LEVANT SUPPLEMENTARY SERIES
VOLUME 8

LANDSCAPES IN TRANSITION

Edited by Bill Finlayson and Graeme Warren

OXBOW BOOKS
Oxford and Oakville

Published jointly by
the Council for British Research in the Levant
and
Oxbow Books, Oxford, UK

ISBN 978-1-84217-416-6

A CIP record for this book is available from the British Library

This book is available direct from:

Oxbow Books, Oxford, UK
(Phone: 01865-241249; Fax: 01865-794449)

and

The David Brown Book Company
PO Box 511, Oakville, CT 06779, USA
(Phone: 860-945-9329; Fax: 860-945-9468)

or from our website

www.oxbowbooks.com

Cover image:
Reconstruction drawing of harvesting in the early Neolithic at Dhra',
near the Dead Sea in Jordan, supplied courtesy of Eric Carlson

Printed in Great Britain by
Information Press, Eynsham
Oxfordshire

Contents

List of Figures and Tables

List of Figures

List of Tables

Acknowledgements

The Landscapes in Transition conference was primarily funded by the Council for British Research in the Levant (CBRL) with further support from UCD Seed Funding (including all transport costs for the UCD participants). Many individual schemes supported particular speakers, and these are acknowledged where appropriate in text. During the conference we were the privileged guests of the Royal Society for the Conservation of Nature Wadi Faynan Ecolodge who helped to support the conference through the special rate they gave us. We are very grateful to their staff for looking after us so well. We also stayed at the Petra Inn and would like to thank Mohammad Tweisi for his help here.

CBRL staff in Amman were very helpful throughout, especially Mrs Nadja Qaisi who organised accommodation and transport for the participants.

Prof Zeidan Kafafi of Yarmuk University gave a tour of his excavations at the famous site of 'Ain Ghazal in Amman. Dr Hans-Georg Gebel gave a guided tour of his excavations at Ba'ja, while Moritz Kinzel led us around the University of Copenhagen's excavations at the Shkarat Msaiad. Our thanks to all.

We are very grateful to Prof Alasdair Whittle for refereeing the entire volume and for his useful comments. Finally we would like to thank all of the participants for not only giving stimulating papers, but for enthusiastically entering into the spirit of the meeting and actively debating the issues as we travelled through the Jordanian landscape.

List of Contributors

The editors

Bill Finlayson
CBRL,
PO Box 519,
Jubaiha 11941,
Amman, Jordan
b.finlayson@cbrl.org.uk

Graeme Warren
School of Archaeology, UCD,
Newman Building,
Belfield,
Dublin 4, Ireland
graeme.warren@ucd.ie

The authors

Eleni Asouti
School of Archaeology, Classics and Egyptology,
University of Liverpool,
Hartley Building,
Brownlow Street,
Liverpool L69 3BX
E.Asouti@liverpool.ac.uk

Anna Belfer-Cohen
Prehistoric Archaeology,
The Institute of Archaeology,
Hebrew University,
Jerusalem 91905, Israel
belferac@mscc.huji.ac.il

Marion Benz
Department of Near Eastern Archaeology,
Albert-Ludwigs Universität,
79085 Freiburg, Germany
marion.benz@orient.uni-freiburg.de

Stuart Black
School of Human & Environmental Sciences,
University of Reading,
Whiteknights,
PO Box 277,
Reading RG6 6AB
s.black@reading.ac.uk

Amy Bogaard
School of Archaeology,
36 Beaumont Street,
Oxford OX1 2PG
amy.bogaard@arch.ox.ac.uk

Dana Campbell
School of Archaeology, Classics and Egyptology,
University of Liverpool,
Brownlow Street,
Liverpool L69 3BX
danac@liverpool.ac.uk

Chantal Conneller
School of Arts, Histories and Cultures,
University of Manchester,
Oxford Road,
Manchester M13 9PL
Chantal.conneller@manchester.ac.uk

Vicki Cummings
School of Forensic and Investigative Science,
University of Central Lancashire,
Preston PR1 2HE
VCummings1@uclan.ac.uk

Andrew Fairbairn
Room 703,
Michie Building,
University of Queensland,
Brisbane,
St Lucia QLD 4072,
Australia
a.fairbairn@uq.edu.au

Duncan Garrow
School of Archaeology, Classics and Egyptology,
University of Liverpool,
Hartley Building,
Brownlow Street,
Liverpool L69 3GS
Duncan.Garrow@liverpool.ac.uk

Nigel Goring-Morris
Prehistoric Archaeology,
The Institute of Archaeology,
Hebrew University,
Jerusalem 91905, Israel
goring@mscc.huji.ac.il

Valasia Isaakidou
Department of Archaeology,
Northgate House,
Sheffield S1 4ET
valasia_isaakidou@yahoo.co.uk

Thomas Kador
UCD School of Archaeology,
Newman Building,
Belfield,
Dublin 4, Ireland
thomas.kador@ucd.ie

Carole McCartney
University of Cyprus,
Archaeological Research Unit,
12 Gladstonos,
Nicosia, Cyrpus
carolemcc@cytanet.com.cy

Lisa Maher
Leverhulme Centre for Human Evolutionary Studies,
University of Cambridge,
Fitzwilliam Street,
Cambridge CB2 1QH
l.maher@human-evol.cam.ac.uk

Sturt W. Manning
Deptartment of Classics
and
Wiener Laboratory for Aegean and Near Eastern Dendrochronology,
Cornell University,
Ithaca, NY 14853-3201, USA
sm456@cornell.edu

Nicky Milner
Department of Archaeology,
The Kings Manor,
University of York,
York YO1 7EP
nm507@york.ac.uk

Gordon Noble
Department of Archaeology,
University of Aberdeen,
St Mary's,
Elphinstone Road,
Aberdeen AB24 3UF
g.noble@abdn.ac.uk

Claire M. C. Rambeau
School of Human & Environmental Sciences,
University of Reading,
Whiteknights,
PO Box 277,
Reading RG6 6AB
cmc*rambeau*@reading.ac.uk

Stuart Robinson
Royal Society University Research Fellow,
Department of Earth Sciences,
University College London,
Gower Street,
London WC1E 6BT
stuart.robinson@ucl.ac.uk

Bruce W. Sellwood
Deceased, November 2007

David Sewell
13 North Avenue, Leicester
david@sewllberry.org.uk

Alison Sheridan
Archaeology Department,
National Museums Scotland,
Chambers Street
Edinburgh EH1 1JF
a.sheridan@nms.ac.uk

Sarah T. Stewart
Trent University,
Archaeological Research Centre,
1600 West Bank Drive,
Peterborough ON K9J7B8, Canada
sallystewart@trentu.ca

Fraser Sturt
Centre for Maritime Archaeology,
University of Southampton, SO17 1BF
F.Sturt@soton.ac.uk

Richard Tipping
School of Biological and Environmental Sciences,
Stirling University,
Stirling UK FK9 4LA
r.tipping@stir.ac.uk

Paul J. Valdes
School of Geographical Sciences,
University of Bristol,
University Road,
Bristol BS8 1SS
p.j.valdes@bristol.ac.uk

Trevor Watkins
School of History, Classics, and Archaeology,
Old High School,
Infirmary Street,
Edinburgh EH1 1LT
Trevor.Watkins@ed.ac.uk

1. Landscapes in Transition: Introduction

Bill Finlayson and Graeme Warren

This publication presents papers arising from a workshop held in Jordan in spring 2007. Contributors were invited to present papers focusing on landscapes at the time of agricultural transition in Europe and Southwest Asia. In particular, we set out to examine two key questions: *What is it that makes landscapes different?* and *How do we study landscapes in this period?* Differing contributions address these questions to different degrees, with more emphasis being placed on the latter. Our substantive discussion of the key themes raised by the volume is reserved for the conclusion; here we outline the background to and structure of this volume. An additional concern of the meeting was the bringing together of mostly British-based researchers to compare and contrast approaches on the transition taken within the two different areas, allowing a wider debate and encouraging some cross-fertilization.

Background

It is uncontroversial to argue that the concept of landscape has enjoyed considerable success in archaeology in recent years, as noted by Barrett: 'it is central to the archaeological programme as a whole because the history of human life is about ways of inhabiting the world' (Barrett 1999, 30). Numerous volumes attest to the popularity of the concept among researchers working in a range of different periods and locations (e.g. Barker and Mattingly 1999; Bender 2000; Bottema 1999; Chapman and Gearey 2000; Cummings and Whittle 2004; Darvill 1999; Edmonds 1999; Tilley 1994; Ucko and Layton 1999; Walsh 1999) and clearly demonstrate the great diversity of research encompassed within 'landscape archaeology'. To a great extent this diversity of practice arises from the nature of the concept itself – landscape, in Gosden and Head's discussion, is a 'usefully ambiguous term' (Gosden and Head 1994), and one should not anticipate a one-size-fits-all methodology. However, ambiguity can be a problem as well as a strength, and this is particularly important where different traditions of archaeological interpretation meet, as, for example, in the transition from hunting and gathering to farming. This transition has been understood as a major division in archaeological practice, and attitudes to 'landscape' across the transition reflect this dichotomy. This in turn creates a weakness in our accounts of the processes involved in the transition. Ingold notes that a central difficulty with landscapes is that they are a modern concept and many traditional societies see themselves as part of and within landscapes, contradictory to our entire notion of the landscape (Ingold 2000). This tension may be most visible at the hunting and gathering to farming transition, where almost all archaeological models assume that this is the key point where people came out of their landscape and began to be creators of it (e.g. Hodder 1990). This separation of social landscape from physical landscape is a central issue in transition studies, especially given the major climatic and environmental changes that were happening at about the same time, yet these are simply facets of a single landscape, and need to be studied together (cf. Trigger 2000), reflecting the complex relationship people have with the world (Descola 1992). This volume directly arises from an exploration of these landscapes, taking case studies from the contrasting regions of the Levant, where the transition to farming was indigenous, and the northwest of Europe, where the transformation was initiated externally. The contrast between the two regions provides valuable comparisons between archaeological traditions and bodies of evidence.

The workshop

The workshop took place in Jordan on 26–30 March 2007. The first four days took place at the Royal Society for the Conservation of Nature's Wadi Faynan Eco-Lodge (Figure 1.1). Field trips provided a key aspect of the workshop, and sites including Dhra', WF16 and Ghuwayr 1 were visited from here. We are especially grateful to Zeidan Kafafi for a tour of 'Ain Ghazal and to Hans Gebel and his research team for the tours of Ba'ja and Shkarat Msaid. The final day of the workshop was hosted at Wadi Mussa,

Fig. 1.1. View approaching the Royal Society for the Conservation of Nature's Wadi Faynan Eco-Lodge, our base for most of the workshop.

Fig. 1.2. Workshop participants at the eco-lodge.

with field trips to Beidha and Basta. The workshop was primarily funded by the Council for British Research in the Levant with support from University College Dublin Seed Funding, although many institutions supported individual speakers (Figure 1.2). Pre-circulated papers were presented at the workshop, with extensive discussion of all papers. All contributors were then given the chance to make alterations to their papers in advance of submission to this volume.

The structure of this volume

It is in the nature of papers focusing on landscapes that they transcend categories. At the workshop, papers were placed into one of eight sessions: Landscape Change; Taskscape/Technology; Landscapes of Settlement; Landscape and Climate; Landscapes and Time; Worldview, Contact and Colonization; Scale and Regions; and Ritual and Routine.

For this volume, the structure is simplified into three parts: Changing Landscapes: Process and Scale; Moving Landscapes: Worldviews and Contact; and Landscapes of Settlement. Considerable overlap between these themes is present.

Bringing researchers working in Northwest Europe and Southwest Asia together always raises some points of terminology. The European terminology of Palaeolithic, Mesolithic and Neolithic was applied to Southwest Asian material, but as greater knowledge and understanding has developed over time the terminologies have diverged. In Europe, 'Mesolithic' is reserved for Holocene hunter-gatherers and 'Neolithic' is associated with a distinctive package of economic and material culture traits. In Southwest Asia, the Natufian was initially interpreted as being 'Mesolithic', but this late Pleistocene culture is now understood as being the final Epipalaeolithic of the southern Levant. 'Neolithic' has also altered, as various apparently transitional cultures lacking the key material culture trait of ceramics became identified as 'Pre-Pottery Neolithic' (PPN). The earliest PPN cultures (PPNA) emerge just before or just after the beginning of the Holocene, leaving no room for a Mesolithic, although there remains some question as to whether the earliest manifestation of the PPNA is really Neolithic at all!

Part One: Changing Landscapes: Process and Scale

Papers in this section address related issues of change over time at varied scales and incorporate reviews of the nature of Late Pleistocene/Early Holocene physical landscape and climate changes in the two geographic focal points for the volume. This includes discussion of the possible relationships between climate change, environmental change and human activity as well as the challenges that integrating this data presents. Further papers outline the synergies and conflicts between broad and small-scale accounts of historical change and the central role of the landscape in this. In the Levant there has been a great amount of new information coming from the northern Levant and from Anatolia that has changed our perceptions of what is happening in the traditionally more intensively studied southern Levant, and this in turn has been encouraging people to look at a very large region in attempts to understand the process that is happening. Within the British Isles there is debate over the significance of regional diversity. The results of dietary analysis from skeletal material are sometimes taken to indicate that the overall change of economy was far more significant than any local patterns.

Nigel Goring-Morris and Anna Belfer-Cohen open the contributions with a review of the changing landscapes and cultures of Southwest Asia from the Late Glacial into the Holocene. This account stresses the 'fluid nature of cultural dynamics' in a context of varied landscapes. Key themes include the identification of solutions to stresses associated with increased proximity caused by sedentism, with many of these seeming to lie in the elaboration of ritual at a landscape level. They argue for the persistence of worldviews through long periods of cultural and landscape change.

Fraser Sturt discusses the distinctions between varied scales of analysis of the transition in Northwest Europe. In particular, he is concerned to avoid fragmentation of data through the adoption of a 'rhythmanalytical' approach and, utilizing musical metaphors, he outlines the differences between 'big beat' and 'bebop' approaches to change in the past. A case study from Eastern England is used to demonstrate the need to understand change at multiple scales (spatial and temporal), and that transition is an ever-present facet of landscapes.

Lisa Maher returns our attention to Southwest Asia, with an emphasis on the significance of scale and dichotomies in our models. In a discussion with strong links to a number of papers in Moving Landscapes (below) Maher argues that our notion of core geographical areas is challenged by recent data. She is also critical of the emphasis within Epipalaeolithic research on the Natufian as a key point of change in the record, arguing for longer-term precedents for supposedly new forms of behaviour, especially mortuary practice.

Nicky Milner's paper examines questions of temporality and subsistence, asking what a rapid or slow change might mean in different regions. Stressing the need to think beyond simplistic discussions of either 'slow' or 'rapid' change, her paper argues that a much greater emphasis on local variability – or 'messiness' – is required. Bayesian dating, and further advances in understanding reservoir effects, offer some methodological support here, but the critical requirement is for a focused landscape archaeology that accepts and engages with local variation.

Stuart Robinson and colleagues emphasize the long term with a synopsis of a recently published review of terrestrial and marine climate data from c. 25,000 to 5,000 years ago. Many contributions to the debate on the transition to agriculture, in both of our study areas, have stressed the possible role of climate change in structuring human cultural change. This chapter includes explicit discussion of the challenges involved in integrating different data sources, from the problems of interpolated dates through to the importance of demonstrating cause and effect.

Richard Tipping's discussion of Northwest Europe is an example of how one might progress in this regard. Tipping reviews a range of palaeoclimatic data, arguing for a significant climate change coincident with the transition to agriculture in Britain and Ireland. He then models how these changes may have impacted on the environment and human behaviour, including the conceptual. This latter includes discussion of the evidence for natural events being 'memorialized'.

Marion Benz's contribution continues the consideration of the long-term interplay between environmental and cultural processes. In particular, her argument, based on an ethnographic database of hunter-gatherer societies, suggests that traditional hunter-gatherer practices, in a context of a changing environment, led to the development of sedentism and the consequent transformation of the landscape. She stresses the importance of 'path dependency' in this context: i.e., once a process has begun, it can be very difficult to reverse it.

Part Two: Moving Landscapes: Worldviews and Contact

The caricatures that archaeologists often employ in discussions of hunter-gatherers and farmers include the ways in which worldviews changed radically at the point of transition. These caricatures include powerful ideas of the wild and tame, as well as conceptions of people as belonging, or being embedded in the world as opposed to being external from and altering the world. This perspective appears to be an area with much held in common between East and West. Apart from the issue of food acquisition, which is in some cases more blurred, a conceptual difference is often presented as the biggest divide between the two modes of life. Yet the relationship between hunter-gatherers and farmers (and the transitional states in between) is often ignored. In the Levant the focus is always on the transition of hunter-gatherers into farmers and the subsequent spread of those farmers, with their ideas, into a landscape that often appears to be thought of as empty, whereas in Northwest Europe a continued polarization of debate sets colonizers against indigenes. The contributions here examine these themes in more detail. In particular, a number of these papers stress the role of movement in spreading ideas and the problematics of archaeological understandings of these 'moving landscapes'.

Alison Sheridan opens this section with a review of the evidence for Neolithization in Britain and Ireland. Sheridan stresses the continental context for many of the novel features of the Neolithic package or packages, arguing for the significance of direct colonization as a mechanism for change. Issues of interactions between hunter-gathers and farmers across water and on dry land are foregrounded by these interpretations, and returned to in papers throughout this section.

Trevor Watkins' paper offers further exploration of a key theme in accounts of the transition in Southwest Asia: namely that increased sedentism leads to increased social stress caused by unprecedented levels of proximity and population density. Watkins' discussion here is very critical of any climate-based argument, and focuses instead on how 'hunter-gatherer groups changed their nature' through the development of increased cognitive skills and symbolic cultural faculties. The symbolic construction of the landscape by the first sedentary communities is argued to be different in kind from anything that had preceded it.

Vicki Cummings' contribution takes a rather different tack in examining change and continuity in understandings of the landscape among the Mesolithic and Neolithic inhabitants of Britain and Ireland. Her argument, based on an analysis of the landscape setting of early Neolithic monuments, is 'that Neolithic farmers did not have a radically different view of the landscape than that of preceding Mesolithic hunter-gatherers and that it is the nature of the evidence (and those studying it) that creates the contrast'. Cummings, in common with many of the contributors from Northwest Europe, stresses the real problems with the extant data for the transition in this region, especially given a tendency to fragment the data by focusing on particular aspects of it.

Gordon Noble's discussion of 'islanding' the transition in Northwest Europe furthers concerns already raised about issues of scale and the problematics of considering what may or may not have been a core or peripheral region in the past. Noble's primary focus is on using developments in island archaeology to reconsider large-scale narratives of change, with case studies drawn from across Britain, Ireland and Denmark. He argues for Britain and Ireland that 'maritime traditions may … be crucial in examining the mechanisms of agricultural uptake in this island world'.

Carole McCartney and colleagues' contribution navigates a similar field. Their discussion places Early Holocene Cyprus in its appropriate Levantine landscapes: those that incorporate the sea. They demonstrate that Cyprus, rather than being isolated, must be understood as part of a network of places. Early colonization of the island has therefore been wrongly cast as an episode of demic diffusion into Europe with subsequent development in isolation and should instead be reconsidered with less of an emphasis on the 'event' of colonization and a better understanding of the processes and networks that linked the coastal landscapes of this region.

Thomas Kador's paper follows on closely from these discussions. Kador examines the late Mesolithic of Ireland, situating himself in the context of a long-standing debate about the degree to which the island might be considered 'isolated' from its neighbours in Britain and continental Europe. Reconsidering a range of data, but especially stone tools, Kador argues that our notions of similarity and difference have often lacked subtlety, and that we have been too quick to compare across national boundaries rather than considering patterns of similarity and difference both within and across geographical entities.

Part Three: Landscapes of Settlement

The final part of this book focuses on varied themes relating to organization, temporalities and changes in the landscape associated with settlement and subsistence. Changing climates, changing settlement patterns and changing methods of exploiting the environment are presumed to have effected an overall change in the landscape. While illustrations in popular books tend to show a transfer from a wild forest landscape to a fully domesticated and tilled one, much of the emphasis in the British academic context has been the monumentalization of the landscape and a very different attitude to the world, with a downplaying of the role of 'economic' change. An alternative major landscape change that is often postulated in the Levant is the steady damage to the environment around 'mega-sites' caused by over-exploitation through farming without soil management, goat herding, and the burning of timber for lime making on a huge scale. The degree of resource use change and its rapidity are central to any conceptions of the landscape in our study areas. At times our understandings of the organization of subsistence, and the necessary relationships between this and the structure of the landscape, are weakly developed. Too often, this debate has been collapsed into a problematic dichotomy between temporary (= mobile) and permanent (= settled). This simple model, itself very flawed, has obscured the need for more developed and dynamic understandings of the landscapes of settlement and the ways in which these landscapes are the basis for ideologies and inequalities. Our contributions demonstrate the significant potential in reconsidering landscapes of settlement.

Eleni Asouti and Andrew Fairbairn argue for a new agenda for archaeobotany in Southwest Asia. They suggest that too much research has been obsessed with evidence for the origins of domesticates rather than directed towards understanding plants as evidence for social processes and landscape relationships. Stressing the problems associated with the uncritical use of analogies, they emphasize the possibilities of reconstructing *practices*, arguing that we should refocus archaeobotany and landscape studies on 'the data-informed reconstruction of the dynamic ecological relationships between communities as socio-economic agents and targeted resources, including the reconstruction of the social practices associated with food production and consumption'.

Dana Campbell's contribution criticizes models that associate the abandonment of the large PPN mega-sites with agricultural collapse caused by environmental degradation, itself linked to population density and the nature of early agriculture. Using 'Ain Ghazal as a test case, Campbell provides reconstructions of the landscape and agricultural systems that demonstrate that the immediate environs of a mega-site could have supported its population, and that the reasons for abandonment should not be seen to lie in ecological degradation. The paper also considers the nature of analogies and reconstructions of early farming systems.

Chantal Conneller's discussion of the taskscape appraises an influential approach to the links between routine technical practices and temporality. Tim Ingold's concept of the taskscape has been influential in hunter-gatherer archaeology in Northwest Europe, although is much less used in Southwest Asia. The concept is powerful, and yct 'the focus on routine tasks produces a sameness (in our accounts?) and also makes it difficult to think about the complex ways in which understandings were generated and inequalities incorporated into these landscapes'. This is especially problematic in considering the taskscape in and around key transitions, especially where these are matched by divisions of archaeological practice.

Amy Bogaard and Valasia Isaakidou transcend our primary case study areas with a paper that moves from Southwest Asia to Central Europe via Greece. They utilize a variety of analogies to help consider the organization of Neolithic agricultural practice, especially in terms of intensive or extensive practice, productivity and settlement size. The three case studies demonstrate the differing ways in which the organization of the routine farming landscape can be understood in terms of settlement size and practice and the ways in which this is therefore evidence of the ideological character of farming.

Duncan Garrow's paper considers the temporality of occupation around the 'transition' in the British Isles. In particular, his case study from Eastern England utilizes settlement, archaeobotanical and ceramic evidence in conjunction with broad analogies in order to suggest some possible models for the duration of settlement. In common with many of the papers in this section, he demonstrates how imprecise assumptions – in this case that lack of definite structural evidence means lack of permanent settlement – cause significant problems in assessing wider discussions of continuity or change at times of transition. In order to understand landscapes in transition at the relevant scales Garrow argues that we must embrace positive and negative evidence, and be both more imaginative and grounded in our approaches to the archaeological material.

Conclusion

Following these papers we offer a review of the key themes arising from the conference. Most of these arose during the course of conference discussions, and the revised papers submitted after the meeting have, to a greater or lesser extent, been modified in light of that debate. The themes cross-cut the sections into which this volume is divided, as the various papers provide a series of differing perspectives on the issues.

Bibliography

Barker, G. and Mattingly, D. (1999) General editors' introduction: The Populus Project, environmental reconstruction in Mediterranean landscape archaeology. Pp. i–vii in G. Barker and D. Mattingly (eds), *The Archaeology of Mediterranean Landscapes 2*. Oxford: Oxbow Books.

Barrett, J. C. (1999) Chronologies of landscape. Pp. 21–30 in R. Layton and P. Ucko (eds), *The Archaeology and Anthropology of Landscape*. London: Routledge.

Bender, B. (2000) Investigating landscape and identity in the Neolithic. Pp. 23–30 in A. Ritchie (ed.), *Neolithic Orkney in its European Context*. Cambridge: MacDonald Institute for Archaeological Research.

Bottema, S. (2000) Landscape archaeology and reconstruction

of the Mediterranean environment based on palynology. Pp. 9–16 in G. Barker and D. Mattingley (eds), *The Archaeology of Mediterranean Landscapes 2*. Oxford: Oxbow Books.

Chapman, H. P. and Gearey, B. R. (2000) Palaeoecology and the perception of prehistoric landscapes: some comments on visual approaches to phenomenology. *Antiquity* 74, 16–319.

Cummings, V. and Whittle, A. (2004) *Places of Special Virtue: Megaliths in the Neolithic Landscapes of Wales*. Oxford: Oxbow Books.

Darvill, T. (1999) The historic environment, historic landscapes, and space-time-action models in landscape archaeology. Pp. 104–18 in P. J. Ucko and R. Layton (eds), *The Archaeology and Anthropology of Landscape*. London: Routledge.

Descola, P. (1992) Societies of nature and the nature of society. Pp. 107–26 in A. Kuper (ed.), *Conceptualising Society*. London: Routledge.

Edmonds, M. (1999) *Ancestral Geographies of the Neolithic: Landscapes, Monuments, and Memory*. London: Routledge.

Gosden, C. and Head, L. (1994) Landscape – a usefully ambiguous concept. *Archaeology in Oceania* 29/3, 113–16.

Hodder, I. (1990) *The Domestication of Europe: Structure and Contingency in Neolithic Societies*. Oxford: Blackwell.

Ingold, T. (2000) *The Perception of the Environment*. London: Routledge.

Tilley, C. (1994) *A Phenomenology of Landscape*. Oxford: Berg.

Trigger, B. G. (2000) Review Article Roger Beefy's Primer. *Cambridge Archaeological Journal* 10/2, 367–9.

Ucko, P. J. and Layton, R. (eds) (1999) *The Archaeology and Anthropology of Landscape: Shaping Your Landscape*. London: Routledge.

Walsh, K. (1999) Mediterranean Landscape Archaeology and Environmental Reconstruction. Pp. 1–8 in G. Barker and D. Mattingley (eds), *The Archaeology of Mediterranean Landscapes 2*. Oxford: Oxbow Books.

Part One

Changing Landscapes: Process and Scale

2. Different Ways of Being, Different Ways of Seeing … Changing Worldviews in the Near East

Nigel Goring-Morris and Anna Belfer-Cohen

This paper outlines the fluid nature of cultural dynamics during the Terminal Pleistocene and Early Holocene in the Levant against the backdrop of the mosaic of different environments and climatic changes in the area. Regional differences in the pace of changes in settlement patterns, site densities and sizes, together with subsistence, are described. Major changes are demonstrated with the beginning of sedentism during the Natufian, as reflected in augmented social interactions, at which time the material evidence for symbolic and ritual activities increased markedly. The longevity of such worldviews is reflected in both the smaller archaic villages of the PPNA and the large village communities of the PPNB and even later. Still, these large PPNB communities should not be viewed simply as early versions of the later Near Eastern village, for these communities represent but an initial, 'first-time' attempt at living in proximity with one another for prolonged periods with subsistence based on domesticates. In such a context the elaboration and intensification of previous ritual activities should be viewed as means of alleviating the accompanying scalar stresses.

By way of a prologue…

Prior to the period of the semi-sedentary Natufian complex, the Levant provides but sporadic and sometimes equivocal evidence for symbolic activities, with most of the material culture remains relating primarily to the mundane realm of mobile foraging bands. Yet those Upper Palaeolithic and early Epipalaeolithic bands clearly had an awareness of self identity, whether at the level of the individual, the family, the band, or the wider mating network – 'us' as opposed to 'them' (Goring-Morris and Belfer-Cohen 2002; Leroi-Gourhan 1981; Lewis-Williams 2002; and papers in Lee and Daly 1999). These social relationships were accompanied by worldviews with a structured codex of non-mundane behaviours. It is this awareness, rather than any particular techno-typological criteria, that is the *raison d'être* behind identifying particular human entities in the Levant.

We use lithics, the most durable material culture remains, for the earlier periods as means to create units of reference rather than being the cause of such divisions. It is ironic that even the most ardent post-processualists, biased against the lithic 'approach', begin their own archaeological discourses by identifying the described phenomena via the 'archaic' lithic categorization. Awareness of the above may ensure progress in prehistoric research in this region since the Levant is unique in its abundance of different paradigmatic frameworks that combine to enrich the sum of research (Clark and Lindly 1991).

The ecological background

A cardinal observation concerning the Near East, and the Levant in particular, is that it comprises a rich mosaic of ecological settings in proximity to one another. Most floral resources are highly seasonal and clumped within the landscape, while faunal species comprise a range of migratory and territorial types; additionally, the Levant provides a major corridor for migratory birds between Africa and Eurasia (Figure 2.1). Local climatic conditions fluctuated markedly over the Terminal Pleistocene/Early Holocene divide throughout the region, undoubtedly dictating local distributions of both plants and animals.

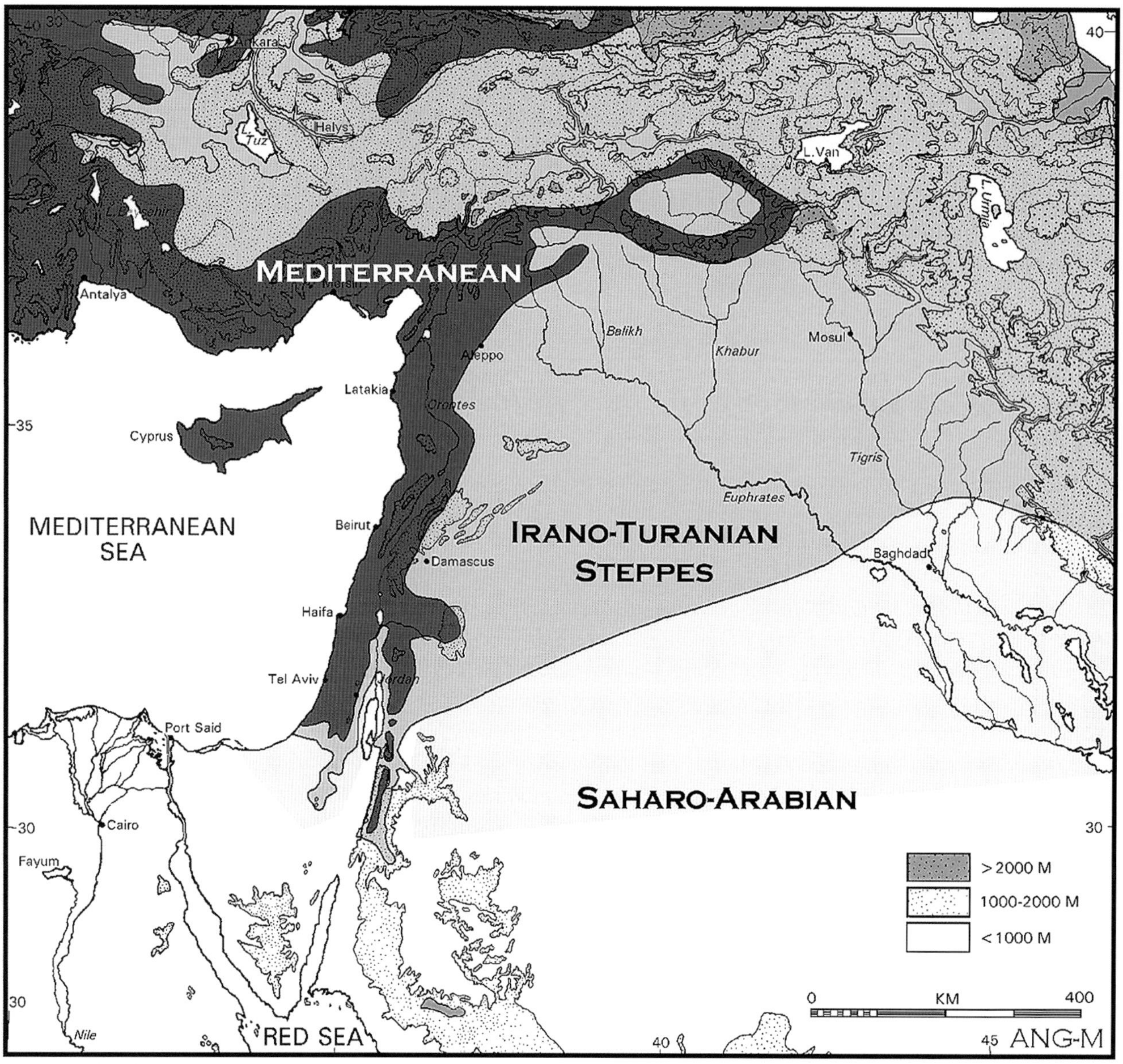

Fig. 2.1. The phytogeographic regions of the Near East. The line between Beirut and Damascus differentiates between southern and northern Levant. Note that some scholars consider that the Upper Tigris is part of Mesopotamia and others that the middle Euphrates is part of south-east Anatolia.

Yet how these environmental changes and the documented shifts in human adaptive systems integrate is notoriously difficult to unravel.

Relative site densities begin to increase markedly only with the beginning of the Epipalaeolithic in the south (see below) and somewhat later in the northern Levant. A steady increase in both the numbers and sizes of sites is observed all through the Epipalaeolithic and into the Early Neolithic (Figure 2.2 – and note that the different entities are adjusted for their respective durations).

Early Epipalaeolithic entities ~22,000–16,000 cal BC (Table 2.1).

The early Epipalaeolithic period corresponds to the Late Glacial Maximum, a time of cold and dry conditions (Enzel *et al.* 2008 and references therein). Sea levels were some 120 m bsl, while Lake Lisan reached high stands, creating eastern and western cultural provinces. On the whole, there is evidence for considerable chronological and geographical variability in the socio-cultural archaeological entities, with much fluidity and dynamic interactions among them (Bar-Yosef and Belfer-Cohen 2002; Goring-Morris and Belfer-Cohen 1997; Goring-Morris *et al.* 2009). There is little if any obvious evidence for sites being specifically located with any but economic criteria in mind: proximity to water sources, clumped exploitable plant resources and distributions of faunal species. Accordingly, many sites are located in ecotones and in low-lying refugia.

West of the Rift Valley postulated Kebaran band ranges,

Period	*Dates cal BC*	*~Duration (yrs)*
Early Epipalaeolithic	22,000–16,000	6,000
Middle Epipalaeolithic	16,000–13,100	2,900
Late Epipalaeolithic	13,100–9,600	3,500
PPNA*	9,600–8,500	1,100
Early PPNB	8,500–8,100	400
Middle PPNB	8,100–7,500	600
Late PPNB	7,500–6,750	750
Final PPNB	6,750–6,400	350
Late Neolithic 1	6,400–5,600	800
Late Neolithic 2 (Early Chalcolithic)	5,600–4,500	1,100

* In the northern Levant, the Round House horizon (in two phases) is estimated at c. 10,750–9,750–8,750 cal BC.

Table 2.1. The dates and duration of the Terminal Pleistocene–Early Holocene prehistoric units in the southern Levant.

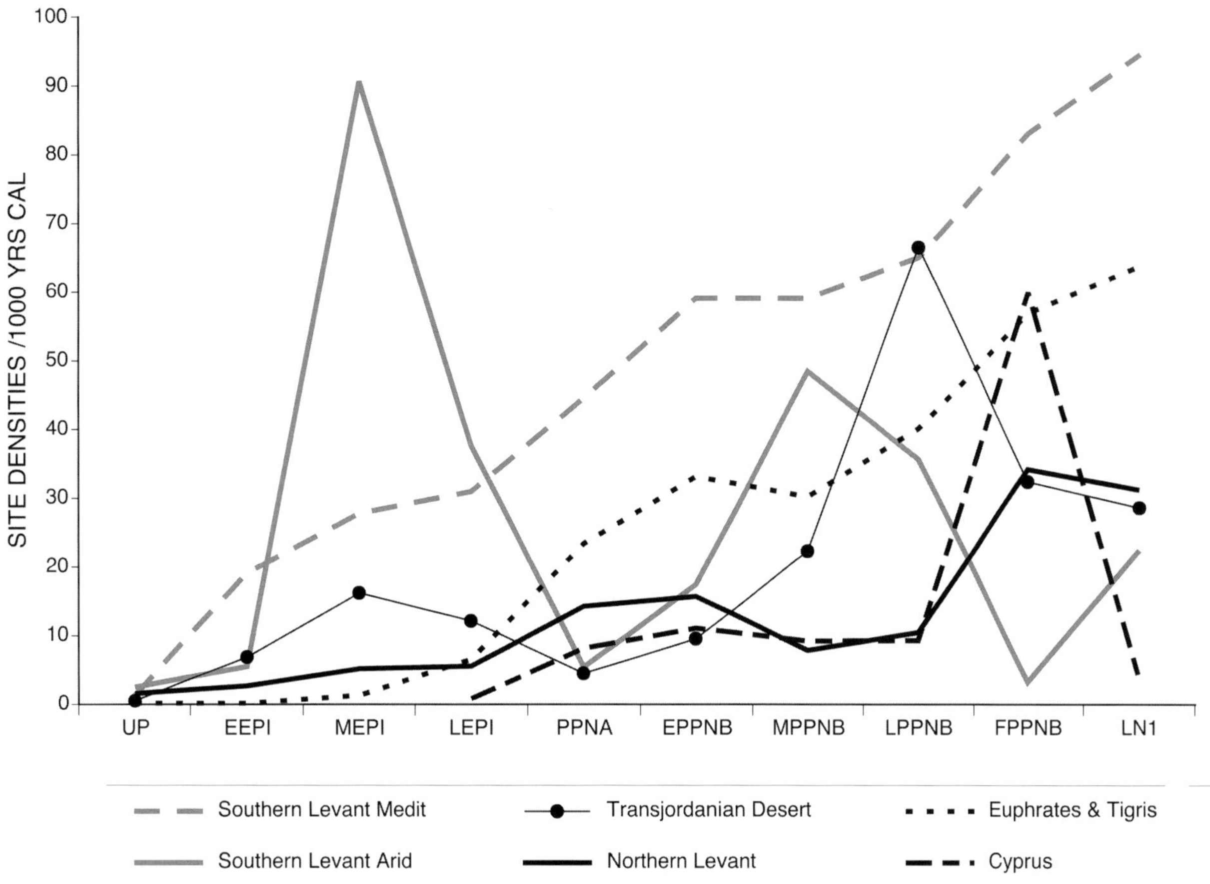

Fig. 2.2. Changing site distribution densities in various areas of the Near East. Note that the differing longevity of each period is taken into account by presenting the relative number of sites/1000 years calibrated for each period (see Table 2.1). Key to periods: UP – Upper Palaeolithic; EEPI – Early Epipalaeolithic; MEPI – Middle Epipalaeolithic; LEPI – Late Epipalaeolithic; PPNA – Pre-Pottery Neolithic A; EPPNB – Early PPNB; MPPNB – Middle PPNB; LEPI – Late Epipalaeolithic; FPPNB – Final PPNB (PPNC); LN1 – Late Neolithic 1 (Pottery Neolithic A).

based on stylistic criteria, appear to have been on the order of 1,200–2,000km², with probable east–west periodic (seasonal?) movements between the coastal plain and the margins of Lake Lisan (Figure 2.3). Sites rarely exceed 150m² and many are considerably smaller, comfortably according with typical ethnographic band sizes (e.g. Lee and De Vore 1976).

By contrast, assumed ranges in eastern Transjordan were probably much larger than in the west and may have been primarily oriented north–south. An additional characteristic

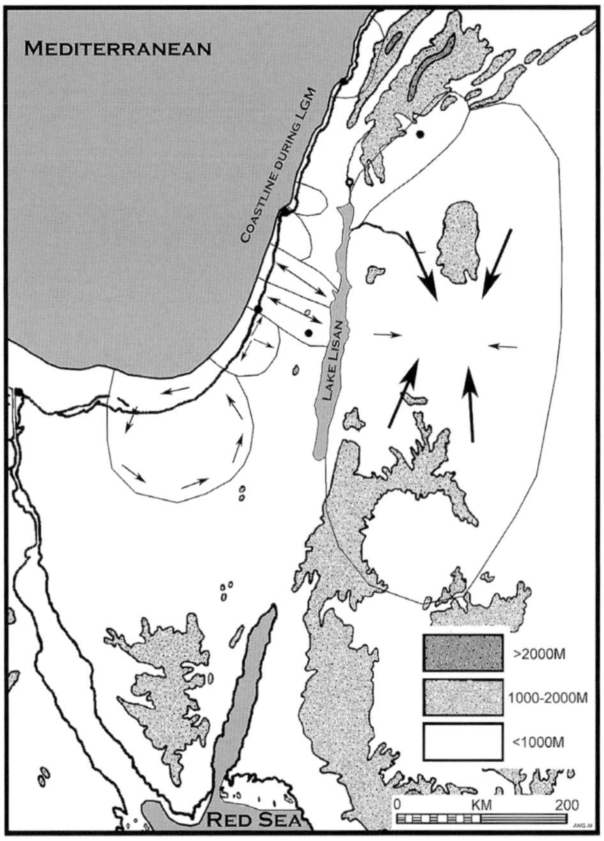

Fig. 2.3. Reconstructed ranges of some early Epipalaeolithic groups in the southern Levant based on the techno-typological and stylistic attributes of lithic assemblages. Note different scales of ranges east and west of the Rift valley.

of the Transjordanian province is the presence of a few major aggregation sites ('microlithic tells') at Jilat 6 and Kharaneh IV on the western margins of the Azraq basin (Garrard 1998; Garrard and Byrd 1992; Muheisen 1988). Undoubtedly these locations attracted bands from far and wide at times of resource abundance and thus served as 'get together' locales (Byrd 2005a; Goring-Morris 1995; Martin 1994).

The few obviously symbolic items recovered from Early Epipalaeolithic contexts (Bar-Yosef 1997; Hovers 1990) are all mobiliary and rarely seem to relate to specific locations; still, it will be interesting to see whether they will be relatively more abundant in the planned excavations at Kharaneh IV (L. Maher and T. Richter pers. comm.).

The Middle Epipalaeolithic entities ~16,000–13,000 cal BC

With the amelioration that heralds the onset of the Middle Epipalaeolithic (the Bølling-Allerød oscillation) conditions improved markedly, enabling populations to increase and expand to the south and east into the (at present) semi-arid and arid regions. Two distinct archaeological complexes can be recognized (Bar-Yosef and Belfer-Cohen 1989; Goring-Morris 1987; Henry 1989). The first is the Geometric Kebaran, seemingly developing directly out of the Kebaran. It is pan-Levantine in scope, while the other, the Mushabian, is restricted to the semi-arid areas of the Negev and Sinai. It appears that, following initial expansion out of the Mediterranean zone, Geometric Kebaran groups underwent contraction if not outright expulsion northwards, back into the Mediterranean zone, in light of the appearance and dominance of the Mushabian (subsequently replaced by the Ramonian) in the Negev and Sinai.

In the Mediterranean zone there may be a tendency towards initial sedentism (Kaufman 1992), although in the southern, marginal, regions bands continued to be highly mobile. Some of the mega-sites in Transjordan appear to have continued to function as previously.

The Late Epipalaeolithic entities ~13,000–9750 cal BC

The Natufian complex is characterized by expansion to the north, east and markedly southwards, especially during its later part. Corresponding to the peak of climatic amelioration (equivalent to the end of the Allerød) the Early Natufian in the favourable areas of the Mediterranean zone is of a scale and complexity hitherto unseen in the archaeological record. Settlements were larger and more sedentary, with a wealth of material remains including numerous obviously ritual items together with indications of symbolic activities. Indeed, it is even possible to recognize certain site localities pertaining not so much to functional economic criteria as to those which mark significant (ritual?) locations in the landscape (Bar-Yosef 2002; Belfer-Cohen and Bar-Yosef 2000; Byrd 2005a; Valla 1999; and papers in Bar-Yosef and Valla 1991 and Delage 2004). The Harifian, restricted to the Negev and Sinai, represents a short-lived final Natufian variant specifically adapted to local conditions. Its demise coincided with forcing conditions of the Younger Dryas, ultimately reaching a critical threshold and leading to abandonment of the region. This probably contributed to increased demographic stress in adjacent Mediterranean areas during the short-lived Khiamian, which perhaps acted as a catalyst in the emergence of the PPNA (Goring-Morris 1991).

Putting subsistence and technology aside, the major increase in obvious symbolic behaviours during the Natufian and onwards reflects, among other factors, the 'scalar stress' deriving from sedentism, with members of larger communities sharing close quarters with one another for extended periods of time. While plant collection may not have increased markedly in terms of actual dietary input, there is substantial evidence that plant collection (e.g. sickles) and processing (e.g. mortars, pestles, etc.) methods became more complex (e.g. Belfer-Cohen and Hovers 2005; Wright 2000). Hunting focused intensively on the acquisition of favoured prey such as the gazelle, but

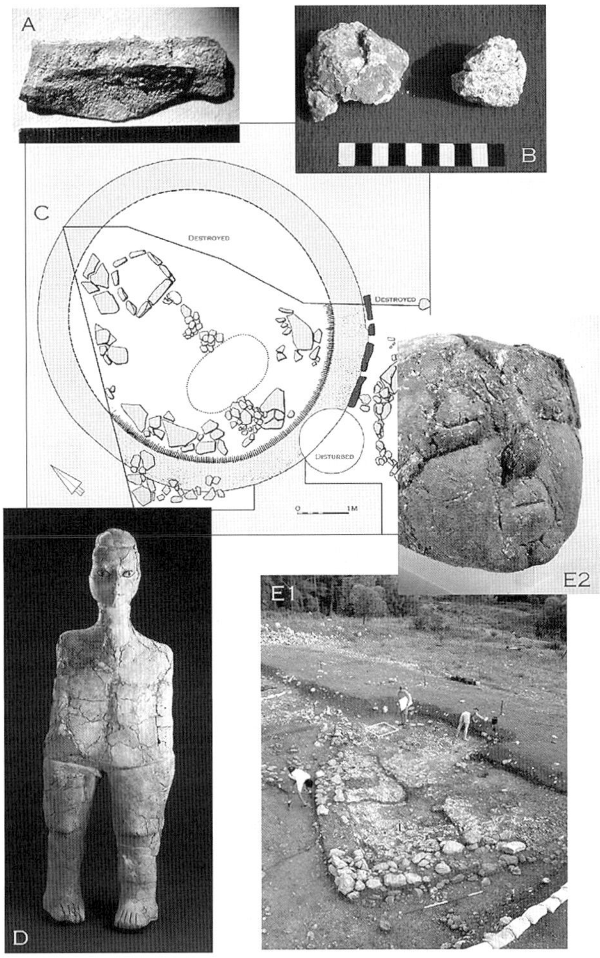

Fig. 2.4. The use of lime plaster: A–B. Middle (Lagama North VIII) and Late Epipalaeolithic (Shunera VII) lime plaster as an adhesive for hafting microliths; C. Natufian lime-plaster bench at Eynan (modified after Perrot and Ladiray 1988); D. PPNB lime-plaster statue from 'Ain Ghazal (courtesy of G. Rollefson); E1–2. PPNB lime-plaster surfaced massive structure and plastered skull from Kfar HaHoresh.

Fig. 2.5. The use of slab pavements in ritual settings: A. Early Natufian Hayonim cave; B. Early Natufian 'Habitation 1' at Eynan (after Perrot and Ladiray 1988); C. Late Natufian Rosh Zin (after Henry 1976); D. 'flagstone' building at PPNB Çayönü (after Ozdogan and Basgelen 1999); E. Mishmar Ha'Emeq (courtesy of O. Barzilai and N. Getzov, IAA); F. structure in cult area of Beidha.

the diet was nevertheless broad, comprising almost every edible animal species available, large and small (e.g. Bar-Oz 2005; Munro 2004).

There are numerous innovations and much variability in the symbolic imagery used through both time and space; yet there is also considerable conservatism displayed in many specific traditions, as indicated by their continuation through to at least the end of the PPNB. Examples demonstrating clear evidence for continuity include lime-plastered installations (Figure 2.4), the use of flagstone-paved structures and pavements (Figure 2.5), the use of baked clay, post-mortem skull removal, the intentional placement of animal bones with human burials (whether dogs, foxes or tortoises), anthropomorphic/zoomorphic arrangements of stones (Figure 2.6), and the use of monoliths (Figure 2.7) (Goring-Morris and Belfer-Cohen 2001; 2002; 2003; 2008).

We may also ask whether the Early Natufian denotes a significantly different way of living as compared with the previously observed patterns, that appear to have centered on the nuclear family and the band. Certainly the large size of many Early Natufian structures raises the issue of the nature of residence patterns. Could they denote residential divisions along the lines of moieties, age or gender partitions? For example, the Wadi Hammeh 27 structures seem much too large for nuclear families (floor spaces on the order of 120–140m^2) (Goring-Morris and Belfer-Cohen 2003; Hardy-Smith and Edwards 2004). The massive stone mortars at Eynan and other sites were probably too large to be owned and used just by nuclear families. Conversely, there is evidence that some areas within settlements were specifically designated for obviously ritual activities. For example, the structures within Hayonim cave are too small for use as habitations, with floor spaces of only c. 2–4m^2.

Fig. 2.6. Anthropomorphic and zoomorphic arrangements: A. Early Natufian Eynan (after Perrot 1966); B. PPNB Kfar HaHoresh, using human long bones (note also post-hole); C1–2. Late Neolithic Biqat Uvda shrine (after Yogev 1980) with slabs to outline leopards and oryx; D1–2. Late Neolithic Khasm e-Tarif shrine in eastern Sinai with adjacent animals (courtesy of U. Avner).

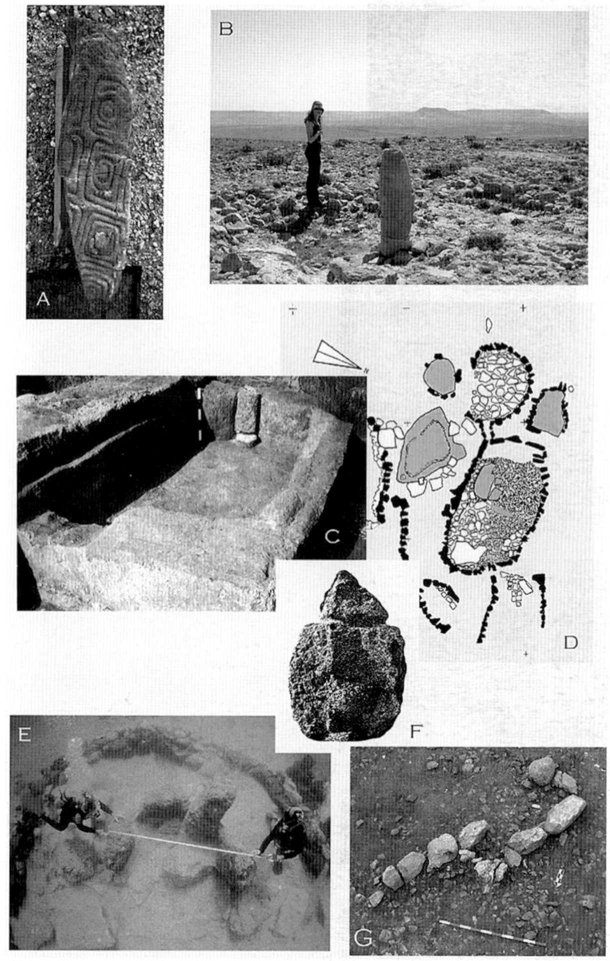

Fig. 2.7. The use of monoliths in symbolic contexts: A. Late Natufian monolith in Locus 4 at Rosh Zin; B. Early Natufian incized monolith from Wadi Hammeh 27 (after Edwards 1991); C. PPNB monolith at Jericho (after Kenyon 1970); D. PPNB monoliths and massive basin at Beidha (modified after Kirkbride 1966); E–F. Final PPNB monoliths and cup-marked basins, 1.60m high anthropomorphic monolith at Atlit Yam (after Galili 2004); G. L1354 at PPNB Kfar HaHoresh: hearth, stone alignment with broken monolith.

As mentioned above, some Late and Final Natufian sites appear to be located in settings that were seemingly non-functional from a purely economic standpoint. Probable examples include Nahal Oren and Sefunim in the Carmel and Nahal Hilazon in western Galilee, as well as some of the Judean desert sites (Figure 2.8). These may rather have served as cemeteries-cum-ritual sites, and perhaps also as boundary markers. Their notable characteristic is the duality of being both known to all who acknowledged their symbolic role, yet secluded in the local landscape, their whereabouts perhaps known only to those who were united by sharing this knowledge.

There is definitely symbolic continuity from earlier times, as people kept tapping the well of their ancestors' beliefs and knowledge – mapping on to time. Thus one can see the pattern of continuity in the enigmatic engravings on the Kebaran Urkan e-Rubb pebble and the Natufian Hayonim cave slabs (Bar-Yosef and Belfer-Cohen 1999; Hovers 1990). Yet the Natufian heralds new ceremonial activities such as rites surrounding death and burial. An example of this is the evidence for ritual activities in the Late Natufian cemetery at Nahal Oren: the large hearth adjacent to the burials, seemingly non-architectural post-holes, paved areas (Figure 2.9) and the breached mortars on top of some burials, not to mention the ceremonial nature of the burials themselves, including post-mortem skull removal (Noy 1989; Stekelis and Yisraely 1963 (and see Figure 2.9)). In some sites bedrock mortars were

Fig. 2.8. Some secluded site settings: A. Final Natufian Hilazon Tachtit in western Galilee (photo by Naftali Hilger, courtesy of L. Grosman); B. PPNB Kfar HaHoresh in lower Galilee; C1–2. PPNB Nahal Hemar cave in the Judean desert (after Bar-Yosef 1985); D1–2. Sefunim cave in the Carmel; E1–3. PPNB Ba'ja in Edom.

Fig. 2.9. Marked graves: A. Final Natufian breached mortar adjacent to headless burial at Nahal Oren (Nahal Oren archive); B. PPNB cache of four skulls adjacent to upright slab at Kfar HaHoresh; C. Child's burial with skull removal and pendant at Kfar HaHoresh; D. Late Neolithic/Early Chalcolithic cache of six skulls adjacent to one of three monoliths at Eilat (courtesy of U. Avner 1991).

occasionally accompanied by runnels and/or shallow cupmarks, as at Saflulim and Upper Besor 6 (Goring-Morris *et al.* 1999; Horwitz and Goring-Morris 2001). Another new phenomenon is the intentional pebble arrangements notable at Eynan and Wadi Hammeh 27, presaging later Neolithic occurrences (Edwards 1991; Perrot and Ladiray 1988).

Some faunal species found within graves from the Natufian onwards most probably functioned in symbolic contexts. There are quite a number of graves containing remains of tortoises, foxes and raptors and their repetitious presence, numbers and placement provide clear evidence of special meaning that, alas, eludes us (Goring-Morris and Belfer-Cohen 2002; Grosman *et al.* 2008).

While there is considerable evidence for groups settling into the local landscape, Natufian exchange networks intensified and expanded markedly, with a wealth of items moving over considerable distances – basalt, flint, molluscs, various exotic minerals and even obsidian (Delage 2001; Valla 1999; Weinstein-Evron *et al.* 2001). While some of these were probably procured for functional and mundane reasons, others were prized for their aesthetic/symbolic properties.

The Neolithic complexes ~9,750–6,150 cal BC

This discussion of the Neolithic starts with the PPNA and runs through to and includes the earliest Late (Pottery) Neolithic Yarmukian. The archaic villages of the southern Levantine PPNA display notable continuity from the Natufian, although there are different variants elsewhere

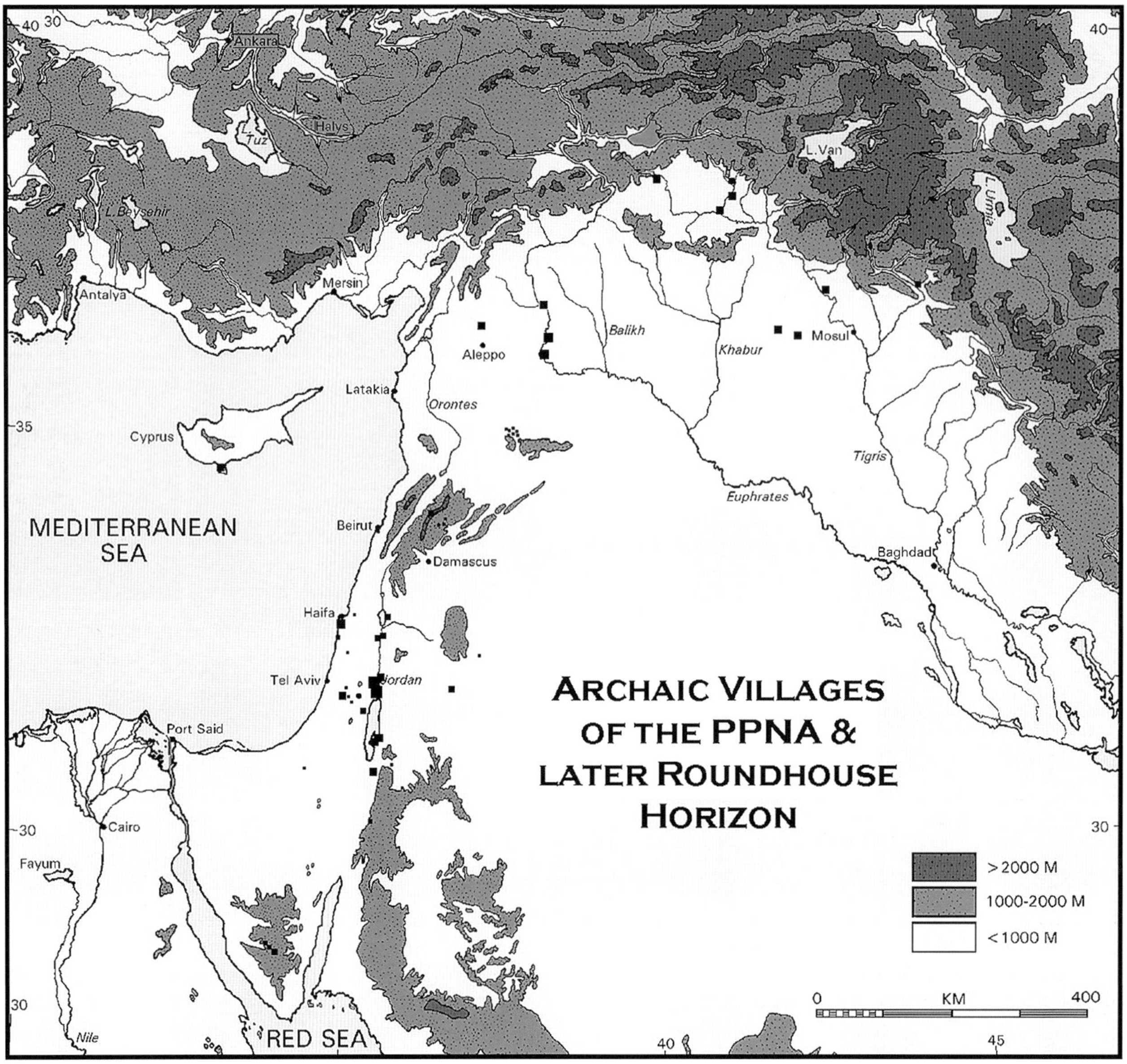

Fig. 2.10. Distribution map of the archaic 'villages' of the PPNA in the Levant.

in the broader Levant (Belfer-Cohen and Bar-Yosef 2000; Cauvin 2000; Kuijt and Goring-Morris 2002; Ozdögan and Basgelen 1999; Simmons 2007, and see Figure 2.10).

The following PPNB (especially its late and final stages) represent the emergence of mega-sites that share enough cultural markers to portray a pan-Levantine *koine* (Bar-Yosef and Belfer-Cohen 1989; Byrd 2005a; Goring-Morris *et al.* 2009 (Figure 2.11)). Still, these village societies should not be viewed simply in terms of being the equivalent of modern peasant village societies (which probably emerged only during the Early Bronze Age in the southern Levant). Agriculture began to be practised more and more intensively in favourable areas during the course of the Neolithic, with reliance on herded animals appearing somewhat later (Horwitz *et al.* 2000; Peters *et al.* 1999; Savard *et al.* 2006). The origins and nature of such processes are under debate, with some arguing for a single centre of innovation in the Middle Euphrates area and subsequent dispersion throughout the Near East, while others assume polycentric developments (e.g. Gopher *et al.* 2001; Lev-Yadun *et al.* 2000 vs Gebel 2004; Willcox 2005). Whatever the case, by the Late and Final stages of this period there was a mosaic of subsistence modes, particularly in the southern Levant. There were thus settled farmers and herders, farmers and hunters, and mobile hunter-gatherers and pastoralists, often over relatively restricted distances. This resulted in a complex array of interactions on all scales (Horwitz *et al.* 2000). A market economy began to emerge, with exchange of sought-after products, whether mundane or luxury items. There may be intimations of the emergence of incipient craft specialization, sometimes around desired resources. Examples include sources of malachite in Wadi Faynan; the flint sources for exquisite naviform blade production;

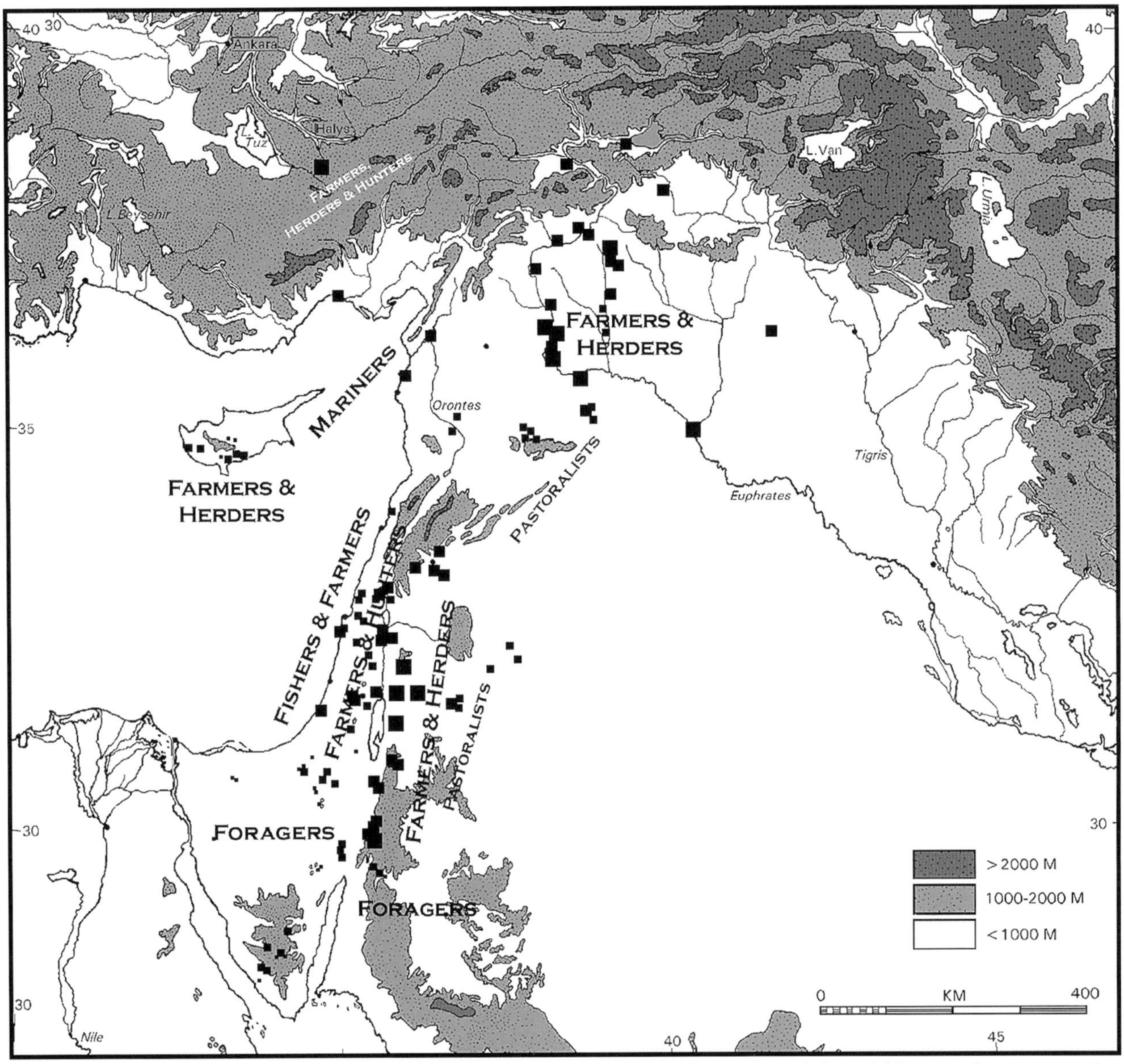

Fig. 2.11. Distribution map of the extent of the PPNB koine *in the Levant, showing variability of different subsistence modes.*

the naviform obsidian workshops in Anatolia; asphalt from the Dead Sea area; and the banded sandstone for bangles in southern Edom (Binder and Balkan-Atli 2001; Gebel and Hermansen 2004; Khalaily *et al.* 2007; Milevski *et al.* 2002). In addition to specialization at the sources, this could also reflect the emergence of itinerant craftsmen along the lines of later tinkers.

Alongside these economic developments were major innovations, regional variability and intensifications in the symbolic imagery used through time and space. It is becoming increasingly clear that most facets of PPNB life were heavily imbued with symbolism, including the supposedly mundane aspects of daily existence (Cauvin 2000; Goring-Morris and Belfer-Cohen 2002; Kuijt 2001; Verhoeven 2004; Watkins 2005). This relates not only to the domestic residence, but also to designated areas within settlements and the wider landscape as a whole. For decades the famous PPNA/B tower and walls of Jericho, whatever their functions, represented an isolated phenomenon. But in recent years exciting new finds, especially in the northern Levant, in the Euphrates–Tigris triangle, as well as in the southern Levant, have transformed our perceptions concerning the manner in which early Neolithic populations viewed their surroundings (Barkai and Liran 2008; Hauptmann 2002; Ozdögan and Basgelen 1999; Peters and Schmidt 2004; Rosenberg and Redding 2000; Schmidt 2005; Stordeur 2004; Watkins 1992, and references therein).

Obvious continuity from the Natufian and PPNA is observed in many specific traditions, such as skull removal and the secluded locations of ceremonial-cum-cemetery sites. Thus there is the cult/cemetery site of Kfar HaHoresh serving nearby lowland settlements; Nahal Hemar cave in the southern Judean desert, with its abundant ritual paraphernalia; and the small Sefunim – 'hidden' – cave in Mt Carmel, with its minor E/MPPNB assemblage overlying earlier Palaeolithic occupations (Figure 2.8; Belfer-Cohen and Goring-Morris 2007). These mostly related to the southeast and southwest of the Mediterranean zone (and see the secluded or remote locations of Ba'ja (Edom) and Khasm et-Tarif (eastern Sinai) – Gebel and Bienert 1997; Goring-Morris 1993). The unique occurrence of the PPNA Jericho tower and surrounding area within a settlement foreshadowed the appearance of sacred precincts, a more common occurrence in the PPNB, often incorporating monoliths (Figure 2.7), flagstone-floored structures and basins (Kuijt 1994; e.g. Beidha – Byrd 2005b; Kirkbride 1966; 'Ain Ghazal – Rollefson 2000; and see Goring-Morris and Belfer-Cohen 2002).

Concurrently we can observe elsewhere in the PPNB *koine* new, awe-inspiring social achievements, with clear regional variability of unprecedented scope. The edifices of PPNA–PPNB Göbekli Tepe, Karahan, and Nevali Çori are situated at watershed locations between the Balikh valley and other tributaries of the Middle Euphrates (Schmidt 2005 and references therein). We can assume, considering the nature, locations and spatial distribution of these sites, that they most probably represent focal points of ancient 'amphictyonies' (Belfer-Cohen and Goring-Morris 2002). Although this conclusion is speculative given the time-depth involved, the available data accord better with this interpretation than with one that assumes that the sites were ritual centres for the neighbouring local population.

While there is considerable continuity from the Natufian through the PPNB (and later) concerning disposal of the dead, the PPNA and especially PPNB burials within settlements often exhibit rather complex circumstances. For example, at 'Ain Ghazal 25 per cent of the MPPNB burials were simply dumped in pits as 'courtyard trash burials' (Rollefson 2000). Others were disposed of more fastidiously, often with later skull removal sometimes entailing subsequent treatment of the skulls. While these rites involved both genders among adults, some children were also treated similarly (Bonogofsky 2004). What sort of hierarchy or ranking does all this ritual activity represent? Does it relate to the status of the individual during life? Or is it only applicable in the afterlife, reflecting communities' rites and beliefs (Kuijt 2001; 2008)?

There are clear-cut human/animal co-associations within some PPNB burials (a phenomenon already known from the Natufian), such as the joint burial of a human and a cat in PPNB Shillourokambos, Cyprus (Vigne and Guilaine 2004). On the mainland, at PPNB Kfar HaHoresh, examples include a headless gazelle carcass with a plastered skull; a skull-less child with pendants in its mouth and a fox mandible on its chest; and a multiple grave containing human remains arranged in the shape of a probable zoomorphic design (Goring-Morris 2005; Horwitz and Goring-Morris 2004). There is also evidence for funerary feasting, as exemplified by parts of a herd of auroch, together with a human burial, representing minimally 500 kg of meat (Goring-Morris and Horwitz 2007). Furthermore, there may be indications for the sacrifice not only of animals but also of humans – see the 'kiva' and accompanying 'sacrificial maiden' at PPNA Jerf el-Ahmar (Stordeur 2000).

Without doubt the social structures of PPNB communities were much more complex than was previously assumed, with symbolic behaviours heavily imbuing all aspects of life. The notable increase of multiple burials (primary and secondary) during the Late PPNB can be tied in with inter-personal and inter-community violence, which is only now beginning to be debated. This may reflect small-scale raiding as settlement densities increased and inequalities emerged on various levels, including differential access to desired resources. Indeed, even the existence of slavery has recently been proposed (Bar-Yosef 2007).

We believe that there is a far greater degree of continuity in the Mediterranean zone from the PPNB into the Pottery Neolithic (PN) entities, namely the Yarmukian/Jericho IX/Lodian, than is frequently portrayed (Garfinkel 1993; Gopher and Gophna 1993). During the Wadi Raba stage, which is commonly viewed today as early Chalcolithic, there is a reversion to smaller communities in the Mediterranean zone (Banning 2003; Garfinkel 1999). In the more arid periphery pastoralists gradually replaced the PPNB foragers but here, too, several themes indicate some continuity in symbolic motifs (Bar-Yosef and Khazanov 1992; Goring-Morris 1993), the meanings of which may well have evolved and changed significantly through time. Examples include the cemetery site of Josephtal (Eilat), with its monolith, sacred post and accompanying skull cache; and sacred locales such as the sanctuaries at Khasm et-Tarif in eastern Sinai and Biqat Uvda in the southern Negev, with their stone arrangements portraying animals and geometric motifs (Avner 2002; Yogev 1983).

Discussion

It seems that the internal – that is, social, cultural, technological and economic – changes at the Terminal Pleistocene–Early Holocene interface nourished each other. New demands in one sphere brought about innovations in another. These involved new techniques and new materials, as well as the production rather than simple procurement of raw materials and resources (e.g. lime plaster); indeed, new concepts were required, which most certainly entailed a new outlook on the landscape itself as an exploitable resource in an active (as opposed to passive) manner. The expansion of communities and the budding-off of new settlements would have necessitated the continuous provision of new agricultural lands. A need arose for substantial, ongoing land clearance (felling and burning) as a result of a lack of crop rotation and manuring (Bogaard this volume). Such activities would have caused considerable modification of the local landscape around settlements. In addition, there were significant changes in the manner in which groups related to the landscape in terms of ownership and rights, including their orderly inter-generational and lateral transfer. Amelioration and climatic conditions in general during the Early and Middle PPNB were instrumental in the initial spread and expansion of the PPNB *koine*, while a relatively abrupt deterioration at the end of the PPNB likely contributed to its demise (Alley *et al.* 1997). But, ultimately, it was the internal processes that were of utmost relevance in both the Neolithic transformation and the demise of the PPNB world.

The initiation of many of the processes associated with Neolithization can be traced back to at least the late Epipalaeolithic (the Natufian) in the southern Levant, as outlined above; they took place selectively, depending on the interaction between internal and external factors. Accordingly, the unfolding picture is one of economically variable, self-sufficient, productive communities, often in close proximity to either pastoral and/or forager communities, with extensive interaction and exchange between them (Figure 2.11). Hunting continued to be important throughout the PPNB, fulfilling both economic and social functions. These inter-relationships stemmed from the need for a demographic/economic 'safety net', as well as the desire for exotic-cum-luxury items, in part perhaps relating to differential status. These new consumer demands were also reflected in incipient specialization on various levels. While the concept of the broader PPNB *koine* cannot be overemphasized, there were still clear boundaries between communities, as portrayed by various elements of the material record.

Among all the changes described above, there appeared the unforeseen consequences of living permanently in large-scale settlements accompanied by major ongoing demographic increase on both community and regional levels. Stress would have occurred in various settings, whether related to hygiene and health issues (contagious diseases), or to the psychological phenomenon of scalar stress that operated on the individual, the communal and the regional levels. There are quite a few indications of such stresses in the archaeological record; one example may be the LPPNB multiple graves, which could hint at either the proliferation of contagious diseases and/or violence.

The end of the PPNB (also termed PPNC) marks the demise of the first thriving Neolithic *koine*. Nevertheless, it should be emphasized that even when major breakdowns occurred, aggravated by an interlude of climatic instability during the FPPNB, not all of the PPNB achievements were lost. By this time, overall population sizes and densities had already passed critical thresholds, ensuring the systematic and orderly inter-generational and lateral transfer of information throughout much, if not most, of the region. Towards the end of the PPNB the cohesion of the large 'mega-site' communities began to break down and they became unable to deal with the disasters that befell them. The absence of viable overarching political and other institutions precluded 'external' help (Goring-Morris and Belfer-Cohen in press). By the Late Neolithic (PN) an interim solution appears to have resolved the problems societies faced; this entailed relating to the landscape in a different manner, by decentralization into smaller and more dispersed groups. However, we believe that the oft-cited differences between the material culture of the PPNB *koine* and subsequent early PN entities have often been overemphasized at the expense of the real level of continuity, which is observable in numerous aspects of the material record and associated belief systems.

Even if one considers the fully fledged Late Neolithic as the concluding chapter of a 'brave new world', the archaeological data clearly present a picture of continuity, with inventions and innovations accumulating on the solid strata of previous knowledge, behaviours and customs, although often with unforeseen, dire consequences. This process should be viewed against the backdrop of real and perceptible environmental changes and the endeavours by communities to achieve an equilibrium between their social correlates and the environment. Ultimately, the Neolithic revolution involved combinations of shifts in the tempo of change, in mindsets, and in the manner in which communities related to themselves and to the physical and cultural landscapes that they inhabited.

Perhaps the circumstances and mechanisms which were instrumental in the 'fall' of the PPNB *koine* sound familiar in the light of present-day global warming; in rather a roundabout way this may indicate that, by Neolithic times, the positive and negative aspects associated with almost, but not quite, 'modern' human society were already present.

Acknowledgements

We are grateful to Bill Finlayson and Graeme Warren for inviting us to participate in a most stimulating conference and accompanying field trips in the Edom region. Both authors acknowledge the support of the Israel Science Foundation funded by the Israel Academy of Sciences (ANG-M – Grants 840/01, 558/04, and 755/07; AB-C – 855/03 and 202/05).

Bibliography

Alley, R. B., Mayewski, P. A., Sowers, T., Stuiver, M., Taylor, K. C. and Clark, P. U. (1997) Holocene climatic instability: a prominent, widespread event 8,200 yr ago. *Geology* 25, 483–6.

Avner, U. (1991) Late Neolithic–Early Chalcolithic burial site in Eilat. *American Journal of Archaeology* 95, 496–7.

Avner, U. (2002) Studies in the Material and Spiritual Culture of the Negev Populations, During the 6th–3rd Millennia B.C. Unpublished PhD thesis, Hebrew University.

Banning, E. B. (2003) Consensus and debate on the Late Neolithic and Chalcolithic of the Southern Levant. *Paléorient* 28, 148–55.

Barkai, R. and Liran, R. (2008) Midsummer sunset at Neolithic Jericho. *Time and Mind: The Journal of Archaeology, Consciousness and Culture* I, 273–84.

Bar-Oz, G. (2005) *Epipalaeolithic Subsistence Strategies in the Levant: A Zooarchaeological Perspective*. The American

School of Prehistoric Research (ASPR) Monograph Series. Boston: Brill Academic Publishers.

Bar-Yosef, O. (1985) *A Cave in the Desert, Nahal Hemar 9,000 Year Old Finds. Catalogue 258.* Jerusalem: Israel Museum.

Bar-Yosef, O. (1997) Symbolic expressions in later prehistory of the Levant: why are they so few? Pp. 161–87 in M. Conkey, O. Soffer, D. Stratmann and N. G. Jablonski (eds), *Beyond Art: Pleistocene Image and Symbol.* Memoirs of the California Academy of Sciences no. 23. San Francisco, CA: California Academy of Sciences.

Bar-Yosef, O. (2002) Natufian. A complex society of foragers. Pp. 91–149 in B. Fitzhugh and J. Habu (eds), *Beyond Foraging and Collecting. Evolutionary Change in Hunter-Gatherer Settlement Systems.* New York: Kluwer Academic/Plenum Publishers.

Bar-Yosef, O. (2007) The emergence of social complexity in the Neolithic of the Near East. *The Annual of the American Schools of Oriental Research* 60/61, 19–40.

Bar-Yosef, O. and Belfer-Cohen, A. (1989) The origins of sedentism and farming communities in the Levant. *Journal of World Prehistory* 40, 447–98.

Bar-Yosef, O. and Belfer-Cohen, A. (1999) Encoding information: unique Natufian objects from Hayonim Cave, western Galilee, Israel. *Antiquity* 73, 402–10.

Bar-Yosef, O. and Belfer-Cohen, A. (2002) Facing environmental crisis – societal and cultural changes at the transition from the Younger Dryas to the Holocene in the Levant. Pp. 55–66 in R. T. J. Cappers and S. Bottema (eds), *The Dawn of Farming in the Near East.* SENEPSE 6. Berlin: *ex oriente.*

Bar-Yosef, O. and Khazanov, A. (eds) (1992) *Pastoralism in the Levant.* Monographs in World Archaeology vol. 10. Madison, WI: Prehistory Press.

Bar-Yosef, O. and Valla, F. (eds) (1991) *The Natufian Culture in the Levant.* Ann Arbor, MI: International Monographs in Prehistory.

Belfer-Cohen, A. and Bar-Yosef, O. (2000) Early sedentism in the Near East – a bumpy ride to village life. Pp. 19–38 in I. Kuijt (ed.), *Life in Neolithic Farming Communities. Social Organization, Identity, and Differentiation.* New York: Kluwer Academic/Plenum.

Belfer-Cohen, A. and Goring-Morris, A. N. (2002) Recent developments in Near Eastern Neolithic research. *Paléorient* 28, 143–8.

Belfer-Cohen, A. and Goring-Morris, A. N. (2007) A new look at old assemblages: a cautionary tale. Pp. 15–24 in L. Astruc, D. Binder and F. Briois (eds), *Technical Systems and Near Eastern PPN Communities.* Antibes: Éditions APDCA.

Belfer-Cohen, A. and Hovers, E. (2005) The groundstone assemblages of the Natufian and Neolithic societies in the Levant – a brief review. *Journal of the Israel Prehistoric Society* 35, 299–308.

Binder, D. and Balkan-Atli, N. (2001) Obsidian and blade technology at Kömürcü-Kaletepe (Central Anatolia). Pp. 1–16 in I. Caneva, C. Lemorini, D. Zampetti and P. Biagi (eds), *Beyond Tools: Redefining the PPN Lithic Assemblages of the Levant.* SENEPSE 9. Berlin: *ex oriente.*

Bonogofsky, M. (2004) Including women and children: Neolithic modeled skulls from Jordan, Israel, Syria and Turkey. *Near Eastern Archaeology* 67, 118–19.

Byrd, B. F. (2005a) Reassessing the emergence of village life in the Near East. *Journal of Archaeological Research* 13, 231–90.

Byrd, B. F. (2005b) *Early Village Life at Beidha, Jordan: Neolithic Spatial Organization and Vernacular Architecture.* Oxford: Oxford University Press.

Cauvin, J. (2000) *The Beginnings of Agriculture in the Near East: A Symbolic Interpretation.* Cambridge: Cambridge University Press.

Clark, G. A. and Lindly, M. (1991) On paradigmatic biases and Paleolithic research tradition. *Current Anthropology* 32, 577–86. Enzel

Delage, C. (2001) Les Ressources Lithiques dans le Nord d'Israel: Le Question des Térritoires d'Approvisionnement Natoufiens Confrontée a l'Hypothese de leur Sedentarité. Unpublished PhD thesis, Université de Paris Panthéon-Sorbonnes.

Delage, C. (ed.) (2004) *The Last Hunter-Gatherers in the Near East.* British Archaeological Reports, International Series 1320. Oxford: BAR.

Edwards, P. C. (1991) Wadi Hammeh 27: an early Natufian site at Pella, Jordan. Pp. 123–48 in O. Bar-Yosef and F. R. Valla (eds), *The Natufian Culture in the Levant.* International Monographs in Prehistory, Archaeological Series 1. Ann Arbor, MI: International Monographs in Prehistory.

Enzel, Y., Amit, R., Dayan, U., Crouvi, O., Kahana, R., Ziv, B. and Sharon, D. (2008) The climatic and physiographic controls of the Eastern Mediterranean over the Late Pleistocene climates in the Southern Levant and its neighboring deserts. *Global and Planetary Change* 60/3–4, 165–92.

Galili, E. (2004) Submerged Settlements of the Ninth to Seventh Millennia BP Off the Carmel Coast. Unpublished PhD thesis, Tel Aviv University.

Garfinkel, Y. (1993) The Yarmukian Culture in Israel. *Paléorient* 19, 115–33.

Garfinkel, Y. (1999) *Neolithic and Chalcolithic Pottery of the Southern Levant.* Qedem vol. 39. Jerusalem: Monographs of the Institute of Archaeology, The Hebrew University of Jerusalem.

Garrard, A. N. (1998) Environment and cultural adaptations in the Azraq Basin: 24,000–7,000 BP. Pp. 139–48 in D. O. Henry (ed.), *The Prehistoric Archaeology of Jordan.* British Archaeological Reports, International Series 705. Oxford: BAR.

Garrard, A. N. and Byrd, B. (1992) New dimensions to the Epipalaeolithic of the Wadi Jilat in central Jordan. *Paléorient* 18, 47–62.

Gebel, H.-G. K. (2004) There was no centre: the polycentric evolution of the Near Eastern Neolithic. *Neo-Lithics* 1/04, 28–32.

Gebel, H.-G. K. and Bienert, H.-D. (1997) Excavating Ba'ja, greater Petra area, southern Jordan. *Neo-Lithics* 1/97, 9–11.

Gebel, H.-G. K. and Hermansen, B. D. (2004) Ba'ja 2003: Summary of the 5th season of excavation. *Neo-Lithics* 2/04, 15–18.

Gopher, A. and Gophna, R. (1993) Cultures of the eighth and seventh millennium BP in the southern Levant: A review for the 1990s. *Journal of World Prehistory* 7, 297–353.

Gopher, A., Abbo, S. and Lev-Yadun, S. (2001) The 'when', the 'where' and the 'why' of the Neolithic Revolution in the Levant. Pp. 49–62 in M. Budja (ed.), *Documenta Praehistoica*, vol. XXVIII. Ljubljana: Oddelek za Arheologijo, Univerza v Ljubljani.

Goring-Morris, A. N. (1987) *At the Edge: Terminal Pleistocene Hunter-Gatherers in the Negev and Sinai.* British Archaeological Reports, International Series 361. Oxford: BAR.

Goring-Morris, A. N. (1991) The Harifian of the Southern

Levant. Pp. 173–234 in O. Bar-Yosef and F. R. Valla (eds), *The Natufian Culture in the Levant*. International Monographs in Prehistory, Archaeological Series 1. Ann Arbor, MI: International Monographs in Prehistory.

Goring-Morris, A. N. (1993) From foraging to herding in the Negev and Sinai: the Early to Late Neolithic transition. *Paléorient* 19, 63–87.

Goring-Morris, A. N. (1995) Complex hunter-gatherers at the end of the Paleolithic (20,000–10,000 BP). Pp. 141–68 in T. E. Levy (ed.), *The Archaeology of Society in the Holy Land*. London: Leicester University Press.

Goring-Morris, A. N. (2005) Life, death and the emergence of differential status in the Near Eastern Neolithic: evidence from Kfar HaHoresh, Lower Galilee, Israel. Pp. 89–105 in J. Clark (ed.), *Archaeological Perspectives on the Transmission and Transformation of Culture in the Eastern Mediterranean*. Oxford: Council for British Research in the Levant and Oxbow Books.

Goring-Morris, A. N. and Belfer-Cohen, A. (1997) The articulation of cultural processes and Late Quaternary environmental changes in Cisjordan. *Paléorient* 23, 71–93.

Goring-Morris, A. N. and Belfer-Cohen, A. (2001) The symbolic realms of utilitarian material culture: the weight of lithics. Pp. 257–71 in I. Caneva, C. Lemorini, D. Zampetti and P. Biagi (eds), *Beyond Tools. Redefining the PPNB Lithic Assemblages of the Levant*. SENEPSE 9. Berlin: *ex oriente*.

Goring-Morris, A. N. and Belfer-Cohen, A. (2002) Symbolic behaviour from the Epipalaeolithic and Early Neolithic of the Near East: preliminary observations on continuity and change. Pp. 67–79 in H.-G. K. Gebel, B. D. Hermansen and C. H. Jensen (eds), *Magic Practices and Ritual in the Near Eastern Neolithic*. SENEPSE 8. Berlin: *ex oriente*.

Goring-Morris, A. N. and Belfer-Cohen, A. (2003) Structures and dwellings in the Upper and Epi-Palaeolithic (ca. 42–10 K BP) Levant: profane and symbolic uses. Pp. 65–81 in O. Soffer, S. Vasil'ev and J. Kozlowski (eds), *Perceived Landscapes and Built Environments*. British Archaeological Reports, International Series 1122. Oxford: BAR.

Goring-Morris, A. N. and Belfer-Cohen, A. (2008) A roof over one's head: developments in Near Eastern residential architecture across the Epipalaeolithic–Neolithic transition. Pp. 239–86 in O. Bar-Yosef and J.-P. Bocquet-Appel (eds), *The Neolithic Demographic Transition and its Consequences*. Springer Verlag: Dortecht Publ.

Goring-Morris, A. N. and Belfer-Cohen, A. (in press) 'Great Expectations', or, the inevitable collapse of the Early Neolithic in the Near East. In M. S. Bandy and J. Fox (eds), *Early Village Society in Global Perspective*. Tucson: University of Arizona Press.

Goring-Morris, A. N. and Horwitz, L. K. (2007) Funerals and feasts in the Near Eastern Pre-Pottery Neolithic B. *Antiquity* 81, 902–19.

Goring-Morris, A. N., Goldberg, P., Goren, Y., Baruch, U. and Bar-Yosef, D. (1999) Saflulim: a Late Natufian base camp in the central Negev highlands, Israel. *Palestine Exploration Quarterly* 131, 1–29.

Goring-Morris, A. N., Hovers, E. and Belfer-Cohen, A. (2009) The dynamics of Pleistocene settlement patterns and human adaptations in the Levant – an overview. Pp. 187–254 in J. J. Shea and D. Lieberman (eds), *Transitions in Prehistory: Papers in Honor of Ofer Bar-Yosef*. Oakville CT: David Brown/Oxbow.

Grosman, L., Munro, N. D. and Belfer-Cohen, A. (2008) A 12,000-year-old Shaman burial from the southern Levant (Israel). *Proceedings of the National Academy of Sciences* 105 (46), 17665–9.

Hardy-Smith, T. and Edwards, P. C. (2004) The garbage crisis in prehistory: artefact discard patterns at the Early Natufian site of Wadi Hammeh 27 and the origins of household refuse disposal strategies. *Journal of Anthropological Archaeology* 23, 253–89.

Hauptmann, H. (2002) Upper Mesopotamia in its regional context during the Early Neolithic. Pp. 263–71 in F. Gerard and L. Thissen (eds), *The Neolithic of Central Anatolia*. Istanbul: Ege Yayinlari.

Henry, D. O. (1976) Rosh Zin: a Natufian settlement near Ein Avdat. Pp. 317–47 in A. E. Marks (ed.), *Prehistory and Paleoenvironments in the Central Negev, Israel. The Avdat/Aqev Area, Part 1*. Dallas, TX: SMU Press.

Henry, D. O. (1989) *From Foraging to Agriculture: The Levant at the End of the Ice Age*. Philadelphia: University of Pennsylvania Press.

Horwitz, L. K. and Goring-Morris, A. N. (2001) Fauna from the Early Natufian site of Upper Besor 6 in the central Negev, Israel. *Paléorient* 26, 111–28.

Horwitz, L. K. and Goring-Morris, A. N. (2004) Animals and ritual during the Levantine PPNB: a case study from the site of Kfar HaHoresh, Israel. *Anthropozoologica* 39, 165–78.

Horwitz, L. K., Tchernov, E., Ducos, P., Becker, C., von den Driesch, A., Martin, L. and Garrard, A. (2000) Animal domestication in the Southern Levant. *Paléorient* 25, 63–80.

Hovers, E. (1990) Art in the Levantine Epi-Palaeolithic: an engraved pebble from a Kebaran site in the lower Jordan Valley. *Current Anthropology* 31, 317–22.

Kaufman, D. (1992) Hunter-Gatherers of the Levantine Epipalaeolithic: the socioecological origins of sedentism. *Journal of Mediterranean Archaeology* 5, 165–201.

Kenyon, K. M. (1970) *The Archaeology of the Holy Land*, 3rd edition. London: Ernest Benn Ltd.

Khalaily, H., Marder, O. and Barzilai, O. (2007) An early Pre-Pottery Neolithic B blade cache from Motza, west of Jerusalem, Israel. Pp. 269–76 in L. Astruc, D. Binder and F. Briois (eds), *Technical Systems and Near Eastern Communities*. Antibes: Éditions APDCA.

Kirkbride, D. (1966) Five seasons at the Prepottery Neolithic village of Beidha in Jordan. *Palestine Exploration Quarterly* 98, 8–72.

Kuijt, I. (1994) Pre-Pottery Neolithic A settlement variability: evidence for sociopolitical developments in the southern Levant. *Journal of Mediterranean Archaeology* 7, 165–92.

Kuijt, I. (2001) Place, death, and the transmission of social memory in early agricultural of the Near Eastern Pre-Pottery Neolithic. Pp. 80–99 in M. S. Chesson (ed.), *Social Memory, Identity, and Death: Anthropological Perspectives on Mortuary Rituals*. Archeological Papers of the American Anthropological Association vol. 10. Naperville, IL: American Anthropological Association.

Kuijt, I. (2008) The regeneration of life: Neolithic structures and symbolic remembering and forgetting. *Current Anthropology* 49/2, 171–97.

Kuijt, I. and Goring-Morris A. N. (2002) Foraging, farming and social complexity in the Pre-Pottery Neolithic of the South-Central Levant: a review and synthesis. *Journal of World Prehistory* 16, 361–440.

Lee, R. B. and Daly, R. (eds) (1999) *The Cambridge Encyclopedia of Hunters and Gatherers*. Cambridge: Cambridge University Press.

Lee, R. B. and De Vore, I. (eds) (1976) *Kalahari Hunter-Gatherers*. Cambridge, MA: Harvard University Press.

Leroi-Gourhan, A. (1981) *The Dawn of European Art. An Introduction to Palaeolithic Cave Painting*. Cambridge: Cambridge University Press.

Lev-Yadun, S., Gopher, A. and Abbo, S. (2000) The cradle of agriculture. *Science* 288/5741, 1602–3.

Lewis-Williams, J. D. (2002) *The Mind in the Cave: Consciousness and the Origins of Art*. London: Thames and Hudson.

Martin, L. A. (1994) Hunting and Herding in a Semi-arid region: An Archaeological and Ethological Study of the Faunal Remains from the Epipalaeolithic and Neolithic of Eastern Jordan. Unpublished PhD thesis, University of Sheffield.

Milevski, I., Marder, O. and Goring-Morris, A. N. (2002) The circulation of asphalt in southern Canaan and Egypt in the Early Bronze Age. Pp. 219–36 in E. C. M. van den Brink and E. Yannai (eds), *In Quest of Ancient Settlements and Landscapes. Archaeological Studies in Honour of Ram Gophna*. Tel Aviv: Ramot Publishing.

Muheisen, M. (1988) The Epipalaeolithic Phases of Kharaneh IV. Pp. 353–67 in A. N. Garrard and H.-G. Gebel (eds), *The Prehistory of Jordan. The State of Research in 1986*. British Archaeological Reports, International Series 396(i). Oxford: BAR.

Munro, N. D. (2004) Zooarchaeological measures of hunting pressure and occupation intensity in the Natufian. Implications for agricultural origins. *Current Anthropology* 45, S5-S33.

Noy, T. (1989) Some aspects of Natufian mortuary behaviour at Nahal Oren. Pp. 53–7 in I. Hershkovitz (ed.), *People and Culture in Change*. British Archaeological Reports, International Series 508(i). Oxford: BAR.

Ozdögan, M. and Basgelen, N. (eds) (1999) *Neolithic in Turkey. The Cradle of Civilization/New Discoveries. Ancient Anatolian Civilizations Series, Vol. 3*. Istanbul: Arkeologie ve Sanat Yayinlari.

Perrot, J. (1966) Le Gisement Natoufien de Mallaha (Eynan), Israel. *L'Anthropologie* 70, 437–83.

Perrot, J. and Ladiray, D. (1988) *Les Hommes de Mallaha (Eynan), Israël*. Mémoires et Travaux du Centre de Recherche Française de Jerusalem, no. 7. Paris: Association Paléorient.

Peters, J. and Schmidt, K. (2004) Animals in the symbolic world of Pre-Pottery Neolithic Gobekli Tepe, south-eastern Turkey: a preliminary assessment. *Anthropozoologica* 39, 179–218.

Peters, J., Helmer, D., Von Den Driesch, A. and Sana Segui, M. (1999) Early animal husbandry in the Northern Levant. *Paléorient* 25, 27–47.

Rollefson, G. O. (2000) Ritual and social structure at Neolithic 'Ain Ghazal. Pp. 163–90 in I. Kuijt (ed.), *Life in Neolithic Farming Communities. Social Organization, Identity, and Differentiation*. New York: Kluwer Academic/Plenum.

Rosenberg, M. and Redding, R. W. (2000) Hallan Cemi and early village organization in Eastern Anatolia. Pp. 39–62 in I. Kuijt (ed.), *Life in Neolithic Farming Communities. Social Organization, Identity, and Differentiation*. New York: Kluwer Academic/Plenum.

Savard, M., Nesbitt, M. and Jones, M. K. (2006) The role of wild grasses in subsistence and sedentism: new evidence from the northern Fertile Crescent. *World Archaeology* 38, 179–96.

Schmidt, K. (2005) 'Ritual centers' and the Neolithisation of Upper Mesopotamia. *Neo-Lithics* 2/05, 13–21.

Simmons, A. H. (2007) *The Neolithic Revolution in the Near East. Transforming the Human Landscape*. Tucson, AZ: University of Arizona Press.

Stekelis, M. and Yisraely, T. (1963) Excavations at Nahal Oren – preliminary report. *Israel Exploration Journal* 13, 1–12.

Stordeur, D. (2000) New discoveries in architecture and symbolism at Jerf el Ahmar (Syria), 1997–1999. *Neo-Lithics* 1/00, 1–4.

Stordeur, D. (2004) Symboles et imaginaire des Premières Cultures Néolithiques du Proche-Orient (Haute et Moyenne Vallée de l'Euphrate). Pp. 15–36 in J. Guilaine (ed.), *Arts et Symboles du Néolithique a la Protohistoire*. Paris: Editions Érrance.

Valla, F. R. (1999) The Natufian: a coherent thought? Pp. 224–41 in W. Davies and R. Charles (eds), *Dorothy Garrod and the Progress of the Palaeolithic*. Oxford: Oxbow Books.

Verhoeven, M. (2004) Beyond boundaries: nature culture and a holistic approach to domestication in the Levant. *Journal of World Prehistory* 18, 179–282.

Vigne, J.-D. and Guilaine, J. (2004) Les premiers animaux de compagnie, 8500 ans avant notre ère? Ou comment j'ai mange mon chat, mon chien et mon renard. *Anthropozoologica* 39, 249–74.

Watkins, T. F. (1992) The beginning of the Neolithic: searching for meaning in material culture change. *Paléorient* 18, 63–75.

Watkins, T. F. (2005) The Neolithic revolution and the emergence of humanity: a cognitive approach to the first comprehensive world-view. Pp. 84–8 in J. Clark (ed.), *Archaeological Perspectives on the Transmission and Transformation of Culture in the Eastern Mediterranean, Levant Supplementary Series 2*. Oxford: Council for British Research in the Levant and Oxbow Books.

Weinstein-Evron, M., Kaufman, D. and Bird-David, N. (2001) Rolling stones: basalt implements as evidence for trade/exchange in the Levantine Epipalaeolithic. *Journal of the Israel Prehistoric Society – Mitekufat Haeven* 31, 25–42.

Willcox, G. (2005) The distribution, natural habitats and availability of wild cereals in relation to their domestication in the Near East: multiple events, multiple centres. *Vegetation History and Archaeobotany* 14, 534–41.

Wright, K. I. (2000) The social origins of cooking and dining in early villages of western Asia. *Proceedings of the Prehistoric Society* 66, 89–121.

Yogev, O. (1983) A fifth millenium B.C.E. sanctuary in the Uvda Valley (Hebrew). *Qadmoniot* 16, 118–22.

3. From Big Beat to Bebop: Settlement Between 6000 and 3000 BC in the Fenland Basin (UK)

Fraser Sturt

In this paper it is argued that archaeology needs to move towards an appreciation of change at a variety of different rates and scales within a number of different mediums. This is demonstrated through an analysis of data pertaining to change within the fenland basin of Eastern England between 6000 and 3000 BC. Here it is shown that in order to understand the variability attested to in the archaeological record we need to remove barriers between different categories of data, approaches to space and sociality. In an effort to achieve this a rhythmanalytical approach is adopted, through which stance a distinction is made between low resolution, 'big beat', traditional approaches to the Mesolithic and Neolithic, and the potential of a more interesting and informative 'bebop', high resolution, account.

Introduction

Whether in the Fertile Crescent or the British Isles, the development and adoption of agriculture required people to improvise and change their daily routines. This process of change has left a significant and varied archaeological record across Europe and beyond. However, in spite of this we often fail to resolve the different strands of evidence into narratives that sufficiently explain both regional variability and apparent broad-scale similarity. Instead of trying to resolve these polyvalent characteristics as a positive feature of the record, debates over how any transition occurred frequently become reduced to polar opposites. In the case of Britain, discussions of settlement have become focused on the presence or absence of permanent structures, rather than what variability in evidence might mean.

I propose that if we wish to understand landscapes in transition we must re-evaluate how we approach the archaeological record and how we conceptualize change. First, it will be argued that we should not break space, time and activity into constituent parts, such as settlement or economy. Rather, we need to consider them as part and parcel of a wider range of interactions and possibilities. In an effort to move beyond the current fragmented attitude to the record, this paper adopts a rhythmanalytical approach. Here a deliberate attempt is made to incorporate as wide a body of data on past spatial conditions and activity as possible, in order to facilitate explanations of human action. Second, the use of a rhythmical mode of investigation ties this study into a broader suite of work, within which music has become a key metaphor for understanding innovation and change. In particular, a contrast will be drawn between melodic, monophonic predictable linear accounts, and harmonic, polyphonic non-linear understandings of change. This largely theoretical discussion is grounded through consideration of the archaeology of the fenland of Eastern England over the period of the Mesolithic–Neolithic transition.

Rhythm and the Mesolithic–Neolithic transition

Despite numerous calls for (and examples of) detailed bottom-up contextual accounts of life in the Mesolithic and Neolithic (Barrett 1997; Edmonds 1997; 1999; King 2001; McFadyen 2006; Thomas 1997; 2001a; 2008; Whittle 2003), current divisions in how we approach different elements of the archaeological record prevent these studies fully integrating the broad spectrum of data available. These divides can be seen to extend beyond the often noted (and increasingly inaccurate) assertion that Mesolithic and Neolithic archaeology are done in very different ways, and thus produce very different interpretations (Bradley 1984; Edmonds 1997; Thomas 1997; 2001b), and into two more troubling areas.

Firstly there is the divide between scientific and humanistic approaches to the past, highlighted by Jones (2002). This sees the constituent parts of changes in time/space, as represented in the archaeological record, broken up and divided among the academic community. Thus advances in humanistic understandings of space to an extent remain separated from advances in areas such as palaeogeography. This is not to say that the products of scientific investigation do not feature in humanistic accounts and vice versa, but that they are given a subordinate place in an interpretative hierarchy within each paradigm. This denies the fact that in order to understand something of perception, ritual, settlement or temporality other broader, biophysical, changes may have been of significance to people in the past.

Secondly, and perhaps more troublingly, is the persistence of a wider, more deliberate, fragmentation in interpretative procedure. Both Barrett (2005) and Thomas (2001b) have argued that current ways of thinking about the Mesolithic and Neolithic require the reduction of complex relationships through essentialist terminology in order to function; tales of settlement, economy, sociality, spatiality and cosmology are told through separate strands of material. Barrett (2005) raises the important point that these are arbitrary categories created in the present and, as such, may not relate to the arenas within which the dynamics of day to day life occurred. This is of particular concern if the object of study is to understand a transitional period. It is quite possible that our terms of settlement, ritual and economy cut through past signatures for change. Similarly, as has been discussed extensively (Armit and Finlayson 1992; Edmonds 1995; Kozlowski 2003; Thomas 2001b), terms such as Mesolithic and Neolithic may do more to cloud our understanding of transitions than shed light on them. In stating this Barrett (2005) has effectively argued that the current practice of archaeology is often robbing of us of meaning rather than creating a plausible interpretation. Thus, as McFadyen (2008) has recently argued, we cannot and should not seek to understand settlement in isolation: it will always be one expression of a wider set of relationships.

In an attempt to move beyond this fragmentation of practice (Jones 2002) and interpretation (Barrett 2005) this paper builds upon the work of Henri Lefebvre, and in particular his concepts of lived space and rhythmanalysis. It is suggested here that people, environment, time and space can never meaningfully be separated, because in so doing we break apart the dynamics that reveal the variety of different ways in which change occurs through time. What I wish to stress in this paper is the need to re-integrate histories of environmental engagement with our accounts of settlement and taskscape (Edmonds 1997).

Lefebvre's work differs from many current accounts of fragmentation of meaning within archaeology (Chapman 2000; Jones 2002; 2005) through the method he proposes, via which we can overcome the problems of disarticulated accounts of life. In Lefebvre's schema, the division of lived experience is considered an unfortunate but established part of current practice. As such, rather than simply critiquing and deconstructing, Lefebvre presents a heuristic device in the form of 'rhythm' through which different strands of data can be reunited.

Rhythmanalysis

Henri Lefebvre stands as one of the most influential writers on the subject of space/time and perception within philosophy, sociology and geography. However, his writings have received considerably less attention within archaeology, with greater time afforded to geographers such as Edward Soja, who were themselves influenced by the work of Lefebvre (see Blake 2002). This is surprising given the degree of harmony that exists between Lefebvre's work and current theories on landscape, space, time and the fragmented nature of archaeology today.

It is not intended to dwell here in any great length upon Lefebvre's wider body of work or the political underpinnings of his theories. Detailed accounts of these can be found in prefaces to modern translations of his more major works (Lefebvre 1991; 1993; 2004) and historical reviews such as that given by Elden (2001). Essentially Lefebvre (1991; 1993) argued that part of the reason why academic theories of space were so far removed from the lived experience of space was not only a failure to see space as praxis, but also the atomization of knowledge and experience that occurred within modern capitalist society.

Central to Lefebvre's theory of the production of space, and how it can be understood, is the concept that space is more than the sum of its parts. Thus if we examine and discuss space in a fragmented manner we can never hope to understand it in its lived form. As such, academia, with its divisions of geography, sociology, architecture and philosophy, could not bring space to mind in a meaningful way for Lefebvre, as they each held part of the puzzle. This, I argue, is exactly what has occurred within archaeology today; social space and physical space have been divided by methodology and outlook. Thus we study settlement via either the permanence of architectural characteristics or the distribution of 'sites', but rarely tie this into other contemporary changes such as shifting landform and ecotones.

To help clarify how we can get back to an understanding of lived space Lefebvre (1993) suggested that we think of it as a trialectic between *le perçu* (perceived time/space), *le conçu* (conceived space) and *le veçu* (experienced time/space). The importance of these ideas lies in the fact that lived space (and the understanding of it) is about more than just appreciating each one of the three parts of the trialectic.

In considering Lefebvre's ideas it quickly becomes clear that our understanding of space is dependent upon our experience of the here and now, the anticipation and imagination of future events and events that will never be, and the memory of the nature of space in the past. It is also about more than human activity alone, in that it includes people's relationships and understandings of broader changes within their surrounds.

Finally, and almost ironically for a researcher who focused on contemporary activity, Lefebvre's work on rhythm forces us to consider the time depth and frequency of activity in the archaeological record. Central to the concept of rhythmanalysis is an appreciation of *measure*, of relative timing, via an understanding of repetition and variation. Where Lefebvre was explicitly trying to reassert the importance of spatiality within a Marxist dialogue that had become overly focused on historicity, archaeology can use it to remember the importance of history within spatiality.

From Big Beat to Bebop: understanding innovation and change

A rhythmanalytical approach to the Mesolithic–Neolithic transition thus requires us to pay close attention to *measure* and evidence for activity types. It is a call to carry out close-grained archaeological research which acknowledges the importance of connections between people and space through time. However, rhythmanalysis in and of itself does not immediately impact on how we understand the process of change, or how we link one regional account to another. It is at this point in considering how change occurs that the thorny issue of scale emerges. In an effort to traverse these complex issues of narratives, scale and change, the Mesolithic–Neolithic transition is considered here through the lens of two musical genres, Big Beat and Bebop. Big Beat is associated with large-scale, low-resolution accounts of pan-European process. Bebop refers to small-scale accounts of regional variation. Neither is held up to be right or wrong: they are simply different readings of the same material.

Big Beat

There is something appealing about Big Beats. They provide an overarching framework that is easy to keep pace with and repeat; to some extent they are almost comforting. Within orthodox accounts of the Mesolithic and Neolithic the Big Beat has long been identified. It is visible in the sweeping arrows of maps of Neolithic process, and in the changes that must have occurred at a broad scale within peoples' day-to-day lives, as new technologies and agriculture were adopted.

Within the Big Beat view of the past there is a fundamental change in the nature of settlement between the Mesolithic and Neolithic. The Mesolithic is about distributed groups of mobile hunter-gatherers living in temporary shelters (Mellars and Reinhardt 1978). These groups may well have operated within some form of seasonal round, revisiting key locations and maintaining a connection with the land. It is important to acknowledge here that the Big Beat view does not preclude a 'social' approach to the Mesolithic – it simply subscribes to a predictable and linear view of process. As Garrow (this volume) notes, this view of the transition sees a gradual shift from a distant otherly past to one within which people become 'more like us'. This approach to the transition can be described as melodic: events played out in a single sequence. The rhythm of this Big Beat view is also well known. It starts off slowly in the Middle East, moves quickly through Europe, and then pauses for a thousand years before a finale of change in the British Isles and Southern Scandinavia c. 4000 BC.

What is important in these accounts is the definitive nature of change between the two periods. Where the Mesolithic is marked by mobility and transitory structures, the Neolithic is connected to permanence and substantial dwellings. Rowley-Conwy (2004) has demonstrated how this view can be maintained via recent results from excavations in the British Isles. He argues that numbers of Early Neolithic house finds are increasing as greater areas are opened up for examination; thus we may have underestimated the role of permanent settlement during the transition and the change within the British Isles may have been closer to that which occurred on the continent than had previously been thought.

The Big Beat model is thus one of conformity and process. The record within the British Isles can be read in such a way that it fits with the sweeping arrows and the broad chronology for change. It is not my intention to dismiss this view of the past, as it has proven to be both stimulating and informative. As Sheridan (2007) has argued, there is a remarkable uniformity in the first pottery that we see within the British Isles and it does all begin to occur at around 4000 BC. Similarly, Rowley-Conwy's (2004) observations are intriguing: some people do appear to have been making permanent house-like structures as soon as farming begins. There is thus a change in rhythm, in lived space and thus in settlement at this point in time. However, it is not the only reading that can be taken from this data. More importantly, it is an account that fails to engage with the majority of the data that we do have from this period.

Bebop

Within the Bebop view of this process of change, the overarching rhythm stays the same. However, rather than being melodic in form, a Bebop view of the past engages with harmonies, of different actions being played out at the same time. Here, it is improvisation upon a theme that provides the 'feel' to the piece of music. Where in Big Beat a central driving force is easily identified through a strong bass line and linear melodic progression, Bebop is more playful, allowing rhythm and harmony to be subverted and reconfigured to create new experiences. It is this ability to engage with bass line rhythms and reconfigure them in new ways that I examine in this section.

It may seem obfuscating to introduce terms such as Bebop, melody and harmony to a discipline that at times feels overly full of empty phrases. However, the shift to a musical metaphor allows engagement with work carried out in organizational studies that uses such metaphors to explain how change and innovation occur (discussed below, see Kamoche *et al.* 2003; Michael 2000). Within a linear, melodic, account of the past it is difficult to envisage any form of transition other than a process of total replacement; one note must follow another without overlap. In this model the Mesolithic of the British Isles falls away as soon as the Neolithic arrives, for the two cannot comfortably co-exist. The rapid appearance of pottery, monuments, domesticates and possibly house structures is taken as a marker for total change. However, as discussed below, we have to be explicit in determining what we mean by 'rapid' and 'total'. We may be discussing changes that occur over a 200–300-year period across the British Isles. Importantly, the Neolithic does not appear to be the same in all of these places at all times. Within East Anglia there is a comparative dearth of monuments when compared to regions such as Wessex (Garrow 2006; Sturt 2006a,b). There is thus variability in terms of 'rapidity' and 'totality' that needs to be taken into consideration.

Work within organizational studies (Kamoche *et al.* 2003; Michael 2000) examining how change and innovation occurs finds linear models to be problematic. Instead, the most frequent metaphor used to communicate innovative states is that of jazz. Within the jazz metaphor harmonics are key; it is quite possible for a series of different notes (practices) to occur at the same time. People experiment with the range of notes available to them to create new musical forms within rhythmical constraints.

This may appear very straightforward, but if we accept that change can (and frequently does) happen in a variety of different ways, employing a suite of different options, we can begin to open ourselves to a Mesolithic–Neolithic transition which is not melodically uniform, but consists of a range of harmonic styles played out in different ways in different parts of the country at different times. Importantly, it makes us think about the unit at which change occurs. Too often we talk in terms of nation states, the British or French Neolithic, rather than units that may have been meaningful in the past. Instead, we must consider how communities might have taken agriculture onboard, adapted to a new geography or responded to new communities of farmers colonizing areas close by. The changes are unlikely to have occurred uniformly at large scales, but probably played out in a variety of different ways as people extemporized around a range of options and possibilities.

As noted above, while the Big Beat vision of the past is correct at a certain level, it also dismisses a large proportion of the data from the Neolithic, and considers uncritically the majority of it from the Mesolithic. It is undeniable that there have been increasing number of house finds within the British Isles, but their distribution is not uniform and their numbers are still very small. Similarly, while distinct changes are visible from 4000 BC onwards, the radiocarbon record paints something less than a uniform picture (see also Milner this volume). While 4000 BC does mark the point of change in some areas, it does not appear so in others. We thus have to acknowledge the possibility that the transition within the British Isles was one of multiple variations upon a theme, of regional renderings of wider processes. In the brief sections below an analysis of data dating to between 6000 and 3000 BC from the fenland basin in Eastern England is offered that demonstrates the importance of appreciating these variations upon a theme, of engaging with rhythm and harmonics.

The fenland and East Anglia: 6000–3000 BC

The fenland is a low-lying area of former coastal wetland in Eastern England (see Figure 3.1). It is a landscape which has undergone a considerable amount of change

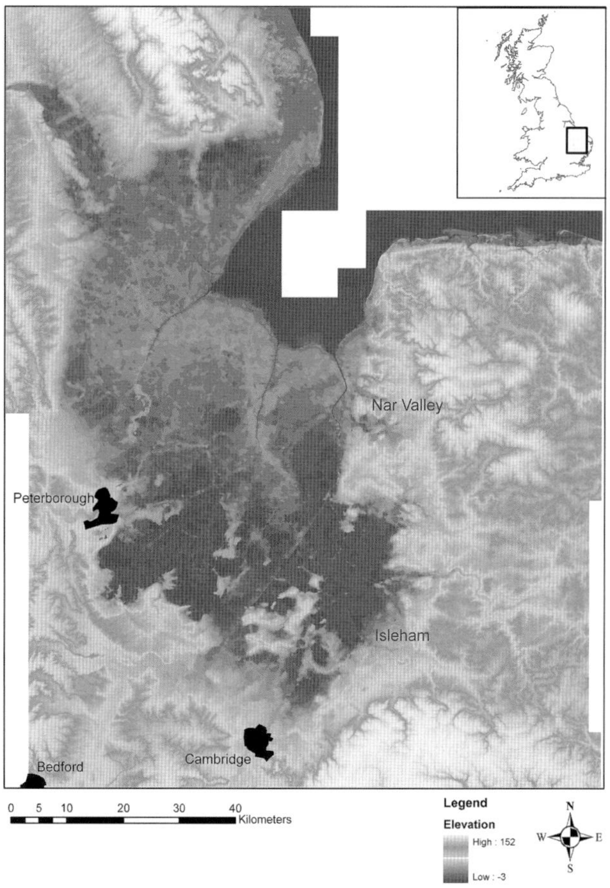

Fig. 3.1. Location map showing the fenland basin and the major sites/areas mentioned in the text. Digital elevation model supplied by Digimap © Crown Copyright/database right 2008. An Ordnance Survey/Edina supplied service.

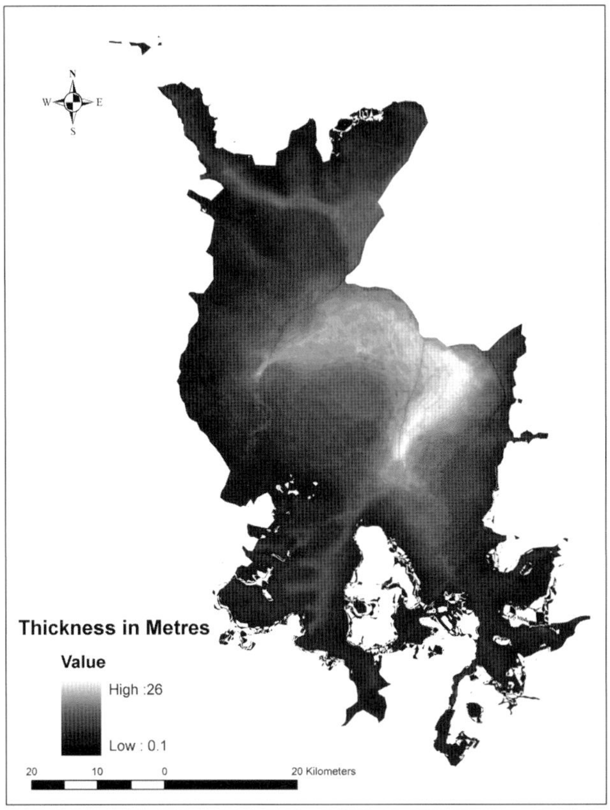

Fig. 3.2. Isopachyte map of Holocene deposits within the fenland basin. This map was created by subtracting the depth of recorded Holocene sediments within boreholes from a digital elevation model of current topography. It is therefore a modelled rather than real surface, as it evens out values between boreholes.

throughout the Holocene. As sea levels rose after the end of the last Ice Age the North Sea Plain was inundated, and so too was the fenland basin. Then, as relative sea-level rise began to slow during the mid Holocene, peat growth saw episodes of regionalized progradation (a seaward expansion of land), at times followed by inundation. This periodic sequence has led to the infilling of the basin and associated preservation of prehistoric land surfaces. The depth of deposits over early Holocene topography can reach over 26 m (Figure 3.2).

The end result of this change is that today's flat agricultural landscape largely conceals the topography and archaeology of the past. Where today we see sugar beet and wheat there once was open sea, mudflats, salt marsh, alder carr and reedswamps. The present fenland is therefore a landscape that is not immediately accessible in terms of its prehistoric characteristics, and there is thus no temptation to make a direct link between how space manifests itself today and the way it would have done so in the past. Rather than being a negative point, in many ways this can be seen as one of the great strengths of the region. Working within this landscape requires constant consideration of how it has changed and when. As Mark Knight has recently described (pers. comm.), working in the fens requires you to think about process at all times, meaning that the dynamics of settlement in this region cannot be extricated from the dynamics of landscape change. Yet how this process of change may have played into histories of settlement and practice during the Early Neolithic has received little direct attention.

Within the fenland, and more broadly East Anglia, there is no record of substantial Early Neolithic domestic structures (as discussed by Garrow *et al.* 2005; Garrow 2006; and this volume). Instead there is evidence for more ephemeral forms of activity: pit sites and lithic scatters (Figures 3.3 and 3.4). While these cannot all be positively categorized as 'settlement', they do represent landscape occupation. It is perhaps in these terms that we need to be thinking if we wish to meaningfully engage with transitions in this region (see Garrow 2006 and this volume). In the absence of permanent structures, despite extensive large-scale programmes of excavation, we have no choice but to consider alternative forms of dwelling practices. As Evans (2004) and McFadyen (2006; 2008) have argued, even a single lithic find indicates some form of connection with a particular place. Similarly, as Garrow (2006; this volume) has demonstrated, pit sites provide a glimpse into a form of occupational practice that may have seen people dwelling in the same spot for months on end, leaving and then returning at a later date. Ephemerality of material remains should not be equated with an ephemeral connection to space.

However, while consideration of the distribution of lithic scatters and findspots from the Mesolithic and Neolithic at this scale forces engagement with data other than that accounted for in the Big Beat approach, a regional view at this level still does not allow us to account for how and why occupation changes through time. What we see at this meso scale is regional trends for scatters and pit sites to be found on sandy soils and in river valleys. This does not give us an understanding of the historicity of space beyond very broad, typologically defined, temporal boundaries. In order to move beyond this I will focus in on the differing records of two smaller areas 20 km apart within the fenland basin (Figure 3.1): Isleham and the Nar Valley.

Isleham

The parish of Isleham lies on the south-eastern edge of the fenland where the counties of Suffolk and Cambridgeshire meet, at the point where the Lower Chalk upland dips down to meet the sand, gravel, greensand and clay of the central basin. The parish includes part of an arc of extensive Early, Middle and Late Neolithic sites along the south-eastern fen-edge. This includes the well-known sites of Shippea Hill, to the north, and the extensive scatters of Soham, to the west (Edmonds *et al.* 1999) (Figure 3.5).

Previous work by the author has focused on modelling the process of landscape change that occurred in this area, alongside survey and investigation of archaeological sites, from the Mesolithic through to the Early Bronze Age (Sturt

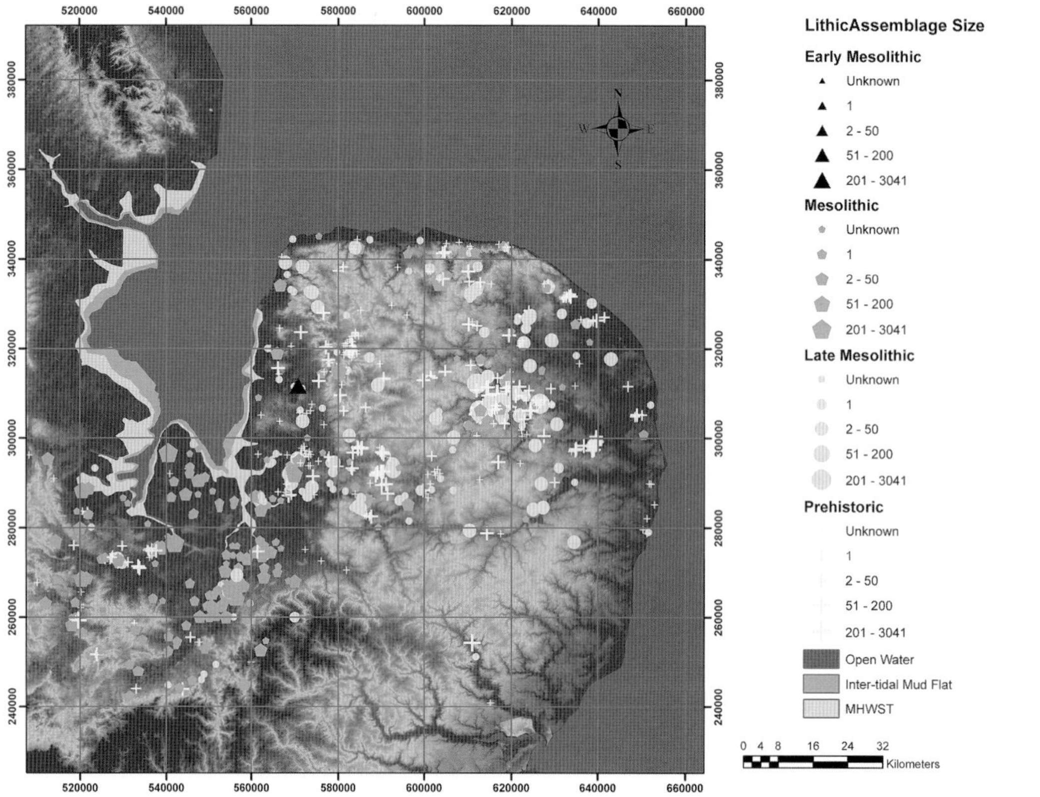

Fig. 3.3. Mesolithic findspots in East Anglia (sea level = 4000 BC).

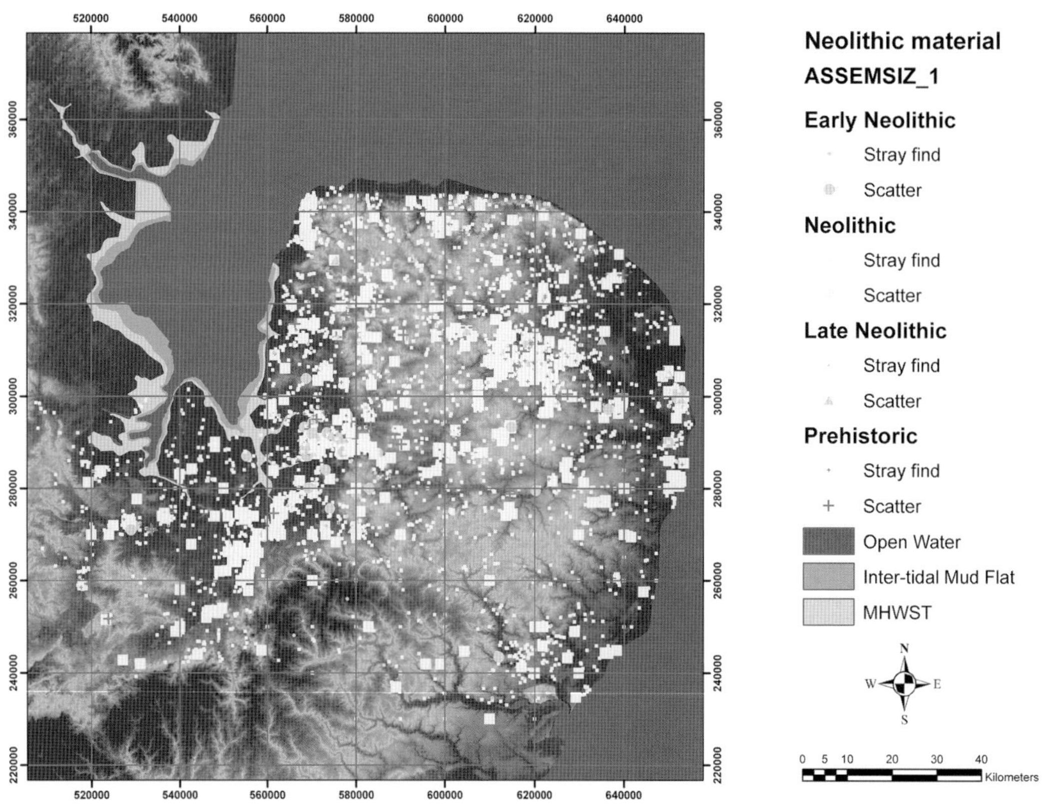

Fig. 3.4. Distribution of Neolithic scatters in East Anglia (sea level = 4000 BC).

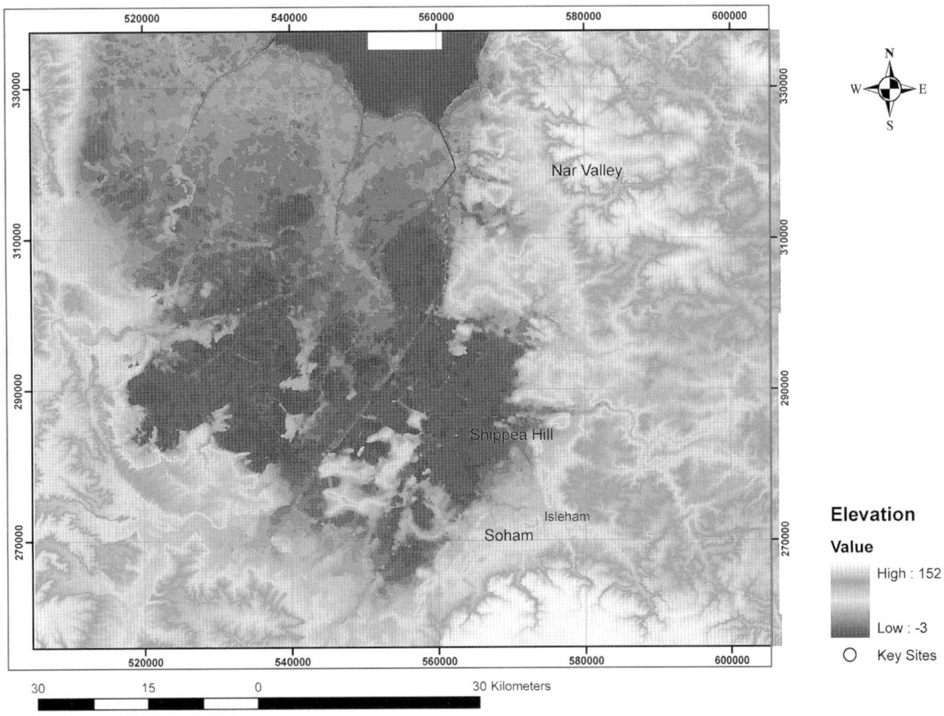

Fig. 3.5. Map showing key sites on the south-east fen-edge.

2006a; 2006b). Here I focus on what the results might mean in terms of occupational histories.

There is some evidence for Late Mesolithic activity in the Isleham area (Figure 3.6). This signature is broadly comparable to that found across the entire fenland region, and is suggestive of a low-density population that made minimal impact on a largely forested environment. The majority of material recovered is found on the banks of, or in close proximity to, river channels, and gives an impression of small groups of people moving down river corridors while living off hunting and gathering activities. At times it is tempting to fit this pattern into a larger-scale model of base camps and hunting camps, but there simply isn't enough data to support such a conclusion in this case. Either the activity of Mesolithic groups extended over much larger areas, and thus can only be understood through a broader regional view, or within this area not enough data survives to provide a detailed image of life in this period.

Conceptualizing Neolithic spatial practices is aided through the more substantial nature of the record. For the Early Neolithic the Isleham data indicates a series of small occupation sites along the edges of the rivers Lark and Snail. Excavation work at one of these sites by the author (Sturt 2006a,b) suggests that these scatters may have been home to groups of people for periods longer than a couple of days. Pits, a post-hole and pottery finds speak of a more substantial settlement that might have been in use for weeks, months or possibly even longer in a similar manner to that discussed by Garrow (2006; this volume). This pit site is likely to have stood at the edge of an estuary in close proximity to salt marsh and reedswamp (Figure 3.6). Given the confined nature of the site it is thought unlikely that arable agriculture occurred here on the island; however, pollen analysis from the Snail palaeochannel (Edmonds *et al.* 2007) does indicate farming within the broader area at this time. The people occupying the small sites along the edges of the channels and on the sand islands may have been carrying out arable farming further upslope and away from the water.

The lack of evidence for Early Neolithic pits at higher elevations could be taken to indicate that within this area the focus of settlement was at the point where dry land, wetland and sea met. However, as with the Mesolithic, there is a danger that the area examined here relates to only part of a wider repertoire of action. It might therefore be ill-conceived to try to paint a complete picture of Early Neolithic land use from this data. For the Early Neolithic it would appear likely that Isleham was a place that was visited and farmed as a part of a larger land/seascape. Although the Isleham pit site in the north of the parish does indicate some form of settlement, it is still small-scale in nature and relatively insubstantial. Perhaps this group of sites forms part of a larger network of practices, incorporating those seen by Garrow (2006) at sites such as Kilverstone.

The Nar Valley

The Nar Valley lies some 20 km to the North of Isleham, yet has a very different archaeological record for activity during this transitional period. The landscape itself is

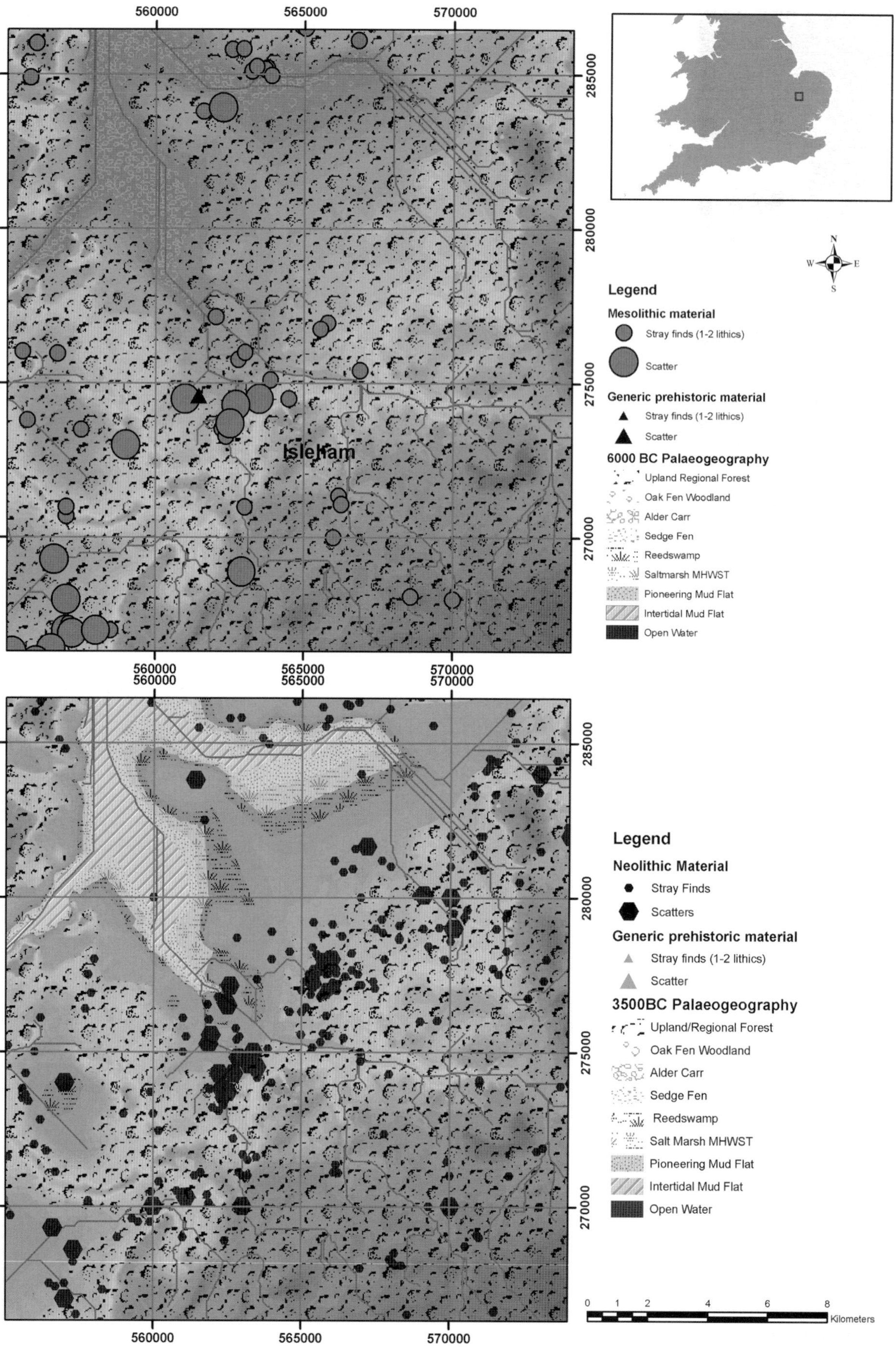

Fig. 3.6. Map showing location of finds in Isleham against a backdrop of modelled palaeoenvironment.

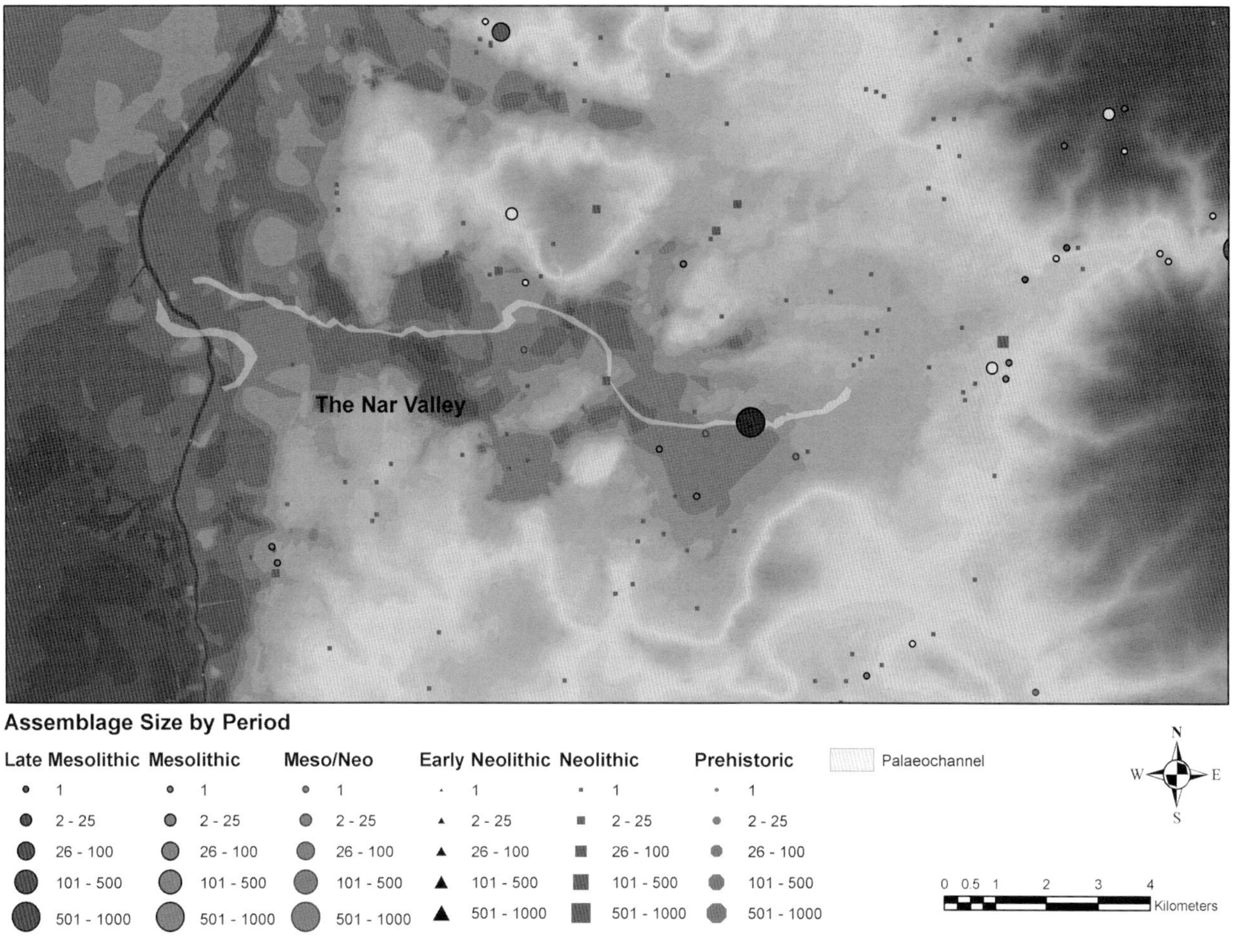

Fig. 3.7. Map showing location of finds in the Nar Valley.

subtly different, with a steeper topography giving way to the deeper parts of the fenland basin. Despite considerable amounts of work in this region (French 1992; Silvester 1988; Sturt 2006a) very little material has been recovered (Figure 3.7). Given the similar research histories of these two regions it is felt that this disparity in the quantity of material found is not a reflection of research bias. Instead, it can be read as a meaningful variation in past practice.

Given the small amount of material recovered from the Nar Valley, it is difficult to model land use based upon this area alone. If this were to be attempted, the themes developed would concern a highly mobile Mesolithic, Neolithic and potentially Bronze Age, as no significant evidence for permanent habitation has been found for any of these periods. For the entirety of the Mesolithic it would appear that this valley was only ever briefly visited by small groups of people. These groups occasionally paused to knap a few flints and perhaps pitched a temporary camp besides the river. Similarly, although there is slightly more evidence for Neolithic activity, it still attests to only a passing engagement. Again, flint was knapped and possibly transient habitation set up, but overall the signature is weak and offers no evidence for substantial investment in this space.

Considering the two areas

It is clear that there is a substantial difference between occupation in Isleham and the Nar Valley during the Late Mesolithic–Early Neolithic. The challenge then becomes to explain this variability. These two regions do not sit isolated from further substantial evidence for Mesolithic and Neolithic activity in the broader East Anglian region (Figures 3.3 and 3.4). Instead they show us the variable engagement with space and possibly differing modes of occupational practice during this period.

The shallowly shelving nature of the south-eastern fen-edge led to the creation of a variety of ecotones along river courses (Figure 3.6). The subtle topography ensured that each of these ecotones was of a substantial size, potentially drawing in a variety of different wildlife. In comparison, the more quickly changing topography of the Nar Valley created a different form of space, with less easy access to a range of resources. The Isleham area appears to have become more attractive to people as sea levels continued to rise. We see a higher number of Mesolithic sites occurring at the same location as Early Neolithic sites in this area when compared with the Nar, and we see an intensification of activity through the Neolithic and into the Bronze Age.

This is interesting because continued sea-level rise reduced the amount of area suitable for arable agriculture, but the connection to place remained. Similarly, Early Holocene sea-level rise will have physically forced Mesolithic hunter-gatherers out of the inundated zones and up the banks of the basin. It would seem plausible that they chose to follow bands of different ecotones up the more shallowly shelving slopes, pausing once the rate of change had slowed (around 4500 BC) and the landscape became more stable.

This history of change, and the presence of a dependable and accessible resource in the form of the fenland areas and marine zone, appears to have been important. As discussed above, the south-eastern fen-edge has remarkably high concentrations of Early Neolithic sites, which, as Evans and Hodder (2006) have noted, stands in contrast to the archaeology of other fenland areas. Although it is always difficult to discuss issues of variability without absolute dates and more intensive excavation, we can begin to think of reasons as to why these differences may exist.

The south-eastern fenland offered a variety of resources to people inhabiting its inundating shores. The rate of change during the mid Holocene was relatively slow, allowing people to slowly adapt to changes in landform. In many ways this was the perfect place to experiment with new forms of practice, to extemporize on a hunting-and-gathering tradition through agricultural activity. Furthermore, there was an historical connection to this space that appeared to hold people to it; settlement practices did not markedly change at this 'transitional' point, but perhaps expanded to include new, agriculturally suited, locations.

The importance of the contrast between Isleham and the Nar Valley is that as a landscape the Nar is as viable for farming as Isleham, but that it has a very different history of occupation. People seem to have had a connection to the south-eastern fen embayment which led to an intensity of occupation and activity not seen further to the north. It would thus seem that histories of settlement location were important. In turn, the locations where people chose to lead their lives shaped the forms of 'Neolithic' that they could live out. We should not be surprised that the Neolithic of a changing fenland is different to that of a Wessex chalk downland, where different harmonies and histories were created. As Lefebvre (2004) has argued, the nature of the spatial relationships that we develop as people shapes who and how we are. Thus the Neolithic of the fenland must, at some level, have been different to that of other areas.

Conclusions

As Garrow (this volume) has argued, rather than looking for landscapes in transition we need to consider all states as transitional. What I hope to have shown through this paper is that there is a wide range of factors that needs to be taken into consideration when examining innovation and transformation. At times we suffer for being too fixated on material culture at the expense of the biophysical record. To people living in the past such distinctions may not have been easy to make, or, as Barrett (2005) has argued, meaningful.

The Neolithic of the south-eastern fen-edge appears to be 'early' and well evidenced. At some levels it does fit into a 'Big Beat' view of the past, within which a package of domesticates, polished stone and pottery swept in and changed people's lives quickly. However, there is also apparent variability. The archaeology of the Neolithic fenland is subtly different to that of wider East Anglia and further afield. It has its own history of change and transition that points to experimentation and extemporization. While it would be comforting to have a series of absolute dates with which to flesh this picture out, the nature of the evidence (lithic scatters and infrequent pits) means that this is unlikely to become available. We need, therefore, to be happy to look at the harmonics displayed through a variety of lines of evidence and pick out the differences. Once we accept this, it becomes possible to see that not all Neolithics need be the same, but that it is possible for them all to be Neolithic: the Big Beat view of sweeping change can and does fit with a Bebop view of regional variation and difference.

Bibliography

Armit, I. and Finlayson, B. (1992) Hunter-gatherers transformed: the transition to agriculture in northern and western Europe. *Antiquity* 66, 664–76.

Barrett, J. C. (1997) Stone Age ideologies. *Analecta Praehistorica Leidensia* 29, 122–9.

Barrett, J. C. (2005) Material culture, humanity and the beginnings of the Neolithic. Pp. 111–24 in T. L. Kienlin (ed.), *Die Dinge als Zeichen: Kulturelles Wissen und materielle Kultur. Internationale Fachtagung an der Johann Wolfgang Goethe-Universitat, Frankfut am Main 3–5 April 2003*. Bonn: Universitatsforchungen zur prahistorischen Archaologie 127.

Blake, E. (2002) Spatiality past and present: an interview with Edward Soja, Los Angeles, 12 April 2001. *Journal of Social Archaeology* 2, 139–58.

Bradley, R. (1984) *The Social Foundations of British Prehistory: Themes and Variations in the Archaeology of Power*. London: Longman Group Ltd.

Chapman, J. (2000) *Fragmentation in Archaeology*. London: Routledge.

Edmonds, M. (1995) *Stone Tools and Society: Working Stone in Neolithic and Bronze Age Britain*. London: B. T. Batsford Ltd.

Edmonds, M. (1997) Taskscape, technology and tradition. *Analecta Praehistorica Leidensia* 29, 99–110.

Edmonds, M. (1999) *Ancestral Geographies of the Neolithic: Landscapes, Monuments and Memory*. London: Routledge.

Edmonds, M., Evans, C. and Gibson, D. (1999) Assembly and collection – lithic complexes in the Cambridgeshire Fenlands. *Proceedings of the Prehistoric Society* 65, 47–82.

Edmonds, M., Gdaniec, K. and Wiltshire, P. (2007) *A Line Across Land: Fieldwork on the Isleham–Ely Pipeline, 1993–4.*

East Anglian Archaeology 121. Cambridge: Cambridge Archaeological Unit.

Elden, S. (2001) Politics, philosophy, geography: Henri Lefebvre in recent Anglo-American scholarship. *Antipode* 33/5, 809–25

Evans, C. and Hodder, I. (2006) *A Woodland Archaeology: Neolithic Sites at Haddenham, Volume 1*. Cambridge: McDonald Institute for Archaeological Research.

Evans, J. G. (2004) Texture and asymmetry in later prehistoric lithics, and their relevance to environmental archaeology. Pp. 215–24 in R. Cleal and J. Pollard (eds), *Monuments and Material Culture: Papers in Honour of an Avebury Archaeologist: Isobel Smith*. Salisbury: Hobnob Press.

French, C. (1992) Fenland research priorities: Late Mesolithic/Early Neolithic transition. *Fenland Research* 7, 2–3.

Garrow, D. (2006) *Pits, Settlement and Deposition during the Neolithic and Early Bronze Age in East Anglia*. British Archaeological Reports, British Series 414. Oxford: BAR.

Garrow, D., Beadsmoore, E. and Knight, M. (2005) Pit clusters and the temporality of occupation: an earlier Neolithic site at Kilverstone, Thetford, Norfolk. *Proceedings of the Prehistoric Society* 71, 139–57

Jones, A. (2002) *Archaeological Theory and Scientific Practice*. Cambridge: Cambridge University Press.

Jones, A. (2005) Lives in fragments: personhood and the European Neolithic. *Journal of Social Archaeology* 5/2, 193–224.

Kamoche, K., Cunha, M. P. E. and da Cunha, J. V. (2003) Towards a theory of organizational improvisation: Looking beyond the jazz metaphor. *Journal of Management Studies* 40, 2023–51.

King, M. (2001) Life and death in the 'Neolithic': dwelling-scrapes in southern Britain. *European Journal of Archaeology* 4/3, 323–45

Kozlowski, S. K. (2003) The Mesolithic: what do we know and what do we believe? Pp. xvii–xxi in L. Larsson, H. Kindgren, K. Knutsson, D. Loeffler and A. Åkerlund (eds), *Mesolithic on the Move: Papers Presented at the Sixth International Conference on the Mesolithic in Europe, Stockholm 2000*. Oxford: Oxbow Books

Lefebvre, H. (1991) *Critique of Everyday Life. Volume 1: Introduction*. London: Verso.

Lefebvre, H. (1993) *The Production of Space*. Oxford: Blackwell.

Lefebvre, H. (2004) *Rhythmanalysis: Space, Time and Everyday Life*. London: Continuum.

McFadyen, L. (2006) Landscape. Pp. 121–38 in C. Conneller and G. Warren (eds), *The Mesolithic of Britain and Ireland: New Approaches*. Stroud: Tempus.

McFadyen, L. (2008) Temporary spaces in the Mesolithic and Neolithic: understanding landscapes. Pp. 121–34 in J. Pollard (ed.), *Prehistoric Britain*. Oxford: Blackwell.

Mellars, P. and Reinhardt, S. C. (1978) Patterns of Mesolithic land-use in southern England: a geological perspective. Pp. 243–93 in P. Mellars (ed.), *The Early Postglacial Settlement of Northern Europe*. London: Duckworth.

Michael, H. Z. (2000) Jazz improvisation and organizing: once more from the top. *Organization Science* 11/2, 227–34.

Rowley-Conwy, P. (2004) How the West was lost: a reconsideration of agricultural origins in Britain, Ireland and Southern Scandinavia. *Current Anthropology* 45 (Supplement), 83–113.

Sheridan, A. (2007) From Picardie to Pickering and Pencraig Hill? New information on the 'Carinated Bowl Neolithic' in northern Britain. Pp. 441–92 in A. Whittle and V. Cummings (eds), *Going Over: The Mesolithic–Neolithic Transition in North-West Europe*. London: British Academy and Oxford University Press.

Silvester, R. J. (1988) *The Fenland Project Number 3: Marshland and the Nar Valley, Norfolk*. Gressenhall: The Norfolk Archaeological Unit.

Sturt, F. (2006a) Landscape and landuse change between 6000–2500 BC on the eastern fen edge. Unpublished PhD thesis, University of Cambridge.

Sturt, F. (2006b) Local knowledge is required: a rhythmanalytical approach to the late Mesolithic and early Neolithic of the East Anglian fenland, UK. *Journal of Maritime Archaeology* 1, 119–39.

Thomas, J. (1997) The materiality of the Mesolithic–Neolithic transition in Britain. *Analecta Praehistorica Leidensi*a 29, 57–64.

Thomas, J. (2001a) Archaeologies of place and landscape. Pp. 165–86 in I. Hodder (ed.), *Archaeological Theory Today*. Cambridge: Polity.

Thomas, J. (2001b) *Understanding the Neolithic*. London: Routledge.

Thomas, J. (2008) The Mesolithic–Neolithic transition in Britain. Pp. 58–89 in J. Pollard (ed.), *Prehistoric Britain*. Oxford: Blackwell

Whittle, A. (2003) The Archaeology of People: Dimensions of Neolithic Life. London: Routledge.

4. People and Their Places at the End of the Pleistocene: Evaluating Perspectives on Physical and Cultural Landscape Change

Lisa A. Maher

Our use of the term landscape in Near Eastern prehistory is somewhat problematic because researchers often dichotomize the landscape into physical and cultural aspects. Furthermore, Epipalaeolithic research, in particular, tends to a) characterize cultural development as occurring within the Mediterranean core versus the arid peripheries and b) focus on the latest phase, the Natufian, as precursors to the Neolithic, such that the cultural landscape is further dichotomized into simple pre-Natufian and complex Natufian groups. Perspectives of landscape that see the physical environment as controlling human behaviour and the cultural landscape as dominated by the Natufian should be abandoned. We need to refocus our field of view to adopt an approach to landscape that recognizes humans' knowledgeable engagement with their surroundings occurring over a longer time depth extending at least throughout the entire Epipalaeolithic period.

Introduction: the landscape of the Levantine Epipalaeolithic (c. 22.5–11.6 ka cal BP)

A recurrent theme in the study of the Levantine Epipalaeolithic (EP)–Neolithic transition is landscape. The term is used frequently, but is usually set up as a series of dichotomous relationships that can hinder our understanding of EP behaviour (Figure 4.1).

The first dichotomy is between the physical and cultural/social landscapes, with the treatment of environment and culture as mutually exclusive categories. Current scenarios generally treat the transition from hunting and gathering to farming in terms of shifts in regional and local environmental conditions and have been environmentally deterministic. The various behavioural expressions of EP groups are attributed to direct responses to climatic conditions (e.g. amelioration and sedentism, stress and mobility). In the cultural landscape social groups are identified by material culture residues. The relative proportions of different microlith tool types are used to determine social and group identity with little understanding of what these artificially constructed typologies 'mean', if anything at all (Olszewski 2001a).

The second dichotomy is within the treatment of the physical landscape. The Levantine palaeoenvironment is divided into two realms: the Mediterranean core which is the homeland for EP groups that 'evolve' unilinearly from Kebaran to Geometric Kebaran to Natufian, and the surrounding periphery zone, characterized as arid and marginal and not contributing to the social development leading to the Neolithic. As these EP groups are perceived as geographically and temporally bounded, our models of culture change leave little room for overlap, exchange or diffusion between them.

The third dichotomy occurs within the cultural landscape. EP entities are defined primarily by their lithic traditions and classified into a variety of different industries, complexes or facies. Pre-Natufian groups consist of a large number of variable Early and Middle EP (and sometimes non-Natufian Late EP) entities lumped together as seasonally mobile hunter-gatherers. The following Natufian is differentiated by the dominance of particular microlith types (e.g. lunates), as well as the sudden appearance of cultural features including sedentism, storage, architecture, art and symbolic expression, and cemeteries. This dichotomous treatment implies that we already have a satisfactory understanding of Early and Middle EP groups, as current research is focused

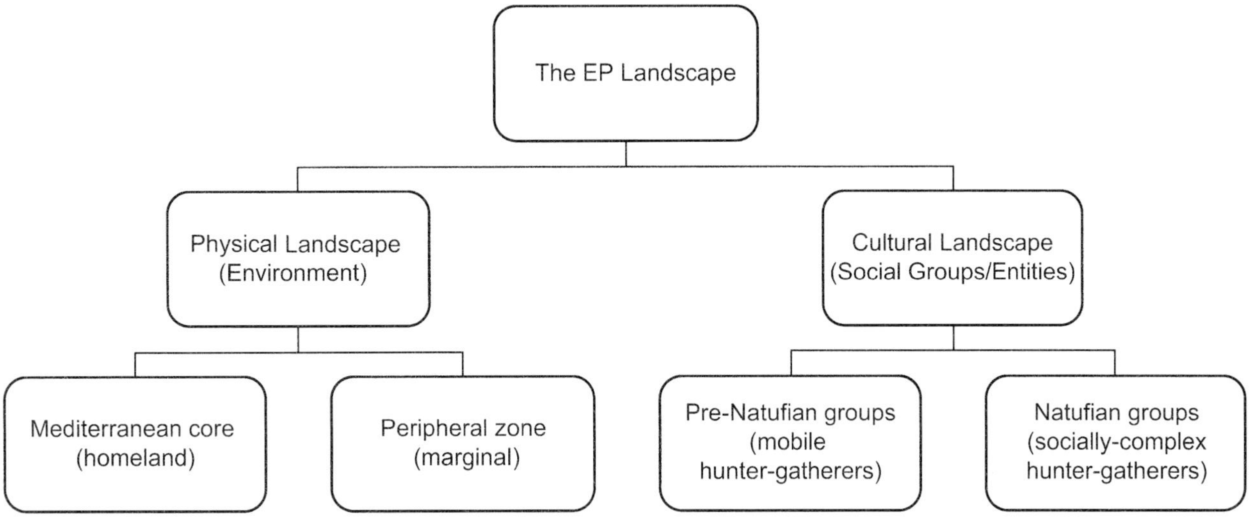

Fig. 4.1. Dichotomous treatment of the EP landscape. At each stage the landscape is more narrowly defined so that geographical spaces become disconnected from the people who inhabit them.

on the latest EP phase, the Natufian, as socially complex precursors to Neolithic farmers and villagers (e.g. Byrd 2005; Verhoeven 2004).

The Landscapes in Transition workshop highlighted several key research themes in the complex and multi-faceted transitions between hunter-gatherers and farmers (Finlayson and Warren this volume). In particular, two themes stand out as directly relevant to this paper: the need for a critical re-assessment of our fragmented approach to physical and social places in the Levantine EP, and the need to consider a perspective with greater time depth that encompasses all of the EP (~23–10 ka BP), and beyond, not just the Natufian.

This paper highlights some of the major theoretical, conceptual and practical issues regarding our approaches to landscape in the EP period. I explore how the two dominant landscape perspectives (physical and cultural) are currently treated as separate spheres of investigation, a situation which in the past has led to a patchy and inconsistent picture of human behaviour during the EP. I suggest that their integration will allow us to move beyond the pervasive environmentally deterministic paradigm. Mortuary behaviour at the Early and Middle EP sites of `Uyun al-Hammam and Kharaneh IV serves as just one brief example of how these landscape dichotomies cannot be maintained in light of the archaeological data. This paper attempts to highlight humans' engagement with their surroundings, rather than focus on simple cause-and-effect.

The physical landscape: space or place?

The physical landscape of the southern Levant has been studied in great detail over recent decades with the aim of reconstructing climatic changes and linking these directly to cultural events. Environmental determinism, explicitly or implicitly, has pervaded most explanatory models for culture change and the origins of agriculture in the Late Pleistocene and Early Holocene (e.g. Bar-Yosef 1987; 1996; Bar-Yosef and Belfer-Cohen 1989; 1992; Bar-Yosef and Meadows 1995; Braidwood 1960; Goring-Morris 1989; Goring-Morris and Belfer-Cohen 1998; Henry 1989; Wright 1993). More recently, less deterministic transitional models have argued that people became farmers through a series of steps, crossing social, economic and technological thresholds until hunting and gathering was no longer an option and there was no turning back (Bar-Yosef and Belfer-Cohen 1998; Belfer-Cohen and Bar-Yosef 2000; Byrd 2005; Kaufman 1986; McCorriston and Hole 1991). However, the physical environment remains the setting for these thresholds and, because several related factors (primarily climate) came together, the adoption of agriculture is argued to have been inevitable, as expressed by Wright: 'the origin of agriculture in the Near East can be attributed to the response of early people to a unique sequence of climatic events from 13,000 to 10,000 years B.P.' (Wright 1993, 466). The emphasis on the environment as the prime motivator in shaping cultural behaviours remains problematic (see also Watkins this volume).

The idea of a physical landscape divorced from cultural events and human agency is further perpetuated by our descriptions of the region. The Levant is described as, and is, an area with dramatic variation (in geology, landform, ecology, elevation) over short distances (e.g. Bar-Yosef 1990; 1998; Bar-Yosef and Belfer-Cohen 1989; Horowitz 1979). Syntheses of the EP often describe the palaeoenvironment in terms of its landforms, ecology, resources available and other geographical facts, then review the relevant cultural entities and introduce the necessary preconditions that allow for sedentism, technological and economic developments, and the adoption of agriculture to proceed (e.g. Bar-Yosef 1990; 1996; Bar-Yosef and Belfer-Cohen 1989; Byrd 2005;

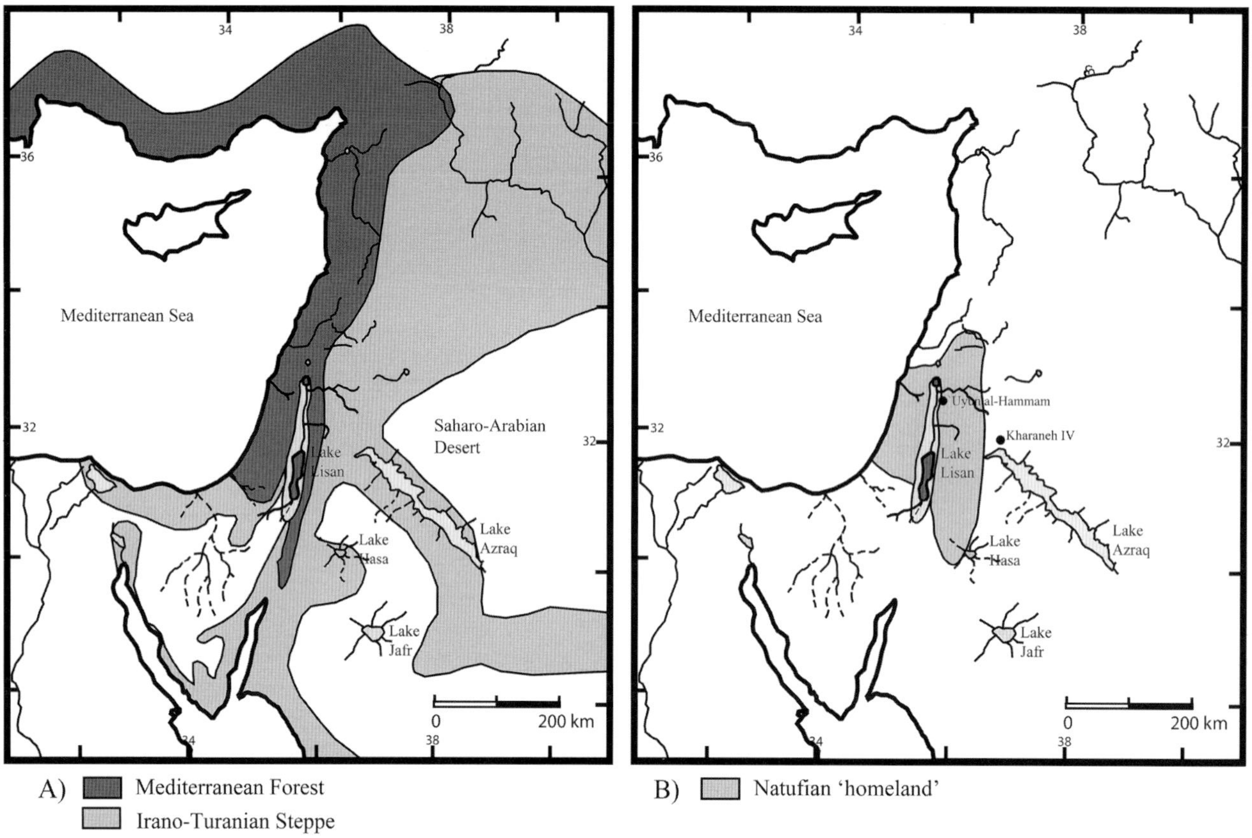

Fig. 4.2. A) reconstructed vegetation zones of the southern Levant differentiated on the basis of a forested Mediterranean core zone and its surrounding, more arid, peripheries (redrawn from Bar-Yosef and Belfer-Cohen 1989). B) the boundaries of the Natufian 'homeland', coinciding largely with the Mediterranean woodland vegetation zone (redrawn from Belfer-Cohen and Bar-Yosef 2000). Note also the locations of the Early and Middle EP sites of `Uyun al-Hammam and Kharaneh IV discussed in the text.

Goring-Morris and Belfer-Cohen 1998; Henry 1983; 1995; Kaufman 1986; Verhoeven 2004). Spatial and temporal changes amongst EP groups are framed in terms of a long trajectory of development in the Mediterranean core area or 'homeland', versus an arid adaptation to the more semi-desert peripheries (Belfer-Cohen and Bar-Yosef 2000, fig. 2; Henry 1995). These syntheses acknowledge ecological complexity, yet their models for cultural development simplify human behaviour within these varied environments to dichotomous core–periphery models.

The physical landscape of the southern Levant is argued to have been a space filled with EP groups whose movements and development were bounded by ecological parameters. It is almost as though this physical landscape is divided into two dichotomous zones, each with its own mutually exclusive trajectory of cultural development (Bar-Matthews and Ayalon 2003; Goring-Morris and Belfer-Cohen 1998; Henry 1995). Corresponding very roughly to modern vegetational boundaries (Bar-Yosef and Belfer-Cohen 1989, fig. 1), the Mediterranean core is characterized as woodland and the peripheries as steppe (Bar-Yosef 1987; Byrd 2005). In the core, abundant springs and rivers dotted a land of oak and pistachio forests full of gazelle, wild cattle, deer, aurochs and wild boar. This is the Natufian 'homeland', where a general abundance of resources promoted a long history of cultural development in which one group evolved into another in the Kebaran–Geometric Kebaran–Natufian sequence (Bar-Yosef 1991b). It is also the space into which most Late EP groups withdrew during the harsh conditions of the Younger Dryas (13–11.6 ka cal BP) because this resource-rich area provided a buffer to the dry climate that made neighbouring areas less inhabitable (Bar-Yosef and Belfer-Cohen 2002; Belfer-Cohen and Bar-Yosef 2000).

The peripheral area contained groups occupying isolated sites tied to oases (Bar-Yosef and Belfer-Cohen 1989) or specialized hunting camps (e.g. Harifian: Goring-Morris 1991 or Hamran: Henry 1995). The significant cultural developments that led to the Neolithic appear to have occured almost exclusively within the core area where conditions were amenable, while groups occupying the peripheries were marginal, their identities not persisting far into the Holocene (Belfer-Cohen and Bar-Yosef 2000; Goring-Morris 1987; 1989; Byrd 2005; Verhoeven 2004). Even substantial multi-phase sites like Kharaneh IV and others in the Azraq Basin (Figure 4.2) are considered marginal in EP–Neolithic transitions. They are classified as unique 'mega-sites' afforded existence owing to their

Approx. Dates cal B.P.	Period	Archaeological Entity	
		Mediterranean Zone	Steppe & Desert Zones
14,600–11,600	Late Epipalaeolithic	Natufian	Harifian
			Natufian
17,500–14,600	Middle Epipalaeolithic	Geometric Kebaran	Ramonian
			Mushabian
			Geometric Kebaran
22,500–17,500	Early Epipalaeolithic	Nizzanan	Nizzanan
		Kebaran	Kebaran
			Nebekian
		Masraqan (Late Ahmarian)	Masraqan (Late Ahmarian)

Table 4.1. The cultural landscape of the EP period with separate cultural trajectories for core and periphery zones (modified after Belfer-Cohen and Goring-Morris 2002).

locations near to Pleistocene lakes. Again, the physical landscape is the setting that determined, forced, or allowed culture change with a modest amount of direct evidence for local palaeolandscape features or their potential relationships to the activities of the people who inhabited the sites. Little remains known of how people may have perceived these spaces and alternative reasons for the repeated, long-term occupation of these mega-sites remain unexplored.

This core–periphery fragmentation of the landscape is extended to the treatment of southern and northern portions of the Levant, with the former being the core of research coverage and the latter the periphery. The Levant is a large region (some 1100 km long and 350 km wide: Bar-Yosef 1998), and the southern part has been much more thoroughly studied. The higher resolution of archaeological, chronological and environmental data from the southern Levant makes comparison between the two regions problematic. Although data from well-dated Epipalaeolithic sites in the northern Levant are increasingly becoming available (Albrecht *et al.* 1992; Gatov and Özdoğan 1994; Olszewski 1993; Otte *et al.* 1995), a lower intensity of research means that archaeological entities cannot yet be traced confidently between the south and the north. Should we continue to consider these two areas as separate, with different patterns and paces of cultural change? By the PPNA there are enough similarities to indicate a sphere of social interaction (Belfer-Cohen and Bar-Yosef 2000), which suggests the possibility of some integration during the preceding EP.

New palaeoenvironmental syntheses (e.g. Robinson *et al.* 2006; this volume) provide us with a more refined picture of the nature and timing of climate changes throughout the EP and Early Neolithic. The emerging picture (see also Watkins, this volume) reinforces a need for a better understanding of both the spatial and temporal variability of the archaeological record, a more complete picture of the palaeoenvironment and more correlation of these two lines of evidence. Watkins (this volume) highlights the theoretical and methodological problems with defining core areas and assigning the emergence of Neolithic societies within them. Similarly, new Early Neolithic landscape-based research in Cyprus is opening our eyes to the presence of probable EP sites on the island, well outside our mainland Levantine 'comfort zone' (McCartney *et al.*, this volume). In light of these new data, we are left with many serious questions regarding our interpretations of the physical landscape. How do we define a core area (and peripheries) and the activities of people in it? Is the core area an environmental space or a social place? Or is it even a necessary or useful concept?

The cultural landscape of the EP: who's who?

The cultural landscape of the Levantine EP is described as an assortment of spatially and temporally overlapping entities defined by relative proportions of microlith types (Table 4.1). Attempts to extricate patterns in human behaviour from the marked variability in the EP demands a consideration of the nature and meaning of lithic variation (Olszewski 2006) and leads us to a number of related questions: does tool style signify group identity? Are these social groups meaningful? At what scale?

To deal with this variability, researchers have tended towards either splitting (Bar-Yosef 1970; Fellner 1995; Goring-Morris 1987; 1996; Goring-Morris and Belfer-Cohen 1998; Henry 1995) or lumping (Byrd 1994; 1998; Gilead 1988; 1991; Neeley and Barton 1994; Olszewski 2006) assemblages; the former dominates the literature, although problems with this type-list approach are becoming visible (Maher 2005a; Neeley and Barton 1994; Olszewski 2001a; 2006; Pirie 2004). Most divisions relate to the presence or absence of the microburin technique and variation within microlithic tools. Different researchers often use different classification schemes, making inter-assemblage comparisons difficult.

Splitters emphasize variability between assemblages,

suggesting that it represents the behaviour of different 'peoples' (Bar-Yosef 1981; 1991a). EP researchers have for a long time argued that stone tool production was a learned tradition allowing for the identification of social groups with a direct link between microlith tool types and group identity (Bar-Yosef 1970; 1981; Fellner 1995; Goring-Morris 1987; Henry 1995). According to splitters, 'variability in chipped stone tool morphology, techniques of manufacture and specific means of hafting reflect, in addition to functional factors, the stylistic traditions of specific groups in the landscape' (Goring-Morris 1996, 130; see also Bar-Yosef 1991a; Goring-Morris 1987; 1995). Henry (1996) argues that a detailed examination of the reduction features of each assemblage clearly indicates that '[t]he techno-typological difference between Geometric Kebaran and Mushabian complexes largely stem from their being produced by different technological systems, not as different stages within a common system' (Henry 1996, 136). Different people were doing different things and social diversity in the EP is the most parsimonious explanation for such high assemblage variation over such short distances (Henry 1996). Thus sites with similarities in lithic assemblages, particularly the predominant microlith form, share a cultural tradition. Those sites with differing lithic assemblages have learned different traditions for tool production because they belong to different groups of people. Explaining lithic variability as material expressions of group identity currently provides some 18 different EP industries in the southern Levant, with 9 of these defined in Jordan alone (Olszewski 2001b).

However, this simplistic connection between microlith types and social identity has been challenged (Byrd 1994; Gilead 1991; Maher 2005a; Neeley and Barton 1994; Pirie 2004; Olszewski 2001a; 2001b; 2006): the distinctions between lithic tool types are not clear-cut or consistent between analysts, and do not follow clear spatial or chronological boundaries. Lumpers argue that over-emphasis on microlith variability creates a confusing, spatially and temporally disjointed, picture of the EP. At its extreme, new archaeological entities are named almost as each new site is excavated. Although lumpers accept that stone tool production is a learned process that conveys inherent social meaning and passes on a tradition of technique and style, they question whether microlith tool types are really indicative of identity. Lumpers suggest that we abandon some of these poorly defined and dated entities in favour of a scheme where the presence of fewer, well-defined groups highlights continuity and interaction (migration, trade and exchange, diffusion) across space and time.

Pirie (2004) critiques the reliance on social explanations for lithic diversity in the EP period and traces the roots of this approach to the methods of researchers who created the typologies. When new material cannot be fitted into existing schemes, a new typology is created. Bar-Yosef (1970) formalized the use of the term Geometric Kebaran for all sites in the Mediterranean core dominated by geometric microliths. Unsatisfied that this term adequately described contemporary assemblages in the Negev and Sinai with different, non-geometric microliths, Goring-Morris (1987) coined the term the Mushabian for these sites. Subsequently, Henry (1995) argued for the uniqueness of contemporary assemblages in southern Jordan and called them the Hamran. Here, the dichotomy of the Mediterranean core and marginal peripheries is corroborated as different social identities are created for each zone (Pirie 2004, 681), where groups in the former exhibit long-term development with increasing complexity to become agriculturalists (Bar-Yosef 1991b), while groups in the latter vanish from the archaeological record (Goring-Morris 1987) or develop a special arid adaptation into something resembling modern-day Bedouin (Henry 1995).

A recent trend in EP literature takes lumping to the extreme. Byrd (1994; 2005) attempts to avoid splitting by using broad descriptive categories. Non-geometric (Neeley *et al.* 1998), non-microlithic (Byrd 1998) or non-Natufian (Byrd 1994) designations lump together cultural-chronological categories into catch-all groups, yet leave us little idea how assemblages within the same group relate to one another. Similarly, Gilead (1988; 1991) argues that the sudden and dramatic behavioural changes evident in the Natufian are the only features that warrant subdivision. He advocates abandoning the terms Early and Middle EP and lumps them together as the latest Upper Palaeolithic or Kebaran, to be followed by the Natufian. Both approaches emphasize the apparent sudden appearance of social complexity in the Natufian (Byrd 2005) and construct a simple pre-Natufian/Natufian dichotomy as the basis for subdividing the EP period, a trend now perpetuated by other researchers (e.g. Verhoeven 2004). This creates a discontinuity between Natufian and other EP cultures and over-emphasizes the Natufian as the immediate precursor to the Neolithic. The variability of Early and Middle EP sites is overlooked in contributing to the development of the Natufian and is only viewed in retrospect from the later Natufian or Neolithic.

A significant implication of defining identity from lithic typology is the inherent, yet overlooked, connection to other aspects of material culture and behaviour. The assignment of an assemblage to the Natufian period is accompanied by the assumption that the 'people' who occupied this site also shared other aspects of 'Natufian' behaviour, such as social complexity and sedentism (Pirie 2004). Some see the Natufian as a crucial point where the economic, social and technological changes underway were pivotal to the Neolithic (Belfer-Cohen and Bar-Yosef 2000). While there is no doubt that Natufian groups played a large role in the intensification of food production, we are becoming increasingly aware that other features, such as sedentism (Belfer-Cohen and Bar-Yosef 2000; Boyd 2006), can be unconnected to agriculture. Moreover, many of the features that have come to characterize the Natufian have earlier roots (Boyd 2006; Maher 2007).

Olszweski (2001b; 2006, 19) addresses the problems in

dealing with variability in EP lithic assemblages by pointing out that the creation of new names for assemblages that do not fit the existing scheme takes little consideration of what these names may or may not actually 'mean' (Olszewski 2001b). According to Olszewski:

> … this debate itself is rooted at least in part in how the distinctions between lithic industries are defined, how they are recognized by various researchers, and how they are applied to particular assemblages. In many cases, it is not clear if the distinctions between assemblages are indeed greater than their overlapping characteristics … In the Levant, which is a very small geographical area, the sheer scale of named Epipalaeolithic industries and their often overlapping characteristics can be quite daunting to sort through, even at times to specialists in the field … For the most part, this plethora has been generated over time by individual scholars rather than by consensus among researchers working in the region. (2006, 20)

An approach to variability that does not dwell on either the minutia of naming entities or abandon naming altogether, but instead recognizes fluidity, diffusion and interaction across space and over time, will give us a more coherent picture of EP *people*, rather than their microliths. As always, this is easier said than done and we are just beginning down this road. It will also dispel the notion of a sudden appearance of social, economic and technological advancement in the Natufian as necessary preconditions for the Neolithic.

Fig. 4.3. Grave IV: the primary, extended interment of an adult female with a flat-sided cobble intentionally propped against the skull.

The Levantine EP landscape

Landscape archaeology in Levantine EP research does not come from the same theoretical milieu as that of British prehistory (e.g. Bradley 1998; Ingold 1993; Knapp and Ashmore 1999; Rossignol and Wandsnider 1992; Thomas 2001; Tilley 1994; Ucko and Layton 1999). Rather, palaeoenvironmental data was, with varying degrees of success, correlated to a confusing assortment of poorly dated and defined cultural phases, industries, complexes and facies that change inconsistently over space and time (Maher 2005a; Olszewski 2006; Pirie 2004), leaving us with an incoherent understanding of the relationships between EP groups and their surroundings and without a clear definition of what the EP landscape is (e.g. Boyd 2004). Is it a physical phenomenon whose spatial and temporal variations can be directly linked as a causal factor to cultural events? Is it a cultural phenomenon where changes in material culture can be used to identify social groups and trace changes in human behaviour? Or, most sensibly, can these dichotomous perspectives be effectively integrated?

I will examine this possibility at two sites in Jordan that, I argue, exemplify the nuanced relationship between people and the landscape and demonstrate that EP people had a very intimate knowledge of their surroundings, filling it with social memory and experiences passed down through several generations. `Uyun al-Hammam is a classic Middle EP site located within the Mediterranean 'homeland'. Kharaneh IV is an Early and Middle EP 'mega-site' located in the peripheral zone of eastern Jordan. Both sites demonstrate substantial occupation of a *place* that extended beyond habitation to include symbolic ties and living–dead interactions as witnessed by mortuary practices. The burial of multiple individuals at both sites suggests their use as burial grounds and reinforces the point that there is no simple pre-Natufian/Natufian division.

`Uyun al-Hammam

Discussed in detail elsewhere (Maher 2005b, 2008), the Geometric Kebaran site of `Uyun al-Hammam is situated in northern Jordan on a large ancient river terrace and is c. 1,500 sq m in size (Figure 4.2). The EP occupation is extremely rich in chipped stone, ground stone and faunal material. The discovery of 7 human burials in 2005, representing at least 10 individuals, more than doubles the number of known human remains from the Middle EP in the entire Levant (Maher 2007). Our preliminary assessment of the skeletons' morphological features, stratigraphic contexts, association with only Geometric Kebaran artefacts and one radiocarbon date from the pelvis of Homo 1 (Grave I; 12,400±180 bp, TO-11704) suggests that they date to sometime within the Geometric Kebaran period.

There are no obvious patterns in body position or orientation, but most of the graves are primary, extended interments of one adult (Maher 2007). There are no formal grave constructions and most of the burials had no discernible burial pit, perhaps being interred in shallow earthen pits. However, several graves had stone inclusions above, below and adjacent to the skeletons (Figure 4.3). All the burials were dug into Geometric Kebaran deposits and the grave fills contained dense concentrations of chipped stone and fauna. Consequently, the identification of grave goods was restricted to objects in direct contact with the bones. A common suite of artefacts was identified, including ground-stone implements (a limestone pounder, several broken basalt pestles, basalt vessel fragments, complete basalt pestles and handstones), chipped stone objects (trapeze/rectangles, cores, blades and endscrapers) and partial animal skeletons (a *Bos*(?) patella, a bone tool, and a canid skull in grave I and a distal phalange from a medium-sized mammal between the clavicle and mandible of the individual in grave IV). The placement of the canid skull in grave I suggests a special animal–human relationship: the animal may have been a pet, or formed personal decoration (Maher 2007). The presence of a worked bone tool and a canid skull in a grave represents the earliest appearance of these features in the EP.

Human remains from this period are extremely rare and the presence of such a large number of individuals at one location enhances our understanding of social, mortuary and ideological practices. The appearance of cemeteries seems to mark a significant feature of EP mortuary practices (Belfer-Cohen 1988; 1991; Byrd and Monahan 1995) in which the continual use of a location – a carefully selected and demarcated physical space – as a burial ground highlights a connection to a place. Byrd (2005, 257) summarizes this connection, stating: '[f]or the first time, Near Eastern hunter-gatherers buried their dead in organized cemeteries within and around these sedentary sites. This development suggests a new way of illustrating ancestral ties to the land … Burials may have served to signify and enforce changes in social relationships brought about by sedentism or to mark social boundaries or belongingness.' Burial grounds could help legitimize residential rights to sites and other landscape features (Byrd and Monahan 1995).

The thickness (more than 1.5 m) and nature of the deposits at `Uyun al-Hammam suggests that it was a favoured locale that was occupied repeatedly (Maher 2005b). After occupation, it appears to have remained a special place in the landscape and was used as a burial ground (Maher 2007, 2008), attesting to its ideological and social importance. Several Natufian base camp sites (Hayonim, Mallaha, Raqefet Cave) also have an underlying Geometric Kebaran occupation, strongly suggesting a pattern of re-occupation of these sites from the Middle to the Late EP. Even the four phases of occupation at Kharaneh IV (Early through Middle EP) attest to its position in the social memory of the people who continued to occupy it.

Fig. 4.4. A view of the Early and Middle EP site of Kharaneh IV, surrounded by a partially destroyed fence.

Fig. 4.5. The flint- and bone-covered surface of Kharaneh IV.

Continuity in site use throughout the EP implies connection to a familiar and favoured 'place'.

At `Uyun al-Hammam we see both repeated occupation of this site over several generations attested to by a high density and diversity of artefacts (lithics, ground stone, shell, fauna) suggesting a multitude of activities occurring on-site, and its subsequent use as a burial ground marking its continued significance in the landscape. It remains unclear whether all the burials were contemporary, but given the current lack of evidence for specific diseases or trauma (J. Stock pers. comm.), and that grave I shows disturbance of an earlier burial by a subsequent interment (Maher 2007), it is unlikely that all the burials were made at once. The site was probably re-used as a formal place in the landscape for the specific purpose of burying the dead.

Kharaneh IV

During the 1980s, Mujahed Muheisen investigated the multi-phase EP site of Kharaneh IV, located 2 km south-west of Qasr Kharaneh (Muheisen 1983; 1988a) (Figure 4.2). Kharaneh IV, which consists of a large, low mound, encompasses more than 21,000 sq m and has an exceedingly high artefact density (Figure 4.4). The entire surface of the mound is paved in flint artefacts and animal bone (Figure 4.5). Test trenches showed that this extremely high artefact content continued to a depth of more than 1.5 m (Muheisen 1988a).

Muheisen documented at least four EP occupation phases containing a complex suite of remains such as post-holes, living floors, hearths, bone, ground stone, lithic tools and debitage, extremely dense faunal material and two human burials (Muheisen 1988a; Rolston 1982). Phases A and B belong to the Kebaran, phase C is transitional, and phase D is Geometric Kebaran (Muheisen 1988a). The four phases span at least 5000 years, with continuity in lithic assemblages over time (Muheisen and Wada 1995). This massive multi-occupational site appears to have been a place of long-term, repeated occupation.

The two burials (both adult males) at Kharaneh IV (phase B) were found within horizons below living floors, one containing hearths. Two stones covering the lower legs and pelvis were placed on top of one burial, and two gazelle horn cores were reportedly found to the left and right of the skull (Muheisen 1988b). The individual was male, suffered from rheumatic arthritis and died between the ages of 35 and 45 years (Rolston 1982). We see a mortuary tradition similar to other Early and Middle EP interments (Maher 2007): the semi-extended burial of adult males in poorly defined pits with a common suite of grave goods. But we are left with questions about the meaning of these. What is the significance of burials with gazelle horns? What are the spatial relationships between burials, living floors and hut structures?

While there is little doubt that the burial (indeed, both burials) carried social significance, the nature of this significance remains unclear. Perhaps the stones were designed to hold down the body and keep it separate from the realm of the living above. Perhaps the connections between the living and the dead at Kharaneh IV were integrated into the daily life of the dwelling's occupants. It is possible that interment served to mark this place in the social (cultural) landscape. The staggering scale and temporal depth of occupation at Kharaneh IV and the diversity of activities apparent throughout the sequence probably signify an attraction to the site for reasons beyond subsistence, although local conditions (water, food supply) surely played a critical role in sustaining occupation at the site. The number of people occupying it at any one time, the number of stays and the duration of each stay remain unknown. We do know that people chose this particular location to aggregate in substantial enough numbers and for long enough to accumulate more than 1.5 m of dense cultural debris spanning several thousand years. From any perspective, the site had important social significance.

EP mortuary practices

The burials at both sites are predominantly single, primary interments of individuals (usually male) in extended (or semi-extended) position with a common suite of grave goods. This pattern appears to be a widespread feature among Early and Middle EP burials (e.g. Ohalo II (Nadel 1994), Neve David (Kaufman and Ronen 1987), Wadi Qadish (Garrard and Yazbeck 2003)) and suggests burial by people with shared mortuary traditions (Maher 2007).

Burials with gazelle horn cores, red ochre and particular human–animal associations are some of the features which characterize the first and sudden appearance of 'complex' burial customs in the Natufian and make them symbolically, ritually and cognitively significant (Byrd and Monahan 1995; Verhoeven 2004, 237–9). The act of burying one's dead on-site suggests ties that go beyond subsistence and habitation. The site becomes a symbolic place and the burial of several individuals strengthens these ties and ensures that the significance of this place spans more than one lifetime/generation. Verhoeven implies that these features first appear in the Natufian period and, according to Byrd and Monahan (1995, 254), 'On-site interment is one of the most prominent changes in social behaviour at Natufian sites. The frequency, patterning, and dating of these burials indicate they are the earliest cemeteries in southwest Asia.' The recognition of this symbolic aspect to landscape in the EP is largely overlooked by Epipalaeolithic and Neolithic researchers. Yet at Kharaneh IV dead people were spatially and symbolically associated with the living in houses and the settlement. At both sites there is a clear association between human and animals; Homo 1 and a canid skull at `Uyun al-Hammam and a human with gazelle horn cores at Kharaneh IV. Both sites contain graves with stones intentionally placed on or adjacent to the bodies, perhaps as grave furnishings or constructions like those at Mallaha (Valla *et al.* 2004) and Hayonim (Belfer-Cohen 1988). The presence of these features and the burial of several individuals at one site (as a burial ground) in the Early and Middle EP suggests that the social and ideological changes linked to Natufian cemeteries are already present in pre-Natufian groups (Maher 2007). A social memory of place was established at `Uyun al-Hammam and Kharaneh IV by the fixing of physical spaces as both recurrent habitation sites and burial grounds, just as in the Natufian. The practices at Kharaneh IV may have signified continued ties between the dead and the living, while at `Uyun al-Hammam importance was placed on maintaining ties to a place after habitation. Furthermore, these changes are not restricted to the Mediterranean core area and imply a local continuity of these practices as 'traditions' (Byrd and Monahan 1995) probably extending throughout the EP, contrasting with the existing simple-pre-Natufian/complex-Natufian dichotomy. Throughout the entirety of the EP, the landscape was filled with socially and culturally constructed places, fixed in time and space by the communities (see Watkins this volume) that inhabited or used them. The patterns emerging in EP burial practices, that may become increasingly apparent as more Early and Middle EP sites are excavated, require us to reconsider the temporal boundaries that we have set for the appearance of features typically used to define Natufian complexity.

Discussion

Byrd (2005) argues for the rapid emergence of sedentary complex hunter-gatherers in the Natufian period during optimal climatic conditions within the resource-rich Mediterranean core area of the southern Levant. Focusing social developments within the Natufian homeland (Figure 4.2), he states:

> The sudden onset of complex, sedentary hunter-gatherers at the start of the Early Natufian was the first step in a process of remarkable changes in settlement pattern, food procurement strategies, community size, social interrelationships, and ideology. These developments … laid the institutional and organizational framework for the dramatic emergence of Early Neolithic village life and intensive cultivation that forever reconfigured how humans interacted with each other and the world around them. (Byrd 1995, 270)

In this paper I have explored EP mortuary practices at two sites to highlight our need for a more integrated approach to landscape that sees the many transitions occurring in the EP with greater time depth. Having briefly reviewed several of the major debates within current EP research, I believe there are several ways to move forward. We need to embrace a more integrated approach to landscape, one that considers the physical spaces people lived in as places imbued with economic, technological, social, and ideological meaning. We need to re-establish an active dialogue among Epipalaeolithic researchers to develop a common terminology for the people and places whose interactions we are interested in investigating. Little new research engages with the Early and Middle EP periods, focusing instead on the later Natufian. Sedentism, architecture, a rich bone tool industry, an emergent and diverse artistic repertoire, intensified plant use and burials with grave goods in formal cemeteries are features used to mark socially complex Natufian groups (Bar-Yosef 1998; Bar-Yosef and Belfer-Cohen 1989; Belfer-Cohen and Bar-Yosef 2000; Byrd 2005; Byrd and Monahan 1995; Verhoeven 2004). These features are considered to be pre-adaptations to, and the Natufian culture a pre-cursor of, Neolithic farming communities. Yet all of the features we interpret as indicators of social complexity appear already established in the Natufian. In order to understand their origins we must look to earlier periods. Consequently, I believe a gap exists in our understanding of a crucial period in prehistory. The cultural features we use to construct the simple-pre-Natufian/complex-Natufian dichotomy are foreshadowed in the Early and Middle EP, pushing back temporal boundaries and highlighting the fact that not all of these traits appeared all at once in the Late EP.

Taking mortuary practices as one example, we see that the nature of burials at Early and Middle EP sites strongly suggests continuity, rather than dichotomy, in the EP. If we shift our focus from concern about defining microlith suites and naming people, or correlating behavioural changes directly with climatic ones, and instead look at the nature of variability throughout the entirety of the EP, we see that EP people thought about and acted within their landscape in ways that hint at the big changes we see later in the Neolithic. In consequence, we need to consider these sites as places, not spaces, marked physically on the ground and in the minds of the people who lived or buried their dead there. I am not arguing that we should view Early and Middle EP groups as socially complex hunter-gatherers equivalent to the Natufian. I am presenting an alternative approach to landscapes that does not focus on dichotomies, but on continuity. `Uyun al-Hammam and Kharaneh IV are both places used for a variety of purposes and over long periods of time. They share many features common to other Early and Middle EP sites, and to later Natufian ones.

Much of the Early EP work I referred to at the beginning of this paper was ground-breaking research that set up a number of indispensable theoretical and methodological frameworks for identifying and interpreting EP variability. Yet it seems that recently there has been a shift away from EP research to other, later, periods. We need to revive interest in the earlier EP by re-conceptualizing the frameworks that determine how we understand issues such as mobility, social interaction and landscapes, and build on the flurry of EP research of the 1980s and 1990s. We now have many sites and data that do not fit the current cultural-chronological schemes and we need to focus on testing whether these sites are truly unique, as we often dismiss them as, or, as I argue, indicative of broader patterns of EP behaviour. This involves integrating a more refined big picture of forager–farmer transitions with data from sites spanning the entire EP located throughout the entire Levant.

Acknowledgements

I would like to thank Bill Finlayson and Graeme Warren for inviting me in to participate in the 'Landscapes in Transition' workshop and all of the participants who made the workshop so thoroughly enjoyable. In particular, I would like to thank Anna Belfer-Cohen, Nigel Goring-Morris, Doug Baird, Trevor Watkins and Stuart Robinson for very helpful and thought-provoking discussions during the workshop, particularly in regards to work at Kharaneh IV. I would also like to thank Kevin Gibbs for his comments on earlier drafts of this paper.

Bibliography

Albrecht, G., Albrecht, B., Berke, H., Burger, D., Moser, J., Rähle, W., Schoch, W., Storch, G., Uerpermann, H. P. and Urban, B. (1992) Late Pleistocene and Early Holocene finds from Öküzini: a contribution to the settlement history of the Bay of Antalya, Turkey. *Paléorient* 18, 123–41.

Bar-Matthews, M. and Ayalon, A. (2003) Climatic conditions in the Eastern Mediterranean during the last glacial (60–10 ky bp) and their relations to the Upper Palaeolithic in the Levant: oxygen and carbon isotope systematics of cave deposits. Pp.

13–18 in A. Belfer-Cohen and N. Goring-Morris (eds), *More Than Meets the Eye: Studies on Upper Palaeolithic Diversity in the Near East*. Oxford: Oxbow Books.

Bar-Yosef, O. (1970) The Epipalaeolithic Cultures of Palestine. Unpublished PhD thesis, Institute of Archaeology, Jerusalem: Hebrew University.

Bar-Yosef, O. (1981) The Epi-palaeolithic complexes in the southern Levant. Pp. 389–408 in J. Cauvin and P. Sanlaville (eds), *Préhistoire du Levant*. Paris: Centre National des Recherches Scientifique.

Bar-Yosef, O. (1987) Late Pleistocene adaptations in the Levant. Pp. 219–36 in O. Soffer (ed.), *The Pleistocene Old World*. New York: Plenum Press.

Bar-Yosef, O. (1990) The last glacial maximum in the Mediterranean Levant. Pp. 58–78 in C. Gamble and O. Soffer (eds), *The World at 18000 BP*. London: Unwin Hyman Ltd.

Bar-Yosef, O. (1991a) Raw material exploitation in the Levantine Epi-Paleolithic. Pp. 235–50 in A. Montet-White and S. Holen (eds), *Raw Material Economies among Prehistoric Hunter-Gatherers*. Lawrence, KS: University of Kansas Press.

Bar-Yosef, O. (1991b) Stone tools and social context in Levantine prehistory. Pp. 371–95 in G. A. Clark (ed.), *Perspectives on the Past: Theoretical Biases in Mediterranean Hunter-Gatherer Research*. Philadelphia, PA: University of Pennsylvania Press.

Bar-Yosef, O. (1996) The impact of the Late Pleistocene–Early Holocene climatic changes on humans in Southwest Asia. Pp. 61–78 in L. Guy Straus, B. Valentin Eriksen, J. M. Erlandson and D. R. Yesner (eds), *Humans at the End of the Ice Age: The Archaeology of the Pleistocene–Holocene Transition*. New York: Plenum Press.

Bar-Yosef, O. (1998) The Natufian culture in the Levant: threshold to the origins of agriculture. *Evolutionary Anthropology* 159–77.

Bar-Yosef, O. and Belfer-Cohen, A. (1989) The origins of sedentism and farming communities in the Levant. *Journal of World Prehistory* 3, 447–98.

Bar-Yosef, O. and Belfer-Cohen, A. (1992) From foraging to farming in the Mediterranean Levant. Pp. 21–48 in A. B. Gebauer and T. D. Price (eds), *Transitions to Agriculture in Prehistory*. Madison, WI: Prehistory Press.

Bar-Yosef, O. and Belfer-Cohen, A. (1998) The Natufian Culture in the Levant: Thresholds to the Origins of Agriculture. *Evolutionary Anthropology* 6, 159–77.

Bar-Yosef, O. and Belfer-Cohen, A. (2002) Facing Environmental Crisis: Societal and Cultural Changes at the Transition from the Younger Dryas to the Holocene in the Levant. Pp. 55–66 in R. T. J. Cappers and S. Bottema (eds), *The Dawning of Farming in the Near East*. Studies in Near Eastern Production, Subsistence and Environment Vol. 6. Berlin: Ex oriente.

Bar-Yosef, O. and Meadows, R. H. (1995) The origins of agriculture in the Near East. Pp. 39–94 in T. D. Price and A. B. Gebauer (eds), *Last Hunters–First Farmers: New Perspectives on the Prehistoric Transition to Agriculture*. Santa Fe, NM: School of American Research Press.

Belfer-Cohen, A. (1988) The Natufian graveyard at Hayonim Cave. *Paléorient* 14, 297–308.

Belfer-Cohen, A. (1991) The Natufian in the Levant. *Annual Review of Anthropology* 20, 167–86.

Belfer-Cohen, A. and Bar-Yosef, O. (2000) Early sedentism in the Near East: a bumpy ride to village life. Pp. 19–37 in I. Kuijt (ed.), *Life in Neolithic Farming Communities: Social Organization, Identity, and Differentiation*. New York: Kluwer Academic/Plenum Publishers.

Belfer-Cohen, A. and Goring-Morris, A. N. (2002) Why Microliths? Microlithization in the Levant. Pp. 57–68 in R. G. Elston and S. L. Kuhn (eds), *Thinking Small: Global Perspectives on Microlith Technologies*. Arlington: American Anthropological Association 12.

Boyd, B. (2004) Agency and landscape: abandoning the 'nature/culture' dichotomy in interpretations of the Natufian and the transition to the Neolithic. Pp. 119–36 in C. Delage (ed.), *The Last Hunter-Gatherer Societies in the Near East*. British Archaeological Reports, International Series 1320. Oxford: BAR.

Boyd, B. (2006) On sedentism in the later Epipalaeolithic (Natufian) Levant. *World Prehistory* 38, 164–78.

Bradley, R. (1998) *The Significance of Monuments: On the Shaping of Human Experience in Neolithic and Bronze Age Europe*. London: Routledge.

Braidwood, R. J. (1960) The agricultural revolution. *Scientific American* 203, 130–52.

Byrd, B. (1994) Late Quaternary hunter-gatherer complexes in the Levant between 20,000 and 10,000 BP. Pp. 205–26 in O. Bar-Yosef and R. Kra (eds), *Late Quaternary Chronology and Paleoclimates of the Eastern Mediterranean*. Tuscon, AZ: Radiocarbon.

Byrd, B. (1998) Spanning the gap from the Upper Paleolithic to the Natufian: the Early and Middle Epipaleolithic. Pp. 64–82 in D. O. Henry (ed.), *The Prehistoric Archaeology of Jordan*. Oxford: Archaeopress.

Byrd, B. (2005) Reassessing the emergence of village life in the Near East. *Journal of Archaeological Science* 13, 231–90.

Byrd, B. F. and Monahan, C. M. (1995) Death, mortuary ritual, and Natufian social structure. *Journal of Anthropological Archaeology* 14, 251–87.

Fellner, R. O. (1995) *Cultural Change and the Epipalaeolithic of Palestine*. BAR International Series 599. Oxford: BAR.

Garrard, A. and Yazbeck, C. (2003) Qadish Valley Prehistory Project, Northern Lebanon: Summary of first two seasons investigations. *Bulletin d'Archéologie et d'Architecture Libanaises* 7, 7–14.

Gatov, I. and Özdoğan, M. (1994) Some Epi-paleolithic sites from NW Turkey. Ağaçlı, Domalı and Gümüşdere. *Anatolica* 20, 97–120.

Gilead, I. (1988) The Upper Palaeolithic to Epi-Palaeolithic transition in the Levant. *Paléorient* 14, 177–82.

Gilead, I. (1991) The Upper Palaeolithic Period in the Levant. *Journal of World Prehistory* 5(2), 105–54.

Goring-Morris, A. N. (1987) *At the Edge. Terminal Pleistocene Hunter-Gatherers in the Negev and Sinai*. BAR International Series 361. Oxford: BAR.

Goring-Morris, A. N. (1989) Developments in Terminal Pleistocene hunter-gatherer socio-cultural systems: a perspective from the Negev and Sinai deserts. Pp. 7–28 in I. Hershkovitz (ed.), *People and Culture in Change. Proceedings of the Second Symposium on Upper Palaeolithic, Mesolithic and Neolithic Populations of Europe and the Mediterranean Basin*. BAR International Series 508. Oxford: BAR.

Goring-Morris, A. N. (1991) The Harifian of the Southern Levant. Pp. 173–216 in O. Bar-Yosef and F. Valla (eds), *The Natufian Culture in the Levant*. Ann Arbor, MI: International Monographs in Prehistory.

Goring-Morris, A. N. (1995) Complex hunter-gatherers at the

end of the Paleolithic (20,000–10,000 BP). Pp. 141–68 in T. E. Levy (ed.), *The Archaeology of the Holy Land*. London: Leicester University Press.

Goring-Morris, A. N. (1996) Square pegs into round holes: a critique of Neeley and Barton. *Antiquity* 70, 130–5.

Goring-Morris, A. N. and Belfer-Cohen, A. (1998) The articulation of cultural processes and the late quaternary environmental changes in Cisjordan. *Paléorient* 23, 71–93.

Henry, D. O. (1983) Adaptive Evolution within the Epipalaeolithic of the Near East. *Advances in World Archaeology* 2, 99–160.

Henry, D. O. (1989) *From Foraging to Agriculture: The Levant at the End of the Ice Age*. Philadelphia, PA: University of Pennsylvania Press.

Henry, D. O. (1995) *Prehistoric Cultural Ecology and Evolution: Insights from Southern Jordan*. New York: Plenum Press.

Henry, D. O. (1996) Functional minimalism versus ethnicity in explaining lithic patterns in the Levantine Epipalaeolithic. *Antiquity* 70, 135–6.

Horowitz, A. (1979) *The Quaternary of Israel*. New York: Academic Press.

Ingold, T. (1993) The temporality of the landscape. *World Archaeology* 25, 152–74.

Kaufman, D. (1986) A reconsideration of adaptive change in the Levantine Epipalaeolithic. Pp. 117–28 in L. Strauss (ed.), *The End of the Palaeolithic in the Old World*. BAR International Series 284. Oxford: BAR.

Kaufman, D. and Ronen, A. (1987) La sepulture kebarienne geometrique de Névé-David Haïfa, Israël. *L'Anthropologie* 91, 335–42.

Knapp, B. and Ashmore, W. (1999) Archaeological landscapes: constructed, conceptualised, ideational. Introduction. Pp. 1–30 in B. Knapp and W. Ashmore (eds), *Archaeological Landscapes: Constructed, Conceptualised, Ideational*. Oxford: Blackwell.

McCorriston, J. and Hole, F. (1991) The ecology of seasonal stress and the origins of agriculture in the Near East. *American Anthropologist* 93, 46–9.

Maher, L. A. (2005a) *The Epipalaeolithic in Context: Palaeolandscapes and Prehistoric Occupation of Wadi Ziqlab, Northern Jordan*. Unpublished PhD thesis, Department of Anthropology, University of Toronto.

Maher, L. A. (2005b) Recent excavations at the Middle Epipalaeolithic encampment of Uyyun al-Hammam, Northern Jordan. *Annual of the Department of Antiquities of Jordan* 49, 101–14.

Maher, L. A. (2007) Microliths and mortuary practices: new perspectives on the Epipalaeolithic in Northern and Eastern Jordan. Pp. 295–301 in T. E. Levy, P. M. M. Daviau, R. W. Younker and M. Shaer (eds), *Crossing Jordan: North American Contributions to the Archaeology of Jordan*. London: Equinox.

Maher, L. A. (2008) 2005 Excavations at the Geometric Kebaran site of `Uyun al-Hammam, al-Koura District, Northern Jordan. *Annual of the Department of Antiquities of Jordan* 51, 163–72.

Muheisen, M. (1983) La Préhistoire en Jordanie. Recherches sur l'Epipaléolithique. L'Example du Gisement de Kharaneh IV. Unpublished PhD thesis, Géologie du Quaternaire et Préhistoire, l'Université de Bordeaux I.

Muheisen, M. (1988a) The Epipalaeolithic phases of Kharaneh IV. Pp. 353–67 in A. Garrard and H. G. Gebel (eds), *The Prehistory of Jordan. The State of Research in 1986*. BAR International Series 396. Oxford: BAR.

Muheisen, M. (1988b) Le Gisement de Kharaneh IV, Note Summaire Sur la Phase D. *Paléorient* 14, 265–9.

Muheisen, M. and Wada, H. (1995) An analysis of the microliths at Kharaneh IV Phase D, Square A20/37. *Paléorient* 21, 75–95.

Nadel, D. (1994) Levantine Upper Palaeolithic–Early Epipalaeolithic burial customs: Ohalo II as a case study. *Paléorient* 20, 113–21.

Neeley, M. and Barton, C. M. (1994) A new approach to interpreting late Pleistocene microlith industries in Southwest Asia. *Antiquity* 68, 275–88.

Neeley, M., Peterson, J., Clark, G., Fish, S. and Glass, M. (1998) Investigations at Tor al-Tareeq: an Epipaleolithic site in the Wadi el-Hasa, Jordan. *Journal of Field Archaeology* 25, 295–317.

Olszewski, D. (1993) Zarzian microliths from Warwasi Rockshelter, Iran: scalene triangles as arrowhead components. Pp. 199–205 in G. L. Peterkin, H. M. Bricker and P. Mellars (eds), *Hunting and Animal Exploitation in the Later Palaeolithic and Mesolithic of Eurasia*. Washington, DC: American Anthropological Association Number 4.

Olszewski, D. (2001a) My 'backed and truncated bladelet', your 'point': terminology and interpretation in Levantine Epipalaeolithic assemblages. Pp. 308–13 in I. Caneva, C. Lemorini, D. Zampetti and P. Biagi (eds), *Beyond Tools: Redefining the PPN Lithic Assemblages of the Levant. Third Workshop on PPN Chipped Lithic Industries*. Ca'Foscari University of Venice: *ex oriente*.

Olszewski, D. (2001b) The Palaeolithic period, including the Epipalaeolithic. Pp. 31–65 in B. McDonald, R. Adams and P. Bienkowski (eds), *The Archaeology of Jordan*. Sheffield: Sheffield Academic Press.

Olszewski, D. (2006) Issues in the Levantine Epipaleolithic: the Madamaghan, Nebekian and Qalkhan (Levant Epipaleolithic). *Paléorient* 32, 19–26.

Otte, M., Yalcinkaya, I., Leotard, J. M., Kartal, M., Bar-Yosef, O., Kozlowski, J., Bayon, I. L. and Marshack, A. (1995) The epi-palaeolithic of Öküzini cave (SW Anatolia) and its mobiliary art. *Antiquity* 69, 931–44.

Pirie, A. (2004) Constructing prehistory: lithic analysis in the Levantine Epipalaeolithic. *Journal of the Royal Anthropological Institute* 10, 675–703.

Robinson, S. A., Black, S., Sellwood, B. W. and Valdes, P. J. (2006) A review of palaeoclimates and palaeoenvironments in the Levant and Eastern Mediterranean from 25,000 to 5000 years BP: setting the environmental background for the evolution of human civilisation. *Quaternary Science Reviews* 25, 1517–41.

Rolston, S. L. (1982) Two prehistoric burials from Kharaneh. *Annual of the Department of Antiquities of Jordan* 26, 221–2.

Rossignol, J. and Wandsnider, L. (1992) *Space, Time, and Archaeological Landscape*. New York: Plenum Press.

Thomas, J. (2001) Archaeologies of place and landscape. Pp. 165–86 in I. Hodder (ed.), *Archaeological Theory Today*. Cambridge: Polity Press.

Tilley, M. (1994) *A Phenomenology of Landscape: Places, Paths, and Monuments*. Oxford: Oxford University Press.

Ucko, P. J. and Layton, R. (1999) *The Archaeology and Anthropology of Landscape: Shaping Your Landscape*. New York: Routledge.

Valla, F. R., Khalaily, H., Valladas, H., Tisnérat-Laborde, N., Samuellian, N., Bocquentin, F., Bridault, A., Simmons, T., Dosseur, G. L., Rosen, A. M., Dubreuil, L., Bar-Yosef Mayer, D. and Belfer-Cohen, A. (2004) Les fouilles de Mallaha en 2000 et 2001: 3ème rapport préliminaire. *Journal of the Israel Prehistoric Society* 34, 49–244.

Verhoeven, M. (2004) Beyond boundaries: nature, culture, and a holistic approach to domestication in the Levant. *Journal of World Prehistory* 18, 179–282.

Wright, H. (1993) Environmental determinism in the Near Eastern prehistory. *Current Anthropology* 34, 458–69.

5. Subsistence at 4000–3700 cal BC: Landscapes of Change or Continuity?

Nicky Milner

Subsistence change at the Mesolithic–Neolithic transition has been much debated, particularly in recent years with the application of stable isotope analysis on human bone. The transition has been characterized by some as rapid, while others argue that the uptake of agriculture was slow. However, when multi-scalar approaches are taken the transition can be argued to be complex and variable. This paper examines the current evidence for the introduction of different domesticated plants and animals in different locations. In addition, the continuation of hunter-gatherer lifestyles are explored. It is argued that there are problems with the data but that methods such as Bayesian modelling are revolutionizing the ways in which the transition can be explored.

Introduction

> The transition from hunting and gathering to agriculture is arguably the most important event in human prehistory, representing a shift from foraging to farming, from food collection to food production, from wild to domestic, that sets the stage for most of the significant subsequent developments in human society (Price 2000, 1).

The transition from foraging to farming has often been hailed as the most profound revolution in human history (e.g. Barker 2006, 414; Price 2000). In terms of subsistence change in Britain and Ireland, the rapidity of the transition has been argued forcefully, particularly in light of recent stable isotope studies (e.g. Richards and Schulting 2003; Rowley-Conwy 2004; *cf.* Thomas 1999). Rowley-Conwy (2004, 106) suggests that archaeologists tend to think of 'slow' in terms of millennia and undoubtedly the introduction of new animals and grain is rapid, if considering change as part of a grand narrative. In Britain, there is little evidence for farming prior to c. 4000 cal BC and by 3700/3600 cal BC there is considerable proof of the widespread presence of domesticated plants and animals. However, 300–400 years is a long time if the transition is to be considered in terms of *human experience*; comparable in terms of time, for example, to the duration of Roman Britain. In this context the transition may be seen as less rapid and more variable depending on location. This is perhaps where some misunderstandings have arisen, particularly with isotope studies, where rapid and slow are words which have been used frequently but with little explanation (Milner *et al.* 2004).

In terms of human experience, it is interesting to consider the ways in which these new plants and animals must have made an impression not only on people but also on the landscape of the islands of Britain and Ireland. Although some people had probably been in contact with communities on the continent before this point, it is perhaps unlikely that most of the population had actually come face to face with the new plants and animals and ways of managing them. Some of the animals, such as sheep and goat, must have seemed completely foreign. In Britain, cattle and pig will have been more familiar, similar in some respects to their wild counterparts, and yet different, particularly in terms of the ways in which they could be controlled. In Ireland, the evidence suggests there were no aurochs in the Mesolithic; in fact, there were no large mammals (elk, red deer or roe deer) (Woodman and McCarthy 2003, 23–33), and therefore the introduction of cattle here must have had a significant impact.

In addition, some of the products, such as milk and wool, would probably have been new to most people. In the Mesolithic there was a huge range of plants that people are likely to have exploited (Mears and Hillman 2007) but the introduction of cereals must have had a significant impact, in that they too would have produced different products: perhaps types of porridge, flour for breads, and maybe even beer (Dineley and Dineley 2000).

We may also consider how these innovations might have affected the landscape and what changes might have occurred. Did domesticated animals like cows need to be penned and did they need to graze in open areas which needed to be created and then maintained? Or were sheep left to wander in already open areas like coastlines where they could graze on coastal plants and seaweeds? Were pigs kept in a similar manner to those in places like New Guinea, where they are semi-feral and allowed to roam freely? To what extent was vegetation cleared for horticulture or larger-scale farming and what sort of size plots may have been needed for these practices?

In order to consider the impact of these new domesticates on both people and the landscapes they inhabit, multi-scalar approaches are necessary which might identify 'variety, messiness and localness' (Cooney 2007, 543; Whittle 2003). However, is this actually possible with the current data sets available? Although the traditional starting point for the transition tends to be seen as 4000 cal BC in Britain, a move towards more detailed chronologies through the use of strict sampling methods and Bayesian statistical frameworks means that the understanding of when innovations occur is becoming more and more refined. Much of this revolutionary dating has been conducted on monuments in the south of Britain, such as the long barrows, which appear to have been constructed from about 3800/3750 cal BC (Bayliss and Whittle 2007; Whittle 2007, 382). Other new practices, such as the use of bowl pottery and stone and flint axes, the construction of 'halls' and the farming of new domesticated plants and cereals, may have appeared earlier. But the crucial period of c. 4000–3700 cal BC, arguably 'the transition', when many of these new introductions are thought to appear in Britain and Ireland, still remains rather unclear (Whittle 2003; 2007). In Ireland, the dates for the early Neolithic are also traditionally seen as from 4000 cal BC (Cooney 2007), but the difference between Britain and Ireland is that there is more evidence – such as cattle dating to the mid 5th millennium cal BC at Ferriter's Cove – to place the beginnings of innovation and change in the latter earlier.

This paper examines this enigmatic period and reviews some of the evidence for subsistence change, with the ultimate aim of assessing how far it is possible to understand the speed of change, the degree of change and the extent to which innovations may have impacted on the landscape. It will first evaluate the evidence for subsistence change, considering a range of different methods and data and their limitations. This will be followed by an assessment of the arguments for a continuation of fishing, hunting and gathering practices into the 4th millennium cal BC.

Subsistence at 4000–3700 cal BC

In recent years, an important source of evidence for dietary change has been the stable isotope analysis of human bone. Both Mesolithic and Neolithic human bones have been analysed in order to determine changes in diet across the transition and the broad picture is a change from marine-dominated diets to terrestrial-based diets (e.g. Schulting and Richards 2002). However, it has been argued that the picture is in fact more complicated than this, and that the stable isotope data do not always accord with the faunal remains (Milner *et al.* 2004; but see also Richards and Schulting 2006; Milner *et al.* 2006).

The main problem with the British and Irish evidence is that there are very few 'Mesolithic' bones dating to the 5th millennium cal BC. These are Cnoc Coig and Caisteal nan Gillean II (both on the island of Oronsay, Scotland), Ferriter's Cove, Rockmarshall and Killuragh cave (all in Ireland). These show a range of values, the latter three suggesting a marine diet, a mixed diet and a terrestrial diet respectively, and those from Oronsay suggesting a predominantly marine diet from Cnoc Coig and rather more mixed ones from Caisteal nan Gillean II (Milner 2006). The 'Neolithic' bones come from a number of caves and tombs across Britain. There are very few with ranges which fall significantly into the 4000–3700 cal BC time period, and many of these have ranges which stretch into the 3700–3400 cal BC period. These dietary signatures are much more terrestrial in nature, though recent work in Denmark suggests that such values may contain a significant freshwater fish component (Bennike *et al.* 2007). Overall, there appears to be a range of diets across the first few centuries of the 4th millennium. The rest of the post-4000 cal BC skeletal material falls into the period post-3800/3700 cal BC (see Milner *et al.* 2004, fig. 1a). However, there are a number of limiting factors with such analyses, including the problems of calibrating dates on bones with marine signatures (see below), and using a linear mixing model (see Evershed 2007 for further details). Consequently, stable isotopes can give us a broad overview of change but determining the rate of change is difficult.

Animal bone assemblages in the Early Neolithic seem to demonstrate a degree of consistency: domestic cattle are usually the dominant species, followed by pig and then sheep/goat in much smaller numbers (Thomas 1999). This certainly appears to be the case for southern Britain (Tresset and Vigne 2007), but it is harder to be certain of the distribution and proportions of domesticates in other parts of the country, where there is less data and less research has been carried out.

Sheep and goat did not have wild ancestors in Britain and Ireland and so their presence in the archaeological record indicates animal husbandry. Sheep/goat has been identified at Early Neolithic sites such as the long barrows post-3800 cal BC (Bayliss and Whittle 2007). However, there are fewer sites which have dates on domesticated animal bone from the first couple of centuries in the

4th millennium cal BC. One notable example is Ascott-under-Wychwood long barrow, which has a pre-barrow occupation site and midden. Six samples have been dated from the midden, three of which are sheep teeth dating to between 3990 and 3780 cal BC (Bayliss *et al.* 2007). Sheep was also found in the pre-cairn phase of Hazleton North long cairn (Levitan 1990, 199) and, although the sheep bones have not been dated, the recent Bayesian modelling suggests the start of the pre-cairn activity at shortly before 3900 cal BC, almost certainly by 3800 cal BC, and that this activity probably lasted a century (Meadows *et al.* 2007, 61). Another site which is often mentioned in the literature is Dalkey Island, with dated sheep bone (Woodman *et al.* 1997). The date when calibrated (OxCal 4.0) is 4036–3654 (OxA-4566, 5050±90BP) and the range of over 350 years should be noted.

Cattle and pig are more complex. They have also been found in the middens at Hazleton North and Ascott-under-Wychwood (Benson and Whittle 2006; Levitan 1990). However, there is known size overlap between wild boar and domestic pigs which means that caution is required when they are found in Early Neolithic sites. There is also known overlap between wild and domestic cattle but, as already mentioned, aurochs were absent in Ireland. The discovery of a number of finds of cattle bone of pre-Neolithic date in Ireland is therefore significant because their presence suggests an early phase of colonization/introduction (Tresset 2003; Tresset and Vigne 2007). A rather puzzling example comes from the site of Sutton, where cattle bone when calibrated using OxCal 4.0 gives very early dates of 5709–5481 (OxA-3691, 6660±80 BP) and 5630–5373 (OxA-3960, 6560±75 BP). However, partly because of the anomalous dates it has been suggested that in fact this might be bear bone (Woodman and McCarthy 2003, 35) and perhaps should be excluded from discussion (Cooney 2007, 548). Kilgreany Cave also contained a cattle bone with what has appeared to be an early date, although when calibrated using OxCal 4.0 it dates to 4234–3798 cal BC (OxA-4269, 5190±80 BP) (Woodman *et al.* 1997, 141): four centuries spanning 4000 cal BC. Perhaps the most well-known early cattle bones are those from the site of Ferriter's Cove. The dated cattle metatarsal (OxA-8775, 5825±50 BP) calibrated using OxCal 4.0 to 4793–4548 cal BC has been of particular interest because it appears 'to be associated with a series of Later Mesolithic phases of activity' (Woodman and McCarthy 2003, 33). However, it is unfortunately not possible to identify whether the cattle were domesticated or wild and therefore whether these were domesticated cows which came from the continent or were perhaps wild cattle or joints of meat which came from Britain (Whittle 2007, 393). In Britain, recent dating programmes have revealed a number of new dates for domesticated cattle which appear in long barrows dating from 3800/3750 cal BC (Bayliss and Whittle 2007). There is also plenty of evidence from slightly later sites such as causewayed enclosures (Bayliss *et al.* 2008; Whittle *et al.* 2008) and these have been connected with consumption, perhaps feasting and sometimes deliberate wasting of certain parts of the animal (Thomas 1999, 27).

Recent advances in ancient DNA (aDNA) research on aurochs and domestic cattle have also provided important insights for the study of the transition. The work to date on samples from prehistoric sites has demonstrated a genetic distinction between *Bos taurus* and *Bos primigenius* within Europe, which excludes an independent domestication of European cattle and suggests an origin in the Near East (Bollongino and Burger 2007). This means that the domesticated cattle found in Britain and Ireland were brought in from the continent on boats. Furthermore, there is a mitochondrial diversity in the cattle from the British Isles that strongly suggests different origins (Tresset and Vigne 2007, 193). Unfortunately, however, aDNA analysis was not possible on the Ferriter's Cove cattle bone.

More recently, innovations in pottery residue analysis have provided a proxy for early cattle domestication. This work has led to the identification of dairy fat residues in pots from a number of Neolithic sites in southern Britain, comparison of which with the faunal remains from these sites suggests that dairying was an integral component of agricultural practices from the onset of the Neolithic (Copley *et al.* 2003; 2005). Although these sites do not all fall into the first centuries of the 4th millennium, pottery from a pre-barrow midden beneath Ascott-under-Wychwood that was analysed for lipids was demonstrated to have contained both meat and dairy fats; as already mentioned, this context does pre-date 3800 cal BC (Bayliss *et al.* 2007, 30). However, it is difficult to say to what *extent* dairying was practised from this evidence: the data do confirm that dairy products were being used early on in southern Britain but it is impossible to be sure whether this is evidence for a widespread dairying economy or milking on a small domestic scale, perhaps even by a minority.

From residue analysis, it would appear that Early Neolithic pottery was used for new forms of consumption practices, and therefore the dating of the pottery is also important in order to answer questions about the introduction of farming practices and changes in subsistence. Sheridan (2003; 2004) has suggested three main strands of movement of groups of farmers from the continent. The first is an Atlantic movement which accords with the small simple passage tombs and small closed polygonal megalithic chambers found near the coast in Wales, western Scotland and Ireland, at perhaps 4200–3800 cal BC (Sheridan 2004, 10). A key example is the tomb at Achnacreebeag on the west coast of Scotland, which, although not dated, has similarities with tombs in north-west France in use about 4000 cal BC (Sheridan 2000). The second is a cross-Channel westward movement from northern Brittany and Normandy to southern England, and the third is a cross-Channel movement from different areas along the Channel bringing Carinated Bowl (CB) pottery to Britain (Sheridan 2004).

The dating of pottery is complicated and usually dates have to be taken on carbonized residues (although it should be noted that such dates can be problematic, depending on what is being dated) or material associated with the pots. From Sheridan's analysis (2007, fig. 6) the earliest dates for CB pottery appear to come from a small number of sites (Coupland, Eweford, Garthdee Road, Carzield, Leven and Dubton), the earliest of which have ranges which suggest a date falling roughly between 4000 and 3700 cal BC, and about another 30 dates which appear to fall roughly in the 3800–3600 cal BC bracket. More recent dates from Coupland (Johnson and Waddington in press) also fall into the 4000–3700 cal BC range. Although Bayesian modelling of the CB phase may provide a better insight, at the moment the only way to examine these dates is by visual inspection, which is not particularly accurate (Bayliss *et al.* 2008, 27). Therefore, it is not possible to say exactly when this CB pottery appeared because the earliest date ranges usually span at least 300 years from c. 4000 to 3700 BC.

Disagreement exists over cereal agriculture: there is much debate over whether it had limited economic importance and was undertaken to produce special or symbolic foods (e.g. Thomas 2004), or whether it was an important part of the subsistence economy (e.g. Rowley-Conwy 2004). Bogaard and Jones (2007) highlight how the former is particularly argued for Britain and the latter for central Europe. However, through a detailed comparison of sites in both Britain and Europe they demonstrate convincing similarities in the evidence in terms of the scale of cereal production and permanence of cultivation plots and suggest that what has been characterized as different practices may represent the same activity. However, much of their evidence again appears to come from sites which probably post-date 3800/3700 cal BC.

In addition, there have been arguments for the introduction of cereals in 5th-millennium contexts and suggestions that this indicated a Neolithic pioneer population, based on cereal-type pollen (e.g. Edwards and McIntosh 1998; Innes *et al.* 2003). However, the evidence is controversial. For instance, a recent critical overview of the evidence for Mesolithic agriculture in and around central Europe suggests that the absolute distinction of wild grasses and cereal is impossible, and such a distinction is therefore highly dubious when very small quantities are identified (Behre 2007). Across this period there is also palynological evidence for the elm decline; the reasons for this are, however, much debated and range from human factors such as clearance for growing cereals and the collection of branches for fodder to environmental drivers such as climate, soil depletion and elm disease.

The evidence for charred cereal grains from securely stratified and well-dated contexts is much more compelling. In an overall analysis of collated dates, Brown (2007) suggests that there was limited cultivation from 3950 to 3800 cal BC with an increase in cultivation from 3800 cal BC. However, once again, dating problems arise. For example, the oak from the destruction level of the timber hall structure at Balbridie has a calibrated mean of between 3900 and 3700 cal BC but the preferred mean of grain, flax and linen is between 3780 and 3640 cal BC (http://www.historic-scotland.gov.uk/index/heritage/archaeology/archaeology-techniques/radiocarbon-dating-search.htm (accessed 22 August 2009)). It is likely that the more reliable dates are those of the grain and that the timber used in the construction is old wood; therefore, it is perhaps safer to rely on dated charred cereal grains.

One date on grain, presented in the appendix to Brown's paper (2007), comes from Ulva cave in the Inner Hebrides, with a date of 3940–3690 BC. However, the report on the cave (Russell *et al.* 1995, 277) suggests that this appears to be from a bulk charcoal sample from a pit containing carbonized cereal grains and should therefore be discounted. Bonsall *et al.* (2002, 7) also collated over 400 dates from Scotland and suggest that the earliest *direct* evidence for agriculture comes from Balbridie and Balfarg Riding School, with dates of 3770–3370 cal BC and 3700–3520 cal BC; and there has since been a charred cereal grain from Claish, Stirling, dated to 3770–3630 cal BC (see Brown 2007 appendix for details of calibrated dates). Lismore fields also has potentially early dates, with charred flax dating to 3950–3640 cal BC and charred cereal grain to 3950–3530 cal BC; however, it should be noted that the ranges span three to four centuries and therefore they may actually date to after 3800 cal BC. It seems that at the moment the evidence for the introduction of grain starts convincingly at about 3800 cal BC.

In summary, the evidence for the introduction of domesticated plants and animals in Britain and Ireland before 3700 cal BC is varied. However, it is important to take into account archaeological visibility and recovery: because the more prominent sites such as the monuments in southern England seem to appear after 3800 cal BC there may be a bias towards these later sites, in terms of archaeological investigations. It is possible that much evidence of farming may have been lost from earlier centuries. At present, there are potential indicators in Ireland for very early domesticates, before 4000 cal BC, but most of these come from shell middens and are not in association with other Neolithic markers like pottery and tombs. The most convincing evidence for the introduction of domesticates in the first couple of centuries in the 4th millennium in Britain is from Ascott-under-Wychwood, in the pre-barrow midden, where there are dates on sheep bone and the use of dairy products in the pottery demonstrating usage perhaps before 3900 cal BC and almost certainly by 3800 cal BC. Other sites, such as Hazleton North, also suggest domesticates in the archaeological record before 3800 cal BC. CB pottery has been argued to arrive at an early date within the 4000–3700 cal BC time span, but the evidence for grain can be argued to appear about 3800 cal BC. It is not always clear whether these different aspects of farming arrived at the same time in each place or whether they were slightly staggered.

Lab code	Radiocarbon bp	Calibrated dates BC	Re-calibrated dates	∂13C
OxA-8014	5495±55	4000–3800	3930–3650	-12.0
OxA-8005	5480±55	4200–4000	4230–3910	-16.0
OxA-8019	5615±45	4200–4000	4060–3770	-12.4
OxA-8004	5740±65	4300–4000	4250–3910	-12.4

Table 5.1. Dates on human bones by Richards and Sheridan (2000) and recalibrated by Gordon Cook (pers. comm.).

Fishing, hunting and gathering at 4000–3700 cal BC

Another way of thinking through the transition is to look at sites that are deemed to be 'Mesolithic' from their material culture but which have 'transitional' or 'Early Neolithic' dates. Examples of these include South Haw, with dates of 4230–3980 cal BC and 3940–3700 cal BC (Chatterton 2007, 70), and March Hill Top, dated to around 4000 cal BC (Spikins 2002, 43). These are all flint scatters with associated hearths which have been dated, but it is not possible to consider subsistence owing to the lack of faunal and floral remains. In some areas there is clear evidence that 5th- and 4th-millennium sites have similar distributions. In the south of Britain, a number of sites including Thrupp, Corporation Farm, Gravelly Guy, Hazleton and Rollright all demonstrate 4th-millennium activity preceded by hunter-gatherer use (Hey and Barclay 2007, 405).

One region which produces further insights is the west coast of Scotland (Armit and Finlayson 1992). Here, a number of shell middens, mainly made up of limpets, also contain other sorts of food remains and sometimes human bone. The key point is that many span the late 5th millennium and early 4th millennium cal BC. Three sites of particular interest are the shell middens on Oronsay, a shell midden in a cleft at Carding Mill Bay and a shell midden under the rock shelter at An Corran. The Oronsay middens are perhaps most well known, particularly for their inclusion of human bone (Meiklejohn *et al.* 2005). The stable isotopes suggest a strong marine diet (Richards and Mellars 1998) but, importantly, one individual appears to date to the first couple of centuries of the 4th millennium cal BC (Richards and Sheridan 2000), evidence which suggests that some people at this time were continuing to fish and hunt seal.

Unlike many of the other dates in this paper, these ones need to be adjusted for marine reservoir effect. Richards and Sheridan (2000) did take this into account when calibrating their dates; however, recent research into the variance in delta-R (the regional difference from the global marine reservoir correction) through time and space (Ascough *et al.* 2007) suggests that these dates should be recalibrated again. When this is carried out it suggests even younger dates for the individuals (calculated by Gordon Cook) (Table 5.1). There are, of course, a number of assumptions and uncertainties associated with corrections, including: 1. it is assumed that an accurate value is available for the delta-R that is applicable to the exact period and geographical area under study; 2. there is an uncertainty as to the stable isotope end members for 100 per cent terrestrial and 100 per cent marine diets; 3. there is an assumption of linearity between the end members; 4. it must be acknowledged that the same percentage marine diet can produce somewhat different stable isotope values depending on which marine resources are consumed; and 5. it must be determined how typical the measured stable isotope values are for the skeletal remains under study. However, despite these potential problems, a correction that takes these into account will produce a calibrated age that is closer to the true age (Gordon Cook pers. comm.).

Carding Mill Bay, near Oban, consisted of a shell midden and a later cist burial (Connock *et al.* 1992). This site is of interest in that it is characteristically 'Mesolithic' but the dates for the midden, collected on a variety of material (including shell, suggesting some marine food consumption at this time), are nearly all post-4000 cal BC (Table 5.2). The human bones have a fairly terrestrial diet signature; however, these are all later in date (after c. 3700 cal BC). Burnt bird and fish bone was also found, but no domesticated plants or animals were present.

The third site of note is An Corran. This is a rock shelter on Skye which contained a shell midden (Saville 1998). What is particularly remarkable here is the length of duration of use. The earliest dates fall in the 7th millennium cal BC and the latest in the 1st millennium cal BC, with dates in every millennium in between (Saville 1998; Milner and Craig 2009). The dates for this site have mainly been taken on bevelled tools. The problem at this site is that very little is known about the accumulation and consumption of marine foods, or indeed the introduction of domesticates. In order to explore this, dates were taken on what had been identified by its small size as domesticated cattle bone. Although these dates should have enabled a better understanding of the introduction of domestic cattle, unfortunately they turned out to be 7th millennium cal BC in date and therefore are most likely to represent small aurochs (Milner and Craig 2009). Although there were some problems with the excavation of this site and the interpretation of the stratigraphy, this highlights the problems of identifying domesticates through size alone. Although further dating work might elucidate the nature of subsistence at this midden through time, it is highly probable that some level of marine consumption continued into the early part of the 4th millennium, if not later.

Lab code	Material	Radiocarbon bp	Calibrated dates BC
OxA-3740	Antler bevel-ended tool	5190±85	4230–3790
GU-2796	Charcoal	5060±50	3970–3710
GU-2899	Shell	5440±50	3970–3660
GU-2898	Shell	5410±60	3960–3660
GU-2797	Charcoal	4980±50	3940–3650
OxA-7664	Human	4830±45	3710–3520
OxA-7663	Human	4800±50	3700–3380
OxA-3739	Bone bevel-ended tool	4765±65	3660–3370
OxA-7665	Human	4690±40	3630–3360
OxA-7890	Human	4330± 60	3350–2700

Table 5.2. Dates from Carding Mill Bay (Schulting and Richards 2002).

It is currently very difficult to be sure how long the seascapes in this part of Scotland continued to be in use but evidence certainly suggests it was a phenomenon that carried on into the 4000–3700 cal BC period. So, how does this accord with other 'Neolithic' evidence in the region? Charred cereal grain has not been found in this part of Scotland at this date and structures like Balbridie, which are often associated with grain, are more common in the east of Scotland, arguably after 3800/3700 cal BC. No domesticated animals dated to this period have been found on these sites either, and small cattle bone needs to be considered with caution. The CB pottery distribution is also confined more to the east and south (Sheridan 2007). However, the earliest forms of monuments, argued to date to the early 4th millennium, are found on this coastline. Overall, there appears to be a little overlap, with Carding Mill Bay being located close to both monuments and pottery, although there is still an area of the west coast, further north and on the islands, where there appears to be less or no Neolithic influence.

There are two ways in which these data could be interpreted, depending on what dates one believes. Through 'eye-balling' the dates it is possible to say that the dates for the tombs and pottery in this area can be interpreted as falling later in the 4000–3700 cal BC period; in this scenario the 'Neolithic' simply appears later, say 3800 cal BC. The other scenario is that at the beginning of the 4th millennium cal BC there are people living in adjacent areas but inhabiting the landscapes in different ways: some are fishing, hunting and gathering, and continuing to use the seascapes and coast; the others are using pottery and building tombs and possibly farming to some degree (though at present there is little direct evidence) (Armit and Finlayson 1992). The research on Achnacreebeag and associated pottery strongly suggests the latter model; in fact, the suggested date for Achnacreebeag is earlier: here late Castellic (i.e. Breton type) pottery is found, suggesting a date somewhere between 4300 and 4000 cal BC. This also accords with the earlier dates from Carrowmore in Ireland (Sheridan 2000).

It is highly likely that there was some interaction between farming communities and fisher-hunter-gatherers, and certainly the stable isotopes on human bones which tend to post-date 3700 cal BC suggest a shift in the diet in these regions, though this does not necessarily mean a wholesale shift in lifestyle. When considering such possible interactions, it is important to consider complexity, as Fewster's (2005) ethnoarchaeological field research in Botswana clearly demonstrates. In this case study a landscape is inhabited by agro-pastoralists and hunter-gatherers. The latter had largely stopped hunting (though they continued to gather), and herded cattle and worked in the fields for the agro-pastoralists in exchange for products such as milk and meat. They also made and traded items such as pestle and mortars. There would appear to be close economic co-operation between groups and at least one marriage, but the relationships were complicated, with the hunter-gatherers being looked down upon by the agro-pastoralists. Of course such a study cannot be used as a direct analogy but it has been used by Fewster to illustrate that material remains are difficult to interpret and models such as availability/substitution tend to obscure the real picture.

Discussion

At the beginning of this paper I asked whether it was possible to try and identify the *speed* of subsistence change, the *degree* of change and the *extent* to which innovations may have impacted on the landscape.

Understanding the speed of change relies heavily on radiocarbon dating. New dating programmes and the application of Bayesian modelling are now being applied to this period, and these are particularly enhancing the dating of monument construction, producing models which suggest they appear from about the 38th century BC in the south of Britain. Some of the dates from domesticated animals found in pre-monument contexts suggest slightly earlier dates for the introduction of these species, perhaps between

4000 and 3800 cal BC in this part of Britain. Cereals seem to appear in Scotland after 3800 cal BC, although there are suggestions that pottery arrives earlier. In Ireland, these introductions all seem to arrive earlier, from 4500 cal BC onwards. However, when considering many of these dates several caveats have to be borne in mind: ranges often span at least one century; in some cases the marine reservoir has to be taken into account (as far as possible); and many dates have not been modelled using Bayesian statistics and instead have to be 'eye-balled' (Bayliss *et al.* 2008, 27). The range of dates can be used to suggest that different introductions happened at different times over a 500-year period c. 4200–3700 cal BC, but, no doubt, as other evidence comes to light these patterns will change further. At least for the time being it can be argued that there is no clear evidence that the speed of subsistence change happened quickly, over a human generation or even a century, right across Britain and Ireland. At the moment the picture looks much more complicated, with areas where certain new foods took off and other areas where change happened more gradually.

The *degree* to which subsistence change occurred is also very hard to determine with any certainty and has been much debated in the literature, often drawing on the same datasets to argue polarized views. For example, Balbridie, a structure which had burnt down and contained vast amounts of carbonized grain (Fairweather and Ralston 1993) is frequently referenced. Rowley-Conwy (2004, 90) sees this as a cereal store and evidence for farming, whereas Thomas (1999, 25) argues against ideas of farmsteads and instead suggests that it was used for specialized storage for what was perceived to be a very special kind of food. In addition, from the plant residues that are found and a consideration of the taphonomic processes Rowley-Conwy argues that the evidence for wild resources suggests a minor input into the economy, whereas Thomas considers that Neolithic diets were dependent on wild resources and that domesticated resources were used for special ceremonies. Again, perhaps by looking in detail at different areas, it may be possible to find evidence for different degrees of subsistence change at certain points in time.

Finally, the ways in which subsistence change impacted on the landscape are largely dependent on the ways in which rapidity of change and degree of change are perceived. In a model where subsistence change is minimal and slow, landscape change is also minimal. For instance, Thomas deconstructs the idea that early agriculturalists in Britain would have been tied to their fields and instead presents a view of small-scale garden horticulture carried out on a sporadic basis. In terms of domesticated animals, he suggests that there is a marked emphasis on cattle in the earlier Neolithic, even though there is variation between sites. It is important to note, however, that most of the evidence comes from monuments such as causewayed enclosures and long barrows, which demonstrates a link between human and cattle remains in 'special sites' (Thomas 1999, 29). It is therefore proposed that the consumption of cattle meat was not an everyday activity but was reserved for particular times and places, such as communal feasts. In sum, Thomas (1999, 29) sees the evidence pointing to a relatively high degree of mobility in lowland Britain, with a partial dependence on cultivated plants and considerable gathering of wild plants. Cattle would have been circulating within the landscape and their slaughter would have taken place at monuments. Even though they may have been consumed only occasionally, the social and symbolic significance of these animals may have been considerable. On the other hand, Rowley-Conwy (2004) argues forcefully for large-scale agriculture and rapid change, with large-scale cereal cultivation and domesticated animals being predominant from early on. Clearly this would have had more of an impact on the landscape in terms of clearance and management.

By taking the view that subsistence change occurred in a number of ways and was complex and messy (Cooney 2007) it is possible to accept a variety of landscape changes. This is reinforced by research into Coleoptera by Robinson (2000, 35), who describes the Neolithic landscape as 'a mosaic of relatively small clearances, grazed park woodland, abandoned clearances in various stages of scrub to woodland succession and woodland which retained an old woodland insect fauna, some of it perhaps even relatively undisturbed primary woodland.' This mosaic effect is also evidenced in the Thames Valley (Hey and Barclay 2007, 403). However, it is also important to note that landscape variability was not new to the Neolithic: the landscape was also varied in the Mesolithic. The extent to which hunter-gatherers either used natural clearings or created and maintained clearings deliberately has been widely discussed (Tipping 2004).

In conclusion, many narratives take simple, large-scale explanations that suggest sudden change or slow change with little in between. There are numerous problems with the data we have, not least with dating, which does not always provide the resolution that is required to understand such changes at the scale of the human generation. However, Bayesian modelling offers some consolation here and is beginning to revolutionize the ways in which the transition is approached. Overall, the data appears to be revealing a complex story of messiness and local character that might be expected and in order to understand this further it is important to begin to think about landscapes at different scales.

Acknowledgements

I thank Graeme Warren and Bill Finlayson for inviting me to contribute to the workshop, and for their helpful comments and useful discussions. I also thank Umberto Albarella, Mark Edmonds, Jon Finch, Gordon Cook, Dale Serjeantson, Alison Sheridan, Clive Waddington and Alasdair Whittle, as well as the workshop participants, for discussions, comments and help on this subject. The viewpoints expressed in this paper are not always in agreement with theirs, and any errors are of course my own.

Bibliography

Armit, I. and Finlayson, B. (1992) Hunter-gatherers transformed: the transition to agriculture in northern and western Europe. *Antiquity* 66, 664–76

Ascough, P. L., Cook, G. T., Dugmore, A. J. and Scott, E. M. (2007) The North Atlantic marine reservoir effect in the Early Holocene: implications for defining and understanding MRE values. *Nuclear Instruments and Methods in Physics Research B* 259, 438–47.

Barker, G. (2006) *The Agricultural Revolution in Prehistory: Why Did Foragers Become Farmers?* Oxford: Oxford University Press.

Bayliss, A. and Whittle, A. (eds) (2007) *Histories of the Dead: Building Chronologies for Five Southern British Long Barrows. Cambridge Archaeological Journal* 17/1 supplement. Cambridge: McDonald Institute for Archaeological Research.

Bayliss, A., Benson, D., Galer, D., Humphrey, L., McFadyen, L. and Whittle, A. (2007) One thing after another: the date of the Ascott-under-Wychwood long barrow. Pp. 29–44 in A. Bayliss and A. Whittle (eds), *Histories of the Dead: Building Chronologies for Five Southern British Long Barrows. Cambridge Archaeological Journal* 17/1 supplement. Cambridge: McDonald Institute for Archaeological Research.

Bayliss, A., Whittle, A. and Healey, F. (2008) Timing, tempo and temporalities in the early Neolithic of southern Britain. Pp. 25–42 in H. Fokkens, B. J. Coles, A. L. Van Gijn, J. P. Kleijne, H. H. Ponjee and C. G. Sloppendel (eds), *Between Foraging and Farming. An Extended Broad Spectrum of Papers Presented to Leendert Louwe Kooijmans*. Analecta Praehistorica Leidensia 40. Leiden: Faculty of Archaeology, Leiden University.

Behre, K.-E. (2007) Evidence for Mesolithic agriculture in and around central Europe? *Vegetation History and Archaeobotany* 16, 203–19

Bennike, P., Fischer, A., Heinemeier, J., Kubiak-Martens, L., Olsen, J., Richards, M. and Robinson, D. E. (2007) The composition of Mesolithic food. Evidence from a submerged settlement. *Acta Archaeologica* 78/2, 163–78.

Benson, D. and Whittle, A. (2006) *Building Memories: The Neolithic Cotswold Long Barrow at Ascott-under-Wychwood, Oxfordshire*. Oxford: Oxbow Books.

Bogaard, A. and Jones, G. (2007) Neolithic farming in Britain and central Europe: contrast or continuity? Pp. 357–76 in A. Whittle and V. Cummings (eds), *Going over: The Mesolithic–Neolithic Transition in North-West Europe*. London: British Academy and Oxford University Press.

Bollongino, R. and Burger, J. (2007) Neolithic cattle domestication as seen from ancient DNA. Pp. 165–88 in A. Whittle and V. Cummings (eds), *Going over: The Mesolithic–Neolithic Transition in North-West Europe*. London: British Academy and Oxford University Press.

Bonsall, C., Anderson, D. E. and Macklin, M. G. (2002) The Mesolithic–Neolithic transition in western Scotland and its European context. *Documenta Praehistorica* 29, 1–20.

Brown, A. (2007) Dating the onset of cereal cultivation in Britain and Ireland: the evidence from charred cereal grains. *Antiquity* 81, 1042–52.

Chatterton, R. (2007) South Haw: an upland Mesolithic site in its local and regional context. Pp. 69–80 in C. Waddington and K. Pedersen (eds), *Mesolithic Studies in the North Sea Basin and Beyond. Proceedings of a Conference held at Newcastle in 2003*. Oxford: Oxbow Books.

Connock, K. D., Finlayson, B. and Mills, C. M. (1992) A shell midden at Carding Mill Bay, near Oban, Scotland. *Glasgow Archaeological Journal* 17, 25–38.

Cooney, G. (2007) Parallel worlds of multi-stranded identities? Considering the process of 'going over' in Ireland and the Irish Sea zone. Pp. 543–66 in A. Whittle and V. Cummings (eds), *Going Over: The Mesolithic–Neolithic Transition in North-West Europe*. London: British Academy and Oxford University Press.

Copley, M. S., Berstan, R., Dudd, S. N., Docherty, G., Mukherjee, A. J., Straker, V., Payne, S. and Evershed, R. P. (2003) Direct chemical evidence for widespread dairying in prehistoric Britain, *Proceedings of the National Academy of Science* 100 (4), 1524–9.

Copley, M. S., Berstan, R., Mukherjee, A. J., Dudd, S. N., Straker, V., Payne, S. and Evershed, R. P. (2005) Dairying in antiquity. III. Evidence from absorbed lipid residues dating to the British Neolithic. *Journal of Archaeological Science* 32/4, 523–46.

Dineley, M. and Dineley, G. (2000) Neolithic ale: barley as a source of malt sugars for fermentation. Pp. 137–54 in A. S. Fairburn (ed.), *Plants in Neolithic Britain and Beyond*. Neolithic Studies Group Seminar Papers 5. Oxford: Oxbow Books.

Edwards, K. J. and McIntosh, C. J. (1998) Improving the detection rate of cereal-type pollen from Ulmus decline and earlier deposits from Scotland. *Pollen et Spores* 30, 179–88.

Evershed, R. P. (2007) Exploiting molecular and isotopic signals at the Mesolithic–Neolithic transition. Pp. 141–64 in A. Whittle and V. Cummings (eds), *Going Over: The Mesolithic–Neolithic Transition in North-West Europe*. London: British Academy and Oxford University Press.

Fairweather, A. D. and Ralston, I. B. M. (1993) The Neolithic timber hall at Balbridie, Grampian Region, Scotland: the building, the date, the plant macrofossils. *Antiquity* 67, 313–23

Fewster, K. (2005) Petso's field: ethnoarchaeology and agency. Pp. 106–25 in N. Milner and P. Woodman (eds), *Mesolithic Studies in the 21st century*. Oxford: Oxbow Books.

Hey, G. and Barclay, A. (2007) The Thames Valley in the late fifth and early fourth millennium cal BC: the appearance of domestication and the evidence for change. Pp. 399–422 in A. Whittle and V. Cummings (eds), *Going Over: The Mesolithic–Neolithic Transition in North-West Europe*. London: British Academy and Oxford University Press.

Innes, J. B., Blackford, J. J. and Davey, P. J. (2003) Dating the introduction of cereal cultivation to the British Isles: early palaeoecological evidence from the Isles of Man. *Journal of Quaternary Science* 18, 603–13

Johnson, B. and Waddington, C. (in press) Prehistoric and Dark Age settlement remains from Cheviot Quarry, Milfield Basin, Northumberland. *Archaeological Journal*.

Levitan, B. (1990) The non-human vertebrate remains. Pp. 199–209 in A. Saville (ed.), *Hazleton North. The Excavation of a Neolithic Long Cairn of the Cotswold–Severn Group*. English Heritage Archaeological Report no 13. London: Historic Buildings and Monuments Commission for England.

Meadows, J., Barclay, A. and Bayliss, A. (2007) A short passage of time: the dating of the Hazleton long cairn revisited. *Cambridge Archaeological Journal* 17/1, 45–64.

Mears, R. and Hillman, G. (2007). *Wild Food*. London: Hodder and Stoughton

Meiklejohn, C., Merrett, D. C., Nolan, R. W., Richards, M. P. and Mellars, P. A. (2005) Spatial relationships, dating and taphonomy of the human bone from the Mesolithic site of Cnoc Coig, Oronsay, Argyll, Scotland. *Proceedings of the Prehistoric Society* 71, 85–106

Milner, N. (2006) Subsistence. Pp. 61–82 in C. Conneller and G. Warren (eds), *Mesolithic Britain and Ireland: New Approaches*. Stroud: Tempus.

Milner, N. and Craig, O. E. (2009) Mysteries of the middens: change and continuity across the Mesolithic-Neolithic transition. Pp. 159–70 in M. J. Allen, N. Sharples and T. O'Connor (eds), *Land and People: Papers in Honour of John G. Evans*. Prehistoric Society Research Paper. Oxford: Oxbow Books.

Milner, N., Craig, O. E., Bailey, G. N., Pedersen, K. and Andersen, S. H. (2004) Something fishy in the Neolithic? A re-evaluation of stable isotope analysis of Mesolithic and Neolithic coastal populations. *Antiquity* 78, 9–22.

Milner, N., Craig, O. E., Bailey, G. N. and Andersen, S. H. (2006) A response to Richards and Schulting. *Antiquity* 80, 456–8.

Price, T. D. (2000) Europe's first farmers: an introduction. Pp. 1–18 in T. D. Price (ed.), *Europe's First Farmers*. Cambridge: Cambridge University Press.

Richards, M. P. and Mellars, P. A. (1998) Stable isotopes and the seasonality of the Oronsay middens. *Antiquity* 72, 178–84.

Richards, M. P. and Schulting, R. J. (2003) Sharp shift in diet at onset of Neolithic. *Nature* 425, 366.

Richards, M. P. and Schulting, R. J. (2006) Against the grain? A response to Milner et al. (2004). *Antiquity* 80, 444–56.

Richards, M. P. and Sheridan, J. A. (2000) New AMS dates on human bone from Mesolithic Oronsay middens. *Antiquity* 74, 313–15.

Robinson, M. A. (2000) Further considerations of Neolithic charred cereals, fruit and nuts. Pp. 85–90 in A. S. Fairburn (ed.), *Plants in Neolithic Britain and Beyond*. Neolithic Studies Group Seminar Papers 5. Oxford: Oxbow Books.

Rowley-Conwy, P. (2004) How the West was lost: a reconsideration of agricultural origins in Britain, Ireland, and Southern Scandinavia. *Current Anthropology* 45 (Supplement), 83–113.

Russell, N. J., Bonsall, C. and Sutherland, D. G. (1995) The exploitation of marine molluscs in the Mesolithic of western Scotland: evidence from Ulva Cave, Inner Hebrides. Pp. 273–88 in A. Fischer (ed.), *Man and the Sea in the Mesolithic. Proceedings of the International Symposium, Kalundborg, Denmark*. Oxford: Oxbow Books.

Saville, A. (1998) An Corran, Staffin, Skye. *Discovery and Excavation in Scotland 1998*, 126–7.

Schulting, R. J. and Richards, M. P. (2002) The wet, the wild and the domesticated: the Mesolithic–Neolithic transition on the west coast of Scotland. *European Journal of Archaeology* 5/2, 147–89.

Sheridan, A. (2000) Achnacreebeag and its French connections: vive the 'auld alliance'. Pp. 1–15 in J. C. Henderson (ed.), *The Prehistory and Early History of Atlantic Europe*. British Archaeological Reports, International Series 861. Oxford: BAR.

Sheridan, A. (2003) French connections I: spreading the *marmites* thinly. Pp. 3–17 in I. Armit, E. Murphy, E. Nelis and D. D. A. Simpson (eds), *Neolithic settlement in Ireland and western Britain*. Oxford: Oxbow Books.

Sheridan, A. (2004) Neolithic connections along and across the Irish Sea. Pp. 9–21 in V. Cummings and C. Fowler (eds), *The Neolithic of the Irish Sea: Materiality and Traditions of Practice*. Oxford: Oxbow Books.

Sheridan, A. (2007) From Picardie to Pickering and Pencraig Hill? New information on the 'Carinated Bowl Neolithic' in northern Britain. Pp. 441–92 in A. Whittle and V. Cummings (eds), *Going Over: The Mesolithic–Neolithic Transition in North-West Europe*. London: British Academy and Oxford University Press.

Spikins, P. (2002) *Prehistoric People of the Pennines. Reconstructing the Lifestyles of Mesolithic Hunter-Gatherers on Marsden Moor*. Leeds: West Yorkshire Archaeology Service.

Thomas, J. (1999) *Understanding the Neolithic*. London: Routledge.

Thomas, J. (2004) Current debates on the Mesolithic–Neolithic transition in Britain and Ireland. *Documenta Praehistorica* 31, 113–30.

Tipping, R. (2004) Interpretative issues concerning the driving forces of vegetation change in the Early Holocene of the British Isles. Pp. 45–53 in A. Saville (ed.), *Mesolithic Scotland and its Neighbours*. Edinburgh: Society of Antiquaries of Scotland.

Tresset, A. (2003) French connections II: of cows and men. Pp. 18–30 in I. Armit, E. Murphy, E. Nelis and D. Simpson (eds), *Neolithic Settlement in Ireland and Western Britain*. Oxford: Oxbow Books.

Tresset, A. and Vigne, J.-D. (2007) Substitution of species, techniques and symbols at the Mesolithic–Neolithic transition in Western Europe. Pp. 189–210 in A. Whittle and V. Cummings (eds), *Going Over: The Mesolithic–Neolithic Transition in North-West Europe*. London: British Academy and Oxford University Press.

Whittle, A. (2003) *The Archaeology of People: Dimensions of Neolithic Life*. London and New York: Routledge.

Whittle, A. (2007) The temporality of transformation: dating the early development of the southern British Neolithic. Pp. 377–98 in A. Whittle and V. Cummings (eds), *Going Over: The Mesolithic–Neolithic Transition in North-West Europe*. London: British Academy and Oxford University Press.

Whittle, A., Bayliss, A. and Healey, F. (2008) The timing and tempo of change: examples from the fourth millennium cal BC in Southern England. *Cambridge Archaeological Journal* 18/1, 65–70.

Woodman, P. C. and McCarthy, M. (2003) Contemplating some awfully interesting vistas: importing cattle and red deer into prehistoric Ireland. Pp. 31–9 in I. Armit, E. Murphy, E. Nelis and D. Simpson (eds), *Neolithic Settlement in Ireland and Western Britain*. Oxford: Oxbow Books.

Woodman, P. C., McCarthy, M. and Monaghan, N. (1997) The Irish Quaternary Fauna Project. *Quaternary Science Reviews* 16, 129–59.

6. A Geological Perspective on Climatic and Environmental Change in the Levant and Eastern Mediterranean from 25,000 to 5000 years BP

Stuart A. Robinson, Stuart Black, Bruce W. Sellwood, Claire M. C. Rambeau and Paul J. Valdes

We recently published an extensive review of terrestrial and marine climatic conditions in the Levant and Eastern Mediterranean from 25,000 to 5,000 cal years BP (Robinson et al. *2006). The aim of our review was to provide a climatic and environmental backdrop against which future studies of climatic, environmental and archaeological change could be compared. Here we provide a brief synopsis of our review, highlighting the major conclusions and including, where appropriate, studies published since our work. We also discuss practical issues surrounding the integration of geological and archaeological data, with particular regard to determining the role (or not) of climate change in driving human behavioural changes.*

Introduction

Robinson *et al.* 2006 presented a comprehensive review of the geological evidence for climatic and environmental change in the Levant and Eastern Mediterranean from 25,000 to 5,000 years before present. This work formed part of the 'Water, Life, Civilisation' project at the University of Reading, UK – a major interdisciplinary initiative to document the past, present and future relationships between climate, water usage and human civilization in the Levant region. The purpose of Robinson *et al.* 2006 was to provide the climatic context for the early history of human civilization in the region, with particular focus on the period with the largest amplitude of climatic change, from the Last Glacial Maximum (LGM) to the Early Holocene. The work provided a comparison between global climate model (GCM) outputs of the LGM and a detailed analysis of the types of geological data available from marine and terrestrial records. We also highlighted the methodological limitations of certain datasets, the geographic areas for which little data exists and the contradictions between some of the datasets.

In this work, we present a synopsis of the major findings of Robinson *et al.* 2006 that incorporates more recent studies and provides an interpretation of the climatic and environmental history of the Levant from a geological proxy perspective. The records discussed here come from throughout the Levant, Eastern Mediterranean and Red Sea (Figure 6.1). We recommend that readers also refer to Robinson *et al.* 2006 for a more detailed and thorough discussion of the proxies, their limitations and the conclusions drawn. It should be stated that all dates discussed below are calendar (calibrated) years (or ka) before present (cal years BP or cal ka BP); calibration details for radiocarbon ages are given in Robinson *et al.* 2006.

What sort of geological records are available?

Terrestrial sedimentary records

Terrestrial sediments can be highly varied in origin, composition and relevance to the wider debate of palaeo-climatic change. At the first order level, the presence

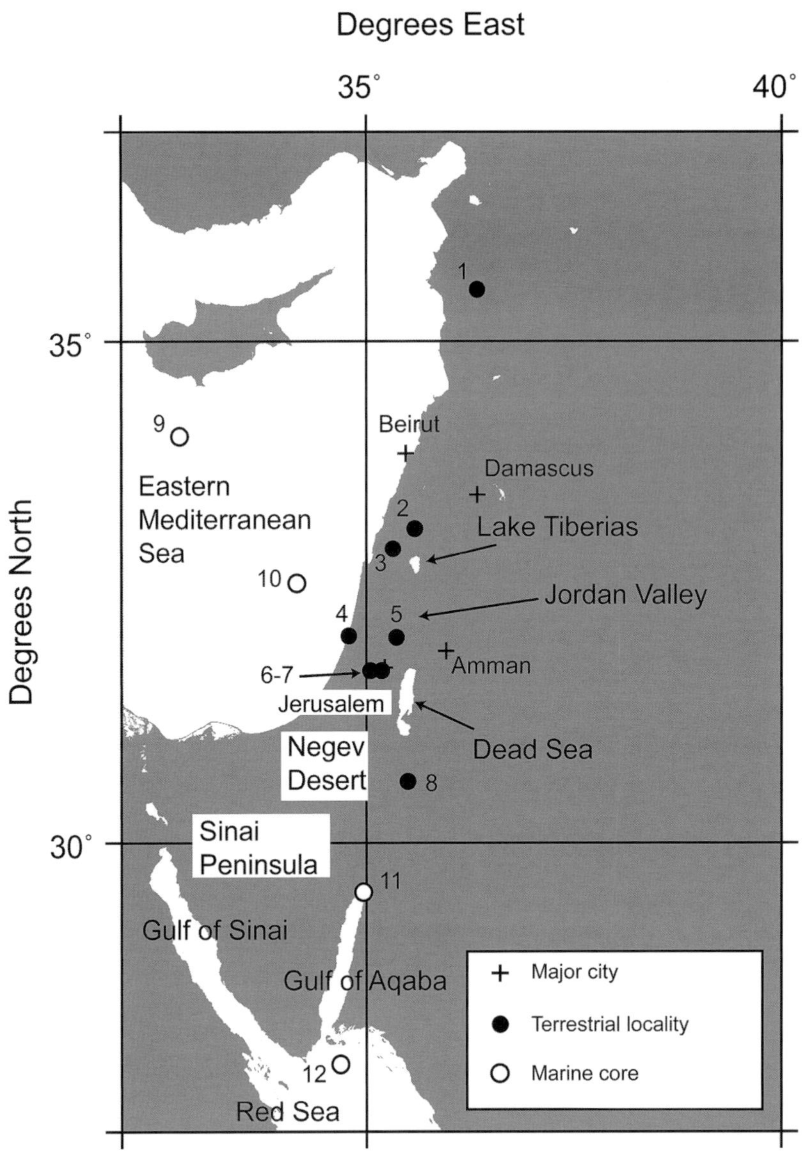

Fig. 6.1. Location map of the Eastern Mediterranean and Levant region. The numbers refer to localities discussed in Robinson et al. *(2006) and in the text: 1 and 2 are pollen records from Ghab Valley and Hula Basin; 3, 5, 6 and 7 are speleothem records from Peqiin Cave, Ma'ale Efrayim Cave, Soreq Cave, and Jerusalem West Cave; 4 and 8 are sedimentological records from the Israeli coastal plain and Wadi Faynan; 9, 10, 11 and 12 are marine records from ODP Site 967, and cores M44-1-KL83, GeoB5804-4, and GeoB5844-2. Map produced with GMT (http://gmt.soest.hawaii.edu/). Figure reproduced with permission from Robinson* et al. *2006.*

of a particular sediment that can be attributed to a specific depositional process is the clearest evidence for reconstructing palaeoenvironments (e.g. fluvial sediments indicate the presence of rivers or streams). However, the use of terrestrial sediments for climate reconstruction can be problematic owing to reasons such as tectonics altering sedimentation patterns, uncertainty over the range of climatic scenarios a sediment type represents, and the non-continuous nature of many terrestrial sedimentary records.

Late Pleistocene–Holocene terrestrial sediments deposited in lakes, as soils and by rivers are common in the Levant and, in some cases, are highly informative for an understanding of the environmental and climatic history of the region (e.g. Goldberg 1994). Lacustrine sediments have proved particularly useful as they provide high-resolution sedimentary archives from which a variety of climatic and environmental information can be gleaned. Much research in the Levant has focused on fluctuations in the levels of the Dead Sea, its late Pleistocene precursor Lake Lisan and Lake Tiberias (e.g. Bartov *et al.* 2002; 2003; Begin *et al.* 1985; Frumkin *et al.* 1994; Hazan *et al.* 2005; Landmann *et al.* 2002; Migowski *et al.* 2006; Neev and Emery 1995; Yechieli *et al.* 1993). The level of these lakes is generally assumed to have been controlled largely by climate change (controlling precipitation, run-off and evaporation) and not by tectonics, even though the Jordan rift basin has been tectonically active throughout the Pleistocene and Holocene (e.g. Marco *et al.* 1996). Lake level curves have generally been constructed by considering the age and altitude of sediments that were deposited at or close to the palaeo-lake shorelines. The presence of hiatuses in sedimentation and thick evaporite deposits mean that some events in the history of Lake Lisan are poorly constrained temporally and poorly understood in terms of lake-level significance.

Large aeolian sand dunes are an obvious indicator of palaeoenvironments and palaeoclimates, indicating arid desert conditions. Goring-Morris and Goldberg (1990) documented a period of dune incursion in the southern Levant which they dated (on the basis of archaeological material) to about the LGM and interpreted as a period of 'hyperaridity'. Palaeosols are also very sensitive indicators of climatic change and, in some cases, can provide quantitative estimates of climatic parameters such as precipitation through their mineralogy, morphology and grain size. In the Levant, the most continuous record of soil formation exists on the Israeli coastal plain (Gvirtzman and Wieder 2001) and provides a 50,000-year record of pedological conditions. However, the distribution of soils in the region is typically less continuous than this, and hence most occurrences of palaeosols provide only a 'snap-shot' view of palaeoenvironment at that locality. In a similar manner, fluvial sediment sequences are generally non-continuous and can be difficult to date accurately owing to an absence of suitable materials. However, in the Wadi Faynan region of southern Jordan McLaren *et al.* (2004) have been able to identify a sequence of fluvial, alluvial and aeolian sediments with sufficient dating to allow them to arrive at some broad conclusions regarding the climate history of this area.

Terrestrial palaeobotanical records

Palaeobotanical records can potentially provide quantitative climatic data through the recognition of climatically sensitive floras. In an area that straddles climatic zones, such as the Levant, it can reasonably be expected that temporal changes in climate will be reflected by temporal changes in vegetation. The best and most informative palaeobotanical records come from palynological studies of high-resolution, continuous and well-dated sediments, such as those recovered by lake and marine cores. Fluvial sediments and palaeosols are much less desirable owing to the possibility of hiatuses and the difficulty of dating such sequences. However, from a palaeoclimate perspective such 'natural' sediment accumulations are preferable to sediments that accumulated at archaeological sites where the potential for anthropogenically induced bias is highest. In the Levant, the best-known and highest-resolution pollen records come from the Ghab Valley in Syria (e.g. van Zeist and Bottema 1982) and the Hula Basin in Israel (e.g. Baruch and Bottema 1991; 1999). The radiocarbon age models for these sites led some authors to suggest that Syria and northern Israel had contrasting climatic histories during the transition from the last glacial to the Early Holocene (e.g. Baruch and Bottema 1991; 1999; Bottema 1995). However, the age models underlying these conclusions have been extremely controversial and are considered by some authors, on the basis of comparisons with marine palynological records in the Eastern Mediterranean and Arabian Sea, to be erroneous (e.g. Meadows 2005; Robinson *et al.* 2006; Rossignol-Strick 1995).

Geochemical records

A variety of geochemical records that utilize biogenic carbonates, abiogenic carbonates, lacustrine sediments and organic matter exist from the Levant. The most informative records come from speleothems, which provide high-resolution, well-dated, continuous, multi-proxy records of climatic and environmental change (e.g. Ayalon *et al.* 1999; Bar-Matthews *et al.* 1997; 1999; 2003; Frumkin *et al.* 1999; 2000; Frumkin and Stein 2004; Vaks *et al.* 2003; 2006). Various climatic and environmental inferences can be drawn from the different types of data generated from speleothems. Oxygen isotopes (expressed in the delta notation $\delta^{18}O$) in speleothem carbonate are controlled by the composition of local precipitation (which is itself the result of a combination of factors including source water, altitude, latitude and air temperature) and evaporation in the soil zone. By combining oxygen-isotope data from speleothem carbonate with hydrogen-isotope data from fluid inclusions in the speleothem it can be possible to make estimates of cave temperatures. Carbon isotopes (expressed as $\delta^{13}C$) in speleothems reflect the composition of soil CO_2 (controlled by the varying proportions of C_3 and C_4 plants in the soil) and dissolution of bedrock carbonate. Strontium isotopes in speleothems are sourced from seawater (directly or through sea-spray), deposition of wind-blown dust in the overlying soil and dissolution of bedrock.

Other terrestrial geochemical records are often less detailed than the speleothems, but can still be informative and useful. The elemental and isotopic chemistry (i.e. Sr-isotopes, Mg/Ca- and Sr/Ca-ratios) of lacustrine sediments can be used to demonstrate different limnological conditions driven by climatic change (Abed and Yaghan 2000; Stein *et al.* 1997). The stable-isotopic ($\delta^{13}C$, $\delta^{18}O$) composition of soil carbonates (calcretes) and molluscs can be used to investigate changes in rainfall source areas, evaporation and vegetational patterns linked to the amount of precipitation (Goodfriend 1991; 1990; 1999; Goodfriend and Magaritz 1988; Magaritz 1986; Magaritz and Heller 1980).

Marine records

It may seem strange to consider marine records of environmental change in a debate that is ostensibly about the links between human behaviour and terrestrial climate change. However, marine records are extremely useful as they provide information over a wider geographic area, can be more continuous than many terrestrial records and are generally easier to date. Additionally, multiple proxies can be used from the same site, using the same age model, to provide insights into many different aspects of palaeoenvironmental change. Unfortunately, one drawback of many marine records is that they were deposited at slow sedimentation rates, and thus some climatic events on the millennial scale or less can be missed.

Many marine cores have been collected in the Eastern Mediterranean and Red Sea, providing an excellent archive of palaeoceanographic information (e.g. Arz *et al.* 2003a; 2003b; Emeis *et al.* 1998; 2000; 2003; Fontugne and Calvert 1992; Hayes *et al.* 2005; Kallel *et al.* 1997a). This information includes changes in physical sedimentation (related to increased continental run-off or changes in aeolian dust supply), changes in the assemblages and geochemistry of foraminifera (related to sea-surface temperature and salinity), changes in the geochemistry of organic matter (related to temperature) and changes in terrestrially sourced pollen assemblages. Fossil corals in the Red Sea provide high-temporal-resolution geochemical records of short intervals of the Holocene (e.g. Moustafa *et al.* 2000), but, unfortunately, a continuous Late Pleistocene–Holocene coral record does not exist in the Levant region.

Geological perspective of climate change in the Levant

Figure 6.2, reproduced from Robinson *et al.* 2006, shows a compilation of the most important marine and terrestrial climatic and environmental datasets from the Levant and Eastern Mediterranean, compared with the GISP2 ice-core temperature record from Greenland. Seven major climatic events are highlighted – Heinrich Event 2 (H2, 23.8 cal ka BP), the Last Glacial Maximum (LGM, 23–19 cal ka BP), Heinrich Event 1 (H1, 16 cal ka BP), the Bølling-Allerød warm interval (B-A, 15–13 cal ka BP), the Younger Dryas (YD, 12.7–11.5 cal ka BP), the Early Holocene Wet Phase (EHWP, 9.5–7 cal ka BP) and a mid-Holocene wet event ('5k', ~5000 cal yrs BP). With the exception of the '5k' event, these periods of time are generally accepted to represent major transitions or extremes in Earth's climate record over the last 25,000 years, and, as such, may be expected to present in the climate records of the Eastern Mediterranean and Levant.

Heinrich Event 2, H2 (~23.8 cal ka BP)

Heinrich Events were times of intense cooling resulting from the addition of freshwater to the North Atlantic inhibiting heat transport from the tropics to northern latitudes (e.g. Bond *et al.* 1992). In the Levant and Eastern Mediterranean the geological evidence suggests that H2 was characterized by a lowering of lake levels and cooling. Bartov *et al.* (2003) suggests that cold-water input into the Mediterranean from the North Atlantic during H2 caused a reduction of evaporation and precipitation in the Eastern Mediterranean region, leading to transient cooling and an evaporation excess over the Levant region. Unfortunately there is not an abundance of data for H2 in the Eastern Mediterranean and Levantine Region.

Last Glacial Maximum (~23 to 19 cal ka BP)

The palaeoclimatic records of the LGM in the Eastern Mediterranean and Levant are in general agreement, suggesting that the region was generally cooler and, possibly, more arid than present (Robinson *et al.* 2006). Lake levels were much higher than present, which Robinson *et al.* (2006) suggested was due to the seasonal (spring) melting of snow in the Jordan River catchment area. Average sea-surface temperatures in the northern Red Sea and Eastern Mediterranean are estimated to have been at least 1 to 2°C colder during the LGM than at present (Emeis *et al.* 2000; 2003; Hayes *et al.* 2005). Average sea-surface salinities were probably elevated at the LGM compared to present values, suggesting increased evaporation and/or decreased freshwater input (e.g. Emeis *et al.* 2000; Kallel *et al.* 1997b). Terrestrial records also point to changes in temperature and aridity: the geochemistry of the speleothem record from Soreq Cave has been interpreted as indicating cooler than modern temperatures above the cave of ~8–12°C during the LGM (McGarry *et al.* 2004) and a mean annual rainfall during the LGM of between 250 and 400 mm (Bar-Matthews *et al.* 1997), considerably less than the 500 mm that presently accumulate there. Speleothem carbon-isotope data for the LGM are consistent with an elevated proportion of C_4 plants growing in the overlying soil, suggesting a cooler and drier climate. Further evidence for dry, cool conditions includes erosion and incursions of sand dunes in the Negev Desert (Goodfriend and Magaritz 1988; Goring-Morris and Goldberg 1990), the precipitation of gypsum in the Dead Sea (Abed and Yaghan 2000), the non-deposition of speleothem carbonate on the eastern side of the central mountain ridge in Israel (suggesting low rainfall: Vaks *et al.* 2003) and elevated non-arboreal pollen values in Syria (see Rossignol-Strick 1995 and Robinson *et al.* 2006).

However, the view presented of a cooler, drier LGM has recently been challenged. Kolodny *et al.* (2005) have presented oxygen-isotopic data from Lake Lisan that broadly show the same pattern as the speleothem data from Soreq Cave (Bar-Matthews *et al.* 2003) and foraminiferal data from the Eastern Mediterranean (Fontugne and Calvert 1992). Kolodny *et al.* (2005) contend that there is a near-constant offset between the oxygen-isotopic composition of sea water, rain water and lake water, suggesting, in their model, that the reason for changes in the oxygen-isotopic composition of speleothem carbonate is not due to changes in the *amount* of rainfall (as suggested by

Bar-Matthews *et al.* 2003) but changes in the isotopic composition of the *source water*, which (ultimately) is the Eastern Mediterranean. This hypothesis is based upon an assumption that the LGM was wetter (more rainfall) which, according to Kolodny *et al.* (2005), is evidenced by the high lake levels. The modelling results presented in Robinson *et al.* 2006 show little change in precipitation patterns across the Levant region during the LGM, with most of the precipitation occurring in the winter months. Modelled (and proxy) temperatures are significantly lower during the LGM, which would have resulted in less evaporation, but still the model predicts that evaporation exceeded precipitation during the LGM. Calculations of the effect of Late Pleistocene–Holocene temperature changes on lake-water oxygen-isotope compositions suggest that most of the changes observed by Kolodny *et al.* (2005) may be explained by the change in temperature (M. Jones pers. comm.). The lower evaporation rates (caused by lower temperatures) suggest that it is not necessary to invoke a wetter than present Levant during the LGM in order to maintain high lake levels (which were established more than 50,000 years ago). Furthermore, spring melting of winter snowfall in the Golan Heights may have contributed to the levels of Lake Lisan (Robinson *et al.* 2006).

A recent study has shown that speleothems in the northern Negev Desert show evidence for non-deposition of carbonate from 25 to 23 cal ka BP and from 13 cal ka BP to present, suggesting climates that were similar or even more arid than at present during these times (Vaks *et al.* 2006). However, Vaks *et al.* (2006) also showed that there were peaks in speleothem growth from 21 to 20 cal ka BP and 19 to 17 cal ka BP, which they suggested were due to more humid conditions, in contrast to the general view of the LGM presented by Robinson *et al.* (2006). As shown by the modelling results, the lower temperatures during the glacial periods would have allowed for a higher precipitation/evaporation (P/E) ratio (although still with an excess of evaporation over precipitation), which may have been sufficiently high to allow speleothem precipitation in the northern Negev without the need for increased rainfall. The carbon-isotope data from the northern Negev speleothems indicate a mostly C_4 type vegetation suggestive of low rainfall and cooler temperatures. At 21 cal ka BP, $\delta^{13}C$ values are very high, suggesting very scarce vegetation owing to arid conditions (Vaks *et al.* 2006), and that it is possible to have speleothem deposition even during arid conditions.

The recent work of Kolodny *et al.* (2005) and Vaks *et al.* (2006) highlight the complexity inherent in determining the climatic conditions of the Levant during the LGM. On the one hand, the oxygen-isotope data from speleothems and Lake Lisan can be interpreted as purely representing changes in the source water (the Eastern Mediterranean) rather than changes in local precipitation. On the other hand, there is clear evidence for a dominance of C4, arid-adapted vegetation and periods of non-deposition of speleothem carbonate, suggesting lower amounts of rainfall. The work of Vaks *et al.* (2006), however, highlights that periods of non-deposition of speleothem carbonate are not synchronous across the region (compare with Vaks *et al.* 2003). Thus it appears that while all authors agree that the LGM was cooler in the Levant, the evidence for changes in precipitation is less conclusive. It may be that the apparent contradictions in evidence arise more from differences in rainfall and temperature patterns, which combine to create differences in the P/E ratio that in turn influence many of the geochemical-proxy records used. Clearly, a greater number of well-dated records are required to determine the precipitation conditions of the Levant during the LGM.

Heinrich Event 1, H1 (16.0 cal ka BP)

As with H2, H1 in the Levant is likely to have been a time of relatively cooler, more arid conditions evidenced by a lowering of lake levels (Bartov *et al.* 2002; 2003), a small positive oxygen-isotope excursion in the Soreq Cave speleothem record (Bar-Matthews *et al.* 1999) and generally low sea surface temperatures [SST] in the Red Sea from ~17 to 15 cal ka BP (Arz *et al.* 2003a; 2003b). Unfortunately the Eastern Mediterranean marine records of SST and sea surface salinities [SSS] are at an insufficient resolution to detect whether H1 had a significant effect. The Red Sea records suggest a small decrease in salinity at ~16 cal ka BP which is difficult to reconcile with the postulation of increased regional evaporation, although long-term sea-level rise and changes in Red Sea circulation may explain some of the salinity decrease.

Bølling-Allerød warm interval (~15 to 13 cal ka BP)

The Bølling-Allerød represents the first sustained phase of warming after the LGM throughout much of the Northern Hemisphere. In the Levant, there is clear evidence from palynology and speleothems that conditions were warmer and wetter than at the LGM, with estimated annual rainfall of between 500 and 750 mm/yr (Rossignol-Strick 1995; Bar-Matthews *et al.* 1997). Carbon-isotope data from speleothems suggest an increase in C_3 plants compared with C_4 plants during the Bølling-Allerød (Bar-Matthews *et al.* 1997; 1999; 2003). The marine records in the Eastern Mediterranean are unclear, probably as a result of age model problems and the low temporal resolution of the samples (a result of low sedimentation rates). In the Red Sea, where sample temporal resolution is higher, SSTs rose dramatically at ~14.5 cal ka BP concomitant with a small decrease in SSSs.

Younger Dryas (~12.7 to 11.5 cal ka BP)

The Younger Dryas is generally considered to have resulted from a rapid input of cold freshwater into the North Atlantic that caused a disturbance in oceanic circulation and Northern Hemisphere climatic regimes (e.g. Alley 2000; Broecker

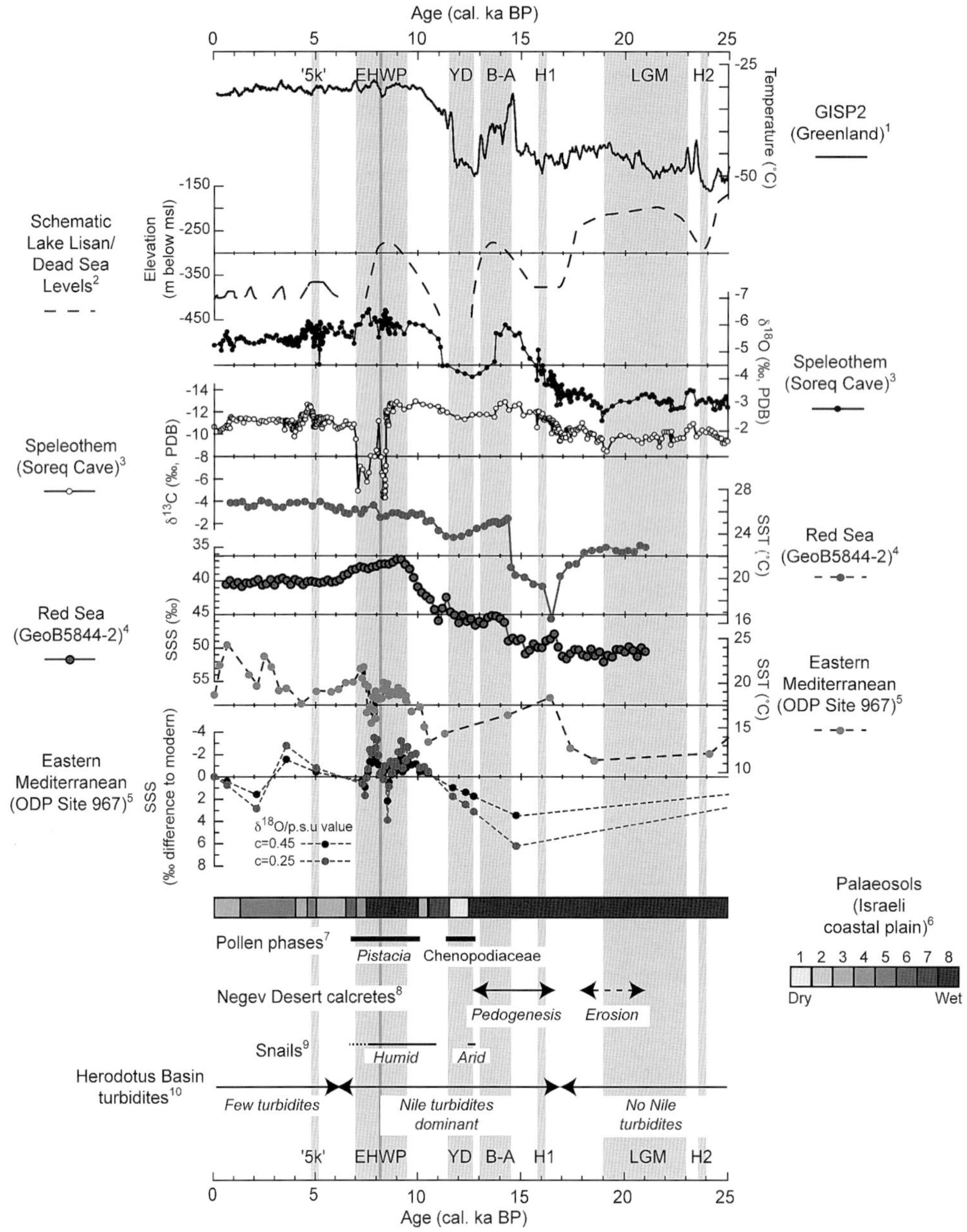

Fig. 6.2. Compilation of terrestrial and marine palaeoclimatic proxy data for the Levant and Eastern Mediterranean, reproduced from Robinson et al. *(2006). Also shown is the ice core record from GISP2 (Greenland). References: 1. Alley (2000); 2. schematic lake level curve from Robinson* et al. *(2006), based on data principally from Bartov* et al. *(2002; 2003), Begin* et al. *(1985), Frumkin* et al. *(1994), Landmann* et al. *(2002), Neev and Emery (1995) and Yechieli* et al. *(1993); 3. Bar-Matthews* et al. *(2003); 4. Arz* et al. *(2003a); 5. Emeis* et al. *(2000; 2003); 6. Gvirtzman and Wieder (2001); 7. Rossignol-Strick (1995); 8. Magaritz (1986), Goodfriend and Magaritz (1988); 9. Magaritz and Heller (1980); Goodfriend (1991; 1990; 1999); 10. Reeder* et al. *(2002). Figure reproduced with permission from Robinson* et al. *2006.*

and Denton 1989). There has been extensive debate about climatic conditions in the Levant during the Younger Dryas, based mainly upon the poorly dated palynological records (e.g. Baruch and Bottema 1991; 1999). However, it is clear from our review of multiple datasets (Figure 6.2) that the period was extremely arid and, most probably, cold compared with the Bølling-Allerød and the Holocene. Extreme aridity is evidenced by massive salt deposition and a lowering of lake levels in Lake Lisan (e.g. Yechieli *et al.* 1993) and deposition of wind-blown sediments on the Israel coastal plain (Gvirtzman and Wieder 2001). The high abundances of Chenopodiaceae and *Artemisia* pollen suggest rainfall of <150 mm per year (Rossignol-Strick 1995). Speleothems record a return towards glacial oxygen- and carbon-isotope values suggesting cooling, aridity and an increase in C_4 vegetation.

The marine record of the Younger Dryas is less conclusive. Cool SSTs (~13°C) are recorded in the Eastern Mediterranean at about 11 cal ka BP (Emeis *et al.* 2000; 2003), which could be part of the Younger Dryas.

Surprisingly, there does not appear to be any change in salinity at this time or a major shift in foraminiferal $\delta^{18}O$ values. However, further west several records show a return towards positive glacial oxygen-isotope values, lower alkenone SSTs, and higher salinities during the Younger Dryas (e.g. Cacho *et al.* 2001; Kallel *et al.* 1997b). In the Red Sea there was a cooling spanning the Younger Dryas, although salinity does not appear to have changed greatly during this interval. Given the dramatic climatic changes recorded by terrestrial proxies it seems surprising that the Eastern Mediterranean marine record does not show a clear climatic response during the Younger Dryas. Higher resolution records from sites with high sedimentation rates are required to determine whether or not the impact of the Younger Dryas on marine environments in the Eastern Mediterranean Sea was significant.

The Early Holocene and Sapropel S1 (~9.5 to 7 cal ka BP)

The Early Holocene appears to have been the wettest phase of the last 25,000 years across much of the Levant and Eastern Mediterranean (Figure 6.2). Increases in *Pistacia* and oak in the pollen records (Rossignol-Strick 1995; Rossignol-Strick 1999), southward migration of the Negev Desert boundary (Goodfriend 1999), meandering streams in southern Jordan (McLaren *et al.* 2004) and the deposition of 'red hamra'-type palaeosols (indicating an excess of precipitation versus evaporation) on the Israeli coastal plain (Gvirtzman and Wieder 2001) all suggest increased rainfall. The carbon- and oxygen-isotopic composition of pedogenic clast coatings from southeastern Turkey also supports a warmer, wetter Early Holocene (Pustovoytov *et al.* 2007). The Dead Sea was recharged to relatively high levels following the massive drawdown during the Younger Dryas (e.g. Frumkin *et al.* 1994), while the oxygen-isotopic composition of speleothems displays a shift towards more negative values, consistent with warmer, wetter conditions; Bar-Matthews *et al.* (2003) estimated rainfall amounts of between 550 and 700 mm/year and McGarry *et al.* (2004) calculated a terrestrial palaeotemperature of ~16°C. There may have been highly localized soil removal caused by the elevated rainfall, evidenced by a remarkably large positive carbon-isotope excursion in Soreq Cave speleothems interpreted by Bar-Matthews *et al.* (1997) as increased bed-rock dissolution caused by decreased soil cover. Elsewhere, speleothem $\delta^{13}C$ records are more typical and are relatively negative during the Early Holocene, compared with the Late Pleistocene or the Late Holocene, suggesting relatively more C_3 vegetation at this time, a situation consistent with warmer and wetter conditions (Frumkin *et al.* 2000).

The marine records of the Eastern Mediterranean and Red Sea also point to warm, wet conditions in the Early Holocene. Sediments rich in organic carbon (Sapropel 1 or S1) were deposited in the Eastern Mediterranean under anoxic conditions between ~9.5 and 7 cal ka BP (e.g. Emeis *et al.* 2003). It has been postulated that the Mediterranean sapropels formed in anoxic conditions that were induced through stratification of the water column driven by a low salinity surface layer (e.g. Kallel *et al.* 1997a; Rohling 1994; Rohling and Hilgen 1991; Rossignol-Strick 1985; Rossignol-Strick *et al.* 1982), which is evidenced by the geochemistry of foraminifera. Geochemical (Scrivner *et al.* 2004) and sedimentary (Reeder *et al.* 2002) data suggest that the source of this low salinity water may have been an enhanced Nile River outflow, driven by increased rainfall in the Sahara (as postulated by Rossignol-Strick 1985; Rossignol-Strick *et al.* 1982). However, further west, geochemical data suggest that influence of the Nile water was minimal (Kallell *et al.* 1997a; Krom *et al.* 1999; Freydier *et al.* 2001) and that the entire central and eastern Mediterranean was likely to have experienced enhanced precipitation. Increased rainfall in the Levant led to an increase in run-off into the Red Sea, as recorded by elevated amounts of terrigenous sediment and a minimum in salinity estimates (Arz *et al.* 2003a; 2003b).

Despite the generalized picture of a warm Early Holocene, there is some evidence, particularly in the Northern Hemisphere, for brief cold events during this period. Rohling and Pälike (2005) have shown that the so-called 8.2 cal ka BP cold event (recognized in the Greenland ice core) is superimposed upon a general background of climatic deterioration in the Northern Hemisphere that started at about 8.6 cal ka BP and lasted until 7.9 cal ka BP. In the Eastern Mediterranean and Levant there is some evidence for climate change during the Early Holocene wet phase that may be related to the events described elsewhere by Rohling and Pälike (2005). Some sapropel records show a brief interruption of anoxic conditions between 7.9 and 7.7 cal ka BP that has been linked to a brief cooling event in the Adriatic Sea (e.g. Myers and Rohling 2000; Rohling *et al.* 1997). Bar-Matthews *et al.* (1999) claim that an excursion in speleothem $\delta^{13}C$ at about 8.2 cal ka BP represents a sudden cooling and decrease in precipitation during the terrestrial 'deluge', which is consistent with a peak in salinity in the Eastern Mediterranean record (see Figure 6.2). However, given the uncertainty in the age models for both speleothems and marine cores, it may prove very difficult to determine conclusively whether cooling and increased salinity in the Eastern Mediterranean about 8000 years ago was related to general cooling or to the '8.2 ka cold event'.

Mid Holocene wet event – ~5000 cal yrs BP

There is some suggestion in the terrestrial data that there may have been a slightly wetter phase in the mid Holocene (~5000 years BP). Lake levels were relatively high at this time (Frumkin *et al.* 1994), precipitation calculated from speleothems rose to >500 mm/year at ~4.7 cal ka BP (Bar-Matthews *et al.* 1997; 2003) and the palaeosol record suggests a slightly wetter climate at about 4.8 cal ka BP (Gvirtzman and Wieder 2001). There is also a suggestion of a brief wetter phase in the snail geochemistry record at ~4.5 cal ka BP (Goodfriend 1991). More recently, Migowski *et al.* (2006) presented a high-resolution, very detailed lake-level curve for the Dead Sea and demonstrated that lake levels were relatively high from ~5.5 to 4.3 cal ka BP, with a possible drop between ~5.4 and 5.0 cal ka BP. This latter drop coincides with a well-known brief climate change event known from evidence across the Eastern Mediterranean and Levant region (see Staubwasser and Weiss 2006).

Linking the geological and archaeological records

A key question for geologists and archaeologists is the role of climate change in driving (or not) changes in human behaviour. This question has been examined and discussed in the Levant for the Late Pleistocene to Early Holocene by many authors (e.g. Bar-Yosef and Belfer-Cohen 1992; Hillman 1996; Migowski *et al.* 2006; Mithen 2003; Moore and Hillman 1992; Staubwasser and Weiss 2006; Benz, this volume; Maher, this volume; Watkins, this volume), and we do not intend to attempt any sort of synthesis of the results here. Rather, we would like to highlight, briefly, a number of issues that always need to be considered when integrating and interpreting geological and archaeological datasets. The common language between these datasets is time, and thus dating is of key importance. As discussed previously, geological datasets utilize a wide variety of dating methods and only through the use of calendar or calibrated ages (in years BP) can these methods be fairly compared. Therefore, it is essential that radiocarbon dates are calibrated before comparison with, say, uranium-series (U-series) ages. The error bars on all dating methods need to be apparent, as without them it is not possible to determine which samples may statistically overlap in age and which do not. This is especially important when trying to determine whether geological or climatic events preceded or succeeded archaeological events. Further complications can potentially arise when archaeological data are compared with a time-series of geological data, such as geochemical data from a speleothem or marine core. With the exception of ice cores, tree-rings and varved sediments, age models for many geological records of climatic and environmental change (including speleothems and many marine cores) rely upon assigning numerical ages to samples based upon a linear interpolation between two 'fixed' age tie-points that have been dated by radiocarbon, U-series or some other method. Thus, considerable care must be taken when comparing such records with other datasets, be they geological or archaeological, as much of the data is not actually dated, but is simply assigned an age through interpolation.

A final point relates to the role of climate in affecting human behaviour. Cause and effect is an inherently difficult concept to prove in the geological record and although examples may exist of two events occurring in a sequence and thus suggesting cause and effect, the reality is that we can only state with certainty what the evidence demonstrates and hypothesize about how the events are connected. However, as the weight of evidence increases, so the case for cause and effect can be more robustly made. Whichever view is taken of the role of climate in human behaviour, it is essential that testable hypotheses are constructed that can be confirmed or disproved by well-dated geological and archaeological records. With respect to the Levant, the climatic and environmental perturbations over the last 25,000 years have been significant and variable. Much research (and debate) has focused on the importance of the Younger Dryas as a driver for a sedentary lifestyle, with a variety of viewpoints being presented (e.g. Moore and Hillman 1992; Hillman 1996; Benz, this volume; Maher, this volume; Watkins, this volume). Some authors have pointed to the succession of archaeological change in the Late Pleistocene as evidence that the Younger Dryas was not a significant event from a human behaviour perspective (e.g. Maher, this volume; Watkins, this volume). However, it should be remembered that the Late Pleistocene–Early Holocene was a period of significant climate variability as the Earth transited from a glacial to interglacial state, and that the Younger Dryas was itself just part of this succession. When comparing archaeological and palaeoclimate datasets it is important to remember that both represent dynamic, constantly changing situations. As discussed by Maher in this volume, the transition in the Near East from so-called 'hunter-gatherers' to 'farmers' can be argued to last from the Late Pleistocene, at a point close to the LGM, through to the Early Holocene, and thus it is important when considering cause and effect not only to look at the detailed evidence for the relative timing of events but also to step back and look at the bigger long-term picture. It is certainly quite striking that the most significant human behavioural changes in the Late Pleistocene appear to commence close to the LGM and the onset of the most dramatic climate transition of the last 100,000 years. While it may be tantalizing to draw conclusions regarding environmental forcing of human behaviour based upon the apparent co-incidence in timing of climatic and archaeological change, the evidence to test such hypotheses will ultimately need to come from stratigraphically well-constrained, integrated archaeological and palaeoclimatic studies.

Conclusions

A wide range of geological data is available to be used to interpret the climatic history of the Eastern Mediterranean and Levant. However, the interpretation of this data is not without problems and, for some time periods, appears to be contradictory. This highlights the difficulty of relying on a limited number of datasets that are restricted to a small geographic area: our understanding will improve with more data that covers a greater area. Ideally, these datasets need to be high-resolution and extremely well dated. Despite the difficulties in the spatial distribution and the interpretation of the data, it appears that the major Northern Hemisphere climatic events of the last 25,000 years can be recognized in the Eastern Mediterranean and Levant. The LGM was probably colder than at present, which would have had a significant effect on evaporation. In our opinion, the amount and pattern of precipitation during the LGM remains unresolved. The Bølling-Allerød was relatively warm and wet, whereas the succeeding Younger-Dryas marked a return to more glacial conditions. The Early Holocene marked the warmest and wettest phase of the last 25,000 years in the region. A second wet phase appears to have occurred about 5000 years ago.

Acknowledgements

We thank the Leverhulme Trust (UK) for funding of the 'Water, Life, Civilisation' project (Grant F/00239/R). More details on this project are available at http://www.waterlifecivilisation.org/. Stuart Robinson is funded by a Royal Society University Research Fellowship. We are grateful to Y. Bartov, S. Calvert, K. C. Emeis, M. Geyh, A. Hayes, M. Jones and S. Mithen for sharing their data with us and for useful discussion. B. Finlayson and G. Warren are thanked for inviting Stuart Robinson and Claire Rambeau to participate in the 'Landscapes in Transition' workshop. Useful discussion with N. Goring-Morris, L. Maher and T. Watkins is gratefully acknowledged. Unfortunately, Bruce Sellwood passed away after the submission of this paper; his support and enthusiasm was invaluable to this project and will be sorely missed by us all.

Bibliography

Abed, A. M. and Yaghan, R. (2000) On the palaeoclimate of Jordan during the last glacial maximum. *Palaeogeography, Palaeoclimatology, Palaeoecology* 160, 23–33.

Alley, R. B. (2000) The Younger Dryas cold interval as viewed from central Greenland. *Quaternary Science Reviews* 19, 213–26.

Arz, H. W., Pätzold, J., Müller, P. J. and Moammar, M. O. (2003a) Influence of northern hemisphere climate and global sea level rise on the restricted Red Sea marine environments during termination I. *Paleoceanography* 18, doi:10.1029/2002PA000864.

Arz, H. W., Lamy, F., Pätzold, J., Müller, P. J. and Prins, M. (2003b) Mediterranean moisture source for an Early-Holocene humid period in the Northern Red Sea. *Science* 300, 118–21.

Ayalon, A., Bar-Matthews, M. and Kaufman, A. (1999) Petrography, strontium, barium and uranium concentrations, and strontium and uranium isotope ratios in speleothems as palaeoclimatic proxies: Soreq Cave, Israel. *The Holocene* 9, 715–22.

Bar-Matthews, M., Ayalon, A. and Kaufman, A. (1997) Late Quaternary paleoclimate in the Eastern Mediterranean region from stable isotope analysis of speleothems at Soreq Cave, Israel. *Quaternary Research* 47, 155–68.

Bar-Matthews, M., Ayalon, A., Kaufman, A. and Wasserburg, G. J. (1999) The Eastern Mediterranean paleoclimate as a reflection of regional events: Soreq Cave, Israel. *Earth and Planetary Science Letters* 166, 85–95.

Bar-Matthews, M., Ayalon, A., Gilmour, M., Matthews, A. and Hawkesworth, C. J. (2003) Sea–land oxygen isotopic relationships from planktonic foraminifera and speleothems in the Eastern Mediterranean region and their implication for paleorainfall during interglacial intervals. *Geochimica et Cosmochimica Acta* 67, 3181–99.

Bartov, Y., Stein, M., Enzel, Y., Agnon, A. and Reches, Z. (2002) Lake levels and sequence stratigraphy of Lake Lisan, the late Pleistocene precursor to the Dead Sea. *Quaternary Research* 57, 9–21.

Bartov, Y., Goldstein, S. L., Stein, M. and Enzel, Y. (2003) Catastrophic arid episodes in the Eastern Mediterranean linked with the North Atlantic Heinrich events. *Geology* 31, 439–42.

Baruch, U. and Bottema, S. (1991) Palynological evidence for climatic changes in the Levant ca. 17,000–9,000 B.P. Pp. 11–20 in O. Bar-Yosef and F. R. Valla (eds), *The Natufian Culture in the Levant*. Ann Arbor, MI: International Monographs in Prehistory.

Baruch, U. and Bottema, S. (1999) A new pollen diagram from Lake Hula: vegetational, climatic and anthropogenic implications. Pp. 5–86 in H. Kawanabe, G. W. Coulter and A. C. Roosevelt (eds), *Ancient Lakes: Their Cultural and Biological Diversity*. Ghent: Kenobi Productions.

Bar-Yosef, O. and Belfer-Cohen, A. (1992) From foraging to farming in the Mediterranean Levant. Pp. 21–48 in A. B. Gebauer and T. D. Price (eds), *Transitions to Agriculture in Pre-history*. Madison, WI: Prehistory Press.

Begin, Z. B., Broecker, W., Buchbinder, B., Druckman, Y., Kaufman, A., Magaritz, M. and Neev, D. (1985) Dead Sea and Lake Lisan levels in the last 30,000 years: a preliminary report. *Israel Geological Survey Report* 29/85, 1–18.

Bond, G., Heinrich, H., Broecker, W., Labeyrie, L., McManus, J., Andres, J., Huon, S., Jantschik, R., Clasen, S., Simet, C., Tedesco, K., Klas, M., Bonani, G. and Ivy, S. (1992) Evidence for massive discharges of icebergs into the North Atlantic Ocean during the last glacial period. *Nature* 360, 245–9.

Bottema, S. (1995) The Younger Dryas in the Eastern Mediterranean. *Quaternary Science Reviews* 14, 883–91.

Broecker, W. S. and Denton, G. H. (1989) The role of ocean-atmosphere reorganizations in glacial cycles. *Geochimica et Cosmochimica Acta* 53, 2465–501.

Cacho, I., Grimalt, J. O., Sierro, F. J., Shackleton, N. and Canals, M. (2001) Evidence for enhanced Mediterranean thermohaline circulation during rapid climate coolings. *Earth and Planetary Science Letters* 183, 417–29.

Emeis, K.-C., Schulz, H.-M., Struck, U., Sakamoto, T., Doose, H., Erlenkeuser, H., Howell, M. W., Kroon, D. and Paterne, M. (1998) Stable isotope and temperature records of sapropels from ODP Sites 964 and 967: constraining the physical environment of sapropel formation in the Eastern Mediterranean Sea. *Proceedings of the Ocean Drilling Program, Scientific Results* 160, 309–31.

Emeis, K.-C., Struck, U., Schulz, H.-M., Bernasconi, S., Sakamoto, T. and Martinez-Ruiz, F. (2000) Temperature and salinity of Mediterranean Sea surface waters over the last 16,000 years: constraints on the physical environment of S1 sapropel formation based on stable oxygen isotopes and alkenone unsaturation ratios. *Palaeogeography, Palaeoclimatology, Palaeoecology* 158, 259–80.

Emeis, K.-C., Schulz, H., Struck, U., Rossignol-Strick, M., Erlenkeuser, H., Howell, M. W., Kroon, D., Mackensen, A., Ishizuka, S., Oba, T., Sakamoto, T. and Koizumi, I. (2003) Eastern Mediterranean surface water temperatures and $\delta^{18}O$ composition during deposition of sapropels in the late Quaternary. *Paleoceanography* 18, doi: 10.1029/2000PA000617.

Fontugne, M. R. and Calvert, S. E. (1992) Late Pleistocene variability of the carbon isotopic composition of organic matter in the Eastern Mediterranean: monitor of changes in carbon sources and atmospheric CO_2 concentrations. *Paleoceanography* 7, 1–20.

Freydier, R., Michard, A., De Lange, G. and Thomson J. (2001) Nd isotopic compositions of Eastern Mediterranean sediments: tracers of the Nile influence during sapropel S1 formation. *Marine Geology* 177, 45–62.

Frumkin, A. and Stein, M. (2004) The Sahara–East Mediterranean dust and climate connection revealed by strontium and uranium isotopes in a Jerusalem speleothem. *Earth and Planetary Science Letters* 217, 451–64.

Frumkin, A., Carmi, I., Zak, I. and Magaritz, M. (1994) Middle Holocene environmental change determined from the Salt Caves of Mount Sodom, Israel. Pp. 315–32 in O. Bar-Yosef and R. S. Kra (eds), *Late Quaternary Chronology and Paleoclimates of the Eastern Mediterranean*. Tucson, AZ: *Radiocarbon* supplement.

Frumkin, A., Ford, D. C. and Schwarz, H. P. (1999) Continental oxygen isotopic record of the last 170,000 years in Jerusalem. *Quaternary Research* 51, 317–27.

Frumkin, A., Ford, D. C. and Schwarz, H. P. (2000) Paleoclimate and vegetation of the last glacial cycles in Jerusalem from a speleothem record. *Global Biogeochemical Cycles* 14, 863–70.

Goldberg, P. (1994) Interpreting late Quaternary continental sequences in Israel. Pp. 89–102 in O. Bar-Yosef and R. S. Kra (eds), *Late Quaternary Chronology and Paleoclimates of the Eastern Mediterranean*. Tucson, AZ: *Radiocarbon* supplement.

Goodfriend, G. A. (1990) Rainfall in the Negev Desert during the Middle Holocene, based on ^{13}C of organic matter in land snail shells. *Quaternary Research* 34, 186–97.

Goodfriend, G. A. (1991) Holocene trends in ^{18}O in land snail shells from the Negev Desert and their implications for changes in rainfall source areas. *Quaternary Research* 35, 417–26.

Goodfriend, G. A. (1999) Terrestrial stable isotope records of Late Quaternary paleoclimates in the Eastern Mediterranean region. *Quaternary Science Reviews* 18, 501–13.

Goodfriend, G. A. and Magaritz, M. (1988) Palaeosols and late Pleistocene rainfall fluctuations in the Negev Desert. *Nature* 332, 144–6.

Goring-Morris, A. N. and Goldberg, P. (1990) Late Quaternary dune incursions in the southern Levant: archaeology, chronology and palaeoenvironments. *Quaternary International* 5, 115–37.

Gvirtzman, G. and Wieder, M. (2001) Climate of the last 53,000 years in the Eastern Mediterranean, based on soil-sequence stratigraphy in the coastal plain of Israel. *Quaternary Science Reviews* 20, 1827–49.

Hayes, A., Kucera, M., Kallel, N., Sbeffi, L. and Rohling, E. J. (2005) Glacial Mediterranean Sea surface temperatures based on planktonic foraminiferal assemblages. *Quaternary Science Reviews* 24, 999–1016.

Hazan, N., Stein, M., Agnon, A., Marco, S., Nadel, D., Negendank, J. F. W., Schwab, M. J. and Neev, D. (2005) The late Quaternary liminological history of Lake Kinneret (Sea of Galilee), Israel. *Quaternary Research* 63, 60–77.

Hillman, G. C. (1996) Late Pleistocene changes in wild plant-foods available to hunter-gatherers of the northern Fertile Crescent: possible preludes to cereal cultivation. Pp. 159–203 in D. J. Harris (ed.), *The Origins and Spread of Agriculture and Pastoralism in Eurasia*. London: UCL Press.

Kallel, N., Paterne, M., Duplessy, J.-C., Vergnaud-Grazzini, C., Pujol, C., Labeyrie, L., Arnold, M., Fontugne, M. and Pierre, C. (1997a) Enhanced rainfall in the Mediterranean region during the last sapropel event. *Oceanologica Acta* 20, 697–712.

Kallel, N., Paterne, M., Labeyrie, L., Duplessy, J.-C. and Arnold, M. (1997b) Temperature and salinity records of the Tyrrhenian Sea during the last 18,000 years. *Palaeogeography, Palaeoclimatology, Palaeoecology* 135, 97–108.

Kolodny, Y., Stein, M. and Machlus, M. (2005) Sea-rain-lake relation in the last glacial East Mediterranean revealed by $\delta^{18}O$-$\delta^{13}C$ in Lake Lisan aragonites. *Geochimica et Cosmochimica Acta* 16, 4045–60.

Krom, M. D., Michard, A., Cliff, R. A. and Strohle, K. (1999) Sources of sediment to the Ionian Sea and western Levantine basin of the Eastern Mediterranean during S-1 sapropel times. *Marine Geology* 160, 45–61.

Landmann, G., Abu Qudaira, G. M., Shawabkeh, K., Wrede, V. and Kempe, S. (2002) Geochemistry of the Lisan and Damya Formations in Jordan, and implications for palaeoclimate. *Quaternary International* 89, 45–57.

McGarry, S., Bar-Matthews, M., Matthews, A., Vaks, A., Schilman, B. and Ayalon, A. (2004) Constraints on hydrological and paleotemperature variations in the Eastern Mediterranean region in the last 140 ka given by the δD values of speleothem fluid inclusions. *Quaternary Science Reviews* 23, 919–34.

McLaren, S. J., Gilbertson, D. D., Gratten, J. P., Hunt, C. O., Duller, G. A. T. and Barker, G. A. (2004) Quaternary palaeogeomorphologic evolution of the Wadi Faynan area, southern Jordan. *Palaeogeography, Palaeoclimatology, Palaeoecology* 205, 131–54.

Magaritz, M. (1986) Environmental changes recorded in the Upper Pleistocene along the desert boundary, Southern Israel. *Palaeogeography, Palaeoclimatology, Palaeoecology* 53, 213–29.

Magaritz, M. and Heller, J. (1980) A desert migration indicator – oxygen isotopic composition of land snail shells. *Palaeogeography, Palaeoclimatology, Palaeoecology* 32, 153–62.

Marco, S., Stein, M., Agnon, A. and Ron, H. (1996) Long-term earthquake clustering: a 50,000-year paleoseismic record in the Dead Sea Graben. *Journal of Geophysical Research* 101, 6179–91.

Meadows, J. (2005) The Younger Dryas episode and the radiocarbon chronologies of the Lake Huleh and Ghab Valley pollen diagrams, Israel and Syria. *The Holocene* 15, 631–6.

Migowski, C., Stein, M., Prasad, S., Negendank, J. F. W. and Agnon, A. (2006) Holocene climate variability and cultural evolution in the Near East from the Dead Sea sedimentary record. *Quaternary Research* 66, 421–31.

Mithen, S. J. (2003) *After the Ice: A Global Human History, 20,000–5000 BC*. London: Weidenfeld and Nicolson.

Moore, A. M. T. and Hillman, G. C. (1992) The Pleistocene to Holocene transition and human economy in Southwest

Asia: the impact of the Younger Dryas. *American Antiquity* 57, 482–94.

Moustafa, Y. A., Pätzold, J., Loya, Y. and Wefer, G. (2000) Mid-Holocene stable isotope record of corals from the northern Red Sea. *International Journal of Earth Sciences* 88, 742–51.

Myers, P. G. and Rohling, E. J. (2000) Modeling a 200-yr interruption of the Holocene Sapropel S1. *Quaternary Research* 53, 98–104.

Neev, D. and Emery, K. O. (1995) *The Destruction of Sodom, Gomorrah, and Jericho, Geological, Climatological and Archaeological Background*. Oxford: Oxford University Press.

Pustovoytov, K., Schmidt, K. and Taubald, H. (2007) Evidence for Holocene environmental changes in the northern Fertile Crescent provided by pedogenic carbonate coatings. *Quaternary Research* 67, 315–27.

Reeder, M. S., Stow, D. A. V. and Rothwell, R. G. (2002) Late Quaternary turbidite input into the east Mediterranean basin: new radiocarbon constraints on climate and sea-level control. Pp. 267–78 in S. J. Jones and L. E. Frostick (eds), *Sediment Flux to Basins: Causes, Controls, and Consequences*. Geological Society of London Special Publication no. 191. London: Geological Society.

Robinson, S. A., Black, S., Sellwood, B. and Valdes, P. J. (2006) A review of palaeoclimates in the Levant and Eastern Mediterranean from 25,000 to 5,000 years BP: setting the environmental background for the evolution of human civilisation. *Quaternary Science Reviews* 25, 1517–41.

Rohling, E. J. (1994) Review and new aspects concerning the formation of Eastern Mediterranean sapropels. *Marine Geology* 122, 1–28.

Rohling, E. J. and Hilgen, F. J. (1991) The Eastern Mediterranean climate at times of sapropel formation: a review. *Geologie en Mijnbouw* 70, 253–64.

Rohling, E. J. and Pälike, H. (2005) Centennial-scale climate cooling with a sudden cold event around 8,200 years ago. *Nature* 434, 975–9.

Rohling, E. J., Jorissen, F. J. and De Stigter, H. C. (1997) 200 year interruption of Holocene sapropel formation in the Adriatic Sea. *Journal of Micropalaeontology* 16, 97–108.

Rossignol-Strick, M. (1985) Mediterranean Quaternary sapropels, an immediate response of the Africa monsoon to variation of insolation. *Palaeogeography, Palaeoclimatology, Palaeoecology* 49, 237–63.

Rossignol-Strick, M. (1995) Sea-land correlation of pollen records in the Eastern Mediterranean for the glacial–interglacial transition: biostratigraphy versus radiometric time-scale. *Quaternary Science Reviews* 14, 893–915.

Rossignol-Strick, M. (1999) The Holocene climatic optimum and pollen records of sapropel 1 in the Eastern Mediterranean, 9000–6000 BP. *Quaternary Science Reviews* 18, 515–30.

Rossignol-Strick, M., Nesteroff, O. P. and Vergnaud-Grazzini, C. (1982) After the deluge: Mediterranean stagnation and sapropel formation. *Nature* 295, 105–10.

Scrivner, A. E., Vance, D. and Rohling, E. J. (2004) New neodymium isotope data quantify Nile involvement in Mediterranean anoxic episodes. *Geology* 32, 565–8.

Staubwasser, M. and Weiss, H. (2006) Introduction: Holocene climate and cultural evolution in late prehistoric–early historic West Asia. *Quaternary Research* 66, 372–87.

Stein, M., Starinsky, A., Katz, A., Goldstein, S. L., Machlus, M. and Schramm, A. (1997) Strontium isotopic, chemical, and sedimentological evidence for the evolution of Lake Lisan and the Dead Sea. *Geochimica et Cosmochimica Acta* 61, 3975–92.

Vaks, A., Bar-Matthews, M., Ayalon, A., Schilman, B., Gilmour, M., Hawkesworth, C. J., Frumkin, A., Kaufman, A. and Matthews, A. (2003) Paleoclimate reconstruction based on the timing of speleothem growth and oxygen and carbon isotope composition in a cave located in the rain shadow in Israel. *Quaternary Research* 59, 182–93.

Vaks, A., Bar-Matthews, M., Ayalon, A., Matthews, A., Frumkin, A., Dayan, U., Halicz, L., Almogi-Labin, A. and Schilman, B. (2006) Palaeoclimate and location of the border between Mediterranean climate region and the Saharo–Arabian Desert as revealed by speleothems from the northern Negev Desert, Israel. *Earth and Planetary Science Letters* 249, 384–99.

Van Zeist, W. and Bottema, S. (1982) Vegetational history of the Eastern Mediterranean and the Near East during the last 20,000 years. Pp. 277–321 in J. L. Bintliff and W. van Zeist (eds), *Palaeoclimates, Palaeoenvironments and Human Communities in the Eastern Mediterranean Region in Later Prehistory*. British Archaeological Reports, International Series 133. Oxford: BAR.

Yechieli, Y., Magaritz, M., Levy, Y., Weber, U., Kafri, U., Woelfli, W. and Bonani, G. (1993) Late Quaternary geological history of the Dead Sea area, Israel. *Quaternary Research* 39, 59–67.

7. The Case for Climatic Stress Forcing Choice in the Adoption of Agriculture in the British Isles

Richard Tipping

Many recently obtained high-temporal-resolution palaeoclimatic records are assembled which demonstrate that the change from a Mesolithic to a Neolithic economy in the British Isles coincided with a period of rapid and significant climatic change. A series of processes, physical, ecological and conceptual, are suggested, some of which are demonstrable, by which the change to agriculture became the preferred choice for hunter-gatherer-fishers.

Introduction: climate change and human decision-making

A recently recognized and unavoidable factor in the relation between humans and their environments is large-scale rapid climate change (Mayewski *et al.* 2004). The astonishing growth of data on Holocene climate change since the 1980s has led to a new paradigm with three key properties (Chambers and Brain 2002). First, some climatic changes were very major events, global in extent though regional in effect. Secondly, some were rapid, even abrupt, meaning that the climate system crossed thresholds over only decades. Third, climate extremes exceeded anything within normal human experience (de Menocal 2001).

The last two properties of the paradigm in particular have led to new speculations on the impacts of climate on human societies because there is a temporal association between periods of rapid climatic and cultural change (Berglund 2003). Mayewski *et al.* (2004) identify six such periods in the Holocene, broadly defined at 7000–6000, 5000–4000, 2200–1800 and 1500–500 BC, and 800–1000 and 1400–1850 AD. This paper develops this idea by a restatement of the 'food crisis' hypothesis at the Mesolithic–Neolithic transition in Northwest Europe, emphasizing that people worry about food more than about other things (Fernandez-Armesto 2000; Schulting 2004). The definition here of the Neolithic is that of Fischer (2002, 344): the introduction of cereal cultivation and/or stock raising.

Some analyses of the role of climatic change in human experience are 'catastrophist' in postulating that climate change was sometimes so rapid, large or extreme that adaptation was impossible (e.g. Weiss and Bradley 2001), but the more catastrophist claims are undoubtedly over-simplistic and unremittingly deterministic (Whittle 2003a). Other interpretations have been concerned to emphasize that people always had choices in how they coped with climatic stress. Most well-explored case studies of climatic impacts have considered scenarios in which the carrying capacity of the environment was lowered and a complex agrarian society fragmented or 'shifted to lower subsistence levels by reducing social complexity ... and reorganizing systems of supply and production' (de Menocal 2001, 572). The change from foraging to farming is ostensibly in the opposite direction, from apparent simplicity to greater complexity in social and economic structure, though we should not assume that early agriculture required a substantially more complex society. One of the most successful applications of a climate stress model to increasing socio-economic complexity is the introduction of agriculture in the Near East (Bar-Yosef and Belfer-Cohen 1992). Brooks (2004) has applied a similar 'oasis' model in explaining climatic forcing in the near-synchronous rise of complex societies in low-latitude, monsoon-affected regions around 4000 BC, and Sandweiss *et al.* (1999) have argued that global climate dislocations at this time impacted on several human societies.

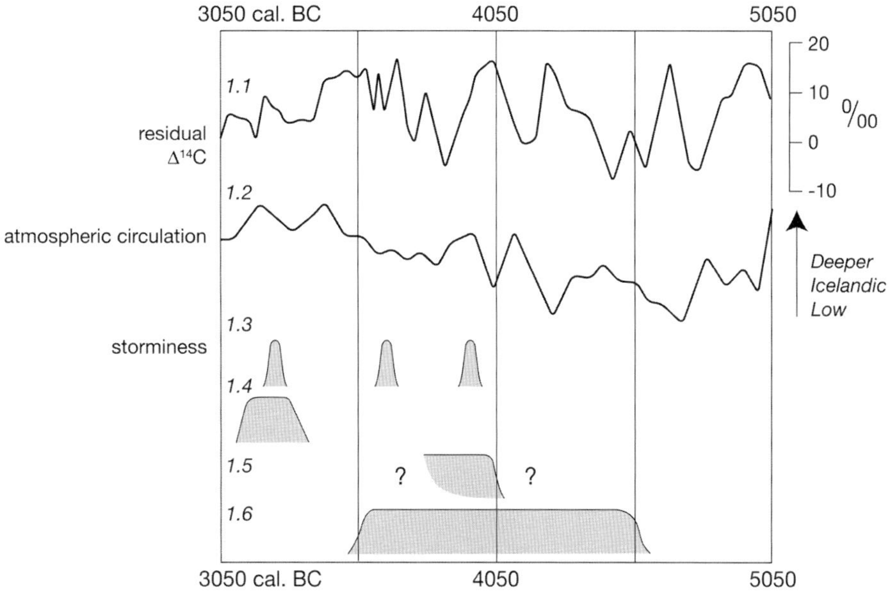

Fig. 7.1. Changes from 5050 to 3050 BC in indicators of change in atmospheric and ocean circulation from major data-sets relating to north-west Europe. Some records have been modified from the original publications to allow comparability: 1.1. residual Δ14C activity (modified from Stuiver and Braziunas 1993) which may represent the global impact of fluctuations in solar activity; 1.2. intensity of North Atlantic atmospheric circulation estimated from the inverse abundance of Na+ (sea salt) in the GISP2 ice core, Greenland, interpreted (Mayewski et al. *2004) as a proxy for the depth/intensity of the Iceland Low atmospheric pressure zone. Deeper periods mean stronger and more zonal winds across the North Atlantic Ocean; 1.3. aeolian sand-blowing events across coastal raised mosses in south-west Sweden (Bjorck and Clemmensen 2004; de Jong* et al. *2006); 1.4. sand-blowing event across peat in north-west Ireland (Caseldine* et al. *2005); 1.5. summary of dune mobilization phases in the UK (Keatinge and Dickson 1979; Mellars 1987; Tooley 1990; Cowie 1996; Gilbertson* et al. *1999); 1.6. sand dune mobilization on the northern Irish coast (Wilson* et al. *2004).*

Baillie (1995), Mayewski *et al.* (1996) and Berglund (2003) noted the temporal association between climatic deterioration at this time and the introduction of farming in the British Isles and Northern Europe respectively, but did not consider the physical, ecological, economic or social mechanisms involved. Bogucki (1998) suggested that agriculture was fully adopted on the Northern European plain to broaden the resource base in response to increased climatic variability. Bonsall *et al.* (2002) have argued that an abrupt change at c. 4000 BC to a drier, more continental climate in the British Isles encouraged the uptake of agricultural resources. Tipping and Tisdall (2004) developed a model, elaborated in this contribution, which examined reasons why resources used by foraging communities might have failed or become much less reliable with climatic deterioration, encouraging those communities to adopt, perhaps only in part, an agrarian 'package'. A new element developed in this paper considers how physical changes to the North Atlantic Ocean might actually have discouraged contact between islands such as the British Isles and the continent before c. 4000 BC.

The model presented here is intended only to explain the adoption of agriculture by hunter-gatherers in the British Isles. In all its details the model is arguably most applicable to the north and west of the British Isles. The model is not applicable to the introduction of agriculture to continental Europe because the timing of this was significantly earlier (Whittle 2003b), although the scale of the climate changes described can in general terms be used to examine the apparent instability of farming populations on the continent closest to Britain at around 4100 BC (Louwe-Kooijmans 1999; Sheridan, this volume). It is considered that the introduction of agriculture to the British Isles and southern Scandinavia was rapid, at 4000–3800 BC (Ashmore 2004; Rowley-Conwy 2004), that there is no evidence in the British Isles for an 'invisible' Neolithic (Tipping 1994; 2004) and that, in contrast to parts of Northwest Europe (Klassen 2002), there is no evidence for contacts between Mesolithic foragers and contemporaneous farmers on the continent (*contra* Thomas 2004). The introduction of agriculture in the British Isles was an event (Schulting and Richards 2002). Of key importance, then, is to explain why this event happened when it did.

Rapid climate change at the Mesolithic–Neolithic event in Northwest Europe

The major climate changes summarized in Figures 7.1 to 7.3 are from a wide range of datasets for Northwest Europe north and west of the Alps, which is modelled to have been for the most part a uniform climatic region (Cheddadi *et al.* 1997), for 5050–3050 BC. At c. 5000 BC extensive areas of intertidal marsh are modelled to have fringed East Anglia and the adjacent continent, including much of the Dogger

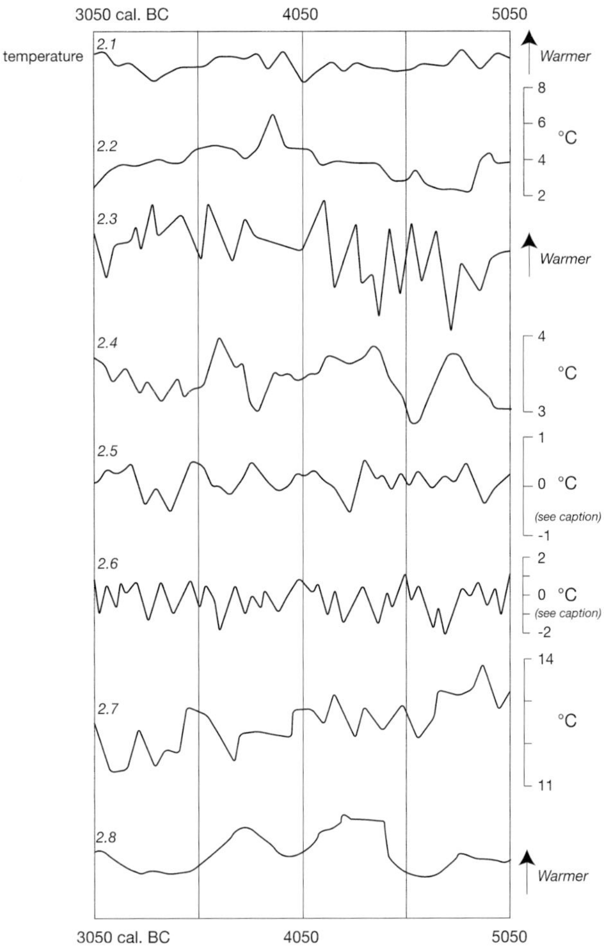

Fig. 7.2. Changes from 5050 to 3050 BC in different indicators of temperature change relating to north-west Europe. Some records have been modified from the original publications to allow comparability: 2.1. $\delta^{18}O$ values in the GRIP ice core, Greenland, from Mayewski et al. *(2004); 2.2. planktonic foraminiferal assemblages in the northern North Atlantic expressed as summer sea surface temperature (°C), from Mayewski* et al. *(2004); 2.3. $\delta^{18}O$ values in a speleothem in western Ireland (McDermott* et al. *2001); 2.4. $\delta^{18}O$ values in a speleothem in western Norway expressed as mean annual air temperature (°C), drawn from Lauritzen and Lundberg (1999); 2.5.* Pinus *tree ring widths in northern Finland expressed as differences in July mean temperature (°C) from 0°C, drawn from Helama* et al. *(2002); 2.6.* Pinus *tree ring widths in northern Finland expressed as July mean temperatures (°C), drawn from Grudd* et al. *(2002); 2.7. pollen spectra in northern Finland expressed as July mean temperatures (°C), drawn from Seppa and Birks (2001); 2.8. peat-stratigraphic analyses from blanket peats in northern Scotland interpreted as a record of relative desiccation, drawn from Tipping and Tisdall (2004).*

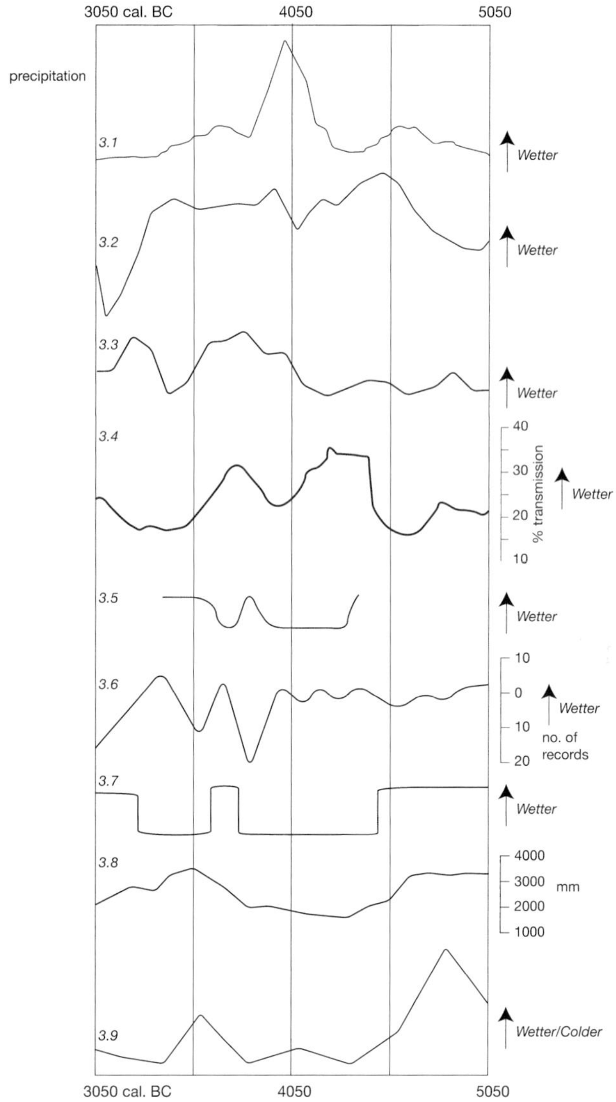

Fig. 7.3. Changes from 5050 to 3050 BC in different indicators of precipitation change relating to north-west Europe. Some records have been modified from the original publications to allow comparability: 3.1. relative deposition rate of oak trees in sediments of the River Main, Germany, suggested to be determined by soil dryness, drawn from Spurk et al. *(2002); 3.2. peat-stratigraphic analyses from a raised moss in north-west England, drawn from Hughes* et al. *(2000); 3.3. peat-stratigraphic analyses from a raised moss in south-east Scotland, drawn from Langdon* et al. *(2003); 3.4. peat-stratigraphic analyses from a fen peat in northern Scotland, drawn from Tipping and Tisdall (2004); 3.5. schematic summary of peat-stratigraphic analyses from a raised moss in the Netherlands (Blaauw* et al. *(2004); 3.6 lake-level fluctuations at several localities in the Alps (numbers of records of high and low lake level), redrawn from Magny (2004); 3.7. lake-level fluctuations at a site in northern Scotland, drawn from Tipping and Tisdall (2004); 3.8. winter precipitation derived from analysis of glacier fluctuations in western Norway, drawn from Nesje* et al. *(2001); 3.9. synthesis of glacier recessions in the Swiss Alps, redrawn from Joerin* et al. *(2006).*

Bank, but by 4000 BC the coastlines were close to those of the present day (Shennan *et al.* 2000).

Palaeoclimate datasets differ in their dating controls and temporal resolutions. This interpretation has tried to focus on the best-constrained and most clearly understood records. Ages are reported in these publications as cal. years BP (these should be referred to for calibration procedures) but are reported here as calibrated years BC. Evidence from datasets that use comparatively imprecise controls like ^{14}C are distinguished in the text by the term c. (circa): other datasets can be resolved to 10–25 years. There is developing in the literature a tendency to 'suck-in' events, to conflate different events into one (Baillie 1991). In addition, a search for cyclicity has meant that many high-resolution records are smoothed when published to 100- or 200-year resolutions.

Four aspects of climate change are explored: a) atmospheric and ocean circulation; b) storminess; c) temperature; and d) precipitation. All use proxy data, of course, and some data are ambiguous as to what they indicate. Tipping and Tisdall (2004) drew on analogies with changes in the most recent climatic excursions c. 1400–1850 AD or at present but this contribution focuses on events that took place at and around the Mesolithic–Neolithic transition. Nonetheless, because many details of climate change in this period are unclear, there is still a need to draw comparisons with possible modern analogues.

A decline in solar output is suggested to have been the most likely cause of the rapid climate change c. 3950 BC (Mayewski *et al.* 2004). Icebergs were transported from the Arctic Ocean after c. 4550 BC, most commonly between c. 4350 and c. 3650 BC. Sea-surface temperatures west of Ireland were sharply reduced (Keigwin and Boyle 2000) and thermohaline circulation in the North Atlantic Ocean was possibly weakened (Broecker 2000). Arctic Ocean currents are modelled to have altered at around 4000 BC because of increased river discharges (Prange and Lohmann 2003). Rates of sedimentation on the Irish Sea floor accelerated after c. 4700 BC (Scourse *et al.* 2002), a phenomenon attributed to a restructuring of continental shelf sea circulation. The same sediment sequence suggests strong seasonal contrasts of up to 15°C in sea-surface temperatures (SST) with winter SST possibly lower than 5°C after c. 4700 BC (Marret *et al.* 2004). After c. 4350 BC the South Jutland current flowing north past Denmark strengthened and outflow from the Baltic Sea was reduced (Gyllencreutz *et al.* 2006). Salinity within the Baltic Sea was rapidly halved at c. 4750 BC but had recovered by c. 4250 BC (Emeis *et al.* 2003).

Polar atmospheric circulation (Fig. 7.1.2) was increasingly strong between 4150 and 3400 BC and mid-latitude westerly storm tracks were more intense, persistent and more strongly meridional (east–west) between 4450 and 3050 BC. Storm events on the Atlantic coast recorded by dune mobilization are most frequent from c. 4000 BC (Figs 7.1.3–7.1.5), though in northern Ireland they are recorded from c. 4500 BC. Dune mobilization events are very poorly dated, however, and several systems in western Scotland were active in the Late Mesolithic period, continuing into the Early Neolithic period, contrasting with more ambiguous biological evidence for climatic quiescence (Andrews *et al.* 1987a; Andrews *et al.* 1987b; Russell *et al.* 1995).

Proxy terrestrial temperature records from Europe (Figs 7.2.1–7.2.8) are rather inconsistent, reflecting different data sources and temporal resolutions. Most reconstructions suggest falls of up to 2°C from c. 4800 BC to c. 4600 BC with an erratic recovery to c. 4200–4000 BC (Figs 7.2.2, 7.2.3, 7.2.6, 7.2.7, 7.2.8). The Late Mesolithic period was generally colder than the Early Neolithic period, with several of the coldest winters in the last 7500 years (Grudd *et al.* 2002). Cheddadi *et al.* (1997) suggest the period around 4050 BC to have been up to 3K warmer than present. The period after c. 4000 BC appears in most reconstructions (except Figs 7.2.7 and 7.2.8) to have been more stable.

Different precipitation proxy data (Fig. 7.3: see also Magny *et al.* 2006) agree more closely (Barber *et al.* 2004). Most records suggest that the climate of Northwest Europe became wetter from c. 5050 to 4550 BC, but then became drier, probably much drier, after c. 4550 BC. Relative dryness may have been most intense or most widely recognized after c. 4250 BC. Seppa and Birks (2001, 535) argued that very dry soils between c. 4400 and 4250 BC should have had 'strong hydrological and ecological consequences'. Recovery had occurred by c. 3900 BC but a further phase of relative dryness occurred in the Early Neolithic period from c. 3850 BC, and in some reconstructions persisted beyond 3050 BC (Figs 7.3.1, 7.3.2, 7.3.4), although in others it is seen to have been interrupted after c. 3650 BC (Figs 7.3.5–7.3.8 inclusive).

Physical and ecological impacts of rapid climate change

Changes in ocean circulation, lower surface temperatures, increased storminess and salinity fluctuations in partly enclosed seas can be expected to have led to at least some of the following physical and ecological changes: changes in spawning and migratory patterns of fish populations; changes in species distributions of coastal faunas; reductions in population numbers or biomass of fish and animal populations; altered coastal geographies; greater wave strengths; increased threats to travel between islands. Few such changes have been demonstrated for this period or, more significantly, ascribed to climatic factors. Only recently have records of dune mobilization been considered to be a measure of storminess rather than relative sea-level change (Orford *et al.* 2000). Other substantial coastal changes in this period, such as the construction of beach ridges, are usually interpreted only in terms of relative sea-level change. Faunal changes signifying a shift from marine to brackish water environments at sites bordering the Baltic Sea, poorly dated but occurring around 4000 BC, have been ascribed to a reduction in tidal amplitude through the submergence of the North Sea (summarized in Fischer 2002, 369) but may be related more directly to atmospherically driven salinity change (Emeis *et al.* 2003).

Climatically induced landscape change away from the coast can be hard to distinguish from anthropogenic change, even within the later Mesolithic period (Tipping 2004). Regionally synchronous change is the most-used distinction. Precipitation is generally thought to have been the most effective climatic factor (Bonsall *et al.* 2002). Rivers in Northwest Europe appear to have flooded more frequently in brief peaks around 4850 BC, and in Central Europe again around 4300 BC and 3970 BC (Macklin *et al.* 2006). British rivers were again more active after c. 3780 BC (Macklin *et al.* 2005; Johnstone *et al.* 2006). Most events, though perhaps not all, relate to phases of increased precipitation (above: Fig. 7.3).

Palaeoecologists have tended to assume a causal relationship between woodland loss and agricultural expansion, but this need not be correct, particularly in periods of rapid climate change. In the boreal forests of Northern Europe many pine trees died between 4150 and 4000 BC, again between 3950 and 3650 BC and at 3600 BC (Zetterberg *et al.* 1996). Tipping *et al.* (2008) describe a collapse of the pine population in northeast Scotland between c. 4250 and 3750 BC associated with otherwise rather poorly defined event(s) between c. 4650 and 4150 BC and c. 3850 and 3650 BC (Bridge *et al.* 1990). Populations of oak trees growing on bog surfaces throughout Northern Europe collapsed for brief periods at 4450 BC and again at 3950 BC (Leuschner *et al.* 2002), events inferred to be related to rising groundwater tables. Few oak trees were preserved on flood plains in northern Germany between 4210 to 3920 BC, possibly because of increased dryness (Spurk *et al.* 2002). Populations of elm trees living alongside these oaks collapsed near-synchronously at 4393–4357 BC (Parker *et al.* 2002). These workers and Cayless and Tipping (2002) argued for a climatic cause for the primary elm decline.

Trees less stressed by climatic extremes, like hazel and birch, seem commonly to have filled gaps created by dying trees, so that the woodland canopy may have remained complete. Very rarely did woodland loss in the Late Mesolithic or Early Neolithic lead to the creation of extensive grasslands (Berglund *et al.* 1996; Richmond 1999; Tipping 1994). The chalk of southern England may be one region where grasslands emerged (Waton 1982), perhaps through precipitation starvation. O'Connell and Molloy (2001) also argued for extensive woodland clearances for pasture 400–500 years long in northwest Ireland from c. 3850 BC in a storm-sensitive region. The apparent completeness of the woodland cover on some exposed Hebridean islands is why Andrews *et al.* (1987a) argued for climatic stability, but other equally exposed localities on this seaboard show that woodland barely took hold in the Holocene (Birks and Madsen 1979; Carter *et al.* 2005; Walker 1984). Blanket peat in upland areas expanded or was established in the Late Mesolithic period to fill the spaces left by dying trees (Tallis 1991). Simmons (1996) argued for moorlands having been created by Late Mesolithic human pressures, but this is debated (Tipping 2004), and the frequency of fires at this time, implied as causal in peat spread, may have been determined by climatic factors (Froyd 2006; Tipping 1996; Tipping and Milburn 2000).

Choices and worldviews: human responses to climate change

Although Bogucki (1998) and Bonsall *et al.* (2002) argued that climate change made farming a more attractive proposition to forager populations by broadening the resource base, most reconstructions emphasize the already wide resource base of later Mesolithic communities (e.g. Mithen 2000), and it is more likely that climate change led to the failure of existing resource strategies or to the conviction among these communities that these were failing. The arguments presented by Sheridan (2007; this volume) for migration of farmers to the British Isles are accepted, but the widespread adoption of agrarian techniques (Jones 2000; Schulting and Richards 2002) is assumed here to have required acceptance and engagement by the native population (Thomas 2004). It is thus argued that the complexity and variability of climate made natural resources used by foraging communities unreliable and unpredictable.

Coastal resources may have become more abundant as a result of coastal erosion accelerating nutrient supply after c. 4700 BC (Andersen 2004, 397), leading to the development of a distinctive shell-midden economy. This may have been sustained after c. 4500 BC by increased storminess as sea-level change slowed. Mithen (2000) suggested major population shifts to these resources on Oronsay at c. 4500 BC, and Andersen (2004, 408) considered that most if not all Late Mesolithic foragers around the Baltic Sea lived on the coast. Deteriorating sea-surface temperature and salinity, increased seasonality and changing ocean currents (above) may, however, have led by c. 4000 BC to a widespread decline in the productivity of these resources (cf. Rowley-Conwy 1984) and the 'slighting of the sea' (Schulting 1998; Schulting and Richards 2002; Richards *et al.* 2003: but see also Milner *et al.* 2004). By denying environmental stress as an explanation of this event because it was so widespread, Thomas (2004, 119) underestimates the hemispheric scale of rapid climate change (Mayweski *et al.* 2004), as might Schulting *et al.* (2004) by assuming the 'gulf stream' to have been a constant ameliorating factor along the Atlantic coast in this period (Broecker 2000; Clark *et al.* 2002; Keigwin and Boyle 2000). The economic significance of the 'slighting of the sea' is uncertain (Fischer 2002, 370; Andersen 2004, 408), but few palaeoeconomic data exist. Middens might then be indicators of famine (Hardy and Wickham-Jones 2003, 375), the 'slighting of the sea' a final failure of the resource, and the development of social mechanisms expressions of concern over resource depletion (e.g. taboos: M. Richards 2003; 2004; Schulting 2004). Natural change in this environment may have been seen as malevolent by Late Mesolithic coastal communities, or at best quixotic and temperamental, and nature as undependable.

Away from the coast there is no evidence that resource strategies changed in the latest Mesolithic, but there is little evidence for resources at all (Warren 2004). We should not assume their constant abundance but the effects of climatic deterioration (above) may not have adversely impacted on these, with the possible exception of migratory fish caught inland (Warren 2005) but coming from the sea. The impact of climate change may have been to promote expansion of grasses beneath thinning woodland canopies (Bonsall *et al.* 2002; Davies *et al.* 2004; Tipping and Tisdall 2004). This may have encouraged the development of agro-pastoralism (Bonsall *et al.* 2002) but would equally have been beneficial to wild animal populations. Change would have allowed both hunters and early pastoralists to thrive, permitting choice. High-resolution faunal analyses in Central Europe suggest that farmers did continue to make choices through resource-switching (Huster-Plogmann *et al.* 1999; Schibler 2006). In these inland and upland landscapes, then, nature simply by changing would have appeared beneficent and generous in spirit. Might this be the context in which woodland glades became charged with special significance (Brown 1999)?

The observation of rapid natural change in their world may have encouraged among forager communities a more ready acceptance of new ideas (Tipping and Tisdall 2004). Mesolithic lives embedded in a world of seasons and ceaseless cycles (e.g. Pollard 1996; Jones 1996), seemingly changeless, were jarred and unsettled by the unavoidable but unexplained evidence of change (Larsson 2003, xxx). Even when resources were not actually threatened, landscape change raised questions and introduced uncertainty. Seasons changed in timing and character, things formerly predictable could no longer be predicted with assurance and were seen to be impermanent. The natural world, and so time, ceased to be perceived in cycles. Mesolithic foragers became more receptive to innovation simply because the natural world appeared differently, as when walking through oak, elm and pine woods that were now decaying; or watching in wonder as fodder for animals magically appeared; or fording rivers made dramatically deeper and more dangerous by frequent floods and in which until recently was reliably found food; or surveying shorelines that in living memory were welcoming and resource-rich; or eyeing islands that were now thought unsafe to reach. Change required no great emotional or philosophical restructuring (Thomas 1999) because it was part of the 'natural order', different only because it was no longer gradual but abrupt and transformative. These people could choose not to change, but increasingly they saw and chose new ways of doing things.

Climatic instability would have had effects on continental farming communities. The periods of crisis in these groups identified by Louwe Kooijmans (1999) may relate to these pressures. The general rise in temperatures after c. 4000 BC (Fig. 7.1.1) and the longer growing season (Cheddadi *et al.* 1997) may have promoted better growing conditions on the continent and in the British Isles, particularly for autumn-sown cereal crops (Bogaard 2004, 164) but the argument in this paper is that the preceding impacts of deteriorating climate did more to prepare the native forager population for an alternative way of life.

In his unease over Sheridan's (2004, 2007) migration hypothesis, Thomas (2004, 118) rightly asks why should 'groups of people ... all have set sail simultaneously' for the British Isles: 'Why did they wait so long, and then all go at once?' Exaggeration apart, these are good questions. The movement of foragers to Oronsay around 4500 BC (Mithen 2000) and the arrival, if securely dated, at Ferriter's Cove in western Ireland of domesticated cattle at around this time (Woodman *et al.* 1999) indicate that sea journeys could be made, but these are separated in time from the second 'wave' of migration by a period which, at least in parts of western Britain, seems to have been one of heightened storminess. Movement in the Mesolithic period between Scottish islands is often assumed to have been routine (Warren 2000; Wickham-Jones 2005), but this may be overstated (Pickard and Bonsall 2004). Such journeys may have been high-risk adventures to collect high-prestige stone from islands (Saville 2003, 346) and we do not know how many such expeditions failed. Might sea travel have become too risky, or been perceived to be, with climatic deterioration? Tipping and Tisdall (2004) suggested that the timing of the introduction of agriculture to the British Isles, and so migration from the continent, was determined by the perception that sea voyages were becoming less climatically hazardous than they had been. The phase of increased storminess identified from the chronology of sand dune mobilization (above: Fig. 7.1) suggests that the earliest Neolithic period was more hostile than the later Mesolithic period, but this chronology is by far the least satisfactorily defined of the records presented in Figs 7.1–7.3. It is critical to define with high precision when storms were most active in this period on the Atlantic coast because the apparent absence of links between Ireland and Scotland in the Mesolithic (Saville 2003; Woodman 2004) can be contrasted with the clear movement between the two in the Early Neolithic (Cooney 2004; Noble 2006; Sheridan 2004). The apparent isolation of foragers on Oronsay at the end of the Mesolithic period can be compared with their return to Islay after c. 4000 BC (Mithen 2000). Raemakers (2003) has argued, for the exposed shorelines of the Netherlands, that our knowledge of settlement in the centuries before 3600 BC is distorted by the loss of sites through coastal erosion, assumed by him to be related to sea-level rise but more likely to have been through heightened storminess. Such phases of increased storminess may have had more than economic impacts.

Might they, for instance, have been memorialized? Portal dolmens look like natural rock formations (C. Richards 2004; Whittle 2004), like tors (Tilley and Bennett 2001), but tors are not found everywhere along the Atlantic coast. Portal dolmens also look like coastal rock formations termed megaclast deposits, enormous stones hurled high on cliff tops onto underlying rocks by truly prodigious storm waves (Williams and Hall 2004). People may have wanted to remember not only natural places (Bradley 2000) but natural events also.

Acknowledgements

I would like to thank Bill Finlayson and Graeme Warren for inviting this contribution.

Bibliography

Andersen, S. H. (2004) Danish shell middens reviewed. Pp. 393–412 in A. Saville (ed.), *Mesolithic Scotland and its Neighbours: The Early Holocene Prehistory of Scotland, its British and Irish Context, and some Northern European Perspectives*. Edinburgh: Society of Antiquaries of Scotland.

Andrews, M. V., Beck, R. B., Birks, H. J. B. and Gilbertson, D. D. (1987a) The past and present vegetation of Oronsay and Colonsay. Pp. 52–77 in P. Mellars (ed.), *Excavations on Oronsay*. Edinburgh: Edinburgh University Press.

Andrews, M. V., Gilbertson, D. D. and Kent, M. (1987b) Storm frequencies along the Mesolithic coastline. Pp. 108–14 in P. Mellars (ed.), *Excavations on Oronsay*. Edinburgh: Edinburgh University Press.

Ashmore, P. (2004) Absolute chronology. Pp. 125–38 in I. A. G Shepherd and G. Barclay (eds), *Scotland in Ancient Europe*. The Neolithic and Early Bronze Age of Scotland in their European Context. Edinburgh: Society of Antiquaries of Scotland.

Baillie, M. G. L. (1991) Suck-in and smear: two related chronological problems for the '90s. *Journal of Theoretical Archaeology* 2, 12–16.

Baillie, M. G. L. (1995) Dendrochronology and past environmental change. *Proceedings of the British Academy* 77, 5–23.

Barber, K. E., Zolitschka, B., Tarasov, P. and Lotter, A. F. (2004) Atlantic to Urals – the Holocene climatic record of mid-latitude Europe. Pp. 417–42 in R. W. Battarbee, F. Gasse and C. E. Stickley (eds), *Past Climate Variability through Europe and Africa*. Dordecht: Kluwer Academic Publishers.

Bar-Yosef, O. and Belfer-Cohen, A. (1992) From foraging to farming in the Mediterranean Levant. Pp. 21–48 in A. B. Gebauer and T. D. Price (eds), *Transitions to Agriculture in Prehistory*. Madison, WI: Prehistory Press.

Berglund, B. E. (2003) Human impact and climate changes – synchronous events and a causal link. *Quaternary International* 105, 7–12.

Berglund, B. E., Birks, H. J. B., Ralska-Jasiewiczowa, M. and Wright, H. E. (1996) *Palaeoecological Events During the Last 15 000 Years: Regional Syntheses of Palaeoecological Studies in Lakes and Mires in Europe*. Chichester: Wiley.

Birks, H. J. B. and Madsen, B. J. (1979) Flandrian vegetational history at Little Loch Roag, Isle of Lewis, Scotland. *Journal of Ecology* 67, 825–42.

Bjorck, S. and Clemmensen, L. B. (2004) Aeolian sediment in raised bog deposits, Halland, SW Sweden: a new proxy record of Holocene winter storminess variation in southern Scandinavia? *The Holocene* 14, 677–88.

Blaauw, M., van Geel, B. and van der Plicht, J. (2004) Solar forcing of climatic change during the mid-Holocene: indications from raised bogs in The Netherlands. *The Holocene* 14, 35–44.

Bogaard, A. (2004) *Neolithic Farming in Central Europe*. London: Routledge.

Bogucki, P. (1998) Holocene climatic variability and early agriculture in temperate Europe: the case of northern Poland. Pp. 77–85 in M. Zvelebil, L. Domańsla and R. Dennell (eds), *Harvesting the Sea, Farming the Forest. The Emergence of Neolithic Societies in the Baltic Region*. Sheffield: Sheffield Academic Press.

Bonsall, C., Macklin, M. G., Anderson, D. E. and Payton, R. W. (2002) Climate change and the adoption of agriculture in north-west Europe. *European Journal of Archaeology* 5, 9–23.

Bradley, R. (2000) *An Archaeology of Natural Places*. London: Routledge.

Bridge, M. C., Haggart, B. A. and Lowe, J. J. (1990) The history and palaeoclimatic significance of subfossil remains of Pinus sylvestris in blanket peats from Scotland. *Journal of Ecology* 78, 77–99.

Broecker, W. S. (2000) Was a change in thermohaline circulation responsible for the Little Ice Age? *Proceedings of the National Academy of Sciences* 97, 1339–442.

Brooks, N. (2004) Cultural responses to aridity in the Middle Holocene and increased social complexity. *Quaternary International* 151, 29–49.

Brown, A. G. (1999) Floodplain vegetation history: clearings as potential ritual spaces? Pp. 49–62 in A. S. Fairbairn (ed.), *Plants in Neolithic Britain and Beyond*. Neolithic Studies Group Seminar Papers 5. Oxford: Oxbow Books.

Carter, S., Dalland, M. and Long, D. (2005) Early land-use and landscape development in Arisaig. Scottish Archaeological Internet Report 15: www.sair.org.uk.

Caseldine, C., Thompson, G., Langdon, P. and Hendon, D. (2005) Evidence for an extreme climatic event on Achill Island, Co. Mayo, Ireland around 5200–5100 cal. yr BP. *Journal of Quaternary Science* 20, 169–78.

Cayless, S. M. and Tipping, R. (2002) Data on mid-Holocene climatic, vegetation and anthropogenic interactions in southern Scotland. *Vegetation History and Archaeobotany* 11, 201–10.

Chambers, F. M. and Brain, S. A. (2002) Paradigm shifts in late-Holocene climatology? *The Holocene* 12, 239–49.

Cheddadi, R., Yu, G., Guiot, J., Harrison, S. P. and Prentice, I. C. (1997) The climate of Europe 6000 years ago. *Climate Dynamics* 13, 1–9.

Clark, P. U., Pisias, N. G., Stocker, T. F. and Weaver, A. J. (2002) The role of the thermohaline circulation in abrupt climate change. *Nature* 415, 863–9.

Cooney, G. (2004) Neolithic worlds: islands in the Irish Sea. Pp. 145–59 in V. Cummings and C. Fowler (eds), *The Neolithic of the Irish Sea. Materiality and Traditions of Practice*. Oxford: Oxbow Books.

Cowie, T. G. (1996) Torrs Warren, Luce Sands, Galloway: a report on archaeological and palaeoecological investigations undertaken in 1977 and 1979. *Transactions of the Dumfries and Galloway Natural History and Antiquarian Society* 71, 11–105.

Davies, A. L., Tisdall, E. and Tipping, R. (2004) Holocene climatic variability and human settlement in the Scottish Highlands: fragility and robustness. Pp. 2–11 in R. A. Housley and G. M. Coles (eds), *Atlantic Connections and Adaptations*. Oxford: Oxbow Books.

De Jong, R., Bjorck, S., Bjorkman, L. and Clemmensen, L. B. (2006) Storminess variation during the last 6500 years as reconstructed from an ombrotrophic peat bog in Halland, southwest Sweden. *Journal of Quaternary Science* 21, 905–19.

De Menocal, P. B. (2001) Cultural responses to climate change during the Holocene. *Science* 292, 667–72.

Emeis, K.-C., Struck, U., Blanz, T., Kohly, A. and Voss, M. (2003) Salinity changes in the central Baltic Sea (NW Europe) over the last 10000 years. *The Holocene* 13, 411–21.

Fernandez-Armesto, F. (2000). *Civilizations*. London: Macmillan.

Fischer, A. (2002) Food for feasting? An evaluation of explanations of the Neolithisation of Denmark and southern Sweden. Pp. 343–93 in A. Fischer and K. Kristiansen (eds), *The Neolithization of Denmark: 150 Years of Debate*. Sheffield, J. R. Collis Publications.

Froyd, C. A. (2006). Holocene fire in the Scottish Highlands: evidence from macroscopic charcoal records. *The Holocene* 16, 235–50.

Gilbertson, D. D., Schwenninger, J.-L., Kemp, R. A. and Rhodes, E. J. (1999) Sand-drift and soil formation along an exposed North Atlantic coastline: 14,000 years of diverse geomorphological, climatic and human impacts. *Journal of Archaeological Science* 26, 439–69.

Grudd, H., Briffa, K. R., Karlén, W., Bartholin, T. S., Jones, P. D. and Kromer, B. (2002) A 7400-year tree-ring chronology in northern Swedish Lapland: natural climatic variability expressed on annual to millennial timescales. *The Holocene* 12, 657–66.

Gyllencreutz, R., Backman, J., Jakobsson, M., Kissel, C. and Arnold, E. (2006) Postglacial palaeoceanography in the Skagerrak. *The Holocene* 16, 975–85.

Hardy, K. and Wickham-Jones, C. R. (2003) Scotland's first settlers: an investigation into settlement, territoriality and mobility during the Mesolithic in the Inner Sound, Scotland: first results. Pp. 369–81 in L. Larsson, H. Kindgren, K. Knutsson, D. Leoffler and A. Åkerlund (eds), *Mesolithic on the Move: Papers presented at the Sixth International Conference on the Mesolithic in Europe, Stockholm 2000*. Oxford: Oxbow Books.

Helama, S., Lindholm, M., Timonen, M., Merilainen, J. and Eronen M. (2002) The supra-long Scots pine tree-ring record for Finnish Lapland: 2. interannual to centennial variability in summer temperatures for 7500 years. *The Holocene* 12, 681–8.

Hughes, P. D. M., Mauquoy, D., Barber, K. E. and Langon, P. G. (2000) Mire-development pathways and palaeoclimatic records from a full Holocene peat archive at Walton Moss, Cumbria, England. *The Holocene* 10, 465–80.

Huster-Plogmann, H., Schibler, J. and Stephan, K. (1999) The relationship between wild animal populations, climatic fluctuations and economic adaptations. A transdisciplinary study on Neolithic sites from Lake Zurich, southwest Germany and Bavaria. Pp. 189–200 in C. Becker, H. Manhart, J. Peters and J. Schibler (eds), *Historia Animalium ex Ossibus. Festschrift fur Angela von den Driesch*. Rahden: Westf.: Leidorf.

Joerin, U. E., Stocker, T. F. and Schlüchter, C. (2006) Multicentury glacier fluctuations in the Swiss Alps during the Holocene. *The Holocene* 16, 697–704.

Johnstone, E., Macklin, M. G. and Lewin, J. (2006) The development and application of a database of radiocarbon-dated Holocene fluvial deposits in Great Britain. *Catena* 66, 14–23.

Jones, A. (1996) Food for thought: material culture and the transformation in food use from the Mesolithic to the Neolithic. Pp. 291–300 in T. Pollard and A. Morrison (eds), *The Early Prehistory of Scotland*. Edinburgh: Edinburgh University Press.

Jones, G. (2000) Evaluating the importance of cultivation and collecting in Neolithic Britain. Pp. 79–84 in A. S. Fairbairn (ed.), *Plants in Neolithic Britain and Beyond*. Neolithic Studies Group Seminar Papers 5. Oxford: Oxbow Books.

Keatinge, T. H. and Dickson, J. H. (1979) Mid-Flandrian changes in vegetation on mainland Orkney. *New Phytologist* 82, 585–612.

Keigwin, L. D. and Boyle, E. A. (2000) Detecting Holocene changes in thermohaline circulation. *Proceedings of the National Academy of Sciences* 97, 1343–6.

Klassen, L. (2002) The Ertebølle culture and Neolithic continental Europe: traces of contact and interaction. Pp. 305–17 in A. Fischer and K. Kristiansen (eds), *The Neolithisation of Denmark: 150 Years of Debate*. Sheffield: J. R. Collis Publications.

Langdon, P. G., Barber, K. E. and Hughes, P. D. M. (2003) A 7500-year peat-based palaeoclimatic reconstruction and evidence for an 1100-year cyclicity in bog surface wetness from Temple Hill Moss, Pentland Hills, southeast Scotland. *Quaternary Science Reviews* 22, 259–74.

Larsson, L. (2003) The Mesolithic of Sweden in retrospective and progressive perspectives. Pp. xxii–xxxii in L. Larsson, H. Kindgren, K. Knutsson, D. Leoffler and A. Åkerlund (eds), *Mesolithic on the Move: Papers presented at the Sixth International Conference on the Mesolithic in Europe, Stockholm 2000*. Oxford: Oxbow Books.

Lauritzen, S.-E. and Lundberg, J. (1999) Calibration of the speleothem delta function: an absolute temperature record for the Holocene in northern Norway. *The Holocene* 9, 659–71.

Leuschner, H. H., Sass-Klassen, U., Jansmam, E., Baillie, M. G. L. and Spurk, M. (2002) Subfossil European bog oaks: population dynamics and long-term growth depressions as indicators of the Holocene hydro-regime and climate. *The Holocene* 12, 695–706.

Louwe Kooijmans, L. P. (1999) Understanding the Mesolithic/Neolithic frontier in the Lower Rhine basin, 5300–4300 cal. BC. Pp. 407–27 in M. Edmonds and C. Richards (eds), *Understanding the Neolithic of North-Western Europe*. Glasgow: Cruithne Press.

McDermott, F., Mattey, D. P. and Hawkesworth, C. (2001) Centennial-scale Holocene climate variability revealed by a high-resolution speleothem $\delta^{18}O$ record from SW Ireland. *Science* 294, 1328–31.

Macklin, M. G., Johnstone, E. and Lewin, J. (2005) Pervasive and long-term forcing of Holocene river instability and flooding in Great Britain by centennial-scale climate change. *The Holocene* 15, 937–43.

Macklin, M. G., Benito, G., Gregory, K. J., Johnstone, E., Lewin,

J., Soja, R., Starkel, L. and Thorndycraft, V. R. (2006) Past hydrological events reflected in the Holocene fluvial record of Europe. *Catena* 66, 145–54.

Magny, M. (2004) Holocene climate variability as reflected by mid-European lake-level fluctuations and its probable impact on prehistoric human settlements. *Quaternary International* 113, 65–79.

Magny, M., Leuzinger, U., Bortenschlager, S. and Haas, J.-N. (2006) Tripartite climate reversal in Central Europe 5600–5300 years ago. *Quaternary Research* 65, 3–19.

Marret, F., Scourse, J. and Austin, W. (2004) Holocene shelf-sea seasonal stratification dynamics: a dinoflagellate cyst record from the Celtic Sea, NW European shelf. *The Holocene* 14, 689–96.

Mayewski, P. A., Buckland, P. C., Edwards, K. J., Meeker, L. D., O'Brien, S. O. and Buckland, P. (1996) Climate change events as seen in the Greenland ice core (GISP2): implications for the Mesolithic of Scotland. Pp. 74–86 in T. Pollard, and A. Morrison (eds), *The Early Prehistory of Scotland*. Edinburgh: Edinburgh University Press.

Mayewski, P. A., Rohling, E. E., Stager, J. C., Karlen, W., Maasch, K. A., Meeker, M., Schneider, R. R. and Steig, E. J. (2004) Holocene climate variability. *Quaternary Research* 62, 243–55.

Mellars, P. (1987) *Excavations on Oronsay*. Edinburgh, Edinburgh University Press.

Milner, N., Craig, O. E., Bailey, G. N., Pederson, K. and Andersen, S. H. (2004) Something fishy in the Neolithic? A re-evaluation of stable isotope analysis of Mesolithic and Neolithic coastal populations. *Antiquity* 77, 9–22.

Mithen, S. (2000) *Hunter-Gatherer Landscape Archaeology: The Southern Hebrides Mesolithic Project 1988–1998*. Cambridge: McDonald Institute for Archaeological Research.

Nesje, A., Matthews, J. A., Dahl, S. O., Berrisford, M. S. and Andersson, C. (2001) Holocene glacier fluctuations of Flatebreen and winter-precipitation changes in the Jostedalsbreen region, western Norway, based on glaciolacustrine sediment records. *The Holocene* 11, 267–80.

Noble, G. (2006) *Neolithic Scotland. Timber, Stone, Earth and Fire*. Edinburgh: Edinburgh University Press.

O'Connell, M. and Molloy, K. (2001) Farming and woodland dynamics in Ireland during the Neolithic. *Biology and Environment: Proceedings of the Royal Irish Academy B* 101, 99–128.

Orford, J. D., Wilson, P., Wintle, A. G., Knight, J. and Braley, S. (2000) Holocene coastal dune initiation in Northumberland and Norfolk, eastern UK: climate and sea-level changes as possible forcing agents for dune initiation. Pp. 197–218 in I. Shennan and J. Andrews (eds), *Holocene Land-Ocean Interaction and Environmental Change around the North Sea*. Geological Society of London Special Publication 166. London: Geological Society.

Parker, A. G., Anderson, D. E., Goudie, A. S., Robinson, M. A. and Bonsall, C. (2002) A review of the mid-Holocene elm decline in the British Isles. *Progress in Physical Geography* 26, 1–45.

Pickard, C. and Bonsall, C. (2004) Deep-sea fishing in the European Mesolithic: fact or fantasy? *European Journal of Archaeology* 7, 273–90.

Pollard, T. (1996) Time and tide: coastal environments, cosmology and ritual practice in early prehistoric Scotland. Pp. 198–210 in T. Pollard and A. Morrison (eds), *The Early Prehistory of Scotland*. Edinburgh: Edinburgh University Press.

Prange, M. and Lohmann, G. (2003) Effects of mid-Holocene river runoff on the Arctic Ocean/sea-ice system: a numerical model study. *The Holocene* 13, 335–42.

Raemakers, D. (2003) Cutting a long story short? The process of Neolithization in the Dutch delta re-examined. *Antiquity* 77, 740–8.

Richards, C. (2004) Labouring with monuments: constructing the dolmen at Carreg Samson, south-west Wales. Pp. 72–80 in V. Cummings and C. Fowler (eds), *The Neolithic of the Irish Sea. Materiality and Traditions of Practice*. Oxford: Oxbow Books.

Richards, M. P. (2003) Explaining the dietary isotope evidence for the rapid adoption of the Neolithic in Britain. Pp. 31–6 in M. Parker Pearson (ed.), *Food, Culture and Identity in the Neolithic and Early Bronze Age*. British Archaeological Report, International Series 1117. Oxford: BAR.

Richards, M. P. (2004) The early Neolithic in Britain: new insights from biomolecular archaeology. Pp. 83–90 in I. A. G. Shepherd and G. Barclay (eds), *Scotland in Ancient Europe. The Neolithic and Early Bronze Age of Scotland in their European Context*. Edinburgh: Society of Antiquaries of Scotland.

Richards, M. P., Schulting, R. J. and Hedges, R. E. M. (2003) Sharp shift in diet at onset of Neolithic. *Nature* 425, 366.

Richmond, A. (1999) *Preferred Economies. The Nature of the Subsistence Base throughout Mainland Britain during Prehistory*. British Archaeological Reports, British Series 290. Oxford: BAR.

Rowley-Conwy, P. (1984) The laziness of the short-distance hunter: the origins of agriculture in western Denmark. *Journal of Anthropological Archaeology* 3, 300–24.

Rowley-Conwy, P. (2004) How the West was lost. A reconsideration of agricultural origins in Britain, Ireland, and southern Scandinavia. *Current Anthropology* 45 (Supplement), 83–113.

Russell, N. J., Bonsall, C. and Sutherland, D. G. (1995) The exploitation of marine molluscs in the Mesolithic of western Scotland: evidence from Ulva Cave, Inner Hebrides. Pp. 273–88 in A. Fischer (ed.), *Man and Sea in the Mesolithic*. Oxbow Monograph 53. Oxford: Oxbow Books.

Sandweiss, D. H., Maasch, K. A. and Anderson, D. G. (1999) Transitions in the Mid-Holocene. *Science* 283, 499–500.

Saville, A. (2003) Indications of regionalisation in Mesolithic Scotland. Pp. 340–50 in L. Larsson, H. Kindgren, K. Knutsson, D. Leoffler and A. Åkerlund (eds), *Mesolithic on the Move: Papers presented at the Sixth International Conference on the Mesolithic in Europe, Stockholm 2000*. Oxford: Oxbow Books.

Schibler, J. (2006) The economy and environment of the 4th and 3rd millennia BC in the northern Alpine foreland based on studies of animal bone. *Environmental Archaeology* 11, 49–64.

Schulting, R. J. (1998) Slighting the sea: stable isotope evidence for the transition to farming in northwestern Europe. *Documenta Prehistorica* 25, 203–18.

Schulting, R. J. (2004) An Irish sea change: some implications for the Mesolithic–Neolithic transition. Pp. 22–8 in V. Cummings and C. Fowler (eds), *The Neolithic of the Irish Sea. Materiality and Traditions of Practice*. Oxford: Oxbow Books.

Schulting, R. J. and Richards, M. P. (2002) The wet, the wild and the domesticated: the Mesolithic–Neolithic transition on the west coast of Scotland. *Journal of European Archaeology* 5, 147–89.

Schulting, R. J., Dupont, C. and Tresset, A. (2004) From harvesting the sea to stock-rearing along the Atlantic façade of north-west Europe. *Environmental Archaeology* 9, 143–54.

Scourse, J. D., Austin, W. E. N., Long, B. T., Assinder, D. J. and Huws, D. (2002) Holocene evolution of seasonal stratification in the Celtic Sea: refined age model, mixing depths and foraminiferal stratigraphy. *Marine Geology* 191, 119–45.

Seppa, H. and Birks, H. J. B. (2001) July mean temperature and annual precipitation trends during the Holocene in the Fennoscandian tree-line area: pollen-based climate reconstructions. *The Holocene* 11, 527–39.

Shennan, I., Lambeck, K., Flather, R., Horton, B., McArthur, J. J., Innes, J. B., Lloyd, J. M., Rutherford, M. M. and Wingfield, R. (2000) Modelling western North Sea palaeogeographies and tidal changes during the Holocene. Pp. 299–319 in I. Shennan and J. Andrews (eds), *Holocene Land-Ocean Interaction and Environmental Change around the North Sea*. Geological Society of London Special Publication 166. London: Geological Society.

Sheridan, A. (2004) Neolithic connections along and across the Irish Sea. Pp. 9–21 in V. Cummings and C. Fowler (eds), *The Neolithic of the Irish Sea. Materiality and Traditions of Practice*. Oxford: Oxbow Books.

Sheridan, A. (2007) From Picardie to Pickering and Pencraig Hill? New information on the 'Carinated Bowl Neolithic' in northern Britain. Pp. 439–90 in A. W. R. Whittle and V. Cummings (eds), *Going Over: The Mesolithic–Neolithic Transition in North-West Europe*. London: British Academy and Oxford University Press.

Simmons, I. G. (1996) *The Environmental Impact of Later Mesolithic Cultures*. Edinburgh: Edinburgh University Press.

Spurk, M., Leuschner, H. H., Baillie, M. G. L., Briffa, K. R. and Friedrich, M. (2002) Depositional frequency of German subfossil oaks: climatically and non-climatically induced fluctuations in the Holocene. *The Holocene* 12, 707–15.

Stuiver, M. and Braziunas, T. F. (1993) Sun, ocean, climate and atmospheric $^{14}CO_2$: an evaluation of causal and spectral relationships. *The Holocene* 3, 289–305.

Tallis, J. H. (1991) Forest and moorland in the south Pennine uplands in the mid-Flandrian period. III. The spread of moorland – local, regional and national. *Journal of Ecology* 79, 401–15.

Thomas, J. (1999) *Rethinking the Neolithic*. Cambridge: Cambridge University Press.

Thomas, J. (2004) Current debates on the Mesolithic–Neolithic transition in Britain and Ireland. *Documenta Praehistorica* 31, 113–30.

Tilley, C. and Bennett, W. (2001) An archaeology of supernatural places: the case of West Penwith. *Journal of the Royal Anthropological Institute* 7, 336–62.

Tipping, R. (1994) The form and fate of Scottish woodlands. *Proceedings of the Society of Antiquaries of Scotland* 124, 1–54.

Tipping, R. (1996) Microscopic charcoal records, inferred human activity and climate change in the Mesolithic of northernmost Scotland. Pp. 39–61 in T. Pollard and A. Morrison (eds), *The Early Prehistory of Scotland*. Edinburgh: Edinburgh University Press.

Tipping, R. (2004) Interpretative issues concerning the driving forces of vegetation change in the early Holocene of the British Isles. Pp. 45–54 in A. Saville (ed.), *Mesolithic Scotland and its Neighbours: The Early Holocene Prehistory of Scotland, its British and Irish Context, and some Northern European Perspectives*. Edinburgh: Society of Antiquaries of Scotland.

Tipping, R. and Milburn, P. (2000) The mid-Holocene charcoal fall in southern Scotland: spatial and temporal variability. *Palaeogeography, Palaeoclimatology, Palaeoecology* 164, 193–209.

Tipping, R. and Tisdall, E. (2004) Continuity, crisis and climate change in the Neolithic and early Bronze Periods of North West Europe. Pp. 71–82 in I. A. G. Shepherd and G. Barclay (eds), *Scotland in Ancient Europe. The Neolithic and Early Bronze Age of Scotland in their European Context*. Edinburgh: Society of Antiquaries of Scotland.

Tipping, R., Ashmore, P., Davies, A., Haggart, A., Moir, A., Newton, A., Sands, R., Skinner, T. and Tisdall, E. (2008) Prehistoric Pinus woodland dynamics in an upland landscape in northern Scotland. *Vegetation History and Archaeobotany* 17, 251–67.

Tooley, M. J. (1990) The chronology of coastal dune development in the United Kingdom. Pp. 81–8 in Th. W. M. Bakker, P. D. Jungerius and J. A. Klijn (eds), *Dunes of the European Coasts*. Cremlingen, Catena Supplement 18.

Walker, M. J. C. (1984) A pollen diagram from St Kilda, Outer Hebrides, Scotland. *New Phytologist* 97, 99–113.

Warren, G. (2000) Seascapes: peoples, boats, and inhabiting the later Mesolithic in western Scotland. Pp. 97–104 in R. Young (ed.), *Mesolithic Lifeways*. Leicester Archaeology Monographs 7. Leicester: University of Leicester, School of Archaeological Studies.

Warren, G. (2004) The start of the Neolithic in Scotland. Pp. 91–103 in I. A. G. Shepherd and G. Barclay (eds), *Scotland in Ancient Europe. The Neolithic and Early Bronze Age of Scotland in their European Context*. Edinburgh: Society of Antiquaries of Scotland.

Warren, G. (2005) *Mesolithic Lives in Scotland*. Stroud: Tempus.

Waton, P. V. (1982). A Palynological Study of the Impact of Man on the Landscape of Central Southern England. Unpublished PhD thesis, University of Southampton.

Weiss, H. and Bradley, R. S. (2001) What drives societal collapse? *Science* 291, 609–10.

Whittle, A. (2003a) Review of Zvelebil, Domanska and Dennell (eds) *Harvesting the Sea, Farming the Forest: the Emergence of Neolithic Societies in the Baltic Region. Environmental Archaeology* 8, 92–3.

Whittle, A. (2003b) *Europe in the Neolithic: the Creation of New Worlds*. Cambridge: Cambridge University Press.

Whittle, A. (2004) Stones that float to the sky: portal dolmens and their landscapes of memory and myth. Pp. 81–90 in V. Cummings and C. Fowler (eds), *The Neolithic of the Irish Sea. Materiality and Traditions of Practice*. Oxford: Oxbow Books.

Wickham-Jones, C. (2005) Summer walkers? – mobility and the Mesolithic. Pp. 30–41 in N. Milner and P. Woodman (eds), *Mesolithic Studies at the Beginning of the 21st Century*. Oxford: Oxbow Books.

Williams, D. M. and Hall, A. M. (2004) Cliff-top megaclast deposits of Ireland, a record of extreme waves in the

North Atlantic – storms or tsunamis? *Marine Geology* 206, 101–17.

Wilson, P., McGourty, J. and Bateman, M. D. (2004) Mid- to late-Holocene coastal dune event stratigraphy for the north coast of Ireland. *The Holocene* 14, 406–16.

Woodman, P. (2004) Some problems and perspectives: reviewing aspects of the Mesolithic period in Ireland. Pp. 285–98 in A. Saville (ed.), *Mesolithic Scotland and its Neighbours: The Early Holocene Prehistory of Scotland, its British and Irish Context, and some Northern European Perspectives*. Edinburgh: Society of Antiquaries of Scotland.

Woodman, P. C., Anderson, E. and Finlay, N. (1999) *Excavations at Ferriter's Cove 1983–95: Last Foragers, First Farmers in the Dingle Peninsula*. Bray: Wordwell.

Zetterberg, P., Eronen, M. and Lindholm, M. (1996) The mid-Holocene climatic change around 3800 B.C.: tree-ring evidence from northern Fennoscandia. *Palaeoklimaforschung* 20, 135–46.

8. Changing Landscapes – Changing Societies? An Anthropological Perspective

Marion Benz

In this article I suggest that Neolithization in the Near East was a long-term, dialectical process between environmental and cultural changes – a process of 'path-dependency'. I show that a different perception of landscapes did not cause the adoption of agriculture, but rather that at the beginning of the Holocene traditional hunter-gatherers' behaviour patterns and preferences led to the durable adoption of a sedentary way of life. In these villages new social rules became necessary which then led in turn to a completely new structuring of the landscape. I argue that, according to an ethno-archaeological model, generalized reciprocity – the principle of sharing with other people and groups – had to be restricted before cultivation could be adopted permanently. Group identities then had to be intensified by visible symbols and territorial markers. The communal buildings of the early Pre-Pottery Neolithic and intensified ritual practices could thus be interpreted as means to counteract the strains caused by the new social order.

Introduction

The differences between foragers' and farmers' perceptions of landscapes have been stressed by several authors. Ideological, religious and cognitive changes have been asserted to explain these differences (Cauvin 1976; Hodder 1990; Watkins 2004). With the discovery of impressive cult centres and communal buildings at sites such as Göbekli Tepe (Schmidt 2006), Jerf el Ahmar (Stordeur *et al.* 2000), 'Abr 3 (Yartah 2004), Sefer Tepe and Karahan Tepe (Çelik 2000; 2006), and Hallan Çemi (Rosenberg and Davis 1992), religious and social factors have been emphasized even more than before (cf. Hayden 1992; Bender 1978). I will classify these models as 'socio-ideological' models where socio-ideological changes, including changes in the perception of landscapes, are thought to be the reason, or at least the trigger, for the development of sedentary life, agriculture and domestication. Because no positive correlation between ecological changes and the development of agriculture has been proven, climatic changes and changes in the natural environment have been regarded as catalysts or have been ignored completely in these models (e.g. Byrd 2005), and these socio-ideological changes sometimes appear as a 'deus ex machina'.

I will argue that Neolithization in the Near East cannot be explained by such a simple correlation or lack of correlation between changing landscapes and cultural changes. Archaeological data for Neolithization are still not dated precisely enough to determine whether there was, or was not, a short period of climatic or ecological stress. For hunter-gatherers one severe drought can lead to resource stress, and the consequences of two or three consecutive years of severe climatic conditions are often hunger, death and migration. However, the level of stress and estimated deprivation depends not only on the objective changes in food supply but also on the culturally determined perception of these changes (Giddens 1993, 628). It is therefore necessary to understand the decision-making and behaviour patterns – in other words the socio-ideological framework – of hunters and gatherers. But it is equally very important to understand climatic and ecological changes.

Landscape – as I understand it – is not a natural phenomenon. Landscapes are always constructions of human agents recognizing them and interpreting them in a certain context and perspective (Schmeling and Schmitz-Emans 2007, 7–12). I will show that the Neolithization of the Near East is a dialectical process in which physical environments influenced the decisions

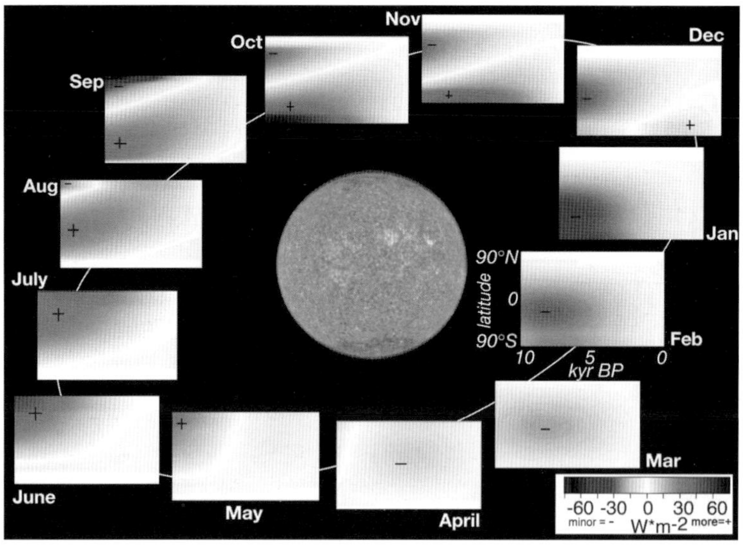

Fig. 8.1. Annual receipt of insolation of the last 10,000 years (Bradley et al. *2003, 133, fig. 6.21). Insolation during the summer months (especially June and July) was even higher at the beginning of the Holocene.*

and behaviour of humans, but where the environment was in turn also modified by the human agents, by their social organization as well as by their cultural traditions and norms. I will therefore try to combine phenomenological (post-processual) and processual concepts of landscapes (Wilkinson 2004, 3–12; Gramsch 2003).

I will suggest that Neolithization was triggered by changing landscapes at the beginning of the Holocene but that the seed for this process had already been set in the ideology, perceptions, preferences and behavioural patterns of hunter-gatherers. Social changes during this process subsequently had consequences that led to a completely changed landscape signature, that of the Pre-Pottery Neolithic A (PPNA), with larger villages and huge communal buildings structuring the landscape.

An ethno-archaeological comparison of 43 recent foraging societies in transition is used here to help deduce possible recurring preferences and behavioural patterns of hunter-gatherers in such a transitional process.

I will first summarize the climatic and ecological changes at the beginning of the Holocene. Then I will present the ethno-archaeological model and conclude by discussing the consequences that persisting behavioural patterns of hunter-gatherers might have had in a changing landscape and how hunter-gatherers changed their environment because of new social norms.

Global and regional proxy data for the climatic development and changing landscapes in the Near East

Global proxies

The most frequently used global proxies are the Greenland and Antarctic ice cores. They show that climate was still quite unstable during the late-glacial Bølling-Allerød warm phase and still cooler than during the Early Holocene. The Younger Dryas (YD) period (10,960/10,860–9620 cal BC) is supposed to have involved a return to ice-age conditions, but, as the process of deterioration that led to the YD had already started in the last phase of the Allerød and lasted for more than a hundred years (Alley *et al.* 1993), people probably did not realize immediately that climatic conditions were worsening.

In contrast, the beginning of the Holocene shows as a very abrupt and qualitative change of climate and landscapes. Ice cores and tree rings show that the climate changed after 9620 cal BC within less than 20 years (Spurk *et al.* 1998). This means that the climate changes occurred within less than one generation and that they were perceptible to people. Global warming was regionally accompanied by a marked qualitative change of the climate. As the arctic thermohaline circulation set in, winter rains occurred on the Eastern Mediterranean coast. In contrast to these humid and mild winters, summers were hotter than today, not only because of global warming, but also because at that time perihelion (earth's nearest position to the sun) occurred during the summer months of the northern hemisphere and aphelion (the most distant point of the earth's orbit to the sun) during the winter months (Bradley *et al.* 2003, 133) (Figure 8.1). In comparison to the Bølling-Allerød interstadial, which was also relatively humid compared to the YD, during the Early Holocene rains seem to be seasonally restricted to the winter months. Although there are so far no direct regional proxies to support this thesis, the reconstruction of global climate development suggests such a seasonally restricted supply of precipitation.

The start of the Holocene was thus characterized by a strong seasonality that had not existed before and has not since then. For North Africa and Eurasia climatologists calculate that summers were about 2–3°C hotter until about 7000 cal BC than they are today, while winters were about 1.5°C colder (Landmann *et al.* 1996; Kneller 1996). How hunter-gatherers may have reacted to these changes will be shown below.

Changing landscapes in the Near East – the regional proxies

While the Younger Dryas is characterized by drier conditions, geomorphological data show an increase in humidity at the beginning of the Holocene (Robinson *et al.* this volume; Kallel *et al.* 2004). Environmental changes are also reflected to different degrees in the pollen records of Lake Huleh, Ghab Valley and Lake Van (for a discussion see Benz 2000, 53–60). The proliferation of Mediterranean vegetation can be observed, with the replacement of perennial plants by annual wild grasses and cereals. Mediterranean trees and shrubs like pistachio, almond, oak and olive trees become more common.

Another important source of data for the changes in the Near East is the settlement patterns of different periods. Whereas during the Natufian small settlements can be observed even in very marginal areas such as the Negev and the Black Desert Basin, during the PPNA settlements seem to cluster next to rivers, lakes, marshes and the coast of the Mediterranean Sea (Goring-Morris 1987; Garrard *et al.* 1994; Betts 1988; Benz 2000, 52f).

From this combined evidence I propose the following conclusions: 1) As the climatic changes from the YD to the Holocene occurred within one generation, people could clearly perceive them. 2) The climatic changes did not only bring a much stronger seasonality, with mild and rainy winters and hot and dry summers, but, owing to increased evaporation, summers became very difficult months for water supply. 3) The concentration of settlements near perennial water resources during the PPNA confirms that water became a critical resource – at least during the summer months. Of course, changed settlement patterns might also be due to changing social or cultural preferences but, in combination with the climatic data, in such a case they might give hints to the quality of climate changes.

This is – in a simplified way – the ecological frame in which some fundamental changes in the behaviour of hunter-gatherers took place. There is some evidence by the Early Natufian for a more sedentary life (Edwards 1991; Valla 1991), but during the Late Natufian, with more than 1000 years of ice-age climate during the YD, hunter-gatherers had returned to a more mobile way of life (Munro 2004). Only during the Early Holocene did they adopt a completely new form of social organization – a permanently sedentary way of life in villages.

In order to understand these decisions and developments, I will now present the results of the analysis of 43 recent foraging societies in transition to agriculture. From this ethnological data I will then develop a model that will be compared to the archaeological data.

The ethno-archaeological model

In order to understand the process of Neolithization 43 recent hunter-gatherer populations from all over the world have been analysed (for the data see Benz 2000). As recent hunter-gatherers survived only in less favourable environments they cannot be compared directly to prehistoric hunter-gatherers (see Lee 1994 and references therein). Nevertheless, given their economic perspective, and probably also their social systems and their ideology, recent hunter-gatherers are closer to prehistoric hunter-gatherers than are we, with our lifestyle of global capitalism, urbanism and monotheistic religions. Based on these 43 ethnographical examples, I will create a model that will hopefully lead to a better contextual – emic – understanding of hunter-gatherers' decisions and behaviours.

The regulation of population density

Most hunter-gatherer populations have a very low population growth rate. As long-term studies of different hunter-gatherer groups have shown, this is due not just to cultural choices but mostly to biological factors (Bentley *et al.* 1993; Blurton Jones *et al.* 1992; Wilmsen 1982; Jones 1980; Howell 1979; Lee 1979). One reason is the relatively low birth rate. Although hunter-gatherers generally prefer to have many children, reduced fertility of women resulting from insufficient nourishment and high mobility appears to keep the rate of reproduction down. Therefore, if mobility was reduced and nourishment for women was stable the biological mechanisms suppressing the birth rate would also be reduced. Adequate nourishment during pregnancy and breastfeeding would also reduce the risk of infant death (Spielmann 1989). Yet this does not mean that infant mortality would automatically decrease. When people live permanently in villages, hygiene is often poor. In addition, close contact with domesticated animals provides new risks of infections which hunter-gatherers had not encountered before. We can therefore suppose:

Ethno-archaeological thesis 1: If there was constantly sufficient nourishment and if mobility was reduced, the birth rate would rise and infant mortality might decrease. As a consequence population would grow, if there were no other limitations – such as diseases attributable to the hygienic conditions in permanent settlements. As hunter-gatherers prefer to have many children, it is unlikely that they would counteract this long-term development by cultural rules.

Group size

For hunter-gatherers the group size seems to be determined, at least in part, by the abundance of resources (Cashdan 1983; Silberbauer 1994). When resources are exhausted, groups usually move or split up. But when resources are abundant, hunter-gatherers often come together in one place (Lee 1979; Hitchcock 1982; 1989). When a larger group remains in one place for a longer period of time – even before natural resources are exhausted – often social conflicts will occur; groups then split up or smaller groups will form separate camps (Guenther 1976; Barnard 1992). Egalitarian social organization is thus also a limiting factor for the group size (R. K. Hitchcock pers. comm.; cf. Marshall 1960; Kent 1995; Lee 1979; Barnard 1986). A comparison of hunter-gatherer groups shows that a viable size for a group is about 30 members.

Ethno-archaeological thesis 2: If the size of a hunter-gatherer group consists of more than 30 members for a longer period of time, social mechanisms must be found to counteract conflicts and to stabilize the group.

Mobility

Mobility is typical for foraging societies. In regions like the Kalahari, but also some regions of the Near East, where the abundance of resources is dependent on highly variable and unpredictable rainfall, mobility seems to be a very efficient way to avoid food shortages. At the same time, nearly all hunter-gatherers would reduce their mobility and put up base camps when resources were abundant (Barnard 1986, 1992). In territories with constantly abundant resources (in South Africa, Australia and Northern America) (cf. Benz 2000) recent hunter-gatherers are more or less sedentary.

But why then do some groups stay in one place even when conflicts occur? The reasons why a more sedentary way of life is adopted for a longer period of time can be divided into two categories: On the one hand, there are external causes, such as changing environment and increased population pressure. If essential resources become very scarce, people might be forced to reduce their mobility or even live permanently in one location. If this is not of their free will, people might perceive sedentism negatively and return to a mobile way of life as soon as possible. On the other hand, observations of recent hunter-gatherers show that villages of farmers, cattle stations or missions, where foragers could obtain food and other valued products easily, were especially attractive when there was a lack of traditional resources. Products from farms or cattle stations were at the beginning included in the diet just like any other natural resource. But once the foragers had become dependent on these farming products – *in theoretical terms: when a dependence on a locally fixed, reliable resource was developed* – they did reduce their mobility.

Thus sedentism might occur without pressure, when hunter-gatherers become dependent on a locally restricted but abundant resource because of changing cultural preferences and demands.

Ethno-archaeological thesis 3: Mobility is abandoned for locally restricted, abundant and calculable resources or because of population pressure or the failure of other resources. If a group becomes dependent on locally restricted resources, permanent settlements will not be abandoned, even when social conflicts and competition over resources occur.

Exploitation of resources

In general, the economic system of mobile hunter-gatherers does not maximize profit but tends to satisfy natural as well as cultural needs. When resources become scarce, hunter-gatherers rarely invest more time and labour in resources. Instead they expand their diet to include less preferred resources, as they aim to get an immediate return. Wild cereals were not preferred resources as they take a lot of time to be processed (Cane 1989). Usually hunter-gatherers do not invest in resources that do not yield an expected return quickly.

Storage is traditionally practised only to compensate for short, predictable failures in resources, to feed hosts at gatherings or in order to barter food for other items (Hitchcock 1989; Cane 1989; Bell 1987; Cashdan 1983).

Another formative principle is generalized reciprocity, the principle of sharing with everybody. Generalized reciprocity is an efficient insurance against food shortages – especially unpredictable ones – using social commitments (cf. Bird-David 1992). It is based on open access to resources, as, for example, when in times of resource shortages in one region a group is allowed to move into the territory of another group and use their resources (Marshall 1960; Barnard 1992). The social networks that are established for this 'social form of storage' must be 'cultivated' and a lot of time is invested in visiting friends. This time investment is often neglected when praising the 'original affluent society' and the relaxed way of life of hunter-gatherers.

If farming is to be adopted permanently, this strong social norm must be reduced. Yet several examples show that in foraging societies generalized reciprocity is suspended only in critical situations (e.g. resource stress or social conflicts) or when the expected delayed return for sharing is not high enough (Lee 1979; Kent 1995; Wiessner 1982; Guenther 1986a; Guenther pers. comm.). Frustration, hopelessness and existential fears might undermine the social norm of sharing as well (Guenter 1986b). The consequence of this reduction is an increase in exclusive rights of use (cf. Benz 2000, 125–31). When resources became scarce, some groups develop a strong commitment to their territory (Heinz 1972). Prestige items are then inherited by parents (Barnard 1992) and staples are hidden in certain places. Endogamy may enforce these exclusive territorial rights.

Ethno-archaeological thesis 4:

a) Storage is practised only when resource failures are calculable and predictable.

b) Generalized reciprocity will be reduced: in times of crises, if the share will be so little that it falls below an acceptable threshold; when the expected delayed social compensation (either in resources or social advantages) will be less than the profit of other alternatives, such as external trade, storage of food or investment in resources; if the chances increase in larger groups of not being detected when acting against the social rules. Such a deviant behaviour normally occurs only in competitive situations or because of conflicts that cannot be avoided by migration.

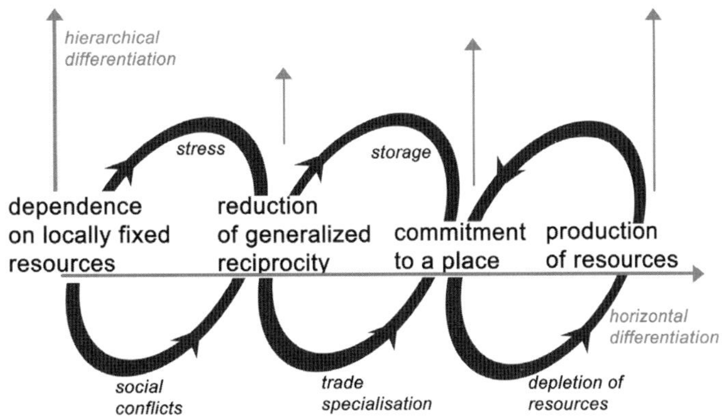

Fig. 8.2. The ethno-archaeological model shows that there are three important thresholds that are passed when moving from a foraging mode of life to the production of resources: a dependence on locally fixed resources; the reduction of the generalized reciprocity; and a commitment to one place. Demographical and biological consequences may then lead into path-dependency. The different stages should not be regarded as evolutionary stages; rather, they represent different forms of social organization that may exist side by side. Hierarchical and horizontal differentiation increase with each stage.

Increasing exclusive rights of use

When generalized reciprocity is reduced territoriality automatically increases, but exclusive rights of use might also be increased for other reasons. In general, hunter-gatherers do have some personal possessions: objects or resources or land in which much labour is invested are owned by a person or a group of people. Hunting equipment, and also locally fixed resources such as places where honey can be obtained, fertile berry grounds, nut trees or land that has been burned, can be owned (Seitz 1984). The adoption of cultivation further increased these claims of ownership and an emotional as well as an economic commitment to one place because of the labour that was invested in the land or in resources (Schapera 1930; Vierich 1982; Lebzelter 1934; Hitchock 1989; Marshall 1960). But it must be stressed that incipient cultivation does not automatically lead to sedentism. In most cases hunter-gatherers do not really care for their fields and often abandon them during the hunting season. While land or plants can be owned, wild animals that migrate from one territory to another cannot be owned.

Ethno-archaeological thesis 5: Exclusive rights of use are increased by labour investments. They most often concern locally restricted resource-rich areas or personal equipment and accessories.

Self-perception

Many ethnographers report that hunter-gatherers who started cultivation or herding long ago still consider themselves as hunters and gatherers (cf. Benz 2000, 134–6). The self-image is often so 'archaic' that some farmers go on their fields with bow and arrow on their back and say that they were 'hunters of the forest'.

Ethno-archaeological thesis 6: In most cases self-perception is changed only a long time after the economical and social systems have been changed.

The ethno-archaeological model

The following model is based on the six theses presented above. People constantly modify their cultural and natural environments to a greater or lesser extent while the always-changing cultural and natural context has an impact on the behaviour and decisions of people. People are therefore constantly confronted with decisions they have to take, yet the consequences of their decisions are often unpredictable.

As long as people's behaviour is not in conflict with environmental and cultural conditions, change might be very subtle. We then speak of balance, when in fact the system is just changing very slowly. However, by taking certain decisions a threshold may be crossed, and then even if the behaviour follows traditional patterns it may lead to completely new conditions.

The consequences of such long-term changes cannot be foreseen. In certain circumstances, returning to former ways of life becomes impossible. Such irreversible long-term processes are described as path-dependency (Witt 1994). In my view, Neolithization is such an irreversible long-term process of path-dependency. Once the system has been changed towards a farming economy returning to the foraging mode of life is very seldom desired, and, if the environment has been changed, it might even have become impossible.

I will now analyse which critical thresholds in economy and society must be crossed during Neolithization, under which conditions such thresholds may be crossed, and what consequences result from the traditional behaviour patterns of hunter-gatherers under such new conditions (Figure 8.2).

1) It has been shown that a commitment to locally fixed resources is the first precondition for farming to become a viable and acceptable way of life (thesis 3). Some kind of territorial commitment must exist. If it does not, foragers tend to leave the place when conflicts occur or

when resources become scarce but they do not invest in resources to increase their yields. Such a dependency might be developed in favourable or unfavourable environmental conditions – be it a resource-rich area that first attracts people or unfavourable conditions that force people to live in one place. If territorial commitment is the result of free choice (abundant resources) it is probably more durable owing to a positive perception, compared with a forced sedentism that will be given up as soon as conditions ameliorate.

2) The second precondition is the reduction of generalized reciprocity (thesis 4). This social obligation is violated only in situations of social conflict or crisis. There must be some pressure before one can decide that the advantage of not sharing compensates for social sanctions. However, in large groups generalized reciprocity may also be reduced because its members do not expect to be caught when acting against the social norm, or because sharing with everybody becomes impossible.

If generalized reciprocity is reduced, the main existing means of insurance against resource failure is lost and must be compensated for in other ways. There are three different solutions: the first one is storage of surplus, the second one is to establish trading networks, and the third one is to invest in resources.

If resources are quite evenly distributed, storage rather than trade might be practised. If the territories differ in their ecological conditions, the group is more likely to resort to trading. In favourable ecological conditions sedentary groups can live in such a situation for a long time without exhausting their resources. As thesis 1 suggests, an increase in population is then to be expected. If population densities increase and sedentary groups grow to more than 30 members, social rules must be developed to appease conflicts (thesis 2) and mechanisms must be found to keep the group together. As the temptation to ignore generalized reciprocity in bigger groups rises, wealth may be accumulated and hierarchies might be established.

The third solution, investment in resources, results in a delayed reward and will thus probably not be the preferred solution. Plants that quickly become ready for harvesting – such as cereals and pulses – were likely to be preferred for cultivation. I therefore suggest that the threshold to investment in resources with delayed returns is a very high one and will only be crossed as a last resort. This means that hierarchically structured sedentary foraging groups may exist in favourable environments alongside groups that practise incipient cultivation in environments where the carrying capacity is transgressed either because population is too dense or because of unfavourable environmental conditions.

3) All three strategies – storage, trade or investment in resources – on the one hand intensify the commitment to one place. On the other hand, they also allow a group to stay longer in one place than a foraging economy does. If we further extrapolate from the reasons hunter-gatherers have for establishing exclusive use rights and possessions (thesis 5), the investment in resources will intensify territoriality and exclusive land rights. The possibility of storing goods then increases social hierarchy within and between groups still further.

Demographic pressure is not a prime motivating factor in this model, but it does determine the speed of transition through different phases. Favourable ecological conditions seem to be attractive conditions for sedentism to be practised for a longer period of time. Ecological conditions may also influence the solutions that are chosen to deal with failures of resources, and they might allow a foraging society to continue their traditional way of hunting and gathering. Yet one of the main social thresholds that must be crossed is the reduction in generalized reciprocity.

Conclusion – Traditional behaviour patterns in a changing landscape

If we now apply this model to prehistoric data in the Near East, we may be able to improve our understanding of the interplay between cultural choices and long-term processes and the ways in which the traditional behaviour of hunter-gatherers in a changing landscape led to the process of Neolithization in the Near East.

The ethno-archaeological model suggests that permanent sedentism was most probably adopted in resource-rich areas. It might even be suggested that our picture of mostly mobile hunter-gatherers in the Upper Palaeolithic is distorted by direct analogies to recent hunter-gatherers living in unfavourable environments. New archaeological data for the Upper Palaeolithic might be interpreted in favour of the existence of more sedentary groups beside mobile foragers (Piperno *et al.* 2004). The climatic amelioration of the Bølling-Allerød period made it possible for some groups during the Early Natufian to live permanently in one location. There are a greater number of archaeological remains of the Natufian compared to the Kebarian, and an increase in population might therefore be suggested, but data from the Kebarian are still too patchy for any conclusion to be drawn (cf. Maher, this volume). Examples are Wadi Hammeh in Jordan, Hatula and Mallaha in Israel and Abu Hureyra on the Euphrates (Valla 1991; Edwards 1991; Lieberman 1994; Moore *et al.* 2000). Yet this incipient trend towards sedentism was frozen by the YD. The spectra of animal bones indicate a return to a more mobile way of life for more than 1000 years (Munro 2004). As the behavioural patterns of recent hunter-gatherers show, farming probably did not start during the YD because the environment was relatively poor and food shortages could not be anticipated.

Compared to the climates of the Bølling-Allerød and the YD, the Early Holocene was a very stable period. There was a strong seasonality in Near Eastern climate and vegetation. There were probably seasonal food and water shortages in summer, but they were predictable. However, the spreading Mediterranean vegetation provided much more food than the steppes of the YD.

The abrupt qualitative change of climate within less than one generation might have had in the long run the following consequences resulting from the traditional behavioural patterns of hunter-gatherers:

1) As thesis 3 of the ethno-archaeological model suggests, hunter-gatherers are attracted by calculable, abundant and nutrient rich resources. The amelioration of climatic conditions and a proliferation of Mediterranean vegetation with nutrient resources such as pistachios, almonds, acorns, pulses, figs and cereals (Byrd 2005, 253) made a sedentary life for some groups attractive. It would have been in accordance with traditional behavioural patterns to stay in a place with abundant resources. Surpluses could easily be gained and stored for the months of predictable resource shortages (thesis 4a). At first, fruit trees and other locally restricted resources might have been more attractive than cereals. But although the Early Holocene seems to be a relatively humid period, water procurement in the hot and dry summer months probably became crucial owing to high evaporation; water supplies were reduced to perennial rivers, lakes and springs.

2) In fact, during the PPNA we can observe a concentration of permanent sites near perennial water sources (Belfer-Cohen, Goring-Morris this volume; Benz 2000, 65ff.). There does not seem to be a decrease in population but rather an aggregation to these favourable areas. Sites became bigger and were occupied for a longer period of time. In addition to permanent settlements, there were also a variety of small temporary sites. In accordance with the ethno-archaeological model, cultivators lived side by side with hunter-gatherers who did not or did not have to cultivate.

3) If nourishment was sufficient in these regions population growth is to be expected (thesis 1) owing to cultural preferences and biological long-term processes.

4) Although cereals are not preferred resources, a dependence on cereals might have developed as the yield per hectare for cereals is very high compared to other plants (Rindos 1984). As cereals grow fast, they fit very well to the ideology of hunter-gatherers who strive for an immediate return on invested work.

5) According to the model, hunter-gatherers invest in resources or practise some kind of storage when shortages are predictable (thesis 4a). Therefore, the strong seasonality and the stability of the Early Holocene climate might have acted as a catalyst for storage being adopted as a sensible practice.

6) If commitment to certain places and the difficulties in water procurement during the summer months 'forced' the inhabitants of these sites to stay there, conflicts over resources might have occurred and generalized reciprocity might have been reduced. Conflicts could no longer be solved by migration. Conflict management and mechanisms to generate a group identity would become indispensable.

The increase of group size in these villages might have been another reason for the reduction of generalized reciprocity. Yet the reduction of generalized reciprocity and the loss of the widely defined network of hunter-gatherers meant the break-down of 'social assurance' against resource shortages. New mechanisms must have been found to keep groups together and commit them to one territory.

Famous communal buildings, such as the tower and walls of Jericho (Naveh 2003) or the structures at Göbekli Tepe, may be interpreted as the remnants of efforts to create and maintain group identities, thus showing the fundamental social changes in Early Holocene societies (Benz 2007). Changing social norms, such as a reduction in generalized reciprocity and changes of conflict management, thus led to a new structuring of the landscape. The social changes led to an appropriation of landscape which had to be demonstrated by monumental markers. The new communal buildings were significant not only as external symbols of group identities, but also as territorial markers. In the same way, the skull cult that reaches its climax during the PPNB, and secondary burials below houses and the open spaces of villages, might be interpreted as an expression of the intention to create relationships between generations and to incorporate even the dead into society. Rituals, particularly secondary burials, are important events for group identity (Kuijt 2000). Ritual caches, which occur on several PPNB sites (Gebel 2002; cf. Astruc *et al.* 2003), may indicate conflicts between the ancient social norm of sharing with everybody and the new social reality that made it impossible to do so. The development of restricted access to buildings and the removal of storage facilities into the houses that has been observed at Beidha (Byrd 1994, 640) support the thesis that generalized reciprocity was reduced.

To sum up, I argue that the cultural preferences of hunter-gatherers, such as the preference to stay in one place when food is easily available, the preference to have many children and the preference for an immediate return, led to path-dependent demographical and biological long-term processes that started with a commitment to one place. Generalized reciprocity in some of these enlarged sedentary groups might then not have been a viable practice any more, and might have been reduced. In consequence, even in times of resource shortages it was no longer possible to move freely and use the resources of any other group. Instead, trading, storage or investment in resources became the only alternatives. Investment in resources then reinforced exclusive rights of use and of property (cf. thesis 5).

To conclude, I want to suggest the hypothesis that storage, trade and investment in the production of resources are the consequences of a commitment to one location and of the reduction of generalized reciprocity. Exclusive group identities are emphasized and maintained by communal buildings and the rituals that are celebrated there. According to the ethno-archaeological model it is not a changed perception of the environment or any ideological or religious change that led to the production of resources, but rather the traditional preferences of hunter-gatherers applied to a rapidly changing landscape that resulted in the long-term process of Neolithization.

Climatic conditions are not a primary factor in our model (cf. Verhoeven 2004, 200ff.), they just form the stage on which the actors play out their drama. By their traditional behaviours hunter-gatherers set the rules. But they were playing in a rapidly changing environment that strongly influenced the storybook, so that the drama in the long run continued in a way probably none of the actors ever had expected. Finally the actors had to change the rules of the game and by so doing they also changed the landscape around them in a way that made a return to a foraging mode of life impossible.

Acknowledgements

I am grateful to Bill Finlayson and Graeme Warren for the wonderful and interesting meeting they organized in Jordan.

Bibliography

Alley, R. B., Meese, D. A., Shuman, C. A., Gow, A. J., Taylor, K. C., Grootes, P. M., White, J. W. C., Ram, M., Waddington, E. D., Mayewski, A. and Zeilinski, G. A. J. (1993) Abrupt increase in Greenland snow accumulation at the end of the Younger Dryas event. *Nature* 362, 527–9.

Astruc, L., Abbès, F., Ibáñez Estévez, J. J. and González Urquijo, J. E. (2003) «Dépots», «réserves» et «caches» de matériel litque taillé au néolithique précéramique au Proche-Orient: quelle gestion de l'outillage? *Paléorient* 29, 59–78.

Barnard, A. (1986) Rethinking Bushman settlement patterns and territoriality. Pp. 41–60 in F. Rottland and R. Vossen (eds), Afrikanische Wildbeuter. *Sprache und Geschichte in Afrika* 7/1.

Barnard, A. (1992) *Hunters and Herders of Southern Africa: A Comparative Ethnography of the Khoisan peoples*. Cambridge: Cambridge University Press.

Bell, D. (1987) Gathered from Kaytey women. Pp. 239–52 in D. J. Mulvaney and J. P. White (eds), *Australians: a Historical Library 1, Australians to 1788*. Sidney: Fairfax, Syme & Weldon.

Bender, B. (1978) Gatherer-hunter to farmer: a social perspective. *World Archaeology* 10, 204–22.

Bentley, G. R., Goldberg, T. and Jasienska, G. (1993) The fertility of agricultural and non-agricultural traditional societies. *Population Studies* 47/2, 260–81.

Benz, M. (2000) *Die Neolithisierung im Vorderen Orient. Theorien, archäologische Daten und ein ethnologisches Modell*. Studies in Early Near Eastern Production, Subsistence, and Environment 7. Berlin: *ex oriente*.

Benz, M. (2007) Zur Bedeutung von Festen während der Neolithisierung im Vorderen Orient. *Ethnographisch Archäologische Zeitschrift* 4, 2006 (2007), 439–62.

Betts, A. V. G. (1988) The Black Desert Survey: summary of prehistoric and palaeoenvironmental research in the Northern Hisma. Pp. 369–91 in A. N. Garrard and H. G. Gebel (eds), *The Prehistory of Jordan*. British Archaeological Reports, International Series 396. Oxford: BAR.

Bird-David, N. (1992) Beyond the 'original affluent society'. *Current Anthropology* 33/1, 25–47.

Blurton Jones, N., O'Connell, J., Hawkes, J., Kamuzora, C. and Smith, L. (1992) Demography of the Hadza, an increasing and high density population of Savanna Foragers. *American Journal of Physical Anthropology* 89, 159–81.

Bradley, R., Briffa, K. R., Cole, J., Hughes, M. K. and Osborn, T. J. (2003) The climate of the last millennium. Pp. 105–41 in K. D. Alverson, R. S. Bradley and Thomas F. Pedersen (eds), *Paleoclimate, Global Change and the Future*. Berlin: Springer.

Byrd, B. (1994) Public and private, domestic and corporate: the emergence of the Southwest Asian village. *American Antiquity* 59/4, 639–66.

Byrd, B. (2005) Reassessing the emergence of early village life in the Near East. *Journal of Archaeological Research* 13/3, 231–90.

Cane, S. (1989) Australian Aboriginal seed grinding and its archaeological record: a case study from the Western Desert. Pp. 99–119 in D. R. Harris and G. C. Hillman (eds), *Foraging and Farming: The Evolution of Plant Exploitation*. London: Unwin Hyman.

Cashdan, E. A. (1983) The effects of food production on mobility in the Central Kalahari. Pp. 311–27 in J. D. Clark and S. A. Brandt (eds), *From Hunters to Farmers*. Berkeley: University of California Press.

Cauvin, J. (1976) *Les Premiers Villages de Syrie-Palestine du IXème au VIIème Millénaire avant J. C.* Lyon: Maison de l'Orient.

Çelik, B. (2000) A new Early-Neolithic settlement: Karahan Tepe. *Neo-Lithics* 2–3, 6–8.

Çelik, B. (2006) A new Pre-Pottery Neolithic site in southeastern Turkey: Sefer Tepe. *Neo-Lithics* 1, 23–5.

Edwards, P. (1991) Wadi Hammeh 27: an Early Natufien site at Pella, Jordan. Pp. 123–48 in O. Bar-Yosef and F. R. Valla (eds), *The Natufian Culture in the Levant*. Ann Arbor, MI: International Monographs in Prehistory.

Garrard, A. N., Baird, D., Colledge, S., Martin, L. and Wright, K. (1994) Prehistoric Environment and Settlement in the Azraq Basin: an interim report on the 1987 and 1988 Excavation Seasons. *Levant* 26, 73–109.

Gebel, H. G. K. (2002) Walls. Loci of forces. Pp. 119–32 in H. G. Gebel, B. Dahl Hermansen and C. Hoffmann Jensen (eds), *Magic Practices and Ritual in the Near Eastern Neolithic*. Studies in Early Near Eastern Production, Subsistence, and Environment 8. Berlin: *ex oriente*.

Giddens, A. (1993) *Sociology*. Cambridge: Polity Press.

Goring-Morris, A. N. (1987) *At the Edge. Terminal Pleistocene Hunter-Gatherers in the Negev and Sinai*. British Archaeological Reports, International Series 361. Oxford: BAR.

Gramsch, A. (2003) Landschaftsarchäologie – ein fachgeschichtlicher Überblick und ein theoretisches Konzept. Pp. 35–54 in J. Kunow and J. Müller (eds), *Landschaftsarchäologie und Geographische Informationssysteme*. Archäoprognose 1. Wünsdorf: Landesamt Brandenburg.

Guenther, M. G. (1976) From hunters to squatters: social and cultural change among the farm San of Ghanzi, Botswana. Pp. 120–33 in R. B. Lee and I. DeVore (eds), *Kalahari Hunter-Gatherers: Studies of the !Kung San and their Neighbours*. Cambridge: Harvard University Press.

Guenther, M. G. (1986a) Acculturation and assimilation of the Bushmen of Botswana and Namibia. Pp. 347–73 in R. Vossen and K. Keuthmann (eds), *Contemporary Studies in Khoisan 1*. Quellen zur Khoisan-Forschung 5.1. Hamburg: Buske.

Guenther, M. G. (1986b) *The Nharo Bushmen of Botswana*. Quellen zur Khoisan-Forschung 3. Hamburg: Buske.

Hayden, B. (1992) Contrasting expectations in theories of domestication. Pp. 11–19 in A. B. Gebauer and T. D. Price (eds), *Transitions to Agriculture in Prehistory*. Madison, WI: Prehistory Press.

Heinz, H. J. (1972) Territoriality among the Bushmen in general and the !Ko in particular. *Anthropos* 67, 405–15.

Hitchcock, R. K. (1982) Patterns of sedentism among the Basarwa of Eastern Botswana. Pp. 223–67 in E. Leacock and R. B. Lee (eds), *Politics and History in Band Societies*. Cambridge: Cambridge University Press.

Hitchcock, R. K. (1989) Settlement, seasonality and subsistence stress among the Tyua of Northern Botswana. Pp. 64–85 in R. Huss-Ashmore and S. H. Katz (eds), *Coping with Seasonal Constraints*. Philadelphia, PA: MASCA.

Hodder, I. (1990) *The Domestication of Europe*. Oxford: Basil Blackwell.

Howell, N. (1979) *Demography of the Dobe !Kung*. New York: Academic Press.

Jones, R. (1980) Hunters in the Australian coastal savanna. Pp. 107–46 in D. R. Harris (ed.), *Human Ecology in Savanna Environments*. London: Academic Press.

Kallel, N., Duplessy, J.-C., Labeyrie, L., Fontugne, M. and Paterne, M. (2004) Mediterranean sea palaeohydrology and pluvial periods during the late Quaternary. Pp. 307–24 in R. W. Battarbee, F. Gasse and C. E. Stickley (eds), *Past Climate Variability through Europe and Africa 6*. Dortrecht: Springer.

Kent, S. (1995) Unstable households in a stable Kalahari community in Botswana. *American Anthropologist* 97/2, 297–312.

Kneller, M. (1996) Ice ages. Pp. 418–23 in S. H. Schneider (ed.), *Encyclopedia of Climate and Weather 1*. New York: Oxford University Press.

Kuijt, I. (2000) Keeping the peace. Ritual, skull caching, and community integration in the Levantine Neolithic. Pp. 137–64 in I. Kuijt (ed.), *Life in Neolithic Farming Communities. Social Organization, Identity, and Differentiation*. New York: Kluwer Academic Publishers.

Landmann, G., Reimer, A. and Kempe, S. (1996) Climatically induced lake level changes at Lake Van, Turkey, during the Pleistocene/Holocene transition. *Global Biogeochemical Cycles* 10/4, 797–808.

Lebzelter, V. (1934) Die Buschmänner Südwestafrikas. *Africa* 7, 70–81.

Lee, R. B. (1979) *The !Kung San*. Cambridge: Cambridge University Press.

Lee, R. B. (1994) Art, science, or politics? The crisis in hunter-gatherer studies. *American Anthropologist* 94/1, 31–54.

Lieberman, D. E. (1994) Seasonality estimates from Hatoula. Pp. 125–31 in M. Lechevallier and A. Ronen (eds), *Le Site de Hatoula en Judée Occidentale. Israel*. Paris: Association Paléorient.

Marshall, L. (1960) !Kung Bushmen bands. *Africa* 30, 325–54.

Moore, A. M. T., Hillman, G. C. and Legge, A. J. (2000) *Village on the Euphrates: From Foraging to Farming at Abu Hureyra*. Oxford: Oxford University Press.

Munro, N. D. (2004) Zooarchaeological measures of hunting pressure and occupation intensity in the Natufian. *Current Anthropology* 45, 5–33.

Naveh, D. (2003) PPNA Jericho: a socio-political perspective. *Cambridge Archaeological Journal* 13/1, 83–96.

Piperno, D. R., Weiss, E., Holst, I. and Nabel, D. (2004) Processing of wild cereals grains in the Upper Palaeolithic revealed by starch grain analysis. *Nature* 430, 670–3.

Rindos, D. (1984) *The Origins of Agriculture: An Evolutionary Perspective*. Orlando, FL: Academic Press.

Rosenberg, M. and Davis, M. K. (1992) Hallan Çemi Tepesi, an early aceramic neolithic site in eastern Anatolia. *Anatolica* 18, 1–18.

Schapera, I. (1930) *The Khoisan Peoples of South Africa*. London: Routledge.

Schmeling, M. and Schmitz-Emans, M. (eds) (2007) *Das Paradigma der Landschaft in Moderne und Postmoderne*. Würzburg: Königshausen and Neumann.

Schmidt, K. (2006) *Sie bauten die ersten Tempel*. München: C. H. Beck.

Seitz, S. (1984) Von der wildbeuterischen zur agrarischen Lebensweise. *Paideuma* 10, 257–74.

Silberbauer, G. B. (1994) A sense of place. Pp. 119–43 in E. S. Burch Jr. and L. J. Ellanna (eds), *Key Issues in Hunter-Gatherer Research*. Oxford: Berg.

Spielmann, K. A. (1989) A review: dietary restrictions on hunter-gatherer women and the implication for fertility and infant mortality. *Human Ecology* 17/3, 321–45.

Spurk, M., Friedrich, M., Hofmann, J., Remmele, S., Frenzel, B., Leuschner, H. H. and Kromer, B. (1998) Revisions and extension of the Hohenheim oak and pine chronologies: new evidence about the timing of the Younger Dryas/Preboreal transition. *Radiocarbon* 40/3, 1107–16.

Stordeur, D., Brenet, M., Der Aprahamian, G. and Roux, J.-C. (2000) Les bâtiments communautaires de Jerf el Ahmar et Mureybet, Horizon PPNA. Syrie. *Paléorient* 26/1, 29–44.

Valla, F. R. (1991) Les Natufiens de Mallaha et l'espace. Pp. 111–22 in O. Bar-Yosef and F. R. Valla (eds), *The Natufian Culture in the Levant*. Ann Arbor, MI: International Monographs in Prehistory.

Verhoeven, M. (2004) Beyond boundaries: nature, culture and a holistic approach to domestication in the Levant. *Journal of World Prehistory* 18/3, 179–282.

Vierich, H. I. D. (1982) Adaptive flexibility in a multi-ethnic setting. Pp. 213–22 in E. Leacock and R. B. Lee (eds), *Politics and History in Band Societies*. Cambridge: Cambridge University Press.

Watkins, T. (2004) Building houses, framing concepts, constructing worlds. *Paléorient* 30/1, 5–24.

Wiessner, P. (1982) Risk, reciprocity and social influences on !Kung San economics. Pp. 61–84 in E. Leacock and R. B. Lee (eds), *Politics and History in Band Societies*. Cambridge: Cambridge University Press.

Wilkinson, T. J. (2003) *Archaeological Landscapes of the Near East*. Tuscon, AZ: The University of Arizona Press.

Wilmsen, E. N. (1982) Studies in diet, nutrition, and fertility among a group of Kalahari Bushmen in Botswana. *Social Science Information* 21/1, 95–125.

Witt, U. (1994) Wirtschaft und Evolution: einige neuere theoretische Entwicklungen. *Wirtschaftswissenschaftliches Studium* 10, 503–12.

Yartah, T. (2004) Tell 'Abr 3, un village du néolithique préceramique (PPNA) sur le Moyen Euphrate. *Paléorient* 30/2, 141–58.

Part Two

Moving Landscapes: Worldviews and Contact

9. The Neolithization of Britain and Ireland: The 'Big Picture'

Alison Sheridan

This contribution offers a model for the Neolithization of Britain and Ireland featuring multiple strands of immigration, from different parts of France to different parts of these islands – at differing scales and for differing reasons – over the course of several centuries from the third quarter of the 5th millennium to the first two centuries of the 4th. A critique of the 'indigenous communities as prime agents of this change' model is offered, and the importance of understanding events in these islands within a broader picture of continental developments is emphasized.

Introduction

Archaeologists have been speculating about the nature of the Mesolithic–Neolithic transition in Britain and Ireland since well before Stuart Piggott proposed his 'multiple colonization' hypothesis in 1954 (Piggott 1954). Over the last two decades the debate has essentially been polarized around the issue of whether the prime agents for the observed changes in economy, lifestyle and material culture were the indigenous population of these islands, actively adopting novel resources and ideas from the continent, or were immigrants. The two points of view entail different perspectives on the nature of the Neolithic economy and lifestyle, and these colour interpretations of the evidence. The proponents of the 'indigenous adoption' model – championed by Julian Thomas (e.g. Thomas 2003; 2004; 2007; 2008) – characterize early farming and its associated lifestyle in Britain and Ireland as being largely mobile, with strong continuity from the traditional hunting, gathering and fishing way of life and the selective uptake of novel practices and traditions. The apparent paucity of house structures over much of Britain is held up as evidence for a non-sedentary way of life, and where house (and house-like) structures exist, their role as settled places for living is downplayed (Thomas 1996). Cereal agriculture is presented as a very minor element in subsistence practices. The proponents of the 'colonization' model portray the change in terms of the introduction of a package (or rather several packages) of radically novel practices, resources, traditions and beliefs, countering claims about the nature of farming and settlement, for example, through a careful reassessment and re-statement of the evidence (e.g. Cooney 2001; Monk 2000; Pailler and Sheridan 2009; M. P. Richards 2004; Richards and Schulting 2006; Rowley-Conwy 2004; Schulting 2000; 2004; Sheridan 2003a; 2003b; 2004; 2005; 2007; in press).

It will come as no surprise to those familiar with the present author's work that her position is firmly within the 'colonization' camp, whereby indigenous acculturation is regarded as a *subsequent development from*, rather than a *prime mover of*, the introduction of novel traditions and practices. (Such a view is not at odds with that of commentators such as Alasdair Whittle (2007; Bayliss *et al.* 2008), who have appealed for models that encompass both immigration and acculturation.) While this contribution starts with a brief critique of the 'indigenous adoption' model, its main thrust is to demonstrate that the Neolithization of Britain and Ireland was by no means a unitary phenomenon, nor did it all happen abruptly around 4000 BC. By sketching the 'big picture', within which more nuanced models of regional processes of change can be situated (e.g. Sturt and Garrow, this volume), it is hoped to broaden and advance the debate, focusing attention on key outstanding questions and encouraging researchers to move on and examine what happened *after* the initial appearance of 'the Neolithic', however it is defined, in Britain and Ireland (cf. Bradley 2007; 2008). Note that the term 'Britain' will here include the Isle of Man. Note also that individual radiocarbon dates cited here (all at 2σ) have been calibrated using OxCal v4.1.

Brief critique of the 'indigenous adoption' model: a plea for informed discourse

> ... it is impossible to imagine how anything other than a colossal invasion of Neolithic people could have completely displaced the indigenes within a couple of centuries ... I submit, then, that small-scale colonization is the least likely explanation for the abrupt beginning of the Neolithic in Britain. (Thomas 2004, 117)

> The principal alternatives are a massive, coordinated seaborne invasion, or the adoption of a Neolithic way of life by Mesolithic hunter-gatherers. The author considers the latter to be more likely. (Thomas 2008, 65)

These two quotes, which demonstrate a retrenchment of a view in the face of all the evidence that has been published in the intervening four years, represent for this author all that is at fault with the 'indigenous adoption' model. To postulate a 'colossal invasion of Neolithic people' as the only alternative to indigenous adoption, and to characterize the process of Neolithization through immigration as complete displacement of the indigenes, is to set up two 'straw man' arguments that betray an ignorance of how culture change has actually worked around the world and over time. As countless examples show, from the 'Founding Fathers' of America to early European traders in Asia, small numbers of immigrants can, sooner or later, have a disproportionate influence on indigenous populations.

The key faults – in this author's opinion – with the 'indigenous adoption' model expounded by Thomas and others are as follows. Firstly, the model is predicated on the assumption that hunter-gatherer-fisher communities in Britain and Ireland were (and had previously been) in contact with Continental farming groups. As pointed out by Pailler and Sheridan (2009), however, there is not one shred of evidence to substantiate this view (and here, the Ferriter's Cove (Co. Kerry) evidence which Thomas cites to support his argument is regarded as an early example of immigration from the Continent, rather than as contact initiated from Ireland: see below). Rather, later Mesolithic communities in Britain and Ireland are remarkable for their insularity: their lithic traditions are markedly different from those on the near Continent, for example (see Kador, this volume). While one cannot rule out the possibility that some short-distance cross-Channel contact took place between Mesolithic groups in southern England and their Continental neighbours, this has left no traces; and regarding farther-flung contacts, to postulate that, say, a group of hitherto insular-orientated Mesolithic hunter-gatherer-fishers from Sligo in north-west Ireland sailed to France and adopted an entire new way of life, including the tradition of constructing causewayed enclosures (see below), stretches credibility and invites the questions 'how?' and 'why?'

Secondly, the 'indigenous adoption' model represents an inappropriate application of a model that had been developed to account for observed changes in hunter-gatherer-fisher communities along the coastal fringe of north-west continental Europe, where these groups were not separated by the sea from their farming neighbours. As Rowley-Conwy (2004) and others (e.g. Louwe Kooijmans 2007) have pointed out, these Ertebølle/Ellerbek/Swifterbant communities had a long history of contact with farming groups further inland, and adopted those elements of the farming lifestyle that suited their own lifestyle and their environment. The archaeological traces of this long relationship are in striking contrast to the appearance of entire packages of novelties in Britain and Ireland.

Thirdly, the model offers no convincing evidence to explain why British and Irish Mesolithic communities should change their lifestyles (including their diet) so profoundly at this time.

Fourthly, and similarly, it fails to take into account the known changes that were occurring on the near Continent during the late 5th and early 4th millennia (see below). In other words, developments in Britain and Ireland are considered from an insular viewpoint, and are not integrated within a broader picture of European developments.

Fifthly, the evidence put forward to support the idea of continuity with pre-existing indigenous ways of life – such as the positioning of some megalithic tombs on shell middens (e.g. Telford 2002) – often ignores the fact that the Mesolithic and Neolithic evidence in question is often separated by well over a millennium, and sometimes by much longer. While one should not underestimate the longevity of traditions, and the potential role of post-colonization contact in the choice of specific landscape locales, nevertheless other factors – including chance, or the desirability of a location for different reasons – could well apply.

Finally, the arguments that have been advanced to play down the role of cereal cultivation and the domestic nature of house structures have already been thoroughly critiqued by others (e.g. Cooney 2001; Monk 2000, Rowley-Conwy 2004), so they will not be repeated here; the question of whether 'halls' are large houses or ritual centres will be covered below. The debate surrounding Richards and Schulting's arguments about the marked Mesolithic–Neolithic dietary change of coastal communities, from predominantly marine to terrestrial (e.g. M. P. Richards 2004; Richards and Schulting 2006; Richards *et al.* 2003; Milner *et al.* 2006), continues, but the basic finding has not been overturned: the 'indigenists'' claims for a substantive continuity of lifestyle are not supported by the available evidence. As for arguments concerning material culture – and here one thinks of Thomas' critique of the current author's work identifying Breton-style pottery at Achnacreebeag, for instance (Thomas 2004) – all too often these betray a depressing lack of familiarity with the British and Continental material in question. While it is perfectly legitimate to argue a position, there is an obligation to be well-informed about the evidence being discussed.

It is with the last point in mind that the author offers the following review of the evidence for Neolithization in Britain and Ireland, pointing out the key developments that enable the 'big picture' narrative to be constructed and refined.

Building and refining the narrative of Neolithization in Britain and Ireland

Our understanding of what happened when, where and why in the process of Mesolithic–Neolithic transition in Britain and Ireland has recently been aided by several key developments on both sides of the Channel.

On this side, the systematic chronological work undertaken by Whittle, Healy and Bayliss regarding the dating of funerary monuments in southern England (Bayliss and Whittle 2007) and of enclosures, mostly but not exclusively in southern England (Bayliss *et al.* 2008; Whittle *et al.* forthcoming), has demonstrated that many of the sites that had traditionally been regarded as belonging to the earliest Neolithic actually date no earlier than the 37th or 38th century BC, and may thus have been constructed by the descendants of the first farming communities. That there was a Carinated Bowl Neolithic (see below) presence in Britain and Ireland prior to that date is indicated by several recently excavated or recently dated sites, including the causewayed enclosure at Magheraboy, Co. Sligo (but see below, p. 98), the large rectangular house at White Horse Stone, Kent, the megalithic monument at Coldrum, also in Kent, and a non-megalithic wooden grave chamber at Yabsley Street, London, all of which have dates for construction or use as early as the 40th century BC (Bayliss *et al.* 2008, 35; Coles *et al.* 2008; Danaher 2007). Peter Woodman's work in Ireland on dating domesticated species' faunal remains from ostensibly 'Mesolithic-style' contexts (Woodman and McCarthy 2003) has helped to clarify the picture, showing where some bones had previously been mis-identified and demonstrating that the only site to have produced pre-Carinated Bowl Neolithic domesticated animal remains is Ferriter's Cove (see below).

On the Continent, our understanding of the processes of late 5th- and early 4th-millennium economic and social change has been transformed by the work of Cassen and colleagues in Brittany (Boujot and Cassen 1992; 1993; Cassen 2000; Cassen and François 2006), Marcigny and colleagues in Normandy (Marcigny *et al.* 2007), Jeunesse and colleagues in the Paris Basin (e.g. Jeunesse 1998), Vanmontfort and colleagues in Belgium (e.g. Vanmontfort 2001, 2004, 2006; Crombé and Sergant 2008), and Louwe Kooijmans and colleagues in the Netherlands (e.g. Louwe Kooijmans 2007). And a remarkable international research project led by Pierre Pétrequin, *Programme JADE*, is shedding significant new light on one of the intriguing elements of Early Neolithic material culture on both sides of the Channel, the 'socially valorised' axeheads of rare Alpine rock types, principally jadeitite and eclogite (Pétrequin *et al.* 2008).

All this, together with the plethora of recently discovered 'Carinated Bowl Neolithic' houses in Ireland (Grogan 2004; Smyth 2007) and other work such as Torben Ballin's and Graeme Warren's careful re-evaluation of aspects of Early Neolithic lithic traditions in Scotland (Ballin 2009; Warren 2004) and Rick Schulting's radiocarbon dating and isotope analysis of human bones from the passage tomb at Broadsands, Devon (Sheridan *et al.* 2008), helps in the creation of the 'big picture' of British and Irish Neolithization, as outlined below.

Neolithization in Britain and Ireland: the 'big picture'

As Stuart Piggott argued over half a century ago in his *Neolithic Cultures of the British Isles* (1954), the Mesolithic–Neolithic transition in Britain and Ireland was not a unitary process of change: we are not dealing with one 'Neolithic', but with several strands, deriving from different areas along the adjacent part of the Continent, and arriving at different times and for different reasons (cf. Kinnes 2004). The principal strands would appear to be as follows (in chronological order): first, there was a possible 'false start', from west or northwest France to southwest Ireland, possibly during the third quarter of the 5th millennium: this is represented by the evidence from Ferriter's Cove, which is reviewed critically below. Secondly, there was a Breton, 'Atlantic' strand, arriving some time between 4300/4200 and 4000 BC at various points on the western coast of Britain and on the coast around the northern half of Ireland. Third, the 'Carinated Bowl Neolithic', or '*trans-Manche* east' strand, from the northernmost part of France, arrived over much of Britain and most of Ireland within the first two centuries of the 4th millennium. And finally, there was a '*trans-Manche* west' strand (or strands), probably from Normandy, to southwest (and parts of southern) England, arriving some time between 4000 and 3800 BC. Contacts between Normandy and the southern/southwestern English coast continued, or were renewed, in subsequent centuries.

Since some of this evidence has already been discussed extensively elsewhere by the current author (e.g. Sheridan 2003a; 2003b; 2004; 2005; 2007; Sheridan *et al.* 2008; Pailler and Sheridan 2009), it is not proposed to describe each strand in detail here. Rather, the main characteristics will be sketched, and the main outstanding research questions highlighted.

1. A possible false start in southwest Ireland during the third quarter of the 5th millennium: the Ferriter's Cove evidence

The discovery of seven bones of domesticated cattle

within a Late Mesolithic hunter-gatherer-fisher settlement at Ferriter's Cove, Co. Kerry, raised the possibility of an early episode of contact with the Continent, since both these species would have had to be imported (Woodman *et al.* 1999, 90, 144–51; Woodman and McCarthy 2003). Initial dating of one of the cattle bones to 5510±70 BP (OxA-3869, 4500–4180 cal BC; Woodman *et al.* 1999, 219) suggested that this had occurred during the third quarter of the 5th millennium; another cattle bone, from a different context, subsequently produced a yet earlier date of 5825±50 BP (OxA-8775, 4790–4550 cal BC: Woodman and McCarthy 2003). It was argued (Woodman and McCarthy 2003; Tresset 2003) that the most plausible area of origin for these animals was west France (Fig. 9.1.1), and a possible explanatory scenario was proposed, featuring a very small-scale immigration of pioneering farmers from France; the hunting of their stock by the indigenous inhabitants of Ferriter's Cove; and the consequent failure of a Neolithic way of life to take root at that point, owing to the absence of a critical mass of people and of domesticates (Sheridan 2003a; Tresset 2003). Since then, the reliability of the earlier cattle date has been thrown into doubt, as the bone had been charred (P. C. Woodman pers. comm.); and given the chronological range afforded by the later date, the candidate area of origin for the cattle and sheep can be expanded to include parts of Brittany (S. Cassen pers. comm.; Tresset and Vigne 2007).

Currently, the Ferriter's Cove evidence remains the earliest directly dated evidence for the presence of domesticated animals on this side of the Channel (not counting dogs). Peter Woodman's aforementioned dating project on other ostensibly early remains of cattle and sheep in Ireland confirmed that none pre-dates the appearance of the Carinated Bowl Neolithic (see below), around or shortly after 4000 BC. Indeed, the most recent discovery of domesticated cattle remains on a hunting/fishing/gathering site (at Lough Kinale II, County Longford) provides another example of their appearance after that of the Carinated Bowl Neolithic (Fredengren 2009; I. Stuijts pers. comm.) This may well relate to the localized acculturation of a 'Mesolithic' group through contact with a 'Carinated Bowl' farming group.

Given that the evidence for a precocious and apparently unsuccessful episode of Neolithization rests on a single acceptable radiocarbon date and on the contextual evidence from just one site, this is clearly not a strong basis from which to argue. Nevertheless, the Ferriter's Cove evidence demands an explanation, and so far the scenario outlined above has not been challenged convincingly. That we are dealing with people coming *from* western/northwest France, and not with Irish Mesolithic groups travelling *to* France, is suggested by: a) the insular nature of Late Mesolithic communities in Ireland (*pace* Kador, this volume); b) the absence of evidence for any Irish Mesolithic-like material in France (although it has to be admitted that nobody has seriously searched for it); and c) a prior history, in the Morbihan area of Brittany, of long-distance seaborne contacts (with Spain) around the mid-5th millennium, which was responsible for the importation of Spanish fibrolite and variscite, and for the copying of Breton-style Alpine axeheads in northwest Spain (Pétrequin *et al.* 2008 and P. Pétrequin pers. comm.). Whether it is feasible to undertake targeted searches around Ferriter's Cove for other evidence relating to our putative early incomers is a moot point; one suspects that any such evidence is most likely to turn up by chance.

2. The Atlantic, 'Breton' Neolithic, arriving c. 4300/4200–4000 BC

This strand of Neolithization is characterized by the appearance of small megalithic closed chambers and simple passage tombs in coastal areas along the Atlantic façade in southwest and northwest Wales, western Scotland and around the northern half of Ireland (for further details, see Sheridan 2003a; 2003b; 2005) (Figs 9.1.2, 9.2). They constitute the earliest megalithic monuments in Britain and Ireland, and the pottery found in at least one of them (at Achnacreebeag, Argyll and Bute) also constitutes the earliest pottery in use in Britain and Ireland. Their structural simplicity has led some of these tombs to be mis-identified as portal tombs (e.g. Carreg Samson, Pembrokeshire: C. Richards 2004), but their affinity with the closed chambers and simple passage tombs of the Morbihan region of southeast Brittany is demonstrable (Pailler and Sheridan 2009). At Achnacreebeag, Graham Ritchie excavated a two-phase monument featuring a closed chamber that had a simple passage tomb added to it (Ritchie 1970). Although the chambers' contents had been almost entirely cleared out, in the passage tomb he found fragments of three distinctive pots, which the current author has identified (with verification from Breton pottery specialists) as being of Breton Middle Neolithic types, namely late Castellic, Pinacle, and a variety that has been found associated with late Castellic pottery in the Morbihan (Fig. 9.3). Despite Thomas' argument that these types of pottery are very diverse and their presence cannot be taken to indicate incomers from the Morbihan (2004), in fact they all form part of the repertoire in use in this part of Brittany during the period 4300/4200–4000 BC (as confirmed, for example, by radiocarbon dates from the Table des Marchands: S. Cassen pers. comm. and see Cassen 2009). Since there is no evidence for external contacts by west-coast Scottish Mesolithic communities, it is hard to explain how Breton-style pottery, in a Breton-style monument, could have appeared in the west of Scotland by means other than a northward movement of people from the Morbihan.

Other pottery has been found in these tombs: a deep, undecorated bowl found at Carreg Samson (Lynch 1975) could form part of the Breton Middle Neolithic repertoire, and stands out as being different from Welsh traditional Carinated Bowl pottery; but as it is of such simple form, it has been hard to persuade sceptics of its likely Breton origins. In Ireland, at the passage tomb cemetery at

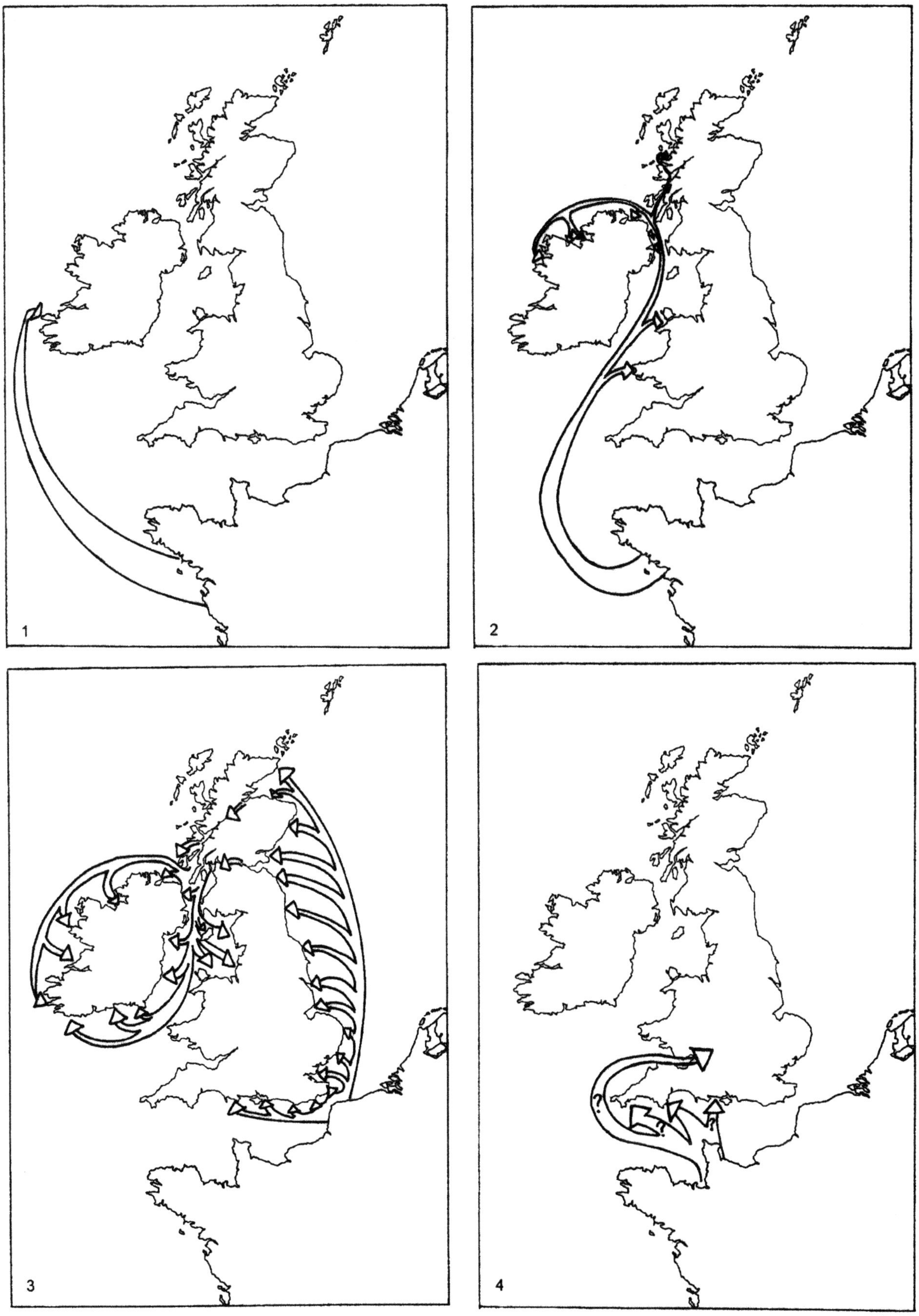

Fig. 9.1. The proposed strands of Neolithization. 1. Northwest France to southwest Ireland (Ferriter's Cove); 2. The Atlantic, 'Breton' Neolithic; 3. The 'Carinated Bowl Neolithic' (overall extent and routes taken from northern France remain open to discussion); 4. The 'Trans-Manche west' strand.

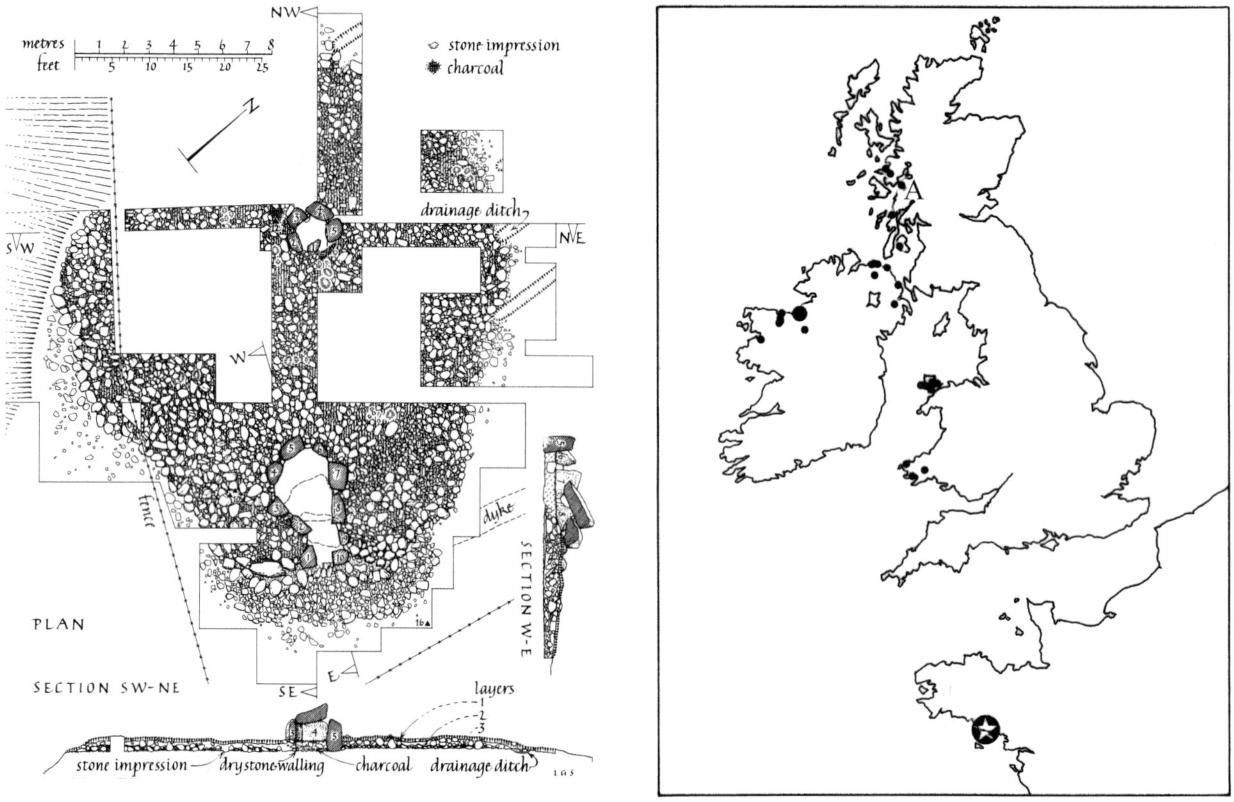

Fig. 9.2. 1. (left). Plan and sections of the closed chamber and simple passage tomb at Achnacreebeag, Argyll and Bute (from Ritchie 1970; Crown copyright). 2. (right) Distribution of Breton-style closed chambers and simple passage tombs in Britain and Ireland. A = Achnacreebeag; star indicates the likely source area in the Morbihan area of Brittany.

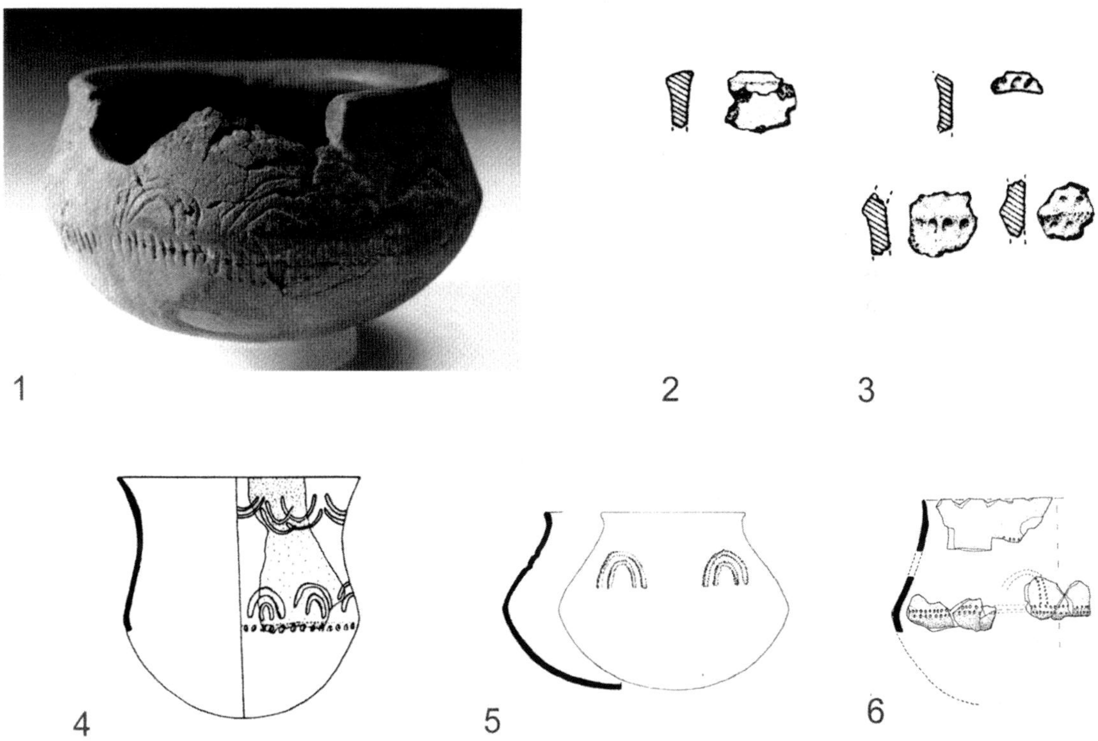

Fig. 9.3. The Breton-style pots from Achnacreebeag (1–3) and some comparanda (4–6): 1. Late Castellic-style bowl with fringed rainbow decoration (photo: NMS); 2–3. Undecorated and decorated bowls (from Ritchie 1970); 4. Early Castellic bowl from Le Castellic, Brittany (from Bailloud 1975); 5. Late Castellic bowl from Vierville, Normandy (from Verron 2000); 6. Late Castellic bowl from Er Grah, Morbihan (from Cassen 2009).

Carrowmore, Co. Sligo, the occasional sherds of coarse decorated 'Carrowkeel' pottery derive, like most or all of the other artefacts from the tombs, from the monuments' secondary use, centuries after their construction (S. Bergh pers. comm.).

Support for the idea that these Breton-style closed chambers and simple passage tombs were constructed broadly within the 4300/4200–4000 BC date bracket is ostensibly provided by a set of dates obtained by Göran Burenhult in the 1990s, which, he argued, relate to the construction and early use of several of the tombs in the Carrowmore cemetery (Burenhult 2001; Sheridan 2003b; note, however, that Stefan Bergh has pointed out (pers. comm.) that the contextual information is not as robust as one might like). A post-4000 BC date for the construction of one simple closed chamber, at Ballintoy, County Antrim, is suggested by the fact that it overlies a layer containing Carinated Bowl pottery (Sheridan 2003b, 11).

The putative immigration evidenced by the appearance of these tombs seems to have had differing outcomes in different areas along the Atlantic façade. In Scotland and Ireland, these simple monuments stand at the beginning of long and complicated sequences of passage tomb development (Sheridan 1986; 2006); and in Scotland, the pottery stands at the beginning of the similarly long and complex sequence of west-coast ceramic developments (Sheridan 2004). This suggests that their builders flourished in these regions. In Wales, by contrast, the passage tomb tradition appears to have died out until its re-introduction, from eastern Ireland, towards the end of the 4th millennium (Lynch 2000).

The possible causes of a small-scale emigration from the Morbihan between 4300/4200 and 4000 BC have been discussed elsewhere. Following Boujot and Cassen (1992; 1993; Cassen 2000), it appears that the shift, at this time, from building massive *tumulus carnacéens* and other monuments to the construction of passage tombs – with the attendant incorporation of parts of some massive decorated menhirs within passage tombs – was part of a broader set of ideological, social and economic changes. These *bouleversements* are perfectly illustrated by the developments at Locmariaquer, where a low earthen barrow (Er Grah) and a stupendously large menhir (le Grand Menhir Brisé) were succeeded by a passage tomb (la Table des Marchands) that incorporates part of a huge decorated menhir, second only in height to the toppled Grand Menhir, as its capstone (Bouillon and Cassen 2008; Cassen 2009; le Roux *et al.* 1996). That some people should choose to leave the Morbihan during this period of change, for a new life in an unknown destination, is consistent with such a reading of the evidence; and, as seen above, there had been a local precedent during previous centuries for the undertaking of long sea journeys. Indeed, if the domestic animals found at Ferriter's Cove were from stock brought from Brittany, this offers a further possible precedent for the undertaking of speculative westward and northward journeys by sea. (Alternatively, given the potential overlap between the Ferriter's Cove date and the estimated date of the 'Atlantic' emigration, it is not impossible that the move to southwest Ireland had actually been part of this strand of Neolithization). Other evidence indicating some kind of movement from Brittany during the late 5th millennium comes from the simple passage tomb at Vierville in Normandy, where the ceramic assemblage includes at least one pot that is believed to have been imported from the Morbihan: a Late Castellic bowl, which incidentally shows marked similarities to its counterpart from Achnacreebeag (Marcigny *et al.* 2007, 143; S. Cassen pers. comm.).

The evidence for this putative Atlantic, 'Breton' strand of Neolithization is – to this author's eyes – persuasive, and yet some commentators have been reluctant to accept it (e.g. A. Whittle pers. comm.) – although none has produced a more plausible explanatory hypothesis. At present the model relies solely on funerary evidence: no associated settlement has been found, nor is there evidence relating to subsistence practices. No bones have survived in the British tombs, and the plentiful deposits of cremated bones found in the Carrowmore cemetery may well relate to secondary tomb use; direct dating of these bones is recommended, to complement Stefan Bergh's current programme of dating bone and antler pins from the tombs, whose results indicate such a scenario (S. Bergh pers. comm.). It may well be that a targeted programme of palaeoenvironmental study and fieldwork in the areas around these monuments might produce evidence relating to their builders (although Burenhult's previous attempt, around Carrowmore, sadly did not produce the kind of information he sought: Burenhult 1984). Perhaps the excavation of a hitherto-uninvestigated Breton-style monument would provide further evidence to support or modify the working hypothesis.

3. The 'Carinated Bowl Neolithic'

This is the most extensively dispersed and best-documented strand of Neolithization (Fig. 9.1.3), even though much remains to be discovered and clarified, particularly as regards its dating, its limits of distribution and its exact geographical area of origin. The 'Carinated Bowl (henceforth CB) Neolithic' is attested over much of Britain and Ireland, and its key features have recently been outlined by this author (Sheridan 2007). These comprise the following 'package' of novelties, which would all have appeared alien and exotic to indigenous Mesolithic communities:

1. A ceramic tradition dominated by the use of gently carinated and S-shaped bowls and jars (of varying sizes and shapes), but also featuring uncarinated, roughly hemispherical bowls and cups and collared jars (Fig. 9.4). Surfaces are generally carefully smoothed and sometimes buffed to a low sheen or burnished to a high sheen; the only decoration, which is rare, is fingertip fluting (usually on the neck and sometimes on the rim) or – more rarely – ripple burnishing. This author has proposed the use of the term 'traditional CB' pottery to distinguish this earliest,

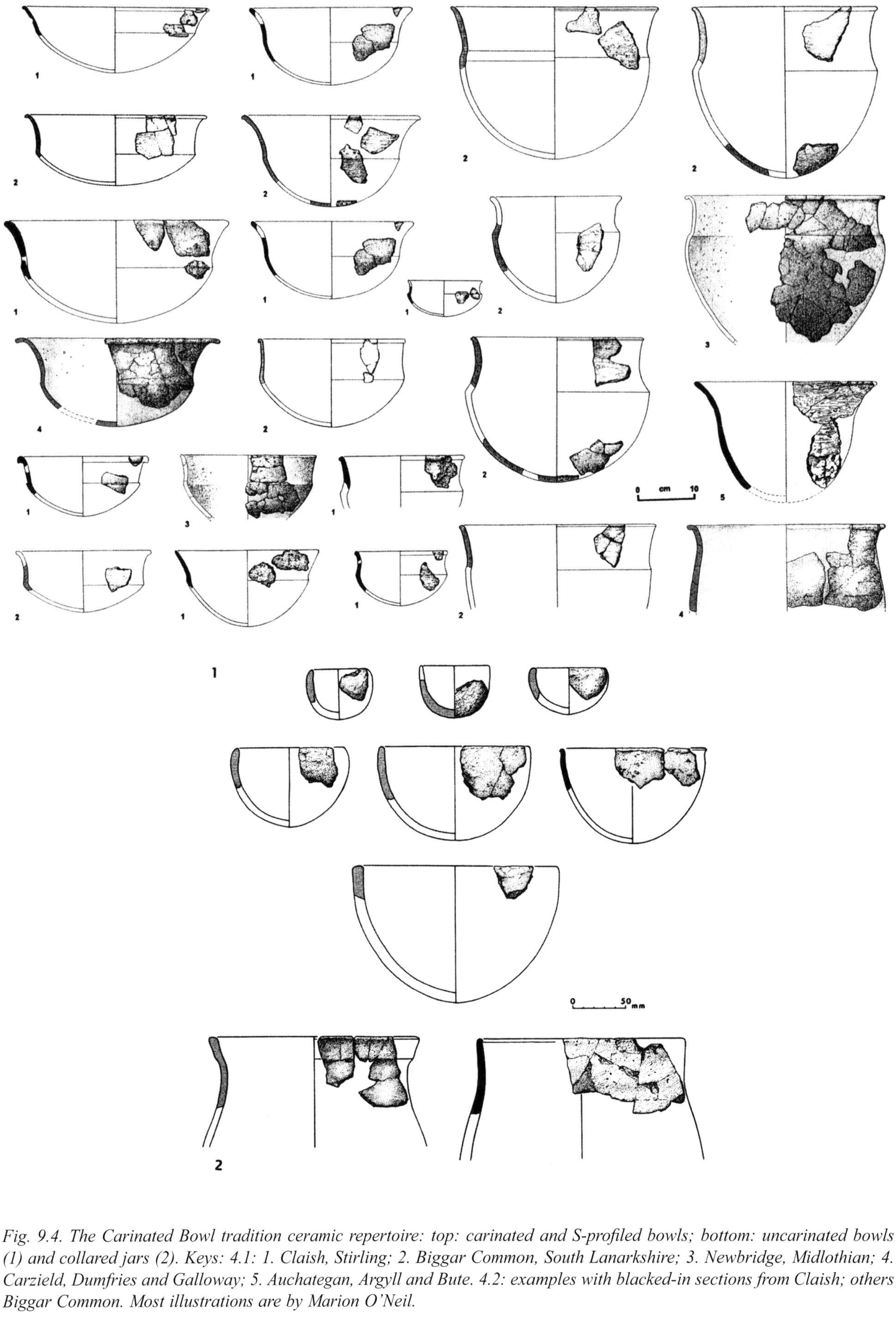

Fig. 9.4. The Carinated Bowl tradition ceramic repertoire: top: carinated and S-profiled bowls; bottom: uncarinated bowls (1) and collared jars (2). Keys: 4.1: 1. Claish, Stirling; 2. Biggar Common, South Lanarkshire; 3. Newbridge, Midlothian; 4. Carzield, Dumfries and Galloway; 5. Auchategan, Argyll and Bute. 4.2: examples with blacked-in sections from Claish; others Biggar Common. Most illustrations are by Marion O'Neil.

Fig. 9.5. Aerial photograph of the Early Neolithic timber 'hall' (c. 20 × 9m) at Warren Field, Crathes, Aberdeenshire, by Moira Greig, courtesy of Aberdeenshire Archaeology.

and markedly homogeneous, version of CB pottery from the various regionally divergent variants ('modified CB') that emerged through 'style drift' – in some cases, at a very early stage (as with the 'North East style' that is found mainly in northeast Scotland).

2. New stone artefact forms (e.g. leaf-shaped arrowheads and plano-convex knives; ground stone axeheads were also a novelty for most of Britain) and a tradition of procuring, working and depositing stone artefacts that markedly differs from Late Mesolithic traditions. (See Warren 2005 for details regarding small lithics in Scotland, and Waddington and Passmore 2004 for Northumberland. The question of the extent to which the working techniques differed in southern England needs to be investigated further.) The movement of various lithic materials – in unworked, partly worked and fully worked states – is well-documented (e.g. Ballin 2009 on pitchstone) and attests to the rapid creation of networks of contacts among the various communities who used CB pottery. Procurement strategies included the opening of flint mines in southern England (Barber *et al.* 1999, 69; Topping 2004) and of quarries – often on mountains – in highland zones (see, for example, Edmonds 2004). The choice of remote or otherwise liminal sources of stone for manufacturing axeheads is likely to have been driven by the ideology surrounding the precious and sacred axeheads made of Alpine rock which the CB users brought with them from the continent as treasured ancestral artefacts (Pétrequin *et al.* 2008).

3. The construction of various non-funerary timber structures (Fig. 9.5). These range from the long rectangular structures up to 24 m long and 11 m wide with convex or straight ends found in Scotland (i.e. the so-called 'halls': Brophy 2007; Murray *et al.* 2009) and southern England (Hey and Barclay 2007) through smaller rectangular, square and oval structures (see Grogan 2004 and Smyth 2007 for reviews of Irish houses, and Kenney and Davidson 2006 for a recently discovered house at Llandygai, Wales) to more slightly constructed buildings, perhaps for less permanent use (e.g. Bolam Lake, Northumberland: Waddington and Davies 2002). There is a marked tendency to construct all but the flimsiest structures on what would have been prime agricultural land. The largest structures relate to the earliest appearance of the CB Neolithic, and much discussion has revolved around their use, with the Scottish 'halls' being regarded by some as being non-residential in nature (see Brophy 2007). In this author's opinion, however, a more plausible explanation is that they were the communal residences of the first generation or two of settlers, built to offer security while their communities became established. (Part of this security included the keeping of large amounts of seed corn, as discovered in the burnt-down remains of the large structure at Balbridie, Aberdeenshire:

Fairweather and Ralston 1993.) These structures would also have constituted powerful statements of identity and presence in the landscape, as would these communities' funerary monuments. Thereafter, individual residential groups were able to 'bud off' to build smaller houses for themselves – a process visible in Aberdeenshire (Murray *et al.* 2009; Sheridan 2007). While these substantial post- or post-and-plank-built structures imply a significant degree of sedentism (which at Crathes Warren Field is borne out by palaeoenvironmental evidence for cereal cultivation in the immediate vicinity: Lancaster 2009; Tipping *et al.* 2009), the flimsier structures suggest that transhumance may have taken place.

4. The keeping of domesticated cattle, sheep (and possibly goats) and pigs; the cultivation of wheat (including the environmentally demanding bread wheat), barley, flax, field bean and probably brassicas, all imported; and the exploitation of a range of wild plant and animal resources (see, e.g., Fairweather and Ralston 1993 and Murray *et al.* 2009). The discovery of an Early Neolithic yew flatbow – which must have been imported from either northwest England or Ireland – in the Moffat Hills at Rotten Bottom, Dumfries and Galloway, strongly points towards the hunting of deer (Sheridan 2007). As with the settlement evidence, the subsistence strategy implies a significant degree of sedentism (necessary for the tilling of crops: see Monk 2000 and Rowley-Conwy 2004), while including some activities that required a more extensive use of the landscape: transhumance, hunting and gathering.

5. At least one CB community seems to have constructed a causewayed enclosure at, or shortly after, the time of arrival (at Magheraboy, Co. Sligo: Danaher 2007). As with the construction of the large communal houses, this implies the presence of a critical mass of people in the area. (The apparent absence of large enclosures from most other areas where CB pottery was used need not indicate failure to reach a critical mass; there would have been other considerations at play in a group's decision over whether or not to build an enclosure.)

6. Funerary practices featuring the use of various non-megalithic monuments, namely: a) rectangular 'linear zone' mortuary structures featuring split oak trunks (some associated with timber façades and long rectangular or trapezoidal post enclosures), many of which were burnt before being covered with long or round mounds (e.g. Eweford West and Pencraig Hill, East Lothian: MacGregor and McLellan 2008); b) cremation sites covered with round or ring cairns (as at Boghead, Moray: Burl 1984); c) a plank-lined rectangular grave (from Yabsley Street, Blackwall, London: Coles *et al.* 2008); d) rectangular mortuary enclosures. Burial in caves was also practised by CB users, as demonstrated at Kilgreany Cave, Co. Waterford (Dowd 2008). Traditional CB pottery has also been associated with simple megalithic structures in Scotland and northern England that appear to represent the translation into stone of timber monuments (as at Cairnholy 1, Dumfries and Galloway, for example). In southeast England, a megalithic tomb at Coldrum, Kent – part of a small, isolated group of megalithic tombs around the Medway estuary – has produced human remains radiocarbon dated to c. 4000 BC (Bayliss *et al.* 2008; Holgate 1981). While no demonstrably CB pottery is known from this tomb (Holgate 1981), the Yabsley Street grave demonstrates that CB pottery was in use in the Thames Valley at a comparable date.

While it is clear that the CB Neolithic is found widely – albeit irregularly – over much of Britain and Ireland, and while its distribution in Scotland and northern England is well documented (Manby *et al.* 2003; Passmore and Waddington 2009; Sheridan 2007), the absence of an overall distribution map hampers our understanding. It appears that parts of eastern and northern Britain (such as Aberdeenshire, Yorkshire and the Milfield Basin in Northumberland) have particular concentrations, while in Wales it has a relatively restricted distribution, mostly in the northwest (Lynch 2000; Peterson 2003); in England, it seems to 'thin out' the further southwest one goes. The southwesterly limit of its distribution is a matter for debate, for while the bowls found beside the wooden trackway, the Sweet Track, in the Somerset Levels (Coles *et al.* 1984, fig. 56) are of forms and fabrics well attested among the CB repertoire, the presence of a black coating on the surface is not a CB feature, and invites comparisons with pottery of similar date (to the 3807/3806 BC of the Sweet Track) further to the southwest (at Penhale Round, Cornwall: J. Nowakowski pers. comm.).

The dating of the CB Neolithic has been, and remains, a matter for debate among those concerned with establishing a firm chronology for the British and Irish Neolithic in general. The number of reliable radiocarbon dates has recently risen steeply in Scotland, northeast England and Ireland (with the Scottish dates being presented in Sheridan 2007 and some of the dates for northeast England published in Passmore and Waddington 2009). These and other dates for Britain and Ireland have recently been Bayesian-modelled by Bayliss *et al.* (and will be published in Whittle *et al.* forthcoming). It appears that there is a very small cluster of dates around 4000 BC from sites in southeast England – note that Bayliss *et al.* are now expressing reservations (A. Bayliss and A. Whittle pers. comm.) about whether the Magheraboy causewayed enclosure in northwest Ireland was also constructed at this time – and then a large number of 39th century BC dates from CB-associated sites around Britain and Ireland. While Bayliss *et al.* interpret this in terms of an initial colonization of southeast England, followed by a spread northwards and westwards from there, it is equally if not more likely that the dates reflect an initial, small-scale emigration to southeast England (with the Magheraboy evidence remaining anomalous) followed by a larger-scale and rapid 'diaspora' from northern France to large parts of Britain and Ireland. Alternatively, it could be that the apparent gap of up to 200 years between the 'first appearance' dates in different areas is partly a function of the Bayesian modelling process, since it requires one to define an end date to the phenomenon being modelled. As noted above, 'traditional CB' pottery developed into its various 'modified' variants at different times in different areas, so defining a meaningful and consistent end point is problematical. It may be that the arbitrary definition of an end point serves to skew the model, and that we are not really dealing with a two-phase CB colonization. No doubt the interpretation of the radiocarbon dates will continue to be debated for some time yet.

The question of the precise area of origin for the CB Neolithic has long been a topic of debate; while Piggott (1954) argued for the far north of France, subsequent commentators (e.g. Kinnes 1988; 2004) have pointed out the absence of exact matches for the pottery, for the use of leaf-shaped arrowheads and for the funerary and non-funerary structures. That this may, however, be a product of an incomplete archaeological record for early 4th-millennium material culture and structures on the far side of the Channel is suggested by the significant advances recently made in Belgium by the work of Bart Vanmontfort (e.g. 2001): here, an entire new regional 'facies' of Middle Neolithic activity has been recognized at and around Spiere, including a possible non-megalithic long barrow at Ottenburg (Vanmontfort 2004, 243). If one entirely 'new' regional group has recently emerged, there is every possibility that others will too. As argued in detail elsewhere (Sheridan 2007, 468–9), CB pottery has many elements in common with northern Chassey and early Michelsberg pottery, and leaf-shaped arrowheads – albeit not identical to CB examples – are a regular feature of Michelsburg assemblages. Furthermore, the past emphasis on highlighting differences between the CB Neolithic and that of the near Continent has masked more significant points of similarity, such as the shared practices of flint mining; of causewayed enclosure use; and, importantly, of establishing and using networks of contacts across which materials, ideas and doubtless people flowed. It seems to this author that the most likely area of origin for the CB Neolithic is the Nord-Pas de Calais region of northern France (possibly extending into northern Picardie), and that it represents one of a number of small regional groups (analogous to Vanmontfort's 'Spiere' group) that emerged at a time of significant social and economic change at the end of the 5th millennium and beginning of the 4th (e.g. Crombé and Sergant 2008; Jeunesse 1998; and Louwe Kooijmans 2007). This view is in accord with Pierre Pétrequin's judgement about the likely proximate source area for Alpine axeheads (Pétrequin *et al.* 2008). Fresh fieldwork, new discoveries and the careful review of existing archive collections should clarify the matter.

4. *The* 'trans-Manche *west' strand*

This strand (Fig. 9.1.4), found in southwest and parts of southern England, involves some elements – closed chambers and simple passage tombs – that are superficially similar to the megalithic monuments that formed part of the Atlantic, 'Breton' Neolithic although, as will be argued below, they relate to a different, and slightly later, set of movements. These movements were arguably between Normandy (and perhaps the Channel Islands and parts of northern Brittany) and southern/southwest England during the first quarter of the 4th millennium, and thus involved shorter distances than those involved in the Atlantic, 'Breton' Neolithic strand. The evidence for this '*trans-Manche* west' strand(s) of Neolithization comprises three main elements: a) the simple, largely drystone-built passage tomb at Broadsands in Devon; b) the drystone-built *rotundae* (closed chambers and simple passage tombs under low round cairns) in the Severn–Cotswold area; and c) several pottery assemblages from southwest and southern England that may, along with the Broadsands pottery, constitute the precursor to the 'Hembury', or 'South-West', style of pottery (Whittle 1977, 77–82). As for possible reasons for this hypothetical immigration(s) from northwest France, the record on the other side of the Channel is not entirely clear although it has been noted that, in Normandy, the second half of the 5th millennium and the early 4th millennium saw a significant expansion in agricultural settlement (including the construction of enclosures) followed, from *c.* 3800 BC, by a period of change when '*les traces archéologiques sont de plus en plus*

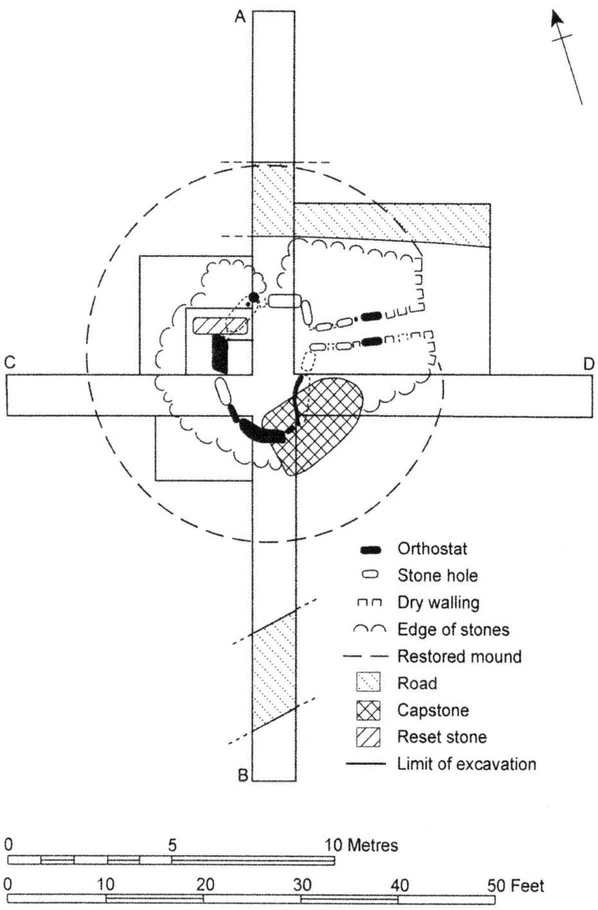

Fig. 9.6. Plan of Broadsands passage tomb, from Radford 1958, redrawn by Floss Wilkins.

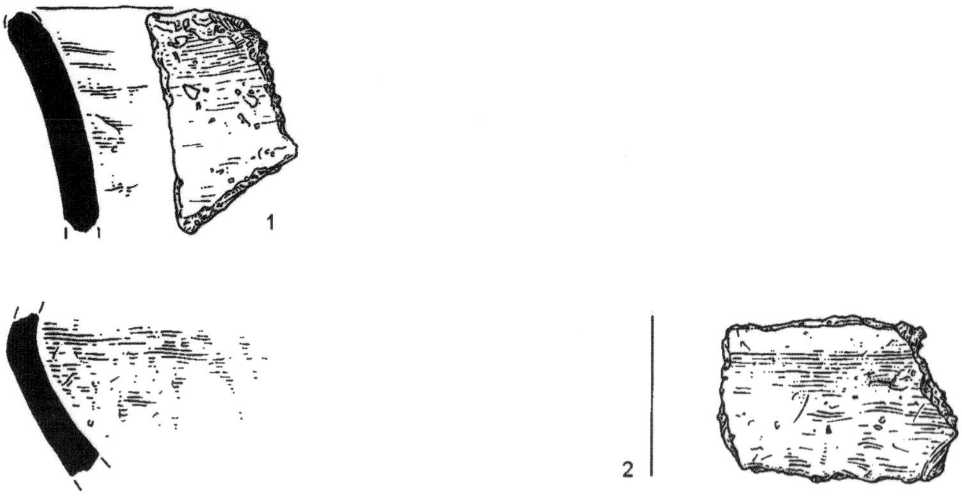

Fig. 9.7. Sherds from two thin fine Early Neolithic carinated bowls from Broadsands passage tomb (scale 2:3). Illustration by Jane Read, courtesy of Henrietta Quinnell.

rares voire inexistantes pour une bonne partie de la Basse-Normandie ... Regroupement des populations, période de trouble, épidémie, problème de reconnaissance des sites ...' (Marcigny *et al.* 2007, 93). Emigration to England may well have formed part of these changes.

The Broadsands passage tomb

This simple passage tomb (Fig. 9.6) on the coast near Paignton, excavated by Courtney Raleigh Radford in 1958 (Radford 1958), has recently been the subject of fresh research by Rick Schulting, this author, Henrietta Quinnell and Roger Taylor (Sheridan *et al.* 2008). It is notable for being unique within the southwestern peninsula, and for its strong affinities, in design and construction, with small drystone-built passage tombs found in Normandy, north Brittany and the Channel Islands (e.g. at Vierville, Normandy; Carn, Brittany; and La Sergenté, Jersey: Verron 1986; 2000; Giot *et al.* 1996; Patton 1993). Broadsands differs from these only in the fact that it had been roofed using capstones, rather than drystone corbelling. Radiocarbon dates recently obtained by Schulting for the human remains indicate that the primary interments were probably deposited between 3845 and 3726 cal BC (at 68 per cent probability, using Bayesian modelling: see Sheridan *et al.* 2008 for details). These provide a *terminus ante quem* for the construction of the tomb. Isotopic analysis indicated that the individuals, like those from similarly dated cave deposits in Devon, had had a terrestrial diet. Sherds from two thin-walled, fine-textured carinated bowls from the tomb (Fig. 9.7) invite comparison not only with other fineware carinated bowls from southwest England (see below), but also with Middle Neolithic II pottery from Normandy, the Channel Islands and northern Brittany (most notably at the simple passage tomb of La Sergenté, Jersey: Patton 1993, fig. iv.3). In terms of northwest French ceramic traditions, we are dealing with pottery that falls within the umbrella term '*Néolithique moyen de l'Ouest*' (Marcigny *et al.* 2007, 92; see also Cassen and François 2006).

The fact that the Broadsands tomb is geographically isolated within the southwest peninsula invites the question of whether we are dealing with single episode of contact – a small group from across the Channel, who lived in Devon and constructed a monument to their dead in the traditional manner – or part of a more extensive (but still probably small-scale) phenomenon of immigration. The evidence reviewed below indicates that the latter may be the case.

'*Rotundae*' in the Severn–Cotswold region

These small drystone-built monuments, featuring closed chambers and simple passage tombs within low round cairns, have been the subject of much debate as regards their dating and significance (Darvill 2004; Smith and Brickley 2006; Pailler and Sheridan 2009; Whittle 2007, 382–3). To cut a long story short, they pre-date Cotswold–Severn megalithic tombs and, in this author's opinion, they should be considered as candidates for further evidence of early 4th-millennium connections with Normandy/the Channel Islands/north Brittany, analogous to the Broadsands tomb. Whittle's argument that they represent 'some kind of monumentalization of those middens which underlie some lateral chambered Cotswold monuments' (2007, 383) is unconvincing. Unfortunately, only a very little pottery has been found in association, and the sherds are too small to be diagnostic. Furthermore, as is the case with Broadsands, there are no known settlements associated with the builders of these tombs; targeted fieldwork may help to remedy this.

The ceramic precursors of 'Hembury'/'South-Western style' pottery

Rosamund Cleal's review (2004) of the earliest pottery

in southern and southwestern England was invaluable in drawing attention to the existence, in this part of Britain, of early 4th-millennium pottery that fits neither within the Carinated Bowl canon (as defined by the current author, e.g. Sheridan 2007) nor within the 'standard' 'Hembury'/'South-Western style' pottery repertoire (as defined by Whittle 1977, 77–82, and Smith 1981). While one may quibble with some of Cleal's claims – Frances Healy (pers. comm.) has argued, for instance, that the assemblage from Flagstones, Dorset, is actually a standard 'Hembury'/'South-Western style' assemblage and that its date of 4960±80 BP (on bulk oak charcoal, HAR-9161, 3950–3640 cal BC) can only be regarded as a *terminus post quem* – the pottery from Coneybury Anomaly, Wiltshire, does nevertheless seem to be genuinely early, and does have elements in common with Middle Neolithic II pottery assemblages across the Channel (such as the vertically perforated lugs on Pot 7: cf. pottery from Louviers in the Eure valley, Normandy: Giligny 2005, fig. 180).

In addition to the Coneybury Anomaly and Broadsands pottery, there are other assemblages that merit reconsideration (see Sheridan *et al.* 2008), not only as potentially having links with French Middle Neolithic pottery in Normandy, the Channel Islands and northern Brittany, but also as constituting the likely precursor for the 'Hembury'/'South-Western style' ceramic tradition (which we know from finds such as Portscatho, Cornwall, to have been in use by the 37th century BC, if not earlier: Jones and Reed 2006, and cf. Whittle *et al.* forthcoming). These include the assemblage from Penhale Round, Cornwall, associated with a fairly substantial sub-rectangular post-built (and presumably roofed) structure, from which short-lived species material produced dates of 5001±75 BP (WK-9839, 3950–3660 cal BC) and 4951±61 BP (WK-9840, 3940–3640 cal BC: J. Nowakowski pers. comm.).

These assemblages are characterized by the presence of thin-walled fineware carinated bowls, whose resemblance to vessels in the Carinated Bowl tradition has caused terminological and conceptual confusion in the past (e.g. Thomas 2004), and indeed still occasions uncertainty in the attribution of individual assemblages (e.g. from the Sweet Track, as mentioned above: Coles *et al.* 1984, fig. 56). Their similarity to carinated bowls in the CB tradition is due to the fact that they share the same ultimate origin in/influence from the northern variant of the widespread *Chasséen* tradition in the Paris Basin (Cassen and François 2006), whose influence extended westwards into Normandy and Armorica and northeastwards into northern France and beyond. However, this author would argue that their appearance in southwest (and parts of southern) England is not due to the CB strand of immigration, but to the '*trans-Manche* west' strand, which arrived at roughly the same time or perhaps slightly later. One distinctive characteristic of some of this pottery – the presence of a deliberate black coating – has also been noted on 'Hembury'/'South-Western style' pottery (e.g. at Carn Brea: Smith 1981) and this is one of several reasons for regarding it as the precursor to that tradition. (The fact that the Sweet Track bowls have this feature makes it possible that they belong to this family of pottery, rather than to the CB tradition, *contra* Sheridan 2007.) Whether this feature is also present on French Middle Neolithic II pottery will require consultation with French ceramic specialists.

Space precludes further discussion of the 'Hembury'/'South-Western style' ceramic tradition, and of its users; suffice it to note that it is recognizably a British regional tradition, albeit one arguably derived from French Middle Neolithic pottery, and whose design was informed by the renewed (or continuing) links with northwest France during the second quarter of the 4th millennium. Its French origins are revealed in the vessel forms discussed above and also in the overall repertoire of shapes (including deep lugged vessels and, at Carn Brea, a flat *plat à pain*: Smith 1981, fig. 74, P153). The subsequent cross-Channel links are manifested by (*inter alia*) the adoption (and adaptation) of the trumpet lug and by the importation of axeheads made of dolerite from Plussulien (Le Roux 1979).

Conclusions

This attempt to sketch a 'big picture' narrative of the Neolithization of Britain and Ireland – and, importantly, to situate it within the context of cross-Channel developments, in all their diversity – has demonstrated the complexity and potential time depth of the phenomenon, and has highlighted areas where we need to find out much more about the individual processes involved. Common to all of the strands is the question of post-colonization relations between incomers and indigenes: is the seemingly rapid disappearance of a lifestyle based solely on hunting, gathering and fishing a genuine phenomenon, and was it (as seems likely) due to ready acculturation to a novel lifestyle that was perceived to promise greater year-round subsistence security (even though this may not have been the case in practice)? And was the disappearance of the 'Mesolithic lifestyle' so rapid? As is suggested by the recent re-calibration of the Oronsay Late Mesolithic human bone dates in the light of new insights into the marine effect on radiocarbon dates (Milner, this volume), it is possible that there was a period of up to several centuries when small groups of indigenous fisher-hunter-gatherers lived not far from, but possibly in unawareness of, small groups of immigrant farmers in the west of Scotland.

The ways in which the various early farming communities adjusted to their environments is also a topic that deserves closer attention (of the kind shown by Sturt's and Garrow's contributions to this volume). Some indication of the trajectories of regional diversification within Scotland, and around the Irish Sea, have already been proposed by this author (Sheridan 1985; 2004; 2006).

It is clear that advances in our understanding will be made only if we frame future research both in terms of 'big picture' issues – and here the big picture encompasses the near continent, not just the British and Irish archipelago – and of fine-grained regional and local studies. There is much to be done (not least in improving our knowledge of the information that is already available to us), but the work will be rewarding.

Acknowledgements

Bill Finlayson and Graeme Warren are thanked for having invited the author to participate in the seminar, and for their patience and kindness. Thanks are also extended to the following: Alasdair Whittle, for his comments; Serge Cassen, for specific details about 5th-millennium Morbihan; Jacky Nowakowski, for providing information on Penhale Round; Rick Schulting and Henrietta Quinnell, for allowing the use of information about Broadsands, and the new drawing of its Early Neolithic pottery by Jane Read, prior to their publication elsewhere; and Hilary Murray, for a similar kindness with regard to Crathes Warren Field.

Bibliography

Bailloud, G. (1975) Les céramiques 'cannelées' du Néolithique morbihannais. *Bulletin de la Société Préhistorique Française* 72, 343–67.

Ballin, T. B. (2009) *Archaeological Pitchstone in Northern Britain: Characterization and Interpretation of an Important Prehistoric Source*. British Archaeological Reports, British Series 476. Oxford: BAR.

Barber, M., Field, D. and Topping, P. (1999) *The Neolithic Flint Mines of England*. London: English Heritage.

Bayliss, A. and Whittle, A. W. R. (eds) (2007) *Histories of the Dead: Building Chronologies for Five Southern British Long Barrows. Cambridge Archaeological Journal* 17/1 supplement.

Bayliss, A., Whittle, A. W. R. and Healy, F. (2008) Timing, tempo and temporalities in the early Neolithic of southern Britain. Pp. 25–42 in H. Fokkens, B. J. Coles, A. L. Van Gijn, J. P. Kleijne, H. H. Ponjee and C. G. Slappendel (eds), *Between Foraging and Farming. An Extended Broad Spectrum of Papers Presented to Leendert Louwe Kooijmans*. Leiden: Leiden University (*Analecta Praehistorica Leidensia* 40).

Bouillon, C. and Cassen, S. (2008) De l'Airbus A380 au site néolithique de la Table des Marchands (Morbihan): limites et perspectives d'un transfert de technologie. Pp. 157–64 in R. Vergnieux and C. Delvoie (eds), *Virtual Retrospect 2007*. Bordeaux: Ausonius.

Boujot, C. and Cassen, S. (1992) Le développement des premières architectures funéraires monumentales en France Occidentale, *Révue d'Archéologie de l'Ouest, Supplément* 5, 195–211.

Boujot, C. and Cassen, S. (1993) A pattern of evolution for the Neolithic funerary structures of the west of France. *Antiquity* 67, 477–91.

Bradley, R. J. (2007) *The Prehistory of Britain and Ireland*. Cambridge: Cambridge University Press.

Bradley, R. J. (2008) The end of the beginning: changing configurations in the British and Irish Neolithic. Pp. 43–9 in H. Fokkens, B. J. Coles, A. L. Van Gijn, J. P. Kleijne, H. H. Ponjee and C. G. Slappendel (eds), *Between Foraging and Farming. An Extended Broad Spectrum of Papers Presented to Leendert Louwe Kooijmans*. Leiden: Leiden University (*Analecta Praehistorica Leidensia* 40).

Brophy, K. (2007) From big houses to cult houses: Early Neolithic timber halls in Scotland. *Proceedings of the Prehistoric Society* 73, 75–96.

Burenhult, G. (1984) *The Archaeology of Carrowmore: Environmental Archaeology and the Megalithic Tradition at Carrowmore, Co. Sligo, Ireland*. Theses and Papers in North-European Archaeology 14. Stockholm: University of Stockholm.

Burenhult, G. (2001) *The Illustrated Guide to the Megalithic Cemetery of Carrowmore, Co. Sligo*, 2nd edition. Tjörnarp: G. Burenhult.

Burl, H. A. W. (1984) Report on the excavation of a Neolithic mound at Boghead, Speymouth Forest, Fochabers, Moray, 1972 and 1974. *Proceedings of the Society of Antiquaries of Scotland* 114, 35–73.

Cassen, S. (2000) *Éléments d'architecture. Exploration d'une tretre funéraire à Lannec er Gadouer (Erdeven, Morbihan). Constructions et reconstructions dans le Néolithique morbihannais. Propositions pour une lecture symbolique*. Chauvigny: Association des Publications Chauvinoises.

Cassen, S. (ed) (2009) *Autour de la Table. Explorations archéologiques et discours savants sur des architectures néolithiques a Locmariaquer, Morbihan (Table des Marchands at Grands Menhir)*. Nantes: Laboratoire de recherches archéologiques, CNRS and université de Nantes.

Cassen, S. and François, P. (2006) Du Chasséen Armoricain à l'Auzay-Sandun: un apport de l'ACR 2003–2006 sur le site de la Table des Marchand (Locmariaquer, Morbihan). *INTERNÉO* 6 (*Actes de la Journée d'information du 18 novembre 2006, Paris*), 77–98.

Cleal, R. (2004) The dating and diversity of the earliest ceramics of Wessex and south-west England. Pp. 164–92 in R. Cleal and J. Pollard (eds), *Monuments and Material Culture*. Salisbury: Hobnob Press.

Coles, J. M., Orme, B. J. and Rouillard, S. E. (1984) *Somerset Levels Papers Number 10*. Cambridge and Exeter: Somerset Levels Project.

Coles, S., Ford, S. and Taylor, A. (2008) An Early Neolithic grave and occupation, and an Early Bronze Age hearth on the Thames foreshore at Yabsley Street, Blackwall, London. *Proceedings of the Prehistoric Society* 74, 215–33.

Cooney, G. (2001) Bringing contemporary baggage to Neolithic landscapes. Pp. 165–80 in B. Bender and M. Winer (eds), *Contested Landscapes: Landscapes of Movement and Exile*. Oxford: Berg.

Crombé, P. and Sergant, J. (2008) Tracing the Neolithic in the lowlands of Belgium: the evidence from sandy Flanders. Pp. 75–84 in H. Fokkens, B. J. Coles, A. L. Van Gijn, J. P. Kleijne, H. H. Ponjee and C. G. Slappendel (eds), *Between Foraging and Farming. An Extended Broad Spectrum of Papers Presented to Leendert Louwe Kooijmans*. Leiden: Leiden University (*Analecta Praehistorica Leidensia* 40).

Danaher, E. (2007) *Monumental Beginnings: The Archaeology of the N4 Sligo Inner Relief Road*. Dublin: Wordwell.

Darvill, T. (2004) *Long Barrows of the Cotswolds and Surrounding Areas*. Stroud: Tempus.

Dowd, M. (2008) The use of caves for funerary and ritual practices in Neolithic Ireland. *Antiquity* 82, 305–17.

Edmonds, M. (2004) *The Langdales: Landscape and Prehistory in a Lakeland Valley*. Stroud: The History Press.

Fairweather, A. M. and Ralston, I. B. M. (1993) The Neolithic timber hall at Balbridie, Grampian Region, Scotland: the building, the date, the plant macrofossils. *Antiquity* 67, 313–23.

Fredengren, C. (2009) Lake-platforms at Lough Kinale – memory, reach and place: a Discovery Programme Project in the Irish midlands. Pp. 882–6 in S. McCartan, R. J. Schulting, G. Warren and P. C. Woodman (eds), *Mesolithic Horizons: Papers Presented at the 7th International Conference on the Mesolithic in Europe, Belfast 2005*. Oxford: Oxbow Books.

Giligny, F. (2005) *Louviers "La Villette" (Eure): Un site néolithique moyen en zone humide*. Rennes: Association pour la Diffusion des Recherches archéologiques dans l'Ouest de la France.

Giot, P.-R., L'Helgouac'h, J. and Monnier, J. (1996) *Préhistoire de la Bretagne*, 3rd edition. Rennes: Ouest-France Université.

Grogan, E. (2004) The implications of Irish Neolithic houses. Pp. 103–14 in I. A. G. Shepherd and G. J. Barclay (eds), *Scotland in Ancient Europe. The Neolithic and Early Bronze Age of Scotland in their European Context*. Edinburgh: Society of Antiquaries of Scotland.

Hey, G. and Barclay, A. (2007) The Thames Valley in the late fifth and early fourth millennium cal BC: the appearance of domestication and the evidence for change. Pp. 399–422 in A. W. R. Whittle and V. Cummings (eds), *Going Over: The Mesolithic–Neolithic Transition in North-West Europe*. Oxford: British Academy and Oxford University Press.

Holgate, R. (1981) The Medway megaliths and Neolithic Kent. *Archaeologia Cantiana* 97, 221–34.

Jeunesse, C. (1998) Pour une origine occidentale de la culture de Michelsberg? Pp. 29–45 in J. Biel, H. Schlichtherle, M. Strobel and A. Zeeb (eds), *Die Michelsberger Kultur und ihre Randgebiete – Probleme der Entstehung, Chronologie unde des Siedlungswesens*. Stuttgart: Theiss (*Materialhefte zur Archäologie in Baden-Württemberg* 43).

Jones, A. M. and Reed, S. J. (2006) By land, sea and air: an Early Neolithic pit group at Portscatho, Cornwall, and consideration of coastal activity during the Neolithic. *Cornish Archaeology* 45, 1–30.

Kenney, J. and Davidson, A. (2006) Neolithic houses – and more besides – at Llandygai. *Current Archaeology* 203, 592–7.

Kinnes, I. A. (1988) The Cattleship Potemkin: the first Neolithic in Britain. Pp. 2–8 in J. C. Barrett and I. A. Kinnes (eds), *The Archaeology of Context in the Neolithic and Bronze Age: Recent Trends*. Sheffield: Department of Archaeology and Prehistory, University of Sheffield.

Kinnes, I. A. (2004) Trans-manche: l'entente cordiale or vive la différence. Pp. 191–5 in J. Cotton and D. Field (eds), *Towards a New Stone Age: Aspects of the Neolithic in South-East England*. Council for British Archaeology Research Report 137. York: CBA.

Lancaster, S. (2009) Palaeoenvironmental synthesis. Pp. 42–50 in H. K. and J. C. Murray and S. Fraser, *A Tale of the Unknown Unknowns: A Mesolithic Pit Alignment and a Neolithic Timber Hall at Warren Field, Crathes, Aberdeenshire*. Oxford: Oxbow Books.

Le Roux, C.-T. (1979) Stone axes of Brittany and the Marches. Pp. 49–56 in T. H. McK. Clough and W. A. Cummins (eds), *Stone Axe Studies, Volume 1*. Council for British Archaeology Research Report 23. London: CBA.

Le Roux, C.-T., L'Helgouac'h, J., Cassen, S., Gauné, F., Lecerf, Y., Le Potier, C., Leroy, D. and Tinevez, J.-Y. (1996) Reprise des fouilles à Locmariaquer (Morbihan). Campagnes 1986–1989. Pp. 440–3 in J.-P. Mohen (ed.), *La Vie Préhistorique*. Dijon: Faton (*Bulletin de la Société préhistorique française*).

Louwe Kooijmans, L. P. (2007) The gradual transition to farming in the Lower Rhine Basin. Pp. 287–309 in A. W. R. Whittle and V. Cummings (eds), *Going Over: The Mesolithic–Neolithic Transition in North-West Europe*. Oxford: British Academy and Oxford University Press.

Lynch, F. M. (1975) Excavations at Carreg Samson, Mathry, Pembrokeshire. *Archaeologia Cambrensis* 124, 15–35.

Lynch, F. M. (2000) The earlier Neolithic. Pp. 42–78 in F. M. Lynch, S. Aldhouse-Green and J. L. Davies, *Prehistoric Wales*. Stroud: Tempus.

MacGregor, G. and McLellan, K. (2008) A burning desire to build: excavations at Eweford West and Pencraig Hill (3950–3380 BC). Pp. 15–45 in O. Lelong and G. MacGregor, *The Lands of Ancient Lothian: Interpreting the Archaeology of the A1*. Edinburgh: Society of Antiquaries of Scotland.

Manby, T. G., King, A. and Vyner, B. E. (2003) The Neolithic and Bronze Ages: a time of early agriculture. Pp. 35–116 in T. G. Manby, S. Moorhouse and P. Ottaway (eds), *The Archaeology of Yorkshire. An Assessment at the Beginning of the 21st Century*. Yorkshire Archaeological Society Occasional Paper 3. Leeds: Yorkshire Archaeological Society.

Marcigny, C., Ghesquière, E. and Desloges, J. (2007) *La Hache et la Meule: Les Premiers Paysans du Néolithique en Normandie (6000–2000 avant notre ère)*. Le Havre: Muséum d'histoire naturelle du Havre.

Miket, R., Edwards, B. and O'Brien, C. (2009) Thirlings: a Neolithic site in Northumberland. Pp. 1–106 in R. Miket, B. Edwards, C. O'Brien, B. Johnson and C. Waddington, *Neolithic and Early Historic Settlement in North Northumberland*. London: Royal Archaeological Institute.

Milner, N., Craig. O. E., Bailey, G. N. and Andersen, S. (2006) A response to Richards and Schulting. *Antiquity* 80, 456–8.

Monk, M. (2000) Seeds and soils of discontent: an environmental archaeological contribution to the nature of the early Neolithic. Pp. 67–87 in A. Desmond, G. Johnson, M. McCarthy, J. Sheehan and E. Shee Twohig (eds), *New Agendas in Irish Prehistory: Papers in Commemoration of Liz Anderson*. Bray: Wordwell.

Murray, H. K., Murray, J. C. and Fraser, S. (2009) *A Tale of the Unknown Unknowns: A Mesolithic Pit Alignment and a Neolithic Timber Hall at Warren Field, Crathes, Aberdeenshire*. Oxford: Oxbow Books.

Pailler, Y. and Sheridan, J. A. (2009) Everything you wanted always to know about … la néolithisation de Grande-Bretagne et de l'Irlande. *Bulletin de la Société préhistorique française* 106/1, 25–56.

Passmore, D. and Waddington, C. (2009) *Managing Archaeological Landscapes in Northumberland. Till–Tweed Studies Volume 1*. Oxford: Oxbow Books and English Heritage.

Patton, M. (1993) *Statements in Stone: Monuments and Society in Neolithic Brittany*. London: Routledge.

Peterson, R. (2003) *Neolithic Pottery from Wales: Traditions of Construction and Use*. British Archaeological Reports, British Series 344. Oxford: BAR.

Pétrequin, P., Sheridan, J. A., Cassen, S., Errera, M., Gauthier, E., Klassen, L., Le Maux, N. and Pailler, Y. (2008) Neolithic Alpine axeheads from the Continent to Great Britain, the Isle of Man and Ireland. Pp. 261–79 in H. Fokkens, B. J. Coles, A. L. Van Gijn, J. P. Kleijne, H. H. Ponjee and C. G. Slappendel (eds), *Between Foraging and Farming. An Extended Broad Spectrum of Papers Presented to Leendert Louwe Kooijmans*. Leiden: Leiden University (*Analecta Praehistorica Leidensia* 40).

Piggott, S. (1954) *Neolithic Cultures of the British Isles*. Cambridge: Cambridge University Press.

Radford, C. A. R. (1958) The chambered tomb at Broadsands, Paignton. *Proceedings of the Devon Archaeological Exploration Society* 5, 147–67.

Richards, C. (2004) Labouring with monuments: constructing the dolmen at Carreg Samson, south-west Wales. Pp. 72–80 in V. Cummings and C. Fowler (eds), *The Neolithic of the Irish Sea: Materiality and Traditions of Practice*. Oxford: Oxbow Books.

Richards, M. P. (2004) The Early Neolithic in Britain: new insights from biomolecular archaeology. Pp. 83–90 in I. A. G. Shepherd and G. J. Barclay (eds), *Scotland in Ancient Europe: The Neolithic and Early Bronze Age of Scotland in their European Context*. Edinburgh: Society of Antiquaries of Scotland.

Richards, M. P. and Schulting, R. J. (2006) Touch not the fish: the Mesolithic–Neolithic change of diet and its significance. *Antiquity* 80, 444–56.

Richards, M. P., Schulting, R. J. and Hedges, R. E. M. (2003) Sharp shift of diet at onset of Neolithic. *Nature* 425, 366.

Ritchie, J. N. G. (1970) Excavation of the chambered cairn at Achnacreebeag. *Proceedings of the Society of Antiquaries of Scotland* 102 (1969–1970), 31–55.

Rowley-Conwy, P. (2004) How the West was lost: a reconsideration of agricultural origins in Britain, Ireland, and southern Scandinavia. *Current Anthropology* 45 (Supplement), 83–113.

Schulting, R. J. (2000) New AMS dates from the Lambourn long barrow and the question of the earliest Neolithic in southern England: repacking the Neolithic package? *Oxford Journal of Archaeology* 19/1, 25–35.

Schulting, R. J. (2004) An Irish sea change: some implications for the Mesolithic–Neolithic transition. Pp. 9–21 in V. Cummings and C. Fowler (eds), *The Neolithic of the Irish Sea: Materiality and Traditions of Practice*. Oxford: Oxbow Books.

Sheridan, J. A. (1985) The Role of Exchange Studies in 'Social Archaeology', with Special Reference to the Prehistory of Ireland from the Fourth to the Early Second Millennium B.C. Unpublished PhD thesis, University of Cambridge.

Sheridan, J. A. (1986) Megaliths and megalomania: an account, and interpretation, of the development of passage tombs in Ireland. *Journal of Irish Archaeology* 3 (1985/6), 17–30.

Sheridan, J. A. (2003a) French Connections I: spreading the *marmites* thinly. Pp. 3–17 in I. Armit, E. Murphy, E. Nelis and D. D. A. Simpson (eds), *Neolithic Settlement in Ireland and Western Britain*. Oxford: Oxbow Books.

Sheridan, J. A. (2003b) Ireland's earliest 'passage' tombs: a French connection? Pp. 9–26 in G. Burenhult and S. Westergaard (eds), *Stones and Bones: Formal Disposal of the Dead in Atlantic Europe during the Mesolithic–Neolithic Interface 6000–3000 BC*. British Archaeological Reports (International Series 1201). Oxford: BAR.

Sheridan, J. A. (2004) Neolithic connections along and across the Irish Sea. Pp. 9–21 in V. Cummings and C. Fowler (eds), *The Neolithic of the Irish Sea: Materiality and Traditions of Practice*. Oxford: Oxbow Books.

Sheridan, J. A. (2005) Les éléments d'origine bretonne autour de 4000 av. J.-C. en Écosse: témoignages d'alliance, d'influence, de déplacement, ou quoi d'autre? Pp. 25–37 in G. Marchand and A. Tresset (eds), *Unité et Diversité des Processus de Néolithisation sur la Façade Atlantique de l'Europe (6e–4e millénaires avant J.-C.)*. Paris: Société Préhistorique Française (Mémoire 36).

Sheridan, J. A. (2006) Creating (and using, amending and reacting to) appropriate dwellings for the dead in Neolithic Scotland. *Journal of Iberian Archaeology* 8, 103–25.

Sheridan, J. A. (2007) From Picardie to Pickering and Pencraig Hill? New information on the 'Carinated Bowl Neolithic' in northern Britain. Pp. 441–92 in A. W. R. Whittle and V. Cummings (eds), *Going Over: The Mesolithic–Neolithic Transition in North-West Europe*. Oxford: British Academy and Oxford University Press.

Sheridan, J. A. (in press) The earliest pottery in Britain and Ireland and its Continental background. In L. P. Louwe Kooijmans, L. Amkreutz and B. Vanmontfort (eds), *Early Pottery Traditions in the Lower Rhine Area*. Leiden: University of Leiden.

Sheridan, J. A., Schulting, R. J., Quinnell, H. and Taylor, R. (2008) Revisiting a small passage tomb at Broadsands, Devon. *Devon Archaeological Society Proceedings* 66, 1–26.

Smith, I. F. (1981) Neolithic pottery. In R. J. Mercer, Excavations at Carn Brea, Illogan, Cornwall, 1970–73 – a Neolithic fortified complex of the third millennium BC. *Cornish Archaeology* 20, 161–85.

Smith, M. and Brickley, M. (2006) The date and sequence of use of Neolithic funerary monuments: new AMS dating evidence from the Cotswold–Severn region. *Oxford Journal of Archaeology* 25/4, 335–55.

Smyth, J. (2007) Neolithic Settlement in Ireland: New Theories and Approaches. Unpublished PhD thesis, University College Dublin.

Telford, D. (2002) The Mesolithic inheritance: contrasting Neolithic monumentality in eastern and western Scotland. *Proceedings of the Prehistoric Society* 68, 289–315.

Thomas, J. S. (1996) Neolithic houses in mainland Britain and Ireland – a sceptical view. Pp. 1–12 in T. Darvill and J. Thomas (eds), *Neolithic Houses in Northwest Europe and Beyond*. Oxford: Oxbow Books.

Thomas, J. S. (2003) Thoughts on the 'repacked' Neolithic revolution. *Antiquity* 77, 67–74.

Thomas, J. S. (2004) Recent debates on the Mesolithic–Neolithic transition in Britain and Ireland, *Documenta Praehistorica* 31, 117–30.

Thomas, J. S. (2007) Mesolithic–Neolithic transitions in Britain: from essence to inhabitation. Pp. 423–39 in A. W. R. Whittle and V. Cummings (eds), *Going Over: The Mesolithic–Neolithic Transition in North-West Europe*. Oxford: British Academy and Oxford University Press.

Thomas, J. S. (2008) The Mesolithic–Neolithic transition in Britain. Pp. 58–89 in J. Pollard (ed.), *Prehistoric Britain*. Oxford: Blackwell.

Tipping, R., Bunting, M. J., Davies, A. L., Murray, H., Fraser, S. and McCulloch, R. (2009) Modelling land use around an early Neolithic timber 'hall' in north-east Scotland from high spatial resolution pollen analysis. *Journal of Archaeological Science* 36, 140–9.

Topping, P. (2004) The South Downs flint mines: towards an ethnography of prehistoric flint extraction. Pp. 177–90 in J. Cotton and D. Field (eds), *Towards a New Stone Age: Aspects of the Neolithic in South-East England*. Council for British Archaeology Research Report 137. York: CBA.

Tresset, A. (2003) French Connections II: of cows and men. Pp. 18–30 in I. Armit, E. Murphy, E. Nelis and D. D. A. Simpson (eds), *Neolithic Settlement in Ireland and Western Britain*. Oxford: Oxbow Books.

Tresset, A. and Vigne, J.-D. (2007) Substitution of species, techniques and symbols at the Mesolithic–Neolithic transition in Western Europe. Pp. 189–210 in A. W. R. Whittle and V. Cummings (eds), *Going Over: The Mesolithic–Neolithic Transition in North-West Europe*. Oxford: British Academy and Oxford University Press.

Vanmontfort, B. (2001) The group of Spiere as a new stylistic entity in the Middle Neolithic Scheldt Basin. *Notae Praehistoricae* 21, 139–43.

Vanmontfort, B. (2004) Converging Worlds. The Neolithisation of the Scheldt Basin during the Late Fifth and Early Fourth Millennium cal BC. Unpublished PhD thesis, University of Leuven.

Vanmontfort, B. (2006) Can we attribute the Middle Neolithic in the Scheldt and Middle Meuse Basins to the Michelsberg cultre? Pp. 109–16 in P. Duhamel (ed.), *Impacts Interculturels au Néolithique Moyen. Du Terroir au Territoire*. Actes du 25e Colloque Interrégional sur le Néolithique, Dijon 20–21 octobre 2001. Dijon: Société Archéologique de l'Est.

Verron, G. (1986) Les civilisations Néolithiques de la Normandie. Pp. 193–206 in J.-P. Demoule and J. Guilaine (eds), *Le Néolithique de la France. Hommage à Gérard Bailloud*. Paris: Picard.

Verron, G. (2000) *La Préhistoire de la Normandie*. Rennes: Ouest-France Université.

Waddington, C. and Davies, J. (2002) An Early Neolithic settlement and Late Bronze Age burial cairn near Bolam Lake, Northumberland: fieldwalking, excavation and reconstruction. *Archaeologia Aeliana* fifth series 30, 1–47.

Waddington, C. and Passmore, D. (2004) *Ancient Northumberland*. London: English Heritage.

Warren, G. (2004) The start of the Neolithic in Scotland. Pp. 91–102 in I. A. G. Shepherd and G. J. Barclay (eds), *Scotland in Ancient Europe. The Neolithic and Early Bronze Age of Scotland in their European Context*. Edinburgh: Society of Antiquaries of Scotland.

Warren, G. (2005) Autre temps, autres lieux: la fin du Mésolithique et le début du Néolithique dans l'Est de l'Écosse. Pp. 13–23 in G. Marchand and A. Tresset (eds), *Unité et Diversité des Processus de Néolithisation sur la Façade Atlantique de l'Europe (6e–4e millénaires avant J.-C.)*. Paris: Société Préhistorique française (Mémoire 36).

Whittle, A. W. R. (1977) *The Earlier Neolithic of Southern England and its Continental Background*. British Archaeological Reports, International Series 35. Oxford: BAR.

Whittle, A. W. R. (2007) The temporality of transformation: dating the early development of the southern British Neolithic. Pp. 377–98 in A. W. R. Whittle and V. Cummings (eds), *Going Over: The Mesolithic–Neolithic Transition in North-West Europe*. London: British Academy and Oxford University Press.

Whittle, A. W. R., Healy, F. M. and Bayliss, A. (forthcoming) *Gathering Time: Dating the Early Neolithic Enclosures of Southern Britain and Ireland*. Oxford: Oxbow Books for English Heritage.

Woodman, P. C. and McCarthy, M. (2003) Contemplating some awful(ly interesting) vistas: importing cattle and red deer into prehistoric Ireland. Pp. 31–9 in I. Armit, E. Murphy, E. Nelis and D. D. A. Simpson (eds), *Neolithic settlement in Ireland and Western Britain*. Oxford: Oxbow Books.

Woodman, P. C., Anderson, E. and Finlay, N. (1999) *Excavations at Ferriter's Cove, 1983–1995: Last Foragers, First Farmers on the Dingle Peninsula*. Bray: Wordwell.

10. Changing People, Changing Environments: How Hunter-Gatherers Became Communities that Changed the World

Trevor Watkins

It is simplistic to imagine that environmental changes that adversely affected the food resources of the hunter-gatherers of the Epipaleolithic and earliest Neolithic of Southwest Asia directly caused the adaptation to the cultivation of plants and the herding of animals. These sophisticated hunters and harvesters, who were trending towards sedentism, exploited their territories quite differently from their mobile hunter-gatherer forebears. We also need to take into account the implications of the new kind of large-scale, permanent community that they constructed and maintained. In effect, communities were creating richly symbolic cultural environments. Thus the environments within which communities operated changed significantly around the end of the Pleistocene or the beginning of the Holocene period, and cultural responses to changes in the bio-environment became the prime among various kinds of environmental challenge and response.

Introduction

We are accustomed to an orthodox view of hunter-gatherers being required to adjust to reduction in food resources as a result of climatic change. That is now the standard account of final Epipalaeolithic hunter-gatherers in the Levantine corridor adopting cultivation in response to the constrictions of the Younger Dryas phase (e.g. Moore and Hillman 1992; Bar-Yosef and Meadow 1995; Bar-Yosef and Belfer-Cohen 2002; Bar-Yosef 2002). An alternative scenario is based on the notion that the environment can 'push' as well as 'pull'. Thus Epipalaeolithic hunter-gatherers in the same Levantine corridor are said to have been able to adopt new economic strategies of harvesting and storing grasses, cereals and legumes when post-Last Glacial Maximum (LGM) climatic amelioration brought about increased bio-diversity, greater bio-mass or different and richer environments. However valid such a simple ecological model may be for earlier periods of pre-*Homo sapiens* prehistory, for the period in which we are interested – in Southwest Asia, the Epipalaeolithic period and into the Neolithic – there were two critically important changes in the ways that hunter-gatherer groups began to operate that were more important than climatic and environmental change in the final Pleistocene.

In the first place, hunter-gatherer groups of the Epipalaeolithic period increasingly focused on harvesting, storing and processing cereals, grasses and legumes; correspondingly, they tended towards sedentism. The range of such groups reduced in extent, as they used stored food resources and relied on a territory that was exploitable from their increasingly permanent home-base. Whereas a mobile hunter-gatherer group might range seasonally over several thousand square kilometres, a sedentary hunter-gatherer group would rely on the intensive exploitation of a territory of a few hundred square kilometres. This change in the way that they moved in, and exploited the food-resources of, the physical environment was more than a matter of geographical range.

Secondly, hunter-gatherer groups changed their nature. In parallel with the trend towards reduced mobility and longer seasonal settlement, transhumance and sedentism, groups became larger. Whereas typical mobile hunter-gatherer groups known from the ethnographic record amount to a small number of tens of persons (except for brief seasonal gatherings, where many small groups may congregate), through the Epipalaeolithic and Early Neolithic periods in Southwest Asia groups not only became more sedentary but also much larger. Communities in the hundreds began to appear in the late Epipalaeolithic period, and permanently co-resident communities of thousands became increasingly common through the Early Neolithic period. These changes in permanence of residence and community size were not simply numerical. The process of producing and maintaining permanent communities that were ten times larger in the Epipalaeolithic, and one hundred times larger in the Early Neolithic, required cultural and symbolizing skills that drove the development of the relevant cognitive skills and symbolic cultural faculties (or, alternately, the development of the capacity for fully symbolic culture opened the way for hunter-gatherer groups to concentrate in larger and larger numbers). Allied to the emergence of larger, permanently co-resident communities was a trend to more and more extensive and intensive networks of social exchange that bound the new communities into powerful interaction spheres.

In what follows, we begin by considering briefly how the subject of environmental relations has developed over recent years. In particular, I am concerned to examine critically the notion that a change such as the adoption of farming, which saw the domestication of both plants and animals, can be explained simply as a response to environmental pressures. Then we can turn to the changing subsistence and settlement strategies of hunter-gatherers of the final Palaeolithic and earliest Neolithic and their changing relations to the territories that they exploited. Thirdly, we can explore the implications of the new kind of community that were constructed by Epipalaeolithic and Early Neolithic hunter-gatherer groups in Southwest Asia. The conclusion of this paper is that the emergence of *Homo sapiens* was predicated on novel cognitive and cultural faculties with symbolic representation, and that a steep upward turn in the graph of cognitive and symbolic cultural abilities coincided with the emergence of novel forms of community that required advanced symbolic modes of construction. Thus, as the cultural environment of everyday life was intensified and deeply enriched, the effective environment within which communities operated changed significantly around the end of the Pleistocene or the beginning of the Holocene period, and cultural responses to changes in the bio-environment became only one among various kinds of environmental challenge and response.

Environmental causes – cultural responses

We can trace back to and beyond Gordon Childe the idea that hunter-gatherers responded to the challenge of environmental pressures driven by climatic change by becoming the first farmers. Childe sought to integrate the thinking of geographers such as Darrell Forde (1934) and the geologist Raphael Pumpelly (1908) on climatic and environmental change and its effects on human affairs with his own ideas on social evolution and a Marxist interpretation of prehistory. A new generation of archaeologists in the 1960s dismissed Childe's work because he wrote 'culture-history', because he was a 'diffusionist' and because his model of desiccation at the end of the Pleistocene had been undermined both by radiometric dating and by improved environmental data. Binford (1968) and Flannery (1969), founder members of the new scientific archaeology, followed Robert Braidwood and H. E. Wright, who had concluded that there was no evidence of climatic change at the time when farming began to be adopted (Braidwood 1960; Braidwood *et al.* 1983; Wright 1968). Binford and Flannery both opted for an alternative ecological variable in the form of levels of human population, and argued for population growth within a scenario of finite environmental resources as the driver for the adaptation of farming. However, there were always those who preferred a climatic change and the reduction of food resources in the environment, and their time came in the 1990s with the identification of the Younger Dryas phenomenon in the final millennium of the Pleistocene period (Alley *et al.* 1993; Berger 1990; Dansgaard *et al.* 1989). It became a dramatically exciting topic among environmental scientists because detailed study of the evidence from Greenland ice cores and the radiocarbon dating evidence showed that its onset and its end were very rapid.

Within Southwest Asia, the focus of researchers had shifted from northeast Iraq and western Iran, where Braidwood and his followers had set it, to the Levant. Gordon Hillman, who had spent many years working on the botanical material recovered in Andrew Moore's excavations at Tell Abu Hureyra, was aware that the site had gone out of occupation around the end of this dramatic climatic reversal (Moore and Hillman 1992; Hillman 1996). He searched for signs of the impact of the Younger Dryas generally in Southwest Asia, and pointed to the decline in tree pollen in the cores from Lake Huleh in northern Israel. The significance of the Mediterranean woodland zone is that it is also the habitat of grasses, cereals and legumes. In a series of maps Hillman modelled the situation at the end of the LGM and the re-colonization of areas by the spreading woodland and associated grasses, cereals and legumes that followed the climatic recovery. Then he modelled the impact of the Younger Dryas based on the Huleh lake core and, most importantly, archaeobotanical evidence from Tell Abu Hureyra, which was sited on the Euphrates in a marginal situation that would be severely affected by any environmental deterioration. In his last map, he showed the remarkably rapid recovery from the Younger Dryas. The version of events written by the major synthesizers of the last two decades is based on this interpretation of the environmental data, and it has become the orthodox account. Robinson and his collaborators have recently reviewed all the data sources, terrestrial from the Levant and marine from the east Mediterranean (Robinson *et al.*, this volume; Robinson *et al.* 2006). As far as the Younger Dryas phenomenon is concerned, they conclude from the terrestrial evidence that there was a significant reduction in rainfall in the southern Levant, but the marine data shows remarkably little evidence for climatic cooling. It is unclear, in their view, whether the ambiguities in the evidence represent a problem with the data or differing responses by different indicators in different environments (Robinson *et al.* 2006, 1537). Building a reconstruction of the changing availability of large-seeded grasses, cereals and legumes in the Levant on such foundations seems risky; and extrapolating to further parts of Southwest Asia (except for the Anatolian plateau, for which there is pollen from lake-bed cores) must be highly speculative. The standard account that supposes that the Younger Dryas caused a severe bottleneck, restricting the availability of storable harvests to the Levant, may be termed the Levantine primacy illusion; the environmental model of a huge Younger Dryas impact cannot be clearly substantiated, and the adoption of cultivation leading to plant domestication cannot be causally linked to climate change.

The phrase Levantine primacy was coined, as far as I am aware, by Patty Jo Watson (1995). The perspective on Southwest Asia taken by the Levantine primacy school is based on the idea that there are environmental reasons why the earliest complex hunter-gatherer societies are to be found in the Mediterranean woodland zone of the Levantine corridor; and that it is within that corridor that complex hunter-gatherers became the earliest cultivators and, later, mixed farmers, again in response to environmental pressures (see also Maher, this volume). The hypothesis is based on the notion that environmental constraints at the LGM and afterwards made the Mediterranean woodland corridor, with its grasses, cereals and legumes, a refugium within which the emergence of more complex strategies of hunter-gatherer subsistence and settlement were possible. It is a very convenient view, because it limits the significant area of Southwest Asia to a relatively small zone of environmental coherence that can also be seen as a single cultural unit. Thus innovations adopted within the Mediterranean corridor are the property of a single cultural group, usually termed successively Natufian, PPNA and PPNB. In consequence, the spread of farming is usually seen as cognate with the spread of the PPNB culture, or of its influence on other, marginal and poorly defined and usually unnamed cultural groups in other parts of the region of Southwest Asia.

However, I assert that a) the environmental evidence for the restrictive view of a Mediterranean woodland refugium is low on data, high on dramatic interpretation and generally unconvincing; b) the archaeological evidence from outside the Levantine corridor contradicts the orthodox interpretation of the environmental data; and c) the hypothesis of Levantine primacy is in fact an untested, but testable, assertion. I have recently rehearsed the criticism of the generally accepted environmental reconstruction in a paper at the 5th International Congress for the Archaeology of the Ancient Near East in Madrid, and give two references (Watkins 2008; 2009a). The archaeological and palaeobotanical evidence that contradicts the Levantine primacy case can be briefly summarized, because so little work has been done across the far southeast of Turkey and the north and northeast of Iraq. At the end of the Epipalaeolithic period there was an occupation of Shanidar cave, in a valley of the Zagros mountains in the far northeast of Iraq, and a small, open village site nearby as Zawi Chemi (Solecki 1981; Solecki 1963; Solecki and Solecki 1983). Both sites produced significant numbers of ground-stone implements for pounding and grinding foodstuffs, and Zawi Chemi contained some circular stone structures built within an accumulation of occupation debris that amounted to a couple of metres or more. At the beginning of the Early Neolithic there were several sedentary hunter-gatherer sites outside the Levantine corridor, three of which (Qermez Dere, Nemrik and M'lefaat) are in north Iraq. These three communities existed at or beyond the very limit of the distribution of the necessary plant resources according to Hillman's map for the period, and yet there is good archaeobotanical evidence from M'lefaat (Savard *et al.* 2003) and Qermez Dere (Nesbitt in Watkins *et al.* 1991, Watkins *et al.* 1995) that pulses, wild grasses and some wild barley were in use, and that an open *Pistacia* woodland environment remained despite the Younger Dryas reversal. In southeast Turkey, east of Diyarbakır, the sedentary village site of Hallan Çemi was, likewise, established at the Epipalaeolithic–Neolithic boundary. There, too, there is evidence of the use of pulses and some wild grasses, while the carbonized remains make it clear that oak–*Pistacia* woodland existed in the area (Rosenberg *et al.* 1995, Rosenberg *et al.* 1998).

While the contradictory evidence is often not recognized, there is a more serious methodological flaw in the orthodox 'Levantine primacy' account of the emergence and development of Neolithic societies in Southwest Asia, for it is an hypothesis that has been elevated to the status of orthodox account without testing. Further work in the Mediterranean woodland zone and its margins (in Jordan, the Palestinian territories, Israel, the Lebanon, western Syria and around the Euphrates in southeast Turkey) does not test the hypothesis. If we wish to define the area in Southwest Asia within which sedentary and semi-sedentary hunter-gatherers operated in the Epipalaeolithic and earliest Neolithic periods, we need to be able to demonstrate its extent by showing which areas were not part of the core area. Following the widely accepted ideas of the philosopher of science Karl Popper, in order to demonstrate a hypothesis we need to apply a test that is capable of falsifying the hypothesis (Popper 1960). How could we falsify the hypothesis? We need to find sites of the appropriate date outside the assumed core area of the Levantine corridor (and not just in the semi-arid areas immediately to the east of the Mediterranean woodland zone); if, like Hallan Çemi and the three sites in northern Iraq, those sites are like those within the core area, then the hypothesis is shown to be false, and it needs revision: that is, the core area needs to be drawn larger. If we find sites of the appropriate date that were formed by non-sedentary, non-storing hunter-gatherers, then we know that we are outside the core area, and we are beginning to define the extent of that core area.

Changing hunter-gatherer subsistence and settlement strategies

Archaeologists and anthropologists began to recognize the social, as well as the economic, implications of the differences between the strategies employed by hunter-gatherers more than 25 years ago. For example, Woodburn wrote of the emergence of investment and delayed return, contrasting with earlier, simpler, hunter-gatherer societies which had operated an immediate return strategy (Woodburn 1968a; 1968b; 1982). Lewis Binford made a distinction between foragers and collectors (Binford 1980; 1990). Foragers operated in small, highly mobile groups, with a strategy that Woodburn would label one of immediate, as opposed to delayed, returns. Collectors, as described by Binford, on the other hand, engage in investment, storage and delayed returns. The French anthropologist Alain Testart wrote of the implications of storage (and inequality) among hunter-gatherers (Testart 1982a; 1982b). He argued that the social implications of storage made those hunter-gatherers who engaged in the practice more similar to farmers than to immediate-return foraging hunter-gatherers. Tim Ingold responded to these discussions, writing about what he defined as practical and social storage (Ingold 1983). Ingold wrote that social storage refers to 'the appropriation of materials in such a way that rights over their future distribution or consumption converge upon a single interest' (Ingold 1983, 561); under these conditions what is stored is considered as property or wealth, and storage becomes part of 'the social relations of distribution'. Ingold (1983, 568) believed that it was the practices of pastoralists and cultivators that, in their different ways, exhibited the characteristics of social storage. In other words, he thought that delayed return and storage as documented in ethnographically observed hunter-gatherer societies do not have a social component of ownership, control and social management. He conceded, however, that the Pacific Northwest Coast peoples seemed to be an exception in this regard, as in so many other ways.

In the Epipalaeolithic of Southwest Asia we see: a) the collection of storable harvests of dry plant foods; b) decreasing group mobility in favour of a trend towards sedentary life; and c) a trend towards larger co-resident group size. We are in fact seeing the emergence of the type of hunter-gatherer subsistence and settlement strategies that Woodburn, Binford and Testart have defined, and 'social storage' as described by Ingold. The social implications of these changes in settlement and subsistence strategy will be discussed below, but we should first consider the implications for the territories exploited by these new hunter-gatherer groups. Let us leave aside two knotty problems. First, it is impossible to estimate the number of locations used by a group of hunter-gatherers over a single year, or over what period (a couple of days, a couple of weeks or several months) any particular location was in use. The ethnographic record tells us that small mobile hunter-gatherer groups might have an annual cycle of 12 (on average, each location representing a month's stay) or even 24 locations each year (moving every two or three weeks) (Binford 1980; 1990; Kelly 1995). Archaeologically, we have a real difficulty in detecting the trend towards sedentism, and there has been a long-running controversy over how to recognize year-round sedentism in the archaeological record. There is little consensus on the question of which groups of Epipalaeolithic hunter-gatherers in Southwest Asia were sedentary or almost sedentary. By the last two millennia of the Epipalaeolithic period, at least in the Mediterranean woodland zone of the Levant and adjacent areas, there are archaeologically and stratigraphically substantial village sites that most archaeologists accept as permanent settlements occupied year-round over many years – and sometimes many centuries or even more than a millennium. On the other hand, at the opposite end of the Epipalaeolithic period the site of Ohalo II, with its unique water-logged deposits and extraordinary organic preservation, exhibits evidence of the harvesting and processing of an extensive range of grasses, cereals and legumes, and seasonality indicators of occupation at every month of the year (Kislev *et al.* 1992; Nadel and Hershkovitz 1991; Piperno *et al.* 2004; Weiss *et al.* 2004). What we can say is that, in some parts of Southwest Asia, at some time in the last 10 millennia of the Pleistocene period, effectively sedentary hunter-gatherer communities became established.

Using the ethnographic evidence, we may suppose that Epipalaeolithic sedentary hunter-gatherer communities intensively used a territory within two hours' walk of their village homes; it makes the arithmetic easy if we calculate a little generously on a 10-km radius, and we end up with a figure of around 320 sq km for the territory around the village site that was regularly, indeed constantly, exploited. Whereas the mobile hunter-gatherer band employed a strategy of seasonal movement from one prime location to another, the sedentary hunter-gatherer community needed to intensify their extraction of resources from within their immediate territory. This total reliance on a relatively small territory that was used year-round and year after year must have changed the way that hunter-gatherer groups related to and perceived their effective environment (Watkins 1997). Flannery proposed the 'broad spectrum' strategy, involving the more intensive pursuit of small mammals, birds, fish, shellfish and amphibians (Flannery 1969), and, while the claim of a broad spectrum revolution has been questioned (Edwards 1989), detailed analysis of faunal remains from a sample of Epipalaeolithic sites from the Levant has shown the shifts within the spectrum that characterize the Epipalaeolithic period (Stiner *et al.* 1999; Stiner *et al.* 2000). The floral evidence was lacking when Flannery first proposed his broad spectrum revolution, but the material from Ohalo II has allowed us a brilliantly detailed insight into the intensive use of plant food resources at the beginning of the Epipalaeolithic period (Weiss *et al.* 2004).

As well as investing in storage (delayed return) and greater labour inputs (in the hunting of small game and birds, and in the processing of many of the stored plant foods that required pounding, grinding or soaking to remove toxins), the (semi-)sedentary hunter-gatherers of the Epipalaeolithic period exchanged the flexibility and risk-reduction of the mobile group for greater risks. Living in larger, permanent groups in one place, and relying on a tightly defined territory immediately accessible from that base, Epipalaeolithic sedentary communities were at risk if changing environmental conditions brought about a reduction in the productivity of their food sources, or if their own population numbers grew over the centuries, or if they chanced to over-hunt important species or harvest too exhaustively. We can see the consequences of trying to maintain this critical equilibrium over the long term in the changes in the spectrum of plant foods at Abu Hureyra on the middle Euphrates in Syria towards the end of the Younger Dryas, which ended in the abandonment of the settlement (difficulties and finally failure that did not affect the nearby site of Tell Mureybet, as it happens) (Hillman 1996; Hillman *et al.* 2001; Moore *et al.* 2000). Similar adjustments, involving greater investment of labour, have been charted through the end of the Epipalaeolithic and the Early Neolithic at Hatula in Israel, as gazelle availability declined and lengthy fishing trips to the Mediterranean shore began (Davis *et al.* 1994).

The symbolic construction of community

Why should hunter-gatherers with tens of millennia of accumulated experience choose to hazard themselves with increased risks? Why should they compound their difficulties by opting for greater concentration on food resources that required more labour investment? And why did they opt for more demanding rules in their social lives, as they turned to the substantial social storage of harvests of legumes, grasses and cereals? Rather than seeking a cause in pressures exerted by changes in the physical environment and its resources, we should look to a different aspect of the environment, the built environment.

It was Peter Wilson who drew attention to the critical significance of the adoption of buildings and village life (Wilson 1988). He distinguishes the 'open society' (the traditional, small-scale, mobile hunter-gatherer band society) and sedentary hunter-gatherer societies, who live in permanent buildings in village societies, which he regards as 'domesticated' societies. For Wilson, domestication is the consequence of living in houses, living in villages. 'Domestication' challenged people's natural, evolved dependence on paying constant attention to one another. As well as presenting a cognitive challenge, domestication could be seen as offering stimulating opportunities for the development of structure, physical and architectural, shaping and modelling the social organization of a new way of life. Wilson reviews many instances in the ethnographic literature of contemporary small-scale sedentary societies that use architecture for the elaboration of thinking about the structure of the world, and the articulation of links between their social world and the cosmos. What Wilson does not do is set this transformation of human psychology in the context of the evolution of human cognitive and cultural faculties.

In a series of conference papers and articles over recent years, I have sought to explore how the ideas of the psychologist Merlin Donald (1991; 2001) on the evolution of human cognition and culture can be set within the archaeological record of the Epipalaeolithic and Neolithic periods of Southwest Asia (Watkins 2003; 2004a; 2004b; 2005; 2006). This task involves recognizing that symbolic culture can operate as a system of symbolic representation generically similar to language. After a first revolution in hominid communication that Donald calls 'mimetic' (not dissimilar to Mithen's 'Singing Neanderthals' (Mithen 2005; 2006)), the second of Donald's stages is the emergence of full modern language, while the third and final stage is the emergence of alphabetic writing systems. We can easily underestimate the massive complexity, subtlety and sophistication of language, and the cognitive power required of our brains to manage speech and the comprehension of utterances that we hear. The critical factor in language as it exists in every human society around the world is that it is a system of symbolic representation. In other words, language consists in more than the accumulation of a series of dictionary definitions of the words used; we put together strings of words so that the meaning of what we say depends on the relations, syntactic and semantic, between those words, and any utterance adds up to much more than the sum of its constituent words (see, for example, Terence Deacon's *The Symbolic Species: The Co-Evolution of Language and the Human Brain* (1997) for a discussion of the complexity of symbolic manipulation involved in ordinary speech). Written language extends the cognitive power of the human brain through the potential of what Donald calls 'external symbolic storage'. The significance of the externalization of memory storage is that the modes of memory are fundamentally different from the brain's memory. For example, the amount of information that can be stored in an external storage system such as written language is virtually unlimited, is more permanent, and can be 'read' by others who see it at some other time or in some other place (telecommunication). Along with Colin Renfrew (1998), I take the view that, between the emergence of full modern language (around 50,000 years ago) and the emergence of writing systems (beginning in some parts of the world around 5000 years ago), there was a critically important intermediate step that took the form of the development of the fully symbolic usage of material culture. Thus the 'revolution in symbols' to which Jacques Cauvin pointed (Cauvin 1994) should be understood as the emergence of the cognitive and cultural ability to create symbolic vocabularies and formulate and 'read' symbolic constructions using material culture (as distinct from spoken or written language).

Early *Homo sapiens* began to use symbolic reference more than 75,000 years ago, and had been carrying out symbolic, ritual actions in the form of ceremonialized burial even earlier. By 50,000 years ago, the linguists tell us, modern humans had evolved fully modern language. From the early Upper Palaeolithic there is evidence of two- and three-dimensional symbolic representation. What is different in the Epipalaeolithic of Southwest Asia is the construction of built environments that were inhabited worlds of symbolism. In the past I have gone along with Jacques Cauvin (1994) in his insistence on a revolution in symbolism at the beginning of the Neolithic, but a good case can be made for symbolic built environments at least as early as the Late Epipalaeolithic period. Two things are important about such symbolically significant built environments. First, as already mentioned, the built environment is inhabited, and it frames the way of life and even the way of thinking of its inhabitants. Architecture can create the arenas within which other objects or actions find their meaningful place. It provides a rich and structured environment within which we think and act. Second, the new kind of environment represented by the early village communities would have had a profound effect on the children who were born and grew up in them. Developmental psychologists emphasize the critical importance of the social and cultural context within which human infants are brought up. Infant experience influences the formation of neural networks, and children brought up in rich cultural environments thrive. The rich symbolic cultural environment of the settlement, therefore, would have become an important driver of rapid cultural development within communities.

Changing people, changing environments

Finally, we return to the title of this contribution to a book entitled *Landscapes in Transition*. The purpose of this essay is to argue that, at least in certain parts of Southwest Asia, there was a steep upward turn in the graph of cognitive and symbolic cultural abilities in the final Pleistocene and earliest Holocene. That coincided with – opened the way for – the emergence of novel forms of community that required advanced symbolic modes of construction. Thus the environments within which communities operated changed significantly, and cultural responses to changes in the bio-environment became less significant than the culturally constructed environment itself. As people and their cognitive faculties changed, they began to construct, both literally and conceptually, their own environments. Thus during the Epipalaeolithic and Early Neolithic periods in at least some parts of Southwest Asia it was the built environment, rich in symbolic values, that became the effective environment, rather than, or as much as, the physical environment from which subsistence resources were taken. Within these settlements, we find striking new uses of architecture, sculpture, modelling and ritualized practices of various kinds; the cultural changes that we can chart may be understood as responses to the cultural environments in which people lived. In these settlements, new, large communities that numbered several hundred and, in the Neolithic period, several thousand inhabitants began to develop rapidly, responding to the rich cultural environment and the density of personal interaction. While we think little of living in urban centres numbering millions of individuals, we should remember that the cognitive and cultural load involved in forming and maintaining communities is very considerable (see Cohen 1985 for a succinct examination of the issue). It was only from the time of the LGM, close to the end of the Pleistocene period, that modern humans had evolved the cognitive and symbolic cultural faculties necessary for the construction of communities that transcended the biological capacity of the human brain (Watkins 2008).

Beyond the settlement, the land that supported its inhabitants constituted an economic resource that belonged exclusively to (constituent groups within) the community. Whether the harvests were of wild plants or cultivated and domesticated crops, the intensively used lands around the settlement were to an extent managed. Further afield there were other communities, and each community was at pains to ensure that it was linked to its neighbours through exchange and sharing (Watkins 2009a). Such networks were in one sense not new, but in scale and intensity they were novel, and they created regional and supra-regional cultural environments. In this regard, we should turn to the work of Clive Gamble on early hominin to Upper Palaeolithic societies and his discussion of human networks (Gamble 1999). Gamble has used recent sociological research on networks, talking in terms of intimate, effective and extensive networks (the last of which is important for us). The individual's intimate network ranges between three and seven, with an average of five, mostly kin. The effective network extends beyond kin and includes those people with whom the individual operates or on whom he/she depends in the course of everyday life. The average size of an effective network is 20 individuals, but may be greater. In mobile hunter-gatherer societies, the effective network equates with the basic 'band'. Bands have a general range of 20 to 40 members. Gamble argues that our early ancestors had intimate and effective networks, but that extensive networks emerged, in Europe at least, only in the Upper Palaeolithic. From a survey of the literature, Gamble suggests that the typical extensive network is in the range of 200–400 persons (Gamble 1999, 60). The formation and maintenance of relations between people who cannot relate to each other face-to-face and, frequently, that is between people who are distant, who rarely see each other, or who may not be known personally to one another, requires the use of symbolic/stylistic signalling. While there is evidence for the movement of stone raw materials over tens of kilometres earlier in the Palaeolithic, the Upper Palaeolithic sees much more intensive movement of high-quality raw materials and also the exchange of items of symbolic value over very long distances. This interpretation of the exchange of marine shells as evidence of new cognitive capacities in modern humans fits with the widely held view among Palaeolithic specialists that the period from about 80,000 to 30,000 years ago saw the emergence of new symbolizing abilities in human cognition and cultural capacity. This capacity allowed humans to build extended networks of relations (of several hundred people) that transcended the fundamental limits of the brain (at around 120–150 persons).

In the Upper Palaeolithic, these new extended networks were spread across the landscape, linking together small, mobile bands. In the Epipalaeolithic period communities began to grow in numbers, so that the extended network became the individual settlement. The far-flung groups who scarcely knew of each other became neighbours. Such large co-resident groups demanded greatly increased cognitive and symbolic cultural skills for the construction and maintenance of the sense of community that enabled them to live together over many years, generations, even centuries. However, these symbolically constructed communities extended their efforts in order to join up into extensive networks of communities that traded materials and goods, replacing the networks of exchange of earlier times. In addition, in subtle and complex ways, they constructed and expressed multi-layered identities through shared symbolic values and practices (Watkins 2008). For us here, thinking about landscapes in transition, the Epipalaeolithic and Early Neolithic communities of Southwest Asia not only transformed the settlement landscape through their changes in subsistence and settlement strategies, but also created symbolic, culturally constructed landscapes that were unlike anything that their Palaeolithic predecessors had known.

Bibliography

Alley, R. B., Meese, D. A., Shuman, C. A., Gow, A. J., Taylor, K. C., Grootes, P. M., White, J. W. C., Ram, M., Waddington, E. D., Mayewski, P. A. and Zielinski, G. A. (1993) Abrupt increase in Greenland snow accumulation at the end of the Younger Dryas event. *Nature* 362, 527–9.

Bar-Yosef, O. (2002) The Natufian culture and the Early Neolithic social and economic trends in Southwestern Asia. Pp. 113–26 in P. Bellwood and C. Renfrew (eds), *Examining the Farming/Language Dispersal Hypothesis*. Cambridge: McDonald Institute for Archaeological Research.

Bar-Yosef, O. and Belfer-Cohen, A. (2002) Facing environmental crisis. Societal and cultural changes at the transition from the Younger Dryas to the Holocene in the Levant. Pp. 55–66 in R. T. J. Cappers and S. Bottema (eds), *The Dawn of Farming in the Near East*. Studies in Early Near Eastern Production, Subsistence and Environment 6. Berlin: *ex oriente*.

Bar-Yosef, O. and Meadow, R. H. (1995) The origins of agriculture in the Near East. Pp. 39–94 in T. D. Price and A. B. Gebauer (eds), *Last Hunters, First Farmers: New Perspectives on the Prehistoric Transition to Agriculture*. Santa Fe, NM: School of American Research Press.

Berger, W. H. (1990) The Younger Dryas cold spell – a quest for causes. *Palaeogeography, Palaeoclimatology, Palaeoecology* 89, 219–37.

Binford, L. R. (1968) Post-Pleistocene adaptations. Pp. 313–41 in L. R. Binford and S. R. Binford (eds), *New Perspectives in Archeology*. Chicago: University of Chicago Press. Reprinted in L. R. Binford (1972) *An Archaeological Perspective*. New York, London: Seminar Press.

Binford, L. R. (1980) Willow smoke and dogs' tails: hunter-gatherer settlement systems and archeological site information. *American Antiquity* 45, 4–20.

Binford, L. R. (1990) Mobility, housing and environment: a comparative study. *Journal of Anthropological Research* 46, 119–52.

Braidwood, L. S., Braidwood, R. J., Howe, B., Reed, C. A. and Watson, P. J. (1983) *Prehistoric Archaeology along the Zagros Flanks*. Oriental Institute Publications no. 105. Chicago: University of Chicago Press.

Braidwood, R. J. (1960) The agricultural revolution. *Scientific American* 203, 130–41. Reprinted in C. C. Lamberg-Karlovsky (ed.), (1972) *Old World Archaeology: The Foundations of Civilization*. Readings from Scientific American. San Francisco: W. H. Freeman.

Cauvin, J. (1994) *Naissance des Divinités, Naissance de l'Agriculture: La Révolution des Symboles au Néolithique*, 2nd edition, revised, 1997. Paris: CNRS Publications. [Published (2000) as *The Birth of the Gods and the Origins of Agriculture*. Cambridge: Cambridge University Press.]

Cohen, A. P. (1985) *The Symbolic Construction of Community*. London: Routledge.

Dansgaard, W., White, J. W. C. and Johnsen, S. J. (1989) The abrupt termination of the Younger Dryas climate event. *Nature*, 339(6225), 532–4.

Davis, S. J. M., Lernau, O. and Pichon, J. (1994) The animal remains: new light on the origin of animal husbandry. Pp. 83–100 in M. Lechevallier and A. Rosen, *Le Gisement de Hatoula en Judée Occidentale, Israël*. Paris: Mémoires et Travaux du Centre de Recherche Français de Jerusalem.

Deacon, T. (1997) *The Symbolic Species: The Co-Evolution of Language and the Human Brain*. New York: W. W. Norton and Co.

Donald, M. (1991) *Origins of the Human Mind: Three Stages in the Evolution of Culture and Cognition*. Cambridge, MA: Harvard University Press.

Donald, M. (2001) *A Mind so Rare*. London: W. W. Norton and Co.

Edwards, P. C. (1989) Revising the broad spectrum revolution: and its role in the origins of Southwest Asian food production. *Antiquity* 63, 225–46.

Flannery, K. V. (1969) The origins and ecological effects of early domestication in Iran and the Near East. Pp. 73–100 in P. J. Ucko and G. W. Dimbleby (eds), *The Domestication and Exploitation of Plants and Animals*. London: Duckham.

Forde, D. (1934) *Habitat, Economy and Society: A Geographical Introduction to Ethnology*. London: Methuen.

Gamble, C. (1999) *The Palaeolithic Societies of Europe*. Cambridge: Cambridge University Press.

Hillman, G. C. (1996) Late Pleistocene changes in wild plant-foods available to hunter-gatherers of the northern Fertile Crescent: possible preludes to cereal cultivation. Pp. 159–203 in D. J. Harris (ed.), *The Origins and Spread of Agriculture and Pastoralism in Eurasia*. London: UCL Press

Hillman, G. C., Hedges, R., Moore, A., Colledge, S. and Pettitt, P. (2001) New evidence of Lateglacial cereal cultivation at Abu Hureyra on the Euphrates. *The Holocene* 11/4, 383–93.

Ingold, T. (1983) The significance of storage in hunting societies. *Man* 18, 553–71.

Kelly, R. L. (1995) *The Foraging Spectrum: Diversity in Hunter-Gatherer Lifeways*. Washington DC: Smithsonian Institution Press.

Kislev, M. E., Nadel, D. and Carmi, I. (1992) Epipalaeolithic (19,000 BP) cereal and fruit diet at Ohalo II, Sea of Galilee, Israel. *Review of Palaeobotany and Palynology* 71, 161–6.

Lee, R. and DeVore, I. (1969) *Man the* hunter. Chicago: Aldine Pub. Co.

Mithen, S. (2005) *The Singing Neanderthals: The Origins of Music, Language, Mind and Body*. London: Weidenfeld and Nicolson.

Mithen, S. (2006) The 'Singing Neanderthals': the origins of music, language, mind and body. *Cambridge Archaeological Journal* 16/1, 97–112.

Moore, A. M. T. and Hillman, G. C. (1992) The Pleistocene to Holocene transition and human economy in Southwest Asia: the impact of the Younger Dryas. *American Antiquity* 57, 482–94.

Moore, A. M. T., Hillman, G. C. and Legge, A. J. (2000) *Village on the Euphrates: From Foraging to Farming at Abu Hureyra*. Oxford: Oxford University Press.

Nadel, D. and Hershkovitz, I. (1991) New subsistence data and human remains from the earliest Levantine Epipalaeolithic. *Current Anthropology* 32, 631–5.

Piperno, D., Weiss, E., Holst, I. and Nadel, D. (2004) Processing of wild cereal grains in the Upper Palaeolithic revealed by starch grain analysis. *Nature* 430, no. 7000, 670–3.

Popper, K. (1960) *The Logic of Scientific Discovery*. London: Hutchinson. [Originally published 1935, Vienna, as *Logik der Forschung*.]

Pumpelly, R. (1908) *Explorations in Turkestan*. Washington, DC: Museum of Natural History.

Renfrew, C. (1998) Mind and matter: cognitive archaeology and external symbolic storage. Pp. 1–6 in C. Renfrew and C. Scarre (eds), *Cognition and Material Culture: the Archaeology of Symbolic Storage*. Cambridge: McDonald Institute for Archaeological Research.

Robinson, S., Black, S., Sellwood, B. and Valdes, P. (2006) A review of palaeoclimates and palaeoenvironments in the Levant and Eastern Mediterranean from 25,000 to 5000 years BP: setting the environmental background for the evolution of human civilisation. *Quaternary Science Reviews* 25, 1517–41.

Rosenberg, M., Nesbitt, R. M., Redding, R. and Strasser, T. F. (1995) Hallan Çemi Tepesi; some preliminary observations concerning Neolithic subsistence behaviors in eastern Anatolia. *Anatolica* 21, 1–12.

Rosenberg, M., Nesbitt, R. M., Redding, R. and Peasnall, B. L. (1998) 'Hallan Çemi, pig husbandry, and post-Pleistocene adaptations along the Taurus-Zagros arc (Turkey). *Paléorient* 24/1, 25–42.

Savard, M., Nesbitt, M. and Gale, R. (2003) Archaeobotanical evidence for early Neolithic diet and subsistence at M'lefaat (Iraq). *Paléorient* 29/1, 93–106.

Solecki, R. L. (1981) *An Early Village Site at Zawi Chemi Shanidar*. Biblioteca Mesopotamica vol. 13. Malibu, CA: Undena Publications.

Solecki, R. L. and Solecki, R. S. (1983) Late Pleistocene – Early Holocene Cultural Traditions in the Zagros and the Levant. Pp. 123–37 in T. C. Young, P. E. L. Smith and P. Mortensen (eds), *The Hilly Flanks and Beyond: Essays on the Prehistory of Southwestern Asia*.

Solecki, R. S. (1963) Prehistory in the Shanidar Valley, Northern Iraq. *Science* 139, no. 3551, 179.

Stiner, M. C., Munro, N. D., Surovell, T. A., Tchernov, E. and Bar-Yosef, O. (1999) Paleolithic population growth pulses evidenced by small animal exploitation. *Science* 283, 190–4.

Stiner, M. C., Munro, N. D. and Surovell, T. A. (2000) The tortoise and the hare: small-game use, the broad-spectrum revolution, and Paleolithic demography. *Current Anthropology* 41, 39–73.

Testart, A. (1982a) *Les Chasseurs-cuilleurs ou l'Origine des Inégalités*. Paris: Société d'Anthropologie.

Testart, A. (1982b) The significance of food storage among hunter-gatherers: residence patterns, population densities and social inequalities. *Current Anthropology* 23, 523–37.

Watkins, T. (1997) The human environment. *Paléorient*, 23/2, 263–75.

Watkins, T. (2003) Memes, memeplexes and the emergence of religion in the Neolithic. Pp. 41–7 in H.-G. Gebel and C. Jensen (eds), *Magic Practices and Ritual in the Near Eastern Neolithic*. Studien zu Produktion, Subsistenz und Umwelt im frühen Vorderasien 8. Berlin: *ex Oriente*.

Watkins, T. (2004a) Building houses, framing concepts, constructing worlds. *Paléorient* 30/1, 5–24.

Watkins, T. (2004b) Architecture and 'theatres of memory' in the Neolithic of southwest Asia. Pp. 97–106 in E. DeMarrais, C. Gosden and C. Renfrew (eds), *Rethinking Materiality: The Engagement of Mind with the Material World*. Cambridge: McDonald Institute for Archaeological Research.

Watkins, T. (2005) The Neolithic revolution and the emergence of humanity: a cognitive approach to the first comprehensive world-view. Pp. 84–8 in J. Clarke (ed.), *Archaeological Perspectives on the Transmission and Transformation of Culture in the Eastern Mediterranean*. Levant Supplementary Series, vol. 2. Oxford: Council for British Research in the Levant and Oxbow Books.

Watkins, T. (2006) Architecture and the symbolic construction of new worlds. Pp. 14–25 in E. Banning and M. Chavan (eds), *Domesticating Space: Construction, Community, and Cosmology in the Late Prehistoric Near East*. Studies in Early Near Eastern Production, Subsistence, and Environment 6. Berlin: *ex oriente*.

Watkins, T. (2008) Supra-regional networks in the Neolithic of Southwest Asia. *Journal of World Prehistory* 21/1, 139–71.

Watkins, T. (2009). Natural environment versus cultural environment: the implications of creating a built environment. Pp. 428–37 in J. Córdoba, M. Molist, C. Pérez, I. Rubio and S. Martínez (eds), *Proceedings of the 5th International Congress on the Archaeology of the Ancient Near East*. Madrid: Universidad Autónoma de Madrid.

Watkins, T. (2009a) Ordering time and space: creating a cultural world. Paper to be published in the Acts of the 5th International Congress on the Archaeology of the Ancient Near East, Madrid.

Watkins, T., Betts, A., Dobney, K. and Nesbitt, M. (1991) *Qermez*

Dere, Tel Afar: Interim Report No. 2, 1989. Edinburgh: Department of Archaeology, University of Edinburgh.

Watkins, T., Dobney, K. and Nesbitt, R. M. (1995) *Qermez Dere, Tel Afar; Interim Report No. 3*. Edinburgh: Department of Archaeology, University of Edinburgh.

Watson, P. J. (1995) Explaining the transition to agriculture. Pp. 21–37 in T. D. Price and A. B. Gebauer (eds), *Last Hunters, First Farmers: New Perspectives on the Prehistoric Transition to Agriculture*. Santa Fe, NM: School of American Research Press.

Weiss, E., Wetterstrom, W., Nadel, D. and Bar-Yosef, O. (2004) The broad spectrum revolution revisited: Evidence from plant remains. *Proceedings of the National Academy of Sciences of the USA* 101, 9551–5.

Wilson, P. J. (1988) *The Domestication of the Human Species*. New Haven, CT, and London: Yale University Press.

Woodburn, J. (1968a) An introduction to Hadza ecology. Pp. 49–55 in Lee and Devore 1969.

Woodburn, J. (1968b) Stability and flexibility in Hadza residential groupings. Pp. 103–10 in Lee and Devore 1969.

Woodburn, J. (1982) Egalitarian societies. *Man* 17, 431–45. Reprinted in J. Gowdy (ed.), (1998) *Limited Wants, Unlimited Means: A Reader on Hunter-Gatherer Economics and the Environment*. Washington, DC: Island Press.

Wright, H. E. (1968) Natural environment of early food production in the Near East. *Science* 161 334–9.

11. Formalizing the Sacred? The Late Mesolithic and Early Neolithic Monumental Landscapes of Britain and Ireland

Vicki Cummings

One of the big unanswered questions in prehistoric archaeology is whether the start of the Neolithic in Britain and Ireland saw the arrival of a new population, bringing with them new forms of material culture from the continent, or whether native hunter-gatherers adopted and adapted Neolithic things into their own worldview. Part of this debate has focused on the role of landscape in the transition process, particularly in relation to monumental architecture. This paper considers the nature of the landscape in Late Mesolithic and Early Neolithic Britain and Ireland, exploring not only the evidence for enduring locations in the Mesolithic and Neolithic but also placing those in a broader context of everyday ritualized activity. I will argue that Neolithic farmers did not have a view of the landscape that was radically different from that of preceding Mesolithic hunter-gatherers and that it is the nature of the evidence (and those studying it) that creates the contrast.

Introduction

In this paper I will consider one of the questions raised by the conference brief: did (Mesolithic) hunter-gatherers have a radically different view of the landscape from that of (Neolithic) farmers, or is it the nature of the evidence that creates the contrast? I will be considering this issue in relation to the Mesolithic–Neolithic transition in Britain and Ireland, with a particular focus on the monumental evidence from western Britain and eastern Ireland. After considering the nature of the evidence, and presenting the results of fieldwork conducted in this area, I raise some areas and points for discussion regarding the importance of the landscape in the transition period. This paper is offered as a discussion piece, and does not seek to definitely 'answer' the questions posed. Instead, I hope it will offer some ways forward and set an agenda for future studies of these questions.

The Late Mesolithic

The archaeology from Late Mesolithic Britain and Ireland has been considered in depth over the years, more recently in relation to the idea that it may represent evidence for 'complex' hunter-gatherers and that the period was somehow a 'precursor' for the subsequent Neolithic (e.g. Cummings 2003; Pollard 2000 and see Warren 2007). There are only limited quantities of evidence from the Late Mesolithic, and the evidence which does exist has been overshadowed by some of the most 'luminous' sites, especially the shell middens of western Scotland. The problem of how Late Mesolithic material has been interpreted has been emphasized: two problems are that material is considered either in very functional or ecological terms, or by Neolithic scholars looking for essentially Neolithic structuring principles in the preceding period (see Conneller and Warren 2006; Warren 2007). Although the Late Mesolithic is problematic, and theoretical considerations could certainly be bolstered by the addition of new sites, as well as the dating of material from known sites (e.g. Chatterton 2006), we can still comment on how Late Mesolithic people conceptualized and interacted with the landscape.

Firstly, what we do have is a range of places in the landscape: these locales range from small lithic scatters to repeatedly visited sites including hearths and, in some cases, structures. There is a temptation to over-emphasize the importance of those sites that were visited on a number of occasions as through them we can discuss the 'persistence of place' in memory and in terms of affinity with particular locations (cf. Barton *et al.* 1995). However, this clearly *was* a feature of a number of Mesolithic locations. Regardless of whether sites were used only once or were used repeatedly, they also show the deliberate and careful choice of where to stay. The analysis of material also enables us to see what activities were undertaken at a site and what material was considered appropriate to be left behind. The most obvious examples of such sites are the shell middens, which are clear evidence for the repeated and prescribed disposal of material in a specific location (Mellars 1987). If we discard the idea that there were separate ritual and domestic spheres in the Mesolithic, and instead the two were enmeshed within the same conceptual understanding of the world, then these sites were potentially used in both ritual *and* domestic spheres: what Bradley (2005) describes as ritualized behaviour. Middens, then, were places which potentially saw a repeated and careful deposition of certain materials at a locale which was returned to on a number of occasions, possibly in a ritualized fashion.

These middens are not found everywhere in Britain and Ireland in the Mesolithic record, although preservation may be a factor in some areas. Nevertheless, they may well be indicative of maritime-based communities who contrast with those communities based inland (R. Schulting pers. comm.). Some middens are early in date, but many of the larger ones clustered around Oban in western Scotland, for example, are later Mesolithic in date (Mellars 1987). The construction of these later middens may have been a localized reaction to something (social, economic or political, perhaps) and it is far too tempting (from a Neolithic point of view at least) to see them as influenced in some way by what was going on in mainland Europe (e.g. the spread of Neolithic material culture along the Atlantic façade and into Belgium/Holland, with the possible early arrival of aspects of this culture into Britain, as at Ferriter's Cove: Woodman *et al.* 1999). It is more likely, however, that they were reactions to internal social politics or ways of doing things. For some hunter-gatherers, the return to place may well have been a way of expressing social identity as well as increasing enduring connections with specific places. These places may have been connected with events, with people, with processes and transformations. Other people may have preferred to keep moving, always picking new locations to dwell, new places to leave deposits. Whether or not these choices had anything to do with subsistence is highly debatable, but they had everything to do with identity and an acute attention to detail with regard to one's place in the cosmos. This key theme of connection with specific place is one to which I will return later.

Secondly, there is growing evidence for what might be called structured deposition, but what can certainly be considered careful and deliberate deposition. Chatterton (2006) details the evidence from Britain and Ireland, emphasizing votive, particularly riverine, deposits of material culture and bodies, as well as the cutting of and deposition of material in pits. There seems to have been deliberate selection of particular materials at different sites, as well as the movement of certain types of raw and worked materials (e.g. Kador 2007). The treatment of human remains – i.e., the circulation of disarticulated human bone – demonstrates the careful working and reworking of bodies in a Late Mesolithic context (e.g. Conneller 2006, 158–60). People therefore had a suite of different materials to draw on (including animal and human bodies) and these materials seem to have been carefully chosen for deposition in very specific parts of the landscape.

There are other strands of evidence to consider. We now have some evidence for cremation from the Mesolithic (Chatterton 2006, 110) which suggests this may have been a common way of disposing of the dead. The formalized cemeteries of Scandinavia are not found in the British record, suggesting that the dead, where deposited, were placed in more ephemeral locations. Places in the landscape were marked out in other ways, though: I have already briefly mentioned middens and there are now a number of examples of the erection of posts (at Stonehenge: Cleal *et al.* 1995 and also see Pitts 2007) which these highlight the potential significance of trees, woodland and clearings (see Gell 1995; Tilley 2007).

The Late Mesolithic world was one of diversity. Some landscapes were dwelt in in rather ephemeral ways. Other landscapes may have been altered and dwelt in regularly and repeatedly by people who left very specific deposits of material culture. Some landscapes were marked out in rather different ways: the construction of a shell midden; the erection of large timber posts. There may of course be a whole range of practices that we simply cannot identify in the archaeological record: the manipulation of woodland seems obvious considering the importance of wood in a Mesolithic economy of substances (cf. Thomas 1999). The significance of rivers and the sea is also likely but only hinted at in the archaeological record (Chatterton 2006; Fowler and Cummings 2003). And what of the mountains, which may have held a very specific place in a Mesolithic worldview, yet did not require active deposition in order for them to be significant (see below)? How were these understood in the Mesolithic? Did they play a part in ritual life, perhaps (see Price 2007)?

We can, and have, obtained further suggestions about the Late Mesolithic from ethnographic analogy (e.g. Cummings and Whittle 2004; Tilley 1994; Barnard 2007). Most historic hunter-gatherer groups invest considerable symbolism in landscape as part of creation mythology, the embodiment of founder ancestors and beings, and the medium for knowledge, stories and morality tales (e.g. Basso 1996; Carmichael *et al.* 1994; Hirsch and O'Hanlon 1995; Kahn 1990). The significance of the landscape to hunter-gatherer groups cannot be overstated, but since landscapes are inherently socially constructed, the question is how we get at these issues in a Late Mesolithic context.

We must be careful, therefore, when talking about the Late Mesolithic, that we do not create something amorphous and uncontextualized. I am uneasy about discussions of Mesolithic lives which remain strangely timeless, ahistorical and not grounded within the cultural and material specificity of this particular context (or, more accurately, sets of contexts). Discussions of place and space, of taskscapes and affinity with place and the engagement and experience of landscape, are all useful and enlightening, but in order to avoid 'banal phenomenological truisms' (Jordan 2003, 130) they need to be explored through the archaeological record in local and regional contexts. As McFadyen (2006, 136) points out, this requires the long-term investment in regional projects. Large-scale projects enable us to address historical context and consider what made people do certain things at certain points. Addressing such issues is not a processual quest for understanding cultural change, but a desire for stories to be enmeshed within specific historical moments *and* trajectories (a trajectory, we should not forget, that ultimately sees hunter-gatherers actively taking on board Neolithic material culture). I do not claim to be able to deliver here a detailed historically contextualized understanding of Late Mesolithic hunter-gatherers in western Britain and eastern Ireland. This is something that needs to be done over the long term, and only with much more detailed datasets can we begin to really get to grips with the kinds of questions that we are currently asking. It is not that we are asking the wrong questions, but that we are going to struggle to answer them if we do not have access to enough relevant data.

Before I turn to the Mesolithic–Neolithic transition, I will outline an ongoing research project in Kintyre, western Scotland (The Southern Kintyre Project: Cummings and Robinson 2007, www.uclan.ac.uk/kintyreonline (accessed 25 August 2009). This project has been designed to address some of the questions raised above: which parts of the landscape were used and occupied by people in the Mesolithic and Neolithic)? What materials did people use and deposit at these locations? Where did these materials come from? What were the different types of deposition at sites? How does the archaeology from this particular area compare with the record in Ireland (just 12 miles away across the sea) and in other parts of western Scotland? In order to address these questions we are finding new sites by detailed fieldwalking and excavation, as well as reassessing old material. We are also considering the local vegetation through environmental reconstruction, as well as exploring characteristics of the landscape: for example, trying to ascertain potential landing places for people who moved around by boat.

Results so far have been intriguing, and suggest that the detailed consideration of individual assemblages, when set against specific research agendas, have the potential to help us answer some of our questions. In southern Kintyre we have found two large Mesolithic scatters (at Macharioch and Machribeg), both from the old dune systems of the Late Mesolithic shoreline. These sites have been located by fieldwalking only, yet each location has produced thousands of pieces of debitage. Each site was therefore either the location of an event involving a massive and intensive episode of flintworking, or was repeatedly visited over time. At Macharioch one flint flake is clearly exotic, appears to be butt-trimmed, and is reminiscent of Bann flakes from across the Irish Sea (Chantal Conneller pers. comm.) The site of Machribeg remains to be analysed but its location is such that there is a clear view out to County Antrim, suggesting a possible relationship that we are keen to explore through the lithics. At both sites we will be analysing the flint to find its original source (did, for example, Mesolithic people utilize Antrim flint? If so, did they travel there? Are depositional practices the same on the western side of the Irish Sea?). There are clearly many questions regarding the differences, as well as the similarities, in identity, practices and deposition between communities in southern Kintyre and those in County Antrim. These will be addressed as the project progresses. It is through studies like these that we can potentially gain insights into the nature of hunter-gatherer communities at particular places and points in time.

The Mesolithic–Neolithic transition

If the evidence for the Late Mesolithic is not as substantial as we might hope, the evidence for the transition period is positively sparse (the transition period here being defined as 4100–3900 BC). There is some evidence that, like the Ertebølle of southern Scandinavia (e.g. Tilley 1996, chapter 1), people in Late Mesolithic Britain and Ireland were in contact with Neolithic people on mainland Europe (e.g. Woodman *et al.* 1999). But the very limited quantity of Neolithic material culture in pre-4000 BC contexts suggests that this kind of contact was unusual or at best sporadic. The real change in Britain and Ireland seems to happen rather abruptly: from about 4000 BC pottery and domesticated animals (especially cattle) appear in the record, along with large numbers of polished stone axes, including exotic jadeite axes from mainland Europe (Whittle 2007). The processes by which this happened have been debated in depth for many decades (for the latest summary see Whittle and Cummings 2007), but it can be boiled down to whether 'the Neolithic' (a term which is itself highly contested) was a result of incoming populations (Sheridan 2004; 2007), the adoption of ideas by a native population (Thomas 2003; 2007) or a mix of the two (e.g. Robb and Miracle 2007).

The big changes in the landscape come a few hundred years later: monuments are not typically found before about 3800 BC (Whittle 2007) and axe factories are not intensively exploited until perhaps as late as 3600 BC (Bradley and Edmonds 1993). There is quite clearly evidence of continuity across the Mesolithic–Neolithic divide in terms of the sites used (shell middens were still in use: Schulting and Richards 2002) and the landscapes in which people dwelt. The way in which people did things in the landscape seemed to continue: there is very little burial evidence from the earliest Neolithic (4000–3800 BC), so we are perhaps looking at cremation and/or deposition in natural features such as rivers. People dug pits and deposited material culture. They also left lithics and presumably organics at dwelling sites. The elm decline also dates to about 3800 BC in many areas, and if this does genuinely represent woodland clearance it seems that, again, this feature of the Neolithic did not really start for a few hundred years after the introduction of Neolithic material culture (of course there is the debate as to whether the elm decline relates to man-made clearance or disease (Edwards and Ralston 1997)). Many monuments seem to have been made initially of wood (the timber phases at Lochhill and Slewcairn, Dumfries and Galloway, and Dooey's Cairn, Co. Antrim, for example: Cooney 2000, 99–103). Wood and woodland must surely have been a key substance and resource in both the Late Mesolithic and very earliest Neolithic (Cummings and Whittle 2003).

One interesting contrast between the Mesolithic and Neolithic is that, overall, there are fewer lithics deposited in the Neolithic than are frequently found at Mesolithic sites (Mithen *et al.* 2007; Josh Pollard pers. comm.), suggesting that at some point it was no longer appropriate to leave large quantities of lithic material in the landscape (this has been noted in Kintyre as well, where Neolithic scatters comprise only a few pieces, while Mesolithic scatters consist of thousands of pieces of debitage). We need some precise dates to record this change in material deposition and to consider this practice in more detail.

Therefore, we have a gap of about 200–300 years before the construction of monuments and the widespread utilization of axe factories, in the past viewed as part and parcel of the Neolithic 'package' and essential for providing us with information on this earliest phase. What may have been as many as 12 generations of people lived in a landscape which, in archaeological terms at least, seems to have been very similar to that of the Late Mesolithic (and see Cummings 2007). On one level, whether these earliest Neolithic people were incomers from Europe or native people who adopted parts of the Neolithic material cultural package may not really matter. This is because, when we come to think of the ways in which the landscape was understood, some ethnographic examples show that incoming populations can adopt the myths, stories and histories of a landscape, as well as patterns of land use, instead of creating new understandings and engagements (for example, Sundstrom 1996). On another level, this issue may be key when thinking about concepts of materiality (and what could be left where) and the ways in which the landscape was inhabited and dwelt in. We are then left, again, wishing for more data to inform our understanding of this key period.

The development of the Neolithic

After about 3800 BC people in many parts (although by no means all) of Britain and Ireland started to construct monuments, leaving permanent and enduring markers in the landscape. At present, there are only a small number of radiocarbon dates for both Irish and Scottish chambered tombs. In many cases these date the use, not the construction, of the monument, and it seems likely that many sites were in use for considerable periods of time. Ideally, we need a whole new series of dates for these sites, comparable with the ones acquired, and modelled, for the Cotswold–Severn sites in southern Britain (cf. Bayliss and Whittle 2007). There is an extensive literature on these chambered tombs and there is no space here to even begin to summarize the range of arguments. However, in recent years archaeologists have become increasingly interested in the landscape setting of these sites (e.g. Tilley 1994), and since this was the overall theme of the conference as well as this paper, I want to move swiftly on to a brief consideration of the landscape settings of these sites, and what they can tell us about how people understood the landscape at this time.

Landscape and the chambered tombs of the Irish Sea zone

There are two main forms of chambered tomb in the Irish Sea zone: portal dolmens and court tombs/Clyde tombs (Figure 11.1). They were all constructed from large slabs

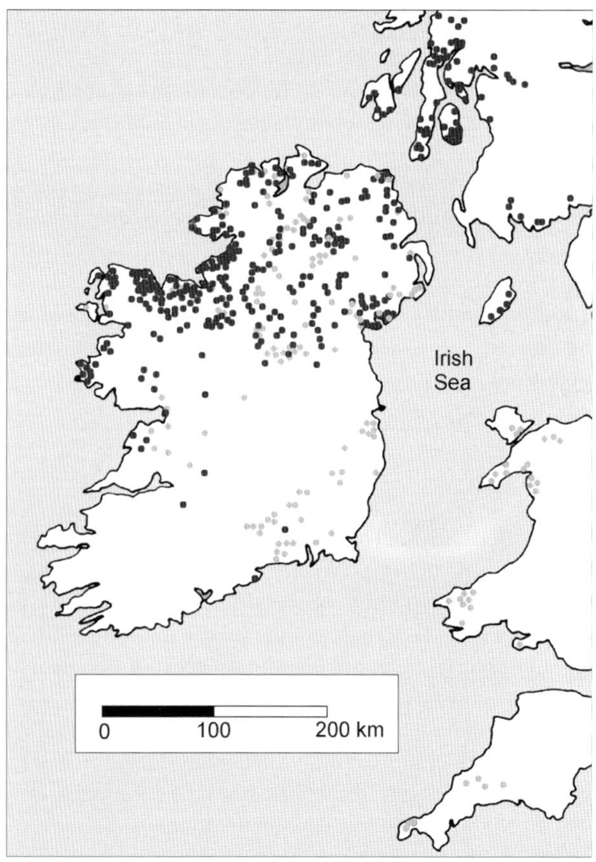

Fig. 11.1. The distribution of chambered tombs in the Irish Sea zone (dark grey = court cairns/Clyde cairns; light grey = dolmens).

Fig. 11.2. Western Scottish Clyde cairns and mountains (Achnagoul and Creag Mhor).

Fig. 11.3. Northeast Irish court cairns and mountains (Dunnaman and Goward).

of rock and contained a standard set of deposits which included human bone, burnt material, pottery and lithics (see Cummings 2009 for a detailed discussion). Some sites have forecourt areas, where people may have gathered and feasted. Many sites were set within long cairns. All sites have been interpreted as representing the construction of a distinctive form of architecture which created a permanent locale in the landscape.

Over the past 10 years I have visited almost all of the chambered tombs found on the shores of the Irish Sea, a total of around 350 sites (Cummings 2009), in order to assess their landscape setting. I had a number of key research questions. What landscape features were visible from these sites and what might their location have meant? Was it possible to identify regional groupings of sites not only through architecture but also through landscape setting? Did these sites represent something different from what had come before, particularly in terms of settings used and referenced?

The result of the examination of the settings of these megaliths was rather surprising. Early Neolithic monuments throughout western Britain and eastern Ireland, the entire Irish Sea zone, were located in very specific and virtually identical parts of the landscape. Monuments were repeatedly positioned in relation to specific landscape features, in particular mountains and water. Although I do not have space here to go into all the details, I do want to briefly summarize some of these relationships.

First, there is a clear connection between monuments and mountains. Megaliths are found only in the areas of western Britain and eastern Ireland which are mountainous. The flatter areas of the Irish Sea zone have notable blanks in the distribution. However, the small number of sites that do exist in flatter areas are very carefully positioned so that distant mountains *are* actually visible. The relationship between megaliths and mountains is very obvious when you are moving around these landscapes. For example, the landscape is quite flat along the east coast of Ireland between Dublin and northern Ireland, and as such is devoid of Early Neolithic megaliths, but as soon as the Mournes and Carlingford Mountain are visible, there is a profusion of monuments. However, megaliths are not built in the mountains themselves, but almost without exception are positioned in the landscape in order to have *views of mountains*. In western Scotland, monuments are situated so there are views of the stunning mountains around Loch Lomond and the visually distinctive mountains of the island of Arran (see Figure 11.2). The Arran mountains are visible from a remarkable number of sites across western Scotland. To the north, Ben Cruachan also

Fig. 11.4. North Welsh dolmens and mountains (Bachwen and Plas Newydd).

Fig. 11.5. View of Arran.

seems significant. In northeastern Ireland, monuments are concentrated in particular around the Antrim Hills to the north and the Mournes to the south (Figure 11.3). A number of megaliths are also carefully positioned so that there are views of Black Mountain, Slieve Gullion and Slemish. A similar relationship is found in Wales, where megaliths cluster around the mountains of north Wales, in particular Snowdonia and the mountains on the Lleyn Peninsula (Figure 11.4).

Overall, then, megaliths are positioned so that there is a view of mountains that are visually very distinctive. In all these areas it is not necessarily the highest mountains, but the ones which have the most distinctive profiles: for example, the mountains of Arran in western Scotland (Figure 11.5), Slieve Binnian in the Mournes in northern Ireland, and Carn Meini in southwest Wales.

So why might mountains have been important? Ethnography studies tell us that mountains often have a central role in mythology and cosmology. Many groups of native North Americans, for example, view mountains as highly symbolic and sacred locales; they are part of creation myths and stories and are key places in the ceremonial use of the landscape. Mountains and hills are often connected with mythical beginnings. The unique properties of mountains and the significance assigned to them in many cultures suggests, then, that mountains could have been considered highly sacred places in the Neolithic too. However, the issue of whether this significance was inherited (from hunter-gatherer populations) and passed on from generation to generation, or created anew, is open to debate.

Many monuments also have views of water, in particular rivers and the sea (Figure 11.6). In western Scotland the distribution of megaliths is coastal, and as such virtually all monuments are positioned with views of the sea or of lochs which are connected to the sea. Many sites are also found alongside rivers. As their name suggests, many of the Clyde tombs have views out over the Firth of Clyde. Other sites are located near streams and you get the sense that these guided people from the sea to the monument along a watercourse, marking out a pathway through the landscape (see Robinson 2007). Similarly, the megaliths of west Wales have a coastal distribution and many sites have clear views of the sea. Again, the sites that are not positioned with a view of the sea overlook rivers (Cummings and Whittle

Fig. 11.6. Monuments and water (Michael's Grave, Bute, and Carreg Samson, Pembrokeshire).

2004). In northeast Ireland almost two-thirds of sites have views of the Irish Sea, even though a considerable number of sites are many miles inland.

Views of mountains and water may not seem remarkable, yet when one contrasts the settings of stone circles in western Britain with their megalithic counterparts, only 7 per cent of stone circles have a view of the sea. It is also interesting that some areas of the Irish Sea zone have no megaliths at all: for the most part these are the areas which afford views of neither visually striking mountains nor the sea. We know these areas were inhabited in the Neolithic, as evidenced by lithic scatters, but they did not see the construction of megalithic architecture. I will address this point again below.

So why do monuments throughout western Britain and eastern Ireland have views of the sea? Just as in the Mesolithic, the sea may have had a highly potent place in worldviews. The sea is a liminal place, and the inter-tidal zone an interface between the land and the sea (Scarre 2002). It provides life but also takes it away. A sea view may also be referencing a (perceived or real) link between Britain and Ireland, and all other parts of the Irish Sea zone. It is interesting that stable isotope analysis shows that people stopped eating things from the sea in the Neolithic (Schulting and Richards 2002), perhaps because it was increasingly connected with the dead (Fowler and Cummings 2003; Thomas 2003). I suggested above that Mesolithic people may have deposited their dead in rivers, and they may also have used the sea in this manner. Perhaps monuments actually represent a real change in how people disposed of the dead, while also making connections with ancestral burial grounds (the sea). Here, then, is evidence to suggest elements of continuity with Late Mesolithic *coastal* populations but also *new* ways of doing things: with people, with places and with things. It might also be possible that a view of the sea was actually concerned with origins, particularly of monument construction and perhaps of the Neolithic itself (whether or not that involved the movement of people or ideas I do not wish to discuss further here).

Let us return to the setting of the chambered tombs of the Irish Sea. Megaliths are located in relation to mountains and the sea, but they are also repeatedly positioned in very similar 'zones' of the landscape. They are never located high in the mountains or on beaches, although there are views of these areas. Inland and mountainous areas are avoided. Instead megaliths are positioned on the side of hills between the mountains and the sea, and this means that in one direction there is a wide view, of mountains and water, and in the other the view is closed or restricted (see Cummings 2009 for full details). Monuments are also carefully orientated in relation to these landscape features, often so that there is a view of water *behind* the monument as you stand at the entrance, and so that there are views of mountains to one side and a closed view to the other. The orientation of the cairn 'opens up' views of specific landscape features on entry or exit and also places the viewer so that long cairn orientation divides the view into two sides. Megaliths are also positioned so that if one approached from downslope, typically from water such as the sea or a river, the monument is highly visible in the landscape (false crested) on approach. If the approach is from inland, the monument remains virtually invisible until you reach the site.

So why were these particular settings appropriate? What might they mean in terms of experiencing and understanding the landscape, or for addressing issues of continuity or change? Elsewhere, Alasdair Whittle and I (Cummings and Whittle 2004) have argued that these places in particular were of significance in a local sense. It is the most visually distinctive mountains that are visible from these sites, places that would have been important for navigating and moving around the landscape and also part of creation mythology and cosmology. The sea was also likely to have been imbued with significances, as both a connecting substance and a highly symbolic natural feature. Furthermore, the locations may well have been important nodes in the landscape long before a monument was constructed (there is clear evidence of this at a number of sites in the form of pre-cairn activity: see Cummings 2009). I am, therefore, strongly in favour of these places having been significant for long periods of time, not just the few hundreds of years between the arrival of Neolithic material culture and the start of monument construction. The places visible from megaliths were steeped in mythology, intimately connected to histories and stories and worldviews that had their origins way back in the Late Mesolithic, if not earlier. These specific places were locales that were appropriate for megalithic architecture *because* of their histories, not as a result of writing new histories onto the landscape. This argument may also go some way towards explaining why there are blank areas in western Britain where no megaliths were constructed. Perhaps these areas did not have the right 'sets' of landscape features and associated mythologies and histories for the construction of such architecture. Or perhaps the people who were living there did not have access to the landscape history which made places so potent. Landscape has agency: it makes people act and respond in particular ways and in some areas this involved megalithic construction, in others it did not.

It is tempting to see that the construction of monumental architecture 200 years into the Neolithic as representing a fundamental shift in terms of landscape and place. It is easy to argue that certain places in the Early Neolithic were now marked out in very distinctive ways. But a consideration of Mesolithic sites shows that this was not the case. Mesolithic examples of altered places include everything from the largest of shell middens in western Scotland to small-scale clearances of woodland. These acts were as much about changing and marking out place as was the construction of a chambered tomb. My research shows that there was only a very limited range of places chosen for the construction of megalithic architecture: only certain landscape features were appropriate to be seen from chambered tombs and only certain parts of the whole of Britain and Ireland saw the construction of megalithic architecture. Vast swathes of this area saw no discernible change from the Mesolithic in terms of the ways that people marked out the landscape. For much of Britain and Ireland, evidence for the Early Neolithic continues to be small-scale deposits of material in pits or on the ground.

Conclusions and future research questions

In this paper I have considered Late Mesolithic and Early Neolithic landscapes. While discussing the Mesolithic material only generally, I have looked at a specific case study from an Early Neolithic context: the possible significances of the setting of chambered tombs. I think this has raised some possibilities for considering my opening question of whether or not Mesolithic hunter-gatherers had a fundamentally different view of landscape from that of Neolithic farmers. This research suggests that was not the case, and that what was different about the two periods is the nature of the evidence we are studying. Chambered tombs are by far the most luminous type of evidence from the Late Mesolithic and Early Neolithic, but, along with shell middens, they skew what may otherwise be considered fairly similar engagements with, and alterations of, landscape. However, I have considered only one of a whole range of aspects of Early Neolithic life. There is certainly much more to be said about the monuments themselves, in terms of deposition, the role of experience, their use and reuse and movements through and around the landscape. All of these issues are addressed elsewhere (Cummings 2009) but cannot be detailed here. Looking beyond the monumental evidence, there are other key issues. How, for instance, did the change in the use of the landscape affect worldviews? Herding cattle has the *potential* to be a radically different engagement with both animals and place than hunting, although again it does not necessarily mean a massive change in terms of belief systems (it is possible that Early Neolithic people were still animistic, and it is also possible that they did not necessarily conceive of a fundamental difference between wild and domestic: see, for example, Barnard 2007 and Kent 1989). There is also much to be said about the procurement of specific stones from specific places and how this also affects and alters perceptions of landscape.

I argue, therefore, that we may still be some way from really being able to get to grips with the questions we want to ask. This is not due to our inability to think beyond the confines of our own historical and cultural context, as there is now a great deal of exciting work on a whole variety of different issues; it seems to be more a limitation of the data we have to work with. We get distracted by the luminous or prolific forms of evidence (here I am as guilty as anyone else, as I have spent most of my research time looking at chambered tombs) when the answers to our questions lie with the more 'mundane' lithic scatters and suites of material left, deliberately and purposefully, by people in both the Mesolithic and Neolithic. I am currently trying to do this in one small part of Scotland, but hopefully the results of this, and other projects underway or completed, can eventually sit against the background painted by the chambered tombs and other forms of evidence, and we may be able to comment further on whether monuments really did represent the formalization of places already sacred in the Mesolithic world.

Acknowledgements

Many thanks to Graeme Warren and Bill Finlayson for their invitation to participate at this event and for their useful editorial comments on this paper.

Bibliography

Barnard, A. (2007) From Mesolithic to Neolithic modes of thought. Pp. 5–20 in A. Whittle and V. Cummings (eds), *Going Over: The Mesolithic–Neolithic Transition in North-West Europe*. London: British Academy and Oxford University Press.

Barton, R., Berridge, P., Walker, M. and Bevins, R. (1995)

Persistent places in the Mesolithic landscape: an example from the Black Mountain uplands of south Wales. *Proceedings of the Prehistoric Society* 61, 81–116.

Basso, K. (1996) Wisdom sits in places: notes on a Western Apache landscape. Pp. 53–90 in S. Feld and K. Basso (eds), *Senses of Places*. Santa Fe, NM: School of American Research Press.

Bayliss, A. and Whittle, A. (eds) (2007) *Histories of the Dead: Building Chronologies for Five Southern British Long Barrows*. *Cambridge Archaeological Journal* 17/1 supplement.

Bradley, R. (2005) *Ritual and Domestic Life in Prehistoric Europe*. London: Routledge.

Bradley, R. and Edmonds, M. (1993) *Interpreting the Axe Trade: Production and Exchange in Neolithic Britain*. Cambridge: Cambridge University Press.

Carmichael, D., Hubert, J., Reeves B. and Schande A. (eds) (1994) *Sacred Sites, Sacred Places*. London: Routledge.

Chatterton, R. (2006) Ritual. Pp. 101–20 in C. Conneller and G. Warren (eds), *Mesolithic Britain and Ireland: New Approaches*. Stroud: Tempus.

Cleal, R., Walker, K. and Montague, R. (1995) *Stonehenge in its Landscape: Twentieth Century Excavations*. London: English Heritage.

Conneller, C. (2006) Death. Pp. 139–64 in C. Conneller and G. Warren (eds), *Mesolithic Britain and Ireland: New Approaches*. Stroud: Tempus.

Conneller, C. and Warren, G. (eds) (2006) *Mesolithic Britain and Ireland: New Approaches*. Stroud: Tempus.

Cooney, G. (2000) *Landscapes of Neolithic Ireland*. London: Routledge.

Cummings, V. (2003) Mesolithic world-views of the landscape in western Britain. Pp. 74–81 in L. Larsson, H. Kindgren, K. Knutsson, D. Leoffler and A. Åkerlund (eds), *Mesolithic on the Move: Papers Presented at the Sixth International Conference on the Mesolithic in Europe, Stockholm 2000*. Oxford: Oxbow Books.

Cummings, V. (2007) From midden to megalith? The Mesolithic–Neolithic transition in western Britain. Pp. 493–510 in A. Whittle and V. Cummings (eds), *Going Over: The Mesolithic–Neolithic Transition in North-West Europe*. London: British Academy and Oxford University Press.

Cummings, V. (2009) *A View from the West: the Neolithic of the Irish Sea Zone*. Oxford: Oxbow Books.

Cummings, V. and Robinson, G. (2007) *Interim Report for the Southern Kintyre Project, 2007 Season*. Specialist Report Number 5. Preston: University of Central Lancashire Studies in Archaeology.

Cummings, V. and Whittle, A. (2003) Tombs with a view: landscape, monuments and trees. *Antiquity* 77, 255–66.

Cummings, V. and Whittle, A. (2004) *Places of Special Virtue: Megaliths in the Neolithic Landscapes of Wales*. Oxford: Oxbow Books.

Edwards, K. and Ralston, I. (1997) *Scotland: Environment and Archaeology, 8000 BC– AD1000*. Chichester: Wiley.

Fowler, C. and Cummings, V. (2003) Places of transformation: building monuments from water and stone in the Neolithic of the Irish Sea. *Journal of the Royal Anthropological Institute* 9, 1–20.

Gell, A. (1995) The language of the forest: landscape and phonological iconism in Umeda. Pp. 232–54 in E. Hirsch and M. O'Hanlon (eds), *The Anthropology of Landscape: Perspectives on Place and Space*. Oxford: Oxford University Press.

Hirsch, E. and O'Hanlon, M. (eds) (1995) *The Anthropology of Landscape: Perspectives on Place and Space*. Oxford: Oxford University Press.

Jordan, P. (2003) Investigating post-glacial hunter-gatherer landscape enculturation: ethnographic analogy and interpretive methodologies. Pp. 128–38 in L. Larsson, H. Kindgren, K. Knutsson, D. Loeffler and A. Åkerlund (eds), *Mesolithic on the Move: Papers Presented at the Sixth International Conference on the Mesolithic in Europe, Stockholm 2000*. Oxford: Oxbow Books.

Kador, T. (2007) Stone age motion pictures: an object's perspective. Pp. 32–44 in V. Cummings and R. Johnston (eds), *Prehistoric Journeys*. Oxford: Oxbow Books.

Kahn, M. (1990) Stone-faced ancestors: the spatial anchoring of myth in Wamira, Papua New Guinea. *Ethnology* 29, 51–66.

Kent, S. (1989) Cross-cultural perceptions of farmers as hunters and the value of meat. Pp. 3–32 in S. Kent (ed.), *The Implications of Sedentism*. Cambridge: Cambridge University Press.

McFadyen, L. (2006) Landscape. Pp. 121–39 in C. Conneller and G. Warren (eds), *Mesolithic Britain and Ireland: New Approaches*. Stroud: Tempus.

Mellars, P. (1987) *Excavations on Oronsay*. Edinburgh: Edinburgh University Press.

Mithen, S., Pirie, A., Smith, S. and Wicks, K. (2007) The Mesolithic–Neolithic transition in western Scotland: a review and new evidence from Tiree. Pp. 511–42 in A. Whittle and V. Cummings (eds), *Going Over: The Mesolithic–Neolithic Transition in North-West Europe*. London: British Academy and Oxford University Press.

Pitts, M. (2007) Why did hunter-gatherers dig row of pits in Scotland? *British Archaeology* 93, 3.

Pollard, J. (2000) Ancestral places in the Mesolithic landscape. *Archaeological Review from Cambridge* 17, 123–38.

Price, B. (2007) Journeying into different realms. Travel, pilgrimage and rites of passage at Graig Lwyd. Pp. 85–101 in V. Cummings and R. Johnston (eds), *Prehistoric Journeys*. Oxford: Oxbow Books.

Robb, J. and Miracle, P. (2007) Beyond 'migration' versus 'acculturation': new models for the spread of agriculture. Pp. 99–116 in A. Whittle and V. Cummings (eds), *Going Over: The Mesolithic–Neolithic Transition in North-West Europe*. London: British Academy and Oxford University Press.

Robinson, G. (2007). *The prehistoric island landscape of Scilly*. British Archaeological Reports, British Series 447. Oxford: BAR.

Scarre, C. (2002) A pattern of islands: the Neolithic monuments of north-west Brittany. *European Journal of Archaeology* 5, 24–41.

Schulting, R. and Richards, M. (2002) The wet, the wild and the domesticated: the Mesolithic–Neolithic transition on the west coast of Scotland. *European Journal of Archaeology* 5, 147–89.

Sheridan, A. (2004) Neolithic connections along and across the Irish Sea. Pp. 9–21 in V. Cummings and C. Fowler (eds), *The Neolithic of the Irish Sea: Materiality and Traditions of Practice*. Oxford: Oxbow Books.

Sheridan, A. (2007) From Pickardie to Pickering and Pencraig Hill? New information on the 'Carinated Bowl Neolithic' in northern Britain. Pp. 441–92 in A. Whittle and V. Cummings (eds), *Going Over: The Mesolithic–Neolithic Transition in North-West Europe*. London: British Academy and Oxford University Press.

Sundstrom, L. (1996) Mirror of heaven: cross-cultural transference of the sacred geography of the Black Hills. *World Archaeology* 28, 177–89.

Thomas, J. (1999) An economy of substances in earlier Neolithic Britain. Pp. 70–89 in J. Robb (ed.), *Material Symbols*. Carbondale, IL: Southern Illinois University Press.

Thomas, J. (2003) Thoughts on the 'repacked' Neolithic revolution. *Antiquity* 77, 67–74.

Thomas, J. (2007) Mesolithic–Neolithic transitions in Britain: from essence to inhabitation. Pp. 423–40 in A. Whittle and V. Cummings (eds), *Going Over: The Mesolithic–Neolithic Transition in North-West Europe*. London: British Academy and Oxford University Press.

Tilley, C. (1994) *A Phenomenology of Landscape*. Oxford: Berg.

Tilley, C. (1996) *An Ethnography of the Neolithic*. Cambridge: Cambridge University Press.

Tilley, C. (2007) The Neolithic sensory revolution: monumentality and the experience of landscape. Pp. 329–46 in A. Whittle and V. Cummings (eds), *Going Over: The Mesolithic–Neolithic Transition in North-West Europe*. London: British Academy and Oxford University Press.

Warren, G. (2007) Mesolithic myths. Pp. 311–28 in A. Whittle and V. Cummings (eds), *Going Over: The Mesolithic–Neolithic Transition in North-West Europe*. London: British Academy and Oxford University Press.

Whittle, A. (2007) The temporality of transformation: dating the early development of the southern British Neolithic. Pp. 377–98 in A. Whittle and V. Cummings (eds), *Going Over: The Mesolithic–Neolithic Transition in North-West Europe*. London: British Academy and Oxford University Press.

Whittle, A. and Cummings, V. (eds) (2007) *Going Over: The Mesolithic–Neolithic Transition in North-West Europe*. London: British Academy and Oxford University Press.

Woodman, P. C., Anderson, E. and Finley, N. (1999) Excavations at Ferriter's Cove 1983–95: Last Foragers, First Farmers in the Dingle Peninsula. Bray: Wordwell.

12. 'Islanding' the Mesolithic–Neolithic Transition: Approaches to Landscapes of Contact and Transformation in Northwest Europe

Gordon Noble

The recent literature on the adoption of agriculture in Northwest Europe leaves the impression that change occurred in uniform ways across large swathes of divergent landscapes. While this issue is the subject of intense debate this often remains at a coarse scale of analysis, examining change within a modern national or even international context. Debate is also polarized between extremes of indigenous adoption and external introduction, resulting in explanations that are applied across-the-board to different contexts and different landscape situations. To explore change at a more regional or local level I suggest that we should begin to 'island' our approaches to the Mesolithic–Neolithic transition. We can do this in at least three ways. Firstly, taking our cue from the discipline of island archaeology, we can treat regions on an equal basis, aware of how modern political conditions affect our concepts of what is either central or marginal. Secondly, we can examine the ways in which change occurred in islands like Britain and Ireland by more fully considering maritime traditions in a region where new ideas were communicated through sea travel. Finally, we can use the concept of an island as metaphor to conceptualize sequences of agricultural adoption. I outline two 'islanded' case studies, one that considers a group of actual islands, those of the British Isles, and another that examines 'islands of agriculture' in the forested landscapes of Jutland, Denmark.

Introduction

Despite a large increase in the identification and publication of Late Mesolithic and Early Neolithic sites in Northwest Europe the nature of the transition between these two 'periods' remains as problematic and contentious as ever. This is in part due to the scale at which this issue has been addressed. Many studies examining the changes between Holocene hunter-gatherers and early farming communities remain at a national or even international scale, producing general models for the ways in which communities in the 5th and 4th millennia BC began to abandon traditional lifestyles and adopt agricultural resources (e.g. Ammerman and Cavalli-Sforza 1984; Douglas Price 2000; Rowley-Conwy 2004; Thomas 1996). These types of approaches are of course useful, but leave the impression that changes in economy and social structure occurred in uniform ways across the variable landscapes of Europe. This is particularly problematic when considering locations such as the fragmented island archipelago(es) of Britain and Ireland, where marked regional traditions of Neolithic material culture, monumentality and settlement suggest that more complex transitions occurred.

Debate concerning the Mesolithic–Neolithic transition has been polarized between approaches advocating gradual adoption by indigenous hunter-gatherers, with a high level of mobility retention and a continuing reliance on largely wild resources (e.g. Bradley 1993; Thomas 1988; 1991; 1999; Whittle 1996), and those arguing for a more rapid and wholesale change, perhaps even involving some form of population movement of already-established farming communities from continental Europe (e.g. Richards and

Hedges 1999; Rowley-Conwy 2004; Schulting and Richards 2002; Sheridan 2000; 2007). These extreme positions leave little common ground. These approaches also tend to generalize data from specific regions or datasets and apply them elsewhere: one example of this is the use of human bone material from the island of Oronsay, a tiny Scottish island, to postulate diet in Mesolithic populations across Britain and Ireland (Schulting and Richards 2002). Considering that Oronsay currently supports a population of five people, the wider applicability of these results remains to be seen. There is also a trend to see areas with long histories of study, often modern political centres, as the norm from which other places are divergent and/or irrelevant (Barclay 2001). However, this misses the point that each region is likely to have witnessed different processes of change and that each must be independently assessed as a means of building up a picture of what is likely to have been a complex process of change. While widespread adoptions of certain forms of material culture did occur as part of this process, there is also much evidence of regionalized traditions and patterns which must not be ignored. In this respect, one of the ways of highlighting diversity is to take our cue from recent developments in island archaeology (e.g. Broodbank 2000; Fitzpatrick 2004; Rainbird 2003). By this, I do not mean that we all necessarily need to explicitly adopt the methodologies of island archaeology, more that we need to be aware of the potential diversity of regionalized sequences of change and to examine the ways in which change may not have been uniform, in that certain areas may have had precocious, early adoptions of agriculture, while others may have been relatively unchanged for some time. We need to regionalize or localize our models of the transition, building them from the bottom up rather than in the generalized, top-down way that tends to happen at present.

Islanding the Mesolithic–Neolithic transition

In order to adopt a more regionalized approach to the Mesolithic–Neolithic transition(s), utilizing approaches from island archaeology may be profitable. There are at least three ways in which island archaeology might impact on our study of the Mesolithic–Neolithic transition.

Firstly, approaches to the study of islands can take many forms, but there is a general consensus in recent literature that islands (and I argue we should extend this to regions generally, whether islands or land-bounded areas) should be considered in their own right, not assuming the general applicability of models developed in specific regions (Broodbank 2000; Fitzpatrick 2004; Rainbird 2003). In this manner, inter-regional pictures can be constructed more fruitfully from the analysis of smaller-scale trends. Regions cannot be treated as 'laboratories' for the study of wider cultural processes (*contra* Evans 1973; Darvill 2000), but must instead be assessed on their own terms, in relation to, but not dependent upon, other regional patterns. The main implication of this is that we must fully consider all sequences of regional change, including areas or regions that have been considered peripheral or marginal, and not use one dataset to characterize another.

Secondly, island archaeology can be used to examine the Mesolithic–Neolithic transition by adopting the actual methodologies of island archaeological analyses. For example, Broodbank has recently studied the ways in which maritime traditions and conditions will have limited or facilitated social change on islands (Broodbank 2000). These considerations seem appropriate when considering Britain and Ireland, an island archipelago which by the beginnings of the Neolithic had been separated from the European mainland for some millennia. Thus both the material attributes and ideas behind farming must have been transmitted through sea travel, yet island archaeological approaches have rarely been considered in models of the Mesolithic–Neolithic transition in this geographical context (Noble 2006a, chapter 2).

Thirdly, island perspectives can be adopted on a metaphorical level as a means for more fully conceptualizing the changes that occurred during the transition to farming. Farming is unlikely to have been adopted or introduced wholesale across every region of Northwest Europe simultaneously; instead, there may have been islands of agricultural uptake, surrounded by people and landscapes relatively untouched by the new ideas, practices and material culture associated with farming.

In order to examine the ways in which the perspectives of island archaeology can be fruitfully applied to the Mesolithic–Neolithic transition I outline two case studies, each at different scales of analysis. One involves the archaeology of a series of actual islands (the British Isles), while the second examines the use of islands as a metaphor in a more land-bounded area in Jutland, Denmark.

The Mesolithic–Neolithic transition in Britain and Ireland

In the British Isles there has been a backlash against the acculturation models for the Mesolithic–Neolithic transition proposed most strongly in the work of Thomas in the 1990s (Thomas 1991; 1996; 1999). In particular, recent work by Sheridan (2000; 2007) and Rowley-Conwy (2004) provides strong criticism of earlier ideas on the nature of the Mesolithic–Neolithic transition. While I would not disagree that aspects of the Neolithic – material culture, monumentality and economy – arrived on the islands more rapidly than has been suggested in the past and that some degree of small-scale movement of farming communities to these islands may have been part of this process of change, we must be wary of the tendency in these revisions to homogenize and typecast the Mesolithic and Neolithic in the Isles and to criticize the necessity of theory and interpretative frameworks in archaeology (See comment by A. Jones in Rowley-Conwy 2004).

The wider-scale patterns among which the earliest traces of the Neolithic occurred in the British Isles should not obscure the fact that if we examine all aspects and areas we find a highly regionalized Neolithic with many temporal patterns of settlement, monumentality, material culture and practice. There are of course aspects which cut across these patterns, such as Carinated Bowl pottery (Sheridan 2007), but we must not favour one form of material culture as a sole characterization of a particular identity or origin: Hodder showed many years ago the problems in assessing group contact, identity and origins from material culture alone (Hodder 1982). We have to acknowledge the limitations that studies of material culture have in assessing the degree to which the Neolithic was a product of colonization or acculturation. Woodman and McCarthy (2003, 36) note the paradox of Later Mesolithic Ireland, where the lithic material suggests insularity, whereas other evidence that is much less likely to survive suggests the presence of much longer-distance contacts. Similarly, in a different geographical context, Robb has highlighted the way in which communities can produce distinctive, and what we can mistakenly identify as insular, forms of material culture in response to social contact as opposed to lack of contact (Robb 2001).

We must be mindful of the likely scale and population size of Late Mesolithic/Early Neolithic communities. The idea of agricultural communities arriving juggernaut-style, devastating everything before them, as implied by Rowley-Conwy (2004), seems unlikely. As I show in my Jutland case study, agriculture and agricultural settlements are likely to have been small islands within a wider environment relatively untouched by the new lifestyles and practices associated with the Neolithic. The indigenous community must therefore have had roles to play in the uptake and trajectory of early farming in the British Isles.

Theoretically informed interpretative frameworks are necessary for contextualizing the nature and significance of the patterns evident at the beginning of the 4th millennium BC. If we invoke colonization, for example, we must have a clear concept of what colonization entails. The introduction of new lifestyles and material culture, or even of new populations, rarely occurs overnight; adoption is almost always dependent upon some form of interaction prior to the main watershed of change (e.g. Anthony 1990). Given the maritime nature of the contact this contact is likely to have been initiated by the coastal indigenous communities of the British Isles, whether or not population movement from the continental landmass was a part of the process of change at a later date. In this respect, actual explanations of the mechanics of change at this period are limited. I do not mean that we have not searched hard enough for origin points or causes for population movement on the continent, but that in an island context the new practices and material culture associated with the Neolithic arrived by sea and we have to consider the ways in which maritime practices and traditions were integrated into the changes we identify now as the Mesolithic–Neolithic transition. This is the first point in which an island archaeology of this period and region would seem useful.

Coastal and island traditions of the British Isles

The earliest examples of domesticated animals found in the British Isles so far are not from the areas of mainland Britain closest to the European mainland, as might have be expected, but from the west coast of Ireland. These were uncovered at Ferriter's Cove in Co. Kerry, in the southwest of Ireland, where cattle and sheep bones have been found within Late Mesolithic occupation deposits (Woodman *et al.* 1999; Woodman and McCarthy 2003, 32). These have been dated to the 5th millennium BC and were associated with contemporary human bones. These date from 500–700 years or more prior to the recognized widespread, rapid changes at the beginning of the 4th millennium in the British Isles. The occurrence of such early domesticates suggests that the individuals who deposited these at Ferriter's Cove may well have been among the first to bring domesticated resources into Britain and Ireland – or to receive them – many centuries before the more wholesale introduction of domesticates elsewhere (Woodman and McCarthy 2003, 33; Tresset 2003). The evidence from Ferriter's Cove has been described as a one-off, abortive attempt at Neolithization (Sheridan 2007), but it is overly premature at present to discount this evidence as irrelevant to the later transformations on these islands. Bone evidence from Mesolithic and Neolithic sites is not exactly abundant and we are missing the contemporary coastline in many areas (Woodman and McCarthy 2003, 32, 36). Without bone preservation, Ferriter's Cove would simply be another artefact scatter, much like the many hundreds found elsewhere. The presence of domesticates in Ireland at such an early date, even in such small quantities, has important implications for contacts between farming communities on continental Europe and the indigenous Mesolithic communities on the islands to the west of the British mainland in the centuries prior to the 4th-millennium changes. Woodman and McCarthy (2003, 35) highlight other possible examples of domesticates in Mesolithic contexts in both Ireland and Brittany. There is also the evidence for pre-elm-decline cereal-type pollen which, while contentious, should not be dismissed outright (e.g. Davey and Innes 2003). We have to think harder about the actual means by which domesticates and other aspects of the Neolithic were introduced to the British Isles.

Sea travel has made only sporadic appearances in accounts of the Mesolithic–Neolithic transition, despite the obvious necessity of advanced levels of seamanship and large craft to carry domesticates from continental Europe. The Atlantic sea routes and the favourable configuration of islands and coast in the west of the British Isles encouraged the development of maritime cultures (Broodbank 2000, 131; Erdoğu 2003, 8), and Henshall (1972, 20) describes this part of the British and Irish archipelago as an indented coastline with a 'maze of lochs, inlets, small islands and skerries', as well as a number of island chains. The tidal streams of Britain and Ireland also favour maritime activity as the Atlantic sea routes channel movement up and down the west coast of Britain (Cunliffe 2001; Davies 1946; Kinnes 1984, 367; Mercer 2003, 69). The island geography, combined with the tidal streams of this region of the British Isles, are thus conducive to the development of maritime activities and arguably make this area more open to change and innovation (Clark 1977). One reason for the early occurrence of domesticates in places like the west coast of Ireland may therefore be the sea traditions of Mesolithic coastal and island communities (Case 1969; Mercer 2003; Noble 2006a, chapter 2; Woodman and McCarthy 2003). The presence of domesticated animal bones in what is not an altogether unusual Mesolithic context suggests that the early contacts with the European mainland were initiated by indigenous hunter-gatherer communities, and that Ireland could have been at the forefront of the adoption of the Neolithic, acting as a dispersion point for material culture and ideas associated with farming.

The early occurrence of traces of the Neolithic in areas that we might characterize as peripheral or marginal has parallels in the Aegean, where distinctive islands, important landmarks for sea travel, have earlier traces of the Neolithic than larger and possibly more productive locations (Erdoğu 2003). It is important not to be too rash in dismissing the possibilities of contacts between the maritime communities of the islands and the continent prior to the 4th millennium BC. Evidence may not occur in the places we might expect, and in this respect we have to be aware of how our modern conceptions of central and marginal affect our understandings of prehistory (Barclay 2001). The Neolithic in the British Isles was not a matter of continental farmers deciding to get on a boat one day and pop over the Channel for a new life. Maritime traditions and skills are complex, restricted and dependent upon generational knowledge and expertise (Noble 2006a, chapter 2; 2006b; Sturt 2005). A deeper understanding of maritime traditions and island archaeology theory should be an important part of new perspectives on the Mesolithic–Neolithic transition in this context.

Regional islands: exploring island dynamism during the Mesolithic–Neolithic transition

In considering the British and Irish Neolithic in the 4th millennium BC we must also be careful not to assume island-wide patterns or traditions once the majority of the traditionally conceived markers of Neolithic economy and ideology were adopted. We must consider carefully the scale at which we analyse this material. As mentioned earlier, there is a tendency to use particular regionalized sets of evidence to make more general cases on a national or even international level. Rowley-Conwy (2004), for example, attempts to return the house to the forefront of Neolithic societies in the British Isles, arguing that the Neolithic was characterized by communities living fully agricultural, sedentary lifestyles and using the explosion of house discoveries in Ireland as one of the supports for this model. The evidence from Ireland should not, however, be transported to other island locations in the British Isles (Bradley 2003). The buildings in Ireland are very different from the ones found in Scotland, for example, and cannot be seen as ubiquitous for the Neolithic in the British Isles. There is no need to use the evidence for these houses (and that is a loaded term) to suggest their presence in every region of the British Isles in the Neolithic. As Garrow (this volume) points out, too much weight has been given to the presence/absence of houses in the Neolithic as a marker of particular lifestyles or origins, without exploring the variability of evidence across time and space. In East Anglia the number of structures that could be termed houses has remained extremely low despite the number of large-scale development-led excavations that has been carried out over the last few decades. In many parts of lowland Scotland the number of identifiable houses also remains low and the interpretation of possible domestic buildings is complex (Barclay *et al.* 2002; Topping 1996). It is not necessary or plausible to cite taphonomic reasons for the sporadic occurrence of what we could term houses in these regions: we should take the negative evidence as being characteristic of a different type of Neolithic. This is not necessarily the mobile Neolithic envisioned by Thomas, but nor is it necessarily the sedentary, house-based communities that Rowley-Conwy and others envision.

In Ireland the rectangular buildings that have been such an influence on the revisionist view of the Neolithic are themselves restricted to a period perhaps as short as a century (Smyth 2007) and do not necessarily date to the very start of the Neolithic; nor are they all the same in form, function or in terms of associated finds (see papers in Armit *et al.* 2003; Smyth 2007). In all regions of the British Isles there are also traditions (much less discussed) of slighter round structures, stake-defined settings, pit complexes and other features, all of which may well describe more widespread (geographically and chronologically), if ephemeral, settlement strategies in the Neolithic (Atkinson 2002; Alexander 2000; Cooney 2000). The small but increasing number of Mesolithic houses from the British Isles and from Scandinavia provides a welcome reminder that the presence or absence of houses does not describe a particular economy or identity (e.g. Andersson *et al.* 2004). These regional patterns and traditions must be incorporated into more nuanced views of the Mesolithic–Neolithic transition and the nature of the Neolithic once established. While traditions such as Carinated Bowl pottery cut across some of these regional patterns, it is important to incorporate diverse regionalized and localized traditions and practices into a more coherent view of the Neolithic(s) of the British Isles. It is important to reconsider the origins of the Neolithic in the British Isles, but not to replace earlier acculturation models with an equally static and all-encompassing colonization and house-based Neolithic model.

Island metaphors

In the second part of this paper I show how ideas of the Mesolithic–Neolithic transition can achieve a greater level of sophistication through the use of the concept of an island as a metaphor for exploring change at the regional and local level. It is important not to forget the probable scale of Neolithic societies and their impacts on the environment and landscape, and the fact that these are likely to have been extremely localized in the early stages of the Neolithic. Here I show again the ways in which the transition to agriculture, this time in Jutland, Denmark, were not uniform across large areas of landscape, but consisted of pockets or 'islands' of agriculture set among a much wider relatively unchanged landscape. These 'islands' where agriculture was first adopted can be seen as pioneer settlements whose influence impacted only on small areas, while the majority of the landscape remained relatively unmodified. I will look at the evidence for agriculture beneath and incorporated within earlier Neolithic long barrows in Denmark. The evidence for settlement and agriculture from these locations is often overlooked as the common focus has been on the final, monumental form of these barrows; however, the evidence from under these barrows can provide significant information on the uptake of farming in Denmark. These barrows seal agricultural traces that may have been among the first adoptions of agricultural practices in a landscape that remained largely unmodified from the forest environment that dominated Mesolithic lives. The findings from barrows show that the transition in this area may have also been an 'islanded' phenomenon. There are likely to have been certain social groups that embraced change before others and thus 'pioneer' landscapes, where agriculture was first adopted, can perhaps be identified. The regionalized patterns and a regional scale of analysis may therefore not be sufficient, as there is also a need to consider more localized sequences of change in the adoption of agriculture.

There has been a growing awareness of the close relationship between earlier Neolithic monumentality in Northwest Europe and settlement both in a physical and metaphorical sense (Bradley 1998; 2002; 2005; Hodder 1990; 1994; Midgley 2005). This includes similarities between the long barrows and related traditions of Northwest Europe and the *Linearbandkeramik* longhouse dwellings of Central Europe. These parallels may have operated on a largely symbolic level, as most longhouses of *Linearbandkeramic* type were no longer constructed by the time long mounds of this type were being constructed. The links between barrows and settlement also include more direct and physical relationships: many long barrows have been shown to overlie previous settlements, including timber dwellings, and everyday building materials were utilized in the construction of these monuments (e.g. Liversage 1981; 1992; Madsen and Jensen 1982). In Denmark, the relationship between earlier Neolithic barrows and settlement has been reinforced by evidence from pollen profiles from buried soils beneath barrows. These show that barrows were, almost without exception, constructed over areas that had been used previously for agriculture and were foci of settlement prior to the construction of the mounds (Andersen 1992, 1994–5). This reinforces excavated evidence from earlier Neolithic barrows in Denmark that often hints at the presence of former domestic dwellings below the mounds, which almost always overlie some form of midden deposit (e.g. Madsen and Jensen 1982; Liversage 1992).

While each pollen profile from the barrows that have been examined in Denmark is slightly different, there is a general consistency in the profiles according to the chronological phases of the Neolithic (Andersen 1992, 1994–5). In the earliest barrows the total percentages of tree pollen obtained are all generally high, indicating that the economy of these sites was clearly adapted to forest life (Andersen 1992; Madsen and Jensen 1982, 84). However, much of the woodland surrounding these barrows was altered by human action. In particular, the profiles show that burning of woodland was common and most profiles contain evidence of both pastoral and cereal farming in the form of barley and wheat grains and agricultural weeds. The profiles include much evidence for what Anderson has termed 'artificial' woodlands generally dominated by birch and hazel – that is, woodlands that regenerate after episodes of clearance (Andersen 1992, 169; 1994–5). The Late Mesolithic and Early Neolithic forest vegetation of Denmark, as shown in regional pollen diagrams, was one dominated by lime woodland. Woodland dominated by trees such as birch and hazel, as occurred in the vicinity of these long barrow sites, was secondary in nature, and 'created more or less intentionally by human action' (Andersen 1992, 165). In the extensive study of barrows constructed throughout the Neolithic in the Vrou area in West Jutland Andersen found that even in the earliest barrows 'natural' or unmodified vegetation was scarcely represented (Andersen 1994–5, 123). All of the pollen analyses from the barrows provide a consistent picture of barrow building being closely associated with areas that had been used for agriculture and places where the forest had been cleared to allow these agricultural practices to occur. However, the type of pollen profiles found under the barrows are poorly represented in the regional pollen sequences obtained from lakes or bogs, which show much less evidence for agricultural and clearance impacts on woodland at this time.

Pollen cores obtained from lakes or bogs provide an amalgam of human activities and vegetation types from a larger area and over a potentially greater time span than those from barrow sites. As a result regional cores are useful in assessing overall regional impacts by human communities, but are not as narrowly focused in terms of time and place as pollen spectra derived from the sealed contexts under barrows (Andersen 1988, 81). As noted above, the pollen compositions found under the barrows are only weakly reflected in these regional pollen diagrams (Andersen 1994–5, 127). Andersen suggests that this probably reflects a low population density in the Earlier Neolithic, but it also implies that the foci of barrow construction were far from randomly chosen, and were in fact exploitation islands in a sea of relatively unmodified woodland: islands of agriculture and secondary woodland set among a wider landscape still largely dominated by original lime woodland. In other words, the places where agriculture was first practised may have been relatively isolated locations, their presence marked simply by differences in the woodland vegetation. The surrounding forest world would have been relatively untouched and traditional forest hunter-gatherer lifestyles may have continued around these farming islands, albeit perhaps in a modified form through contact with farming communities.

Given that landscape change may have been a largely localized process, regional approaches to the Mesolithic–Neolithic transition, while adding sophistication to our current models, may not go far enough; even a regional scale of analysis may not be sufficiently fine-grained to pick up the local sequences of change that would have made up the complicated process of agricultural adoption in certain areas. In this respect, detailed examinations at a site and immediate landscape level are crucial to building wider patterns and adopting an approach that builds bottom-up locally and regionally based models, rather than top-down generalizing models. This places a great emphasis on the importance of developer-funded archaeology, as only large-scale development projects provide the framework and resources necessary in modern archaeology to gain snapshots of entire landscapes at both a local and regional level. The earliest Neolithic sites and the first agricultural practices in a region may have made relatively small-scale impacts on the local environment and landscape. To this end, landscape models at the smaller scales of analysis must also be 'islanded', and the use of islands as a means of conceptualizing change during the Mesolithic–Neolithic transition can be a useful way of creating more complex models of change.

Conclusions

We are a long way from identifying the multiple, tangled processes of change that led to the first agricultural communities in Northwest Europe. We need to resist the temptation to reduce agricultural origins to a simple issue of colonization/acculturation. Our frameworks of analysis must become more detailed and pitched at a variety of scales in order to construct complex models of transition that fully address the changes that occurred. Some of the methodologies and approaches of island archaeology may be a fruitful means of beginning to achieve these aims. In the island geography of Britain and Ireland, the methods and theoretical frameworks of island archaeology allow the consideration of regionalized sequences of change in studies of Neolithization. In addition, our concepts of core/periphery may have to alter before we can achieve more comprehensive views of the early stages of agricultural adoption in Britain and Ireland. A consideration of maritime traditions may also be crucial in examining the mechanisms of agricultural uptake in this island world. We also need new theoretical frameworks for studying material culture as the treatment of similarity or lack of similarity between particular forms of material culture is a dangerous means of assessing levels or lack of social contact in both the Mesolithic and Neolithic. At a more localized scale techniques such as palynology can identify the changes in the landscape that accompanied agricultural adoption. The use of closed contexts such as old ground surfaces beneath barrows can provide the localized pictures of environment and economy that will be crucial in building detailed understandings of Neolithization processes. It is time to abandon generalizing models of change. Instead we must island our approaches to the Mesolithic–Neolithic transition, albeit in models that are able to incorporate similarities between islands as much as differences.

Acknowledgements

The British Academy awarded the author an Overseas Conference Grant in 2007 to attend the conference. Part of this research was undertaken during a postdoctoral fellowship also funded by the British Academy. I gratefully acknowledge the support of the British Academy in both endeavours.

Bibliography

Alexander, D. (2000) Excavations of Neolithic pits, later prehistoric structures and a Roman temporary camp along the line of the A96 Kintore and Blackburn bypass, Aberdeenshire. *Proceedings of the Society of Antiquaries of Scotland* 130, 11–76.

Ammerman, A. J. and Cavalli-Sforza, L. L (1984) *The Neolithic Transition and the Genetics of Populations in Europe*. Princeton: Princeton University Press.

Andersen, S. Th. (1988) Pollen spectra from the double passage-grave, Klekkendehøj on Møn. *Journal of Danish Archaeology* 7, 77–92.

Andersen, S. Th. (1992) Early and Middle Neolithic agriculture in Denmark: pollen spectra from soils in burial mounds of the Funnel Beaker Culture. *Journal of European Archaeology* 1, 153–80.

Andersen, S. Th. (1994–5) Pollen analytical investigations of barrows from the Funnel Beaker and Single Grave cultures in the Vroue area, West Jutland, Denmark. *Journal of Danish Archaeology* 12, 107–32.

Andersson, M., Karsten, P., Knarrström, B. and Svensson, M. (2004) *Stone Age Scania: Significant Places Dug and Read by Contract Archaeology*. Lund: National Heritage Board.

Anthony, D. W. (1990) Migration in archaeology: the baby and the bathwater. *American Anthropologist* 92, 895–914.

Armit, I., Murphy, E., Nelis, E. and Simpson, D. (eds) (2003)

Neolithic Settlement in Ireland and Western Britain. Oxford: Oxbow Books.
Atkinson, J. A. (2002) Excavation of a Neolithic occupation site at Chapelfield Cowie, Stirling. *Proceedings of the Society of Antiquaries of Scotland* 132, 139–92.
Barclay, G. J. (2001) 'Metropolitan' and 'parochial'/'core' and 'periphery': a historiography of the Neolithic of Scotland. *Proceedings of the Prehistoric Society* 67, 1–18.
Barclay, G. J., Brophy, K. and MacGregor, G. (2002) Claish, Stirling: an Early Neolithic structure in its context. *Proceedings of the Society of Antiquaries of Scotland* 132, 65–138.
Bradley, R. (1993) *Altering the Earth: The Origins of Monuments in Britain and Continental Europe*. Society of Antiquaries of Scotland Monograph Series 8. Edinburgh: Society of Antiquaries of Scotland.
Bradley, R. (1998) *The Significance of Monuments*. London: Routledge.
Bradley, R. (2002) *The Past in Prehistoric Societies*. London: Routledge.
Bradley, R. (2003) Neolithic expectations. Pp. 218–22 in I. Armit, E. Murphy, E. Nelis and D. Simpson (eds), *Neolithic Settlement in Ireland and Western Britain*. Oxford: Oxbow Books.
Bradley, R. (2005) *Ritual and Domestic Life in Prehistoric Europe*. London: Routledge.
Broodbank, C. (2000) *An Island Archaeology of the Early Cyclades*. Cambridge: Cambridge University Press.
Case, H. (1969) Neolithic explanations. *Antiquity* 43, 176–86.
Clark, G. (1977) The economic context of dolmens and passage graves in Sweden. Pp. 35–49 in V. Markolic (ed.), *Ancient Europe and the Mediterranean*. Warminster: Amis and Phillips.
Cooney, G. (2000) *Landscapes of Neolithic Ireland*. London: Routledge.
Cunliffe, B. (2001) *Facing the Ocean*. Oxford: Oxford University Press.
Darvill, T. (2000) Neolithic Man in context. Pp. 371–85 in A. Ritchie (ed.), *Neolithic Orkney in its European Context*. Cambridge: McDonald Institute for Archaeological Research.
Davey, P. and Innes, J. (2003) The Early Neolithic and the Manx environment. Pp. 20–7 in I. Armit, E. Murphy, E. Nelis and D. Simpson (eds), *Neolithic Settlement in Ireland and Western Britain*. Oxford: Oxbow Books.
Davies, M. (1946) The diffusion and distribution pattern of the megalithic monuments of the Irish Sea and north channel coastlands. *Antiquaries Journal* 26, 38–60.
Douglas Price, T. (2000) *Europe's First Farmers*. Cambridge: Cambridge University Press.
Erdoğu, B. (2003) Visualizing Neolithic landscape: the early settled communities in Western Anatolia and eastern Aegean islands. *European Journal of Archaeology* 6, 7–23.
Evans, J. D. (1973) Islands as laboratories for the study of culture process. Pp. 517–20 in C. Renfrew (ed.), *The Explanation of Culture Change: Models in Prehistory*. London: Duckworth.
Fitzpatrick, S. M. (2004) *Voyages of Discovery: The Archaeology of Islands*. London: Praeger.
Henshall, A. S. (1972) *The Chambered Tombs of Scotland*. Edinburgh: Edinburgh University Press.
Hodder, I. (1982) *Symbols in Action: Ethnoarchaeological Studies of Material Culture*. Cambridge: Cambridge University Press.
Hodder, I. (1990) *The Domestication of Europe*. Oxford: Blackwell.
Hodder, I. (1994) Architecture and meaning: the example of Neolithic houses and tombs. Pp. 73–86 in M. Parker Pearson and C. Richards (eds), *Architecture and Order: Approaches to Social Space*. London: Routledge.
Kinnes, I. (1984) Microliths and megaliths: monumental origins on the Atlantic fringe. Pp. 367–7 in G. Burenhult (ed.), *The Archaeology of Carrowmore, Co. Sligo*. Stockholm: Institute of Archaeology.
Liversage, D. (1981) Neolithic monuments at Lindebjerg, Northwest Zealand. *Acta Archaeologica* 51, 85–152.
Liversage, D. (1992) *Barkaer: Long Barrows and Settlements*. Copenhagen: University of Copenhagen.
Madsen, T. and Jensen, H. J. (1982) Settlement and land use in Early Neolithic Denmark. *Analecta Praehistorica Leidensia* 15, 63–86.
Mercer, R. J. (2003) The early farming settlement of south western England in the Neolithic. Pp. 56–70 in I. Armit, E. Murphy, E. Nelis and D. Simpson (eds), *Neolithic Settlement in Ireland and Western Britain*. Oxford: Oxbow Books.
Midgley, M. (2005) *The Monumental Cemeteries of Prehistoric Europe*. Stroud: Tempus.
Noble, G. (2006a) *Neolithic Scotland: Timber, Stone, Earth and Fire*. Edinburgh: Edinburgh University Press.
Noble, G. (2006b) Harnessing the waves: monuments and ceremonial complexes in Orkney and beyond. *Journal of Maritime Archaeology* 1, 1–18.
Rainbird, P. (2003) *The Archaeology of Micronesia*. Cambridge: Cambridge University Press.
Richards, M. P. and Hedges, R. E. M. (1999) A Neolithic revolution? New evidence of diet in the British Neolithic. *Antiquity* 73: 891–7.
Robb, J. (2001) Island identities: ritual, travel and the creation of difference in Neolithic Malta. *European Journal of Archaeology* 4, 175–202.
Rowley-Conwy, P. (2004) How the West was lost: a reconsideration of agricultural origins in Britain, Ireland, and southern Scandinavia. *Current Anthropology* 45 (Supplement), 83–113.
Schulting, R. J. and Richards, M. P. (2002) The wet, the wild and the domesticated: the Mesolithic–Neolithic transition on the west coast of Scotland. *European Journal of Archaeology* 5, 147–89.
Sheridan, A. (2000) Achnacreebeag and its French connections: vive the 'auld alliance'. Pp. 1–15 in J. C. Henderson (ed.), *The Prehistory and Early History of Atlantic Europe*. British Archaeological Report, International Series 861. Oxford: BAR.
Sheridan, A. (2007) From Picardie to Pickering and Pencraig Hill? New information on the 'Carinated Bowl Neolithic' in northern Britain. Pp. 439–90 in A. W. R Whittle and V. Cummings (eds), *Going Over: The Mesolithic–Neolithic Transition in North-West Europe*. London: British Academy and Oxford University Press.
Smyth, J. (2007) Neolithic Settlement in Ireland: New Theories and Approaches. Unpublished PhD thesis, University College Dublin.
Sturt, F. (2005) Fishing for meaning: lived space and early Neolithic of Orkney. Pp. 68–80 in V. Cummings and A. Pannett

(eds), *Set in Stone: New Approaches to Neolithic Monuments in Scotland.* Oxford: Oxbow Books.

Thomas, J. (1988) Neolithic explanations revisited: the Mesolithic–Neolithic transition in Britain and southern Scandinavia. *Proceedings of the Prehistoric Society* 54, 59–66.

Thomas, J. (1991) *Rethinking the Neolithic.* Cambridge: Cambridge University Press.

Thomas, J. (1996) *Time, Culture and Identity: An Interpretative Archaeology.* London: Routledge.

Thomas, J. (1999) *Understanding the Neolithic.* London: Routledge.

Topping, P. (1996) Structure and ritual in the Neolithic house: some examples from Britain and Ireland. Pp. 157–70 in T. Darvill and J. Thomas (eds), *Neolithic Houses in Northwest Europe and Beyond.* Oxford: Oxbow Books.

Tresset, A. (2003) French connections II: of cows and men. Pp. 18–29 in I. Armit, E. Murphy, E. Nelis and D. Simpson (eds), *Neolithic Settlement in Ireland and Western Britain.* Oxford: Oxbow Books.

Whittle, A. (1996) *Europe in the Neolithic.* Cambridge: Cambridge University Press.

Woodman, P. and McCarthy, M. (2003) Contemplating some awful(ly interesting) vistas: importing cattle and red deer into prehistoric Ireland. Pp. 31–5 in I. Armit, E. Murphy, E. Nelis and D. Simpson (eds), *Neolithic Settlement in Ireland and Western Britain.* Oxford: Oxbow Books.

Woodman, P., Anderson, E. and Finlay, N. (1999) Excavations at Ferriter's Cove 1983–95: Last Foragers, First Farmers in the Dingle Peninsula. Bray: Wordwell.

13. Reconsidering Early Holocene Cyprus within the Eastern Mediterranean Landscape

Carole McCartney, Sturt W. Manning, David Sewell and Sarah T. Stewart

Recent research in Cyprus has dramatically altered our perception of the Early Holocene occupation of the island. Previously, dominant perspectives defined islands as closed systems that were idealized as isolated, pristine entities. It is now known that foragers camped at Aetokremnos c. 10,000 years ago and the Cypriot Aceramic Neolithic began some two millennia prior to the Khirokitian, who were long supposed to be the original settlers. This new evidence requires new interpretations, ranging from the idea that mainland hunter-foragers marginalized by rising sea levels and/or pressure from emerging farmers were pushed into colonizing Cyprus through to choices made by coastal groups because of perceived opportunities on Cyprus already consistent with their way of life, values and social systems. Critically, such interpretations require landscape analysis within the island, as well as necessitating an understanding of the links between islands and wider regions.

Introduction: the problem

Despite considerable recent debate (Broodbank 2000; 2006; Waldren 2002), the underlying perception in archaeology still regards islands as closed systems, typically isolated and frequently fragile. The effect of this stereotype is clearly shown in the model of island colonization in the Mediterranean, whereby human populations could permanently settle on islands only with the advent of farming technology (Cherry 1981; 1990). Current definitions of the Cypriot Aceramic Neolithic employ the strict forager/farmer dichotomy that derives from this model even though they have evolved to account for startling new evidence. Thus the Khirokitia culture, once seen as an 'original culture' (beginning c. 6000 BC), was thought to represent a population of isolated farmers cut off from the Pre-Pottery Neolithic (PPN) of the Levant (Stanley-Price 1977; Le Brun 1986). Differences between Cyprus and the Levant were emphasized and explained by reference to biogeographic models, employing the 'founder effect' to justify the absence of diagnostic PPN elements (Stanley-Price 1977; Held 1989; 1993). Even then, however, a few researchers promoted the 'antecedent hypothesis', wherein a culture preceding the Khirokitia culture was inferred (Watkins 1973, 47; Todd 1986, 18), and it is this hypothesis which has been proven correct by subsequent discoveries. Foragers camped at Aetokremnos c. 10,500/10,000 BC (Simmons 1999; Simmons and Mandel 2007) and the Cypriot Aceramic or Cypro-PPNB now begins some two millennia prior to the Khirokitian (Swiny 2001; Guilaine and Le Brun 2003; Peltenburg and Wasse 2004). Discussions of this extended Cypriot Neolithic, however, perpetuate the earlier model since true 'colonization' of the island is said to have occurred only with farmers arriving from the Levant with a full agricultural 'package'. (Note that Cypriot sites are normally referred to by both the village name and local toponym: e.g. Kissonerga-Mylouthkia. In this paper only the most common single name, typically the local toponym, will be used: e.g. Mylouthkia. Complete site names can be found in the references cited.)

A problem arises when we attempt close the gap in the archaeological record between Aetokremnos and the Cypro-PPNB. Recent discoveries have now

Cypriot Terminology			
Period	*Phase*	*Dates*	*Comparative mainland period*
Late Epipalaeolithic (pre-Neolithic)	Akrotiri Phase	?↑↑↑↑↑? c. 10,500/10,000 BC	Epipalaeolithic
Early Neolithic	Late Early Neolithic	?↑↑↑↑↑? c. 8900–8600 BC	Late PPNA/Early Neolithic
Cypro-PPNB (Aceramic Neolithic)	Cypro-EPPNB Cypro-MPPNB Cypro-LPPNB Khirokitian	?↑↑↑↑↑? 8200–7900 BC 7900–7500 BC 7500–7000 BC 7000–5500 BC	EPPNB MPPNB LPPNB Final PPNB

Table 13.1. Preliminary sequence of Early Holocene occupation in Cyprus.

demonstrated more significant maritime activity by the Late Epipalaeolithic while an Early Neolithic (PPNA) period on Cyprus is being uncovered (Table 13.1; Ammerman n.d; McCartney *et al.* 2006; 2007; 2008). Providing a new descriptive context for these discoveries, however, is hampered by the idea that a 'colonization' phase separate from such early occupations still needs to be documented and that this could not have taken place until the advent of agriculture with a demic diffusion from the Levant (Peltenburg 2003 with references). To create a new context that incorporates both active maritime interaction in the Eastern Mediterranean (encompassing Cyprus) and a pre-agricultural Neolithic it is necessary to review core assumptions about the Cypriot landscape that have helped to perpetuate the island Neolithic model. We need, in other words, to unpack the island 'black box' and to see Cyprus as a landscape that encompasses land and sea, island and mainland.

Cyprus in the Eastern Mediterranean landscape

To reconsider Cyprus in an Eastern Mediterranean context, it is necessary to review three key assumptions. The first, that Cyprus as an island represents a formidable potential target for colonization, is linked to assumptions concerning the development of maritime technology. Second is the assumption that Cyprus represents a marginal environment poorly suited to the practice of a foraging economy. The third, and related, assumption is linked to questions of mobility versus sedentism: 'colonization' as currently understood demands permanent settlement made possible by farming technology.

Island targets

Held refers to Cyprus as a 'continental' island to which 'oceanic' expeditions were required for colonization (Held 1989; 1993). This model was used to separate Cyprus from the mainland and decrease the likelihood of early colonization. The sea was thus viewed as a barrier to Epipalaeolithic and Early Neolithic groups on the coast of the Near East, who were perceived as reaching, but not passing, the sea's edge. Such modelling may be appropriate to remote Pacific islands (e.g. Irwin 1992), but seems ill-suited to the different geographical circumstances of the Mediterranean Sea. Critically, Cyprus was visible from a large area of the south Anatolian coast, and cloud banks, observations of birds and so on would have made it a 'known' quantity to coastal populations across the Eastern Mediterranean and any groups engaged in maritime activity. Lower sea levels in the Late Pleistocene/Early Holocene would have further integrated areas, opening 'windows' along the coastal littoral, extending both the Cilician coast and the continental shelf of Cyprus and removing a significant component of the 'sea barrier' (Galili and Weinstein-Evron 1985; Gomez and Pease 1992). Following Manning and Hulin (2005), we adapt a well-known map of Mediterranean intervisibility (Chapman 1990, fig. 262 (after Schüle), reproduced in Broodbank 2000, fig. 4 and Horden and Purcell 2000, 127). Figure 13.1 shows the land/sea intervisibility in the Eastern Mediterranean given today's coastlines. The map illustrates that the colonization of Cyprus (like that of Crete and the Aegean islands), did *not* relate to non-visibility or a lack of knowledge of their existence. Instead, for groups engaged in any maritime activity, the abilities to exploit alternative resources available on islands (and the interest in doing so) made the settlement of those islands attractive and possible (Broodbank 2000, 68–143; Finlayson 2004, 17). Any maritime activity off the Eastern Mediterranean coast would have brought Cyprus into play and the circum Eastern Mediterranean coastal mountain chains are likely to have been critical vantage points from which early foragers first glimpsed the island.

High mountains contribute to the area of visibility far more than coastal features of modest height, as shown by the large curves making up the edges of the shaded areas on Fig. 13.1. Areas of overlap show where a sailor could

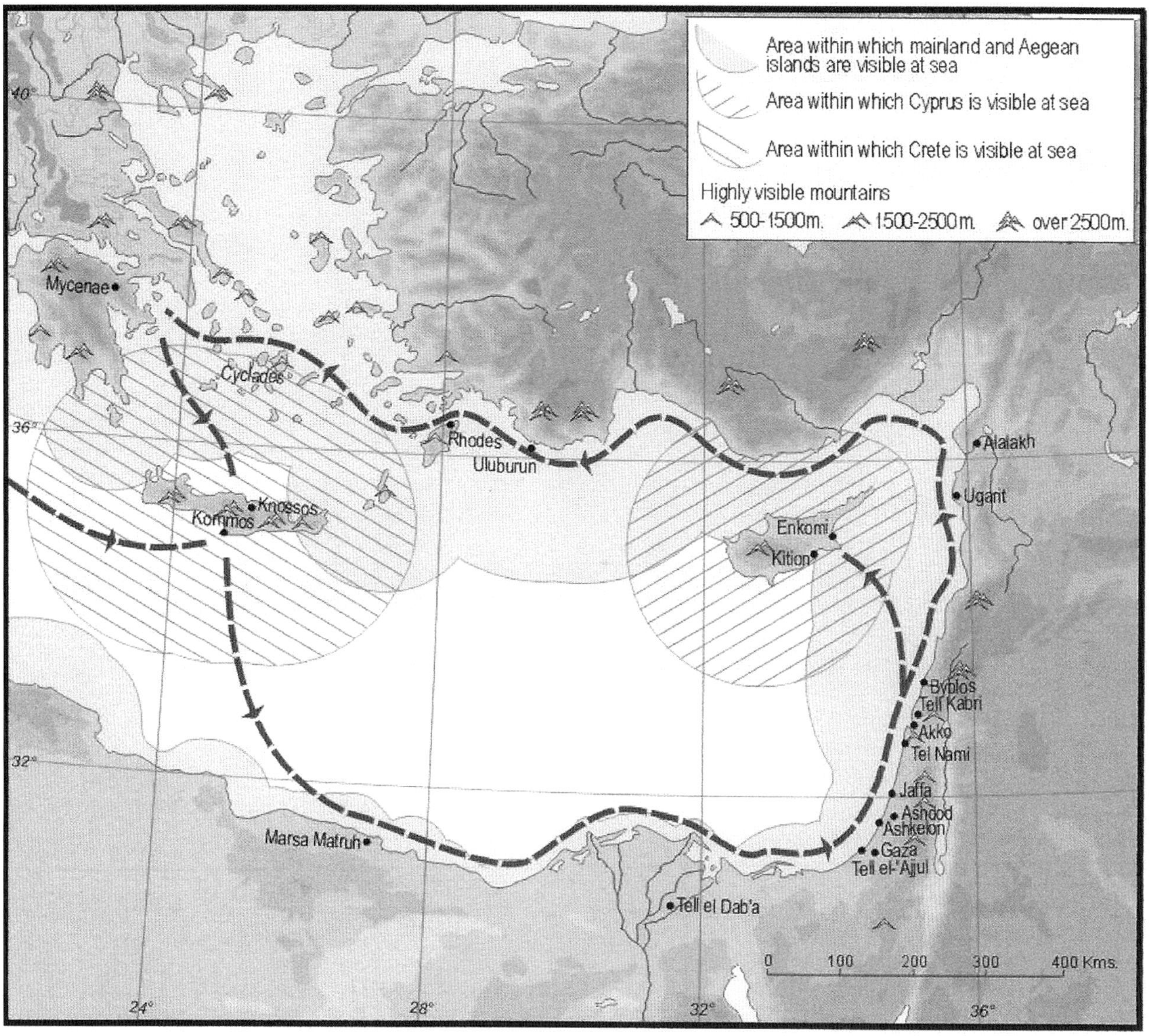

Fig. 13.1. Sea areas from which the mainland and islands are inter-visible.

see both his origin and his destination at the same time. For example, a ship sailing to Cyprus from Byblos would not sight land until about halfway through the voyage, whereas on the return voyage Mt Hermon and Qurnat as-Sawda would be visible from the eastern coast of Cyprus. The approximate distance at which land is first seen by an observer at sea is calculated by the formula: $D = 1.17 \times \text{sqrt}(L) + 1.17 \times \text{sqrt}(O)$.

Where D = distance in nautical miles
L = height above sea level of land in feet
O = height above sea level of observer in feet

This formula takes account of refraction by the atmosphere, but ignores alterations caused by changes in barometric pressure. All calculations were performed in feet and nautical miles, and converted into metric units. It was assumed that the height of the observer was 3 m, a relatively conservative interpretation considering waves, ship masts, elevated topographic features, the use of trees (etc.) and built structures on land.

Significant and long-distance maritime movements were already taking place by the later Palaeolithic in the Mediterranean as illustrated by Melian obsidian and deep sea fish bones at sites like Franchthi Cave (Perlès 2001; see Broodbank 2006 and Ammerman n.d. for comprehensive reviews). Cyprus clearly did not play a part in this early Aegean trade, as indicated by the complete absence of obsidian from Melos on Cyprus; it was dependent, instead, on the resources of central Anatolia (cf. Briois *et al.* 1997; McCartney and Grautze 2003), but not until the Cypro-PPNB. Obsidian is in fact all but completely absent from Cyprus prior to this time (Simmons 1999; McCartney and Todd 2005; McCartney *et al.* 2006). However, the presence of early coastal foragers at Aetokremnos on Cyprus indicates maritime technology in the Eastern Mediterranean by at least the 11th millennium BC (Simmons 1999;

Ammerman *et al.* 2006; Ammerman n.d.). By the Early Neolithic foragers at Asprokremnos had still not engaged either with farming or the exchange of obsidian, yet they were well connected within multiple interaction spheres in the Levant (McCartney *et al.* 2008; McCartney in press).

According to the perspective that sees Cyprus as a remote island environment, colonizing farmers once possessed of maritime technology lose this ability as they become isolated, or special pleading is required to envision outsiders, 'middlemen', who transported them to new pastures (Peltenburg *et al.* 2001, 56–7; Peltenburg 2003, 97). According to this interpretation conservatism in material culture is promoted through a veiled 'founder effect' perspective that assumes increasing isolation (Guilaine *et al.* 2000; Guilaine and Le Brun 2003; Peltenburg 2003; 2004) with little effort to look for choices in technological practice leading to the active definition of identity (Last 1995; McCartney 2004; McCartney and Todd 2005, 211–27). Change is typically limited to transformations that take place wholly within the island environment, disallowing the potential for the exchange of ideas through stimulus diffusion that may have been learned during interaction on the mainland. Fortunately, current considerations of island occupation now view the sea as a 'highway' and a medium of interaction with nearby mainland environments (Broodbank 2000; 2006; Waldren 2002). Some recent interpretations concerning Neolithic Cyprus have placed the island within such a wider Eastern Mediterranean context, citing the need to maintain social relationships and animal stocks or gain access to resources and new technologies (Horwitz *et al.* 2004; Watkins 2004a, 31; Peltenburg 2004, 19; McCartney 2004). Typically such interaction is seen as Cypriots seeking to gain access to resources not available in their impoverished environment, rather than as evidence for the active focus of continued two-way interaction (cf. Finlayson 2004; McCartney in press). There is still a need to alter the perception of the sea as a geographic barrier to an awareness that the sea *joins* Cyprus to the Levant.

Island biomass

Poor biomass attributed to Mediterranean islands was viewed as an impediment to permanent island settlement prior to the introduction of farming (e.g. Held 1989; 1993; Knapp *et al.* 1994; Peltenburg 2004). Thus Aetokremnos was defined as a brief visitation that served to provide background knowledge for later colonizing farmers who arrived to discover an empty landscape (Peltenburg 2003, 98). The tendency is to interpret prehistoric foraging from a continental perspective, and consequently the economic potential of Cyprus has been little considered or explored. Hunting patterns are characteristically interpreted in terms of large terrestrial ungulates, and marine resources are often viewed as marginal foods to be utilized only when the so-called *preferred* resources became unavailable (Hayden 1981). The currently poor evidence for early foragers utilizing the Levantine coast is likely to be more a reflection of research bias and sea-level rise than a true bias against marine resources, and the recent documentation of coastal sites in Greece clearly illustrates a maritime economy previously ignored (Runnels *et al.* 2006; Sampson and Katsarou 2004).

New evidence, however, has begun to demonstrate that Cyprus was not an empty landscape prior to the Cypro-PPNB (McCartney 2005; McCartney and Todd 2005; McCartney *et al.* 2006; Ammerman *et al.* 2006; Ammerman n.d.). Not only do the newly discovered coastal sites at Aspros and Nissi Beach, which provide parallels to Aetokremnos, point to a more significant Late Epipalaeolithic (or 'pre-Neolithic') period on the island, they have been interpreted as illustrating a parallel maritime lifestyle rather than a stage that disappeared with the advent of cultivation (Ammerman n.d.). These sites broadly correspond to Epipalaeolithic coastal sites on the Eastern Mediterranean mainland (Copeland 1991; Albrecht *et al.* 1992; Bar Yosef 2001), but, unlike the more highly diagnostic Neolithic material culture, the simple microlithic technology found at these sites makes straightforward equations with a single parent culture on the mainland difficult. Indeed, it may be that looking for such simple origins obscures an underlying maritime complexity. What we now understand is that the beginning of seafaring opened up the whole of the Mediterranean basin and this seems to have occurred during (or by) the onset of climate change in the Younger Dryas (Broodbank 2006; Ammerman n.d.); it is in the context of such an already interactive seascape that we must evaluate the developments of the ensuing Holocene. Furthermore, we now know that it was not (precocious) farmers who initiated interaction between the Levant and Cyprus during the Early Neolithic. Evidence from Asprokremnos (the only currently excavated site that belongs to this period) includes a hunted fauna and a material culture characterized by hunting tools with significant symbolic value, including arrowheads and shaft-straighteners (McCartney *et al.* 2007; 2008).

On Cyprus the local ecology may provide additional clues to the character of a maritime foraging economy. For example, the location of currently known early coastal sites in Cyprus in proximity to sea-turtle nesting sites is unlikely to be fortuitous (Fig. 13.2). Such ancient nesting zones are abundant in the Eastern Mediterranean and provide not only a food resource but also possibilities for interaction across the wider region, demonstrating a rich foraging environment along the Mediterranean littoral. The turtles (and their eggs), available primarily during the summer, are complemented during spring and autumn by increases in fish, while molluscs, crustaceans, sea plants, birds and seals are available throughout the year. Though some resources may have declined during the Younger Dryas, certain species such as cuttlefish, which are cold-blooded, would have been relatively unaffected (and lower sea levels would have brought deep-water fish within closer reach). Though the abundance of such resources no doubt

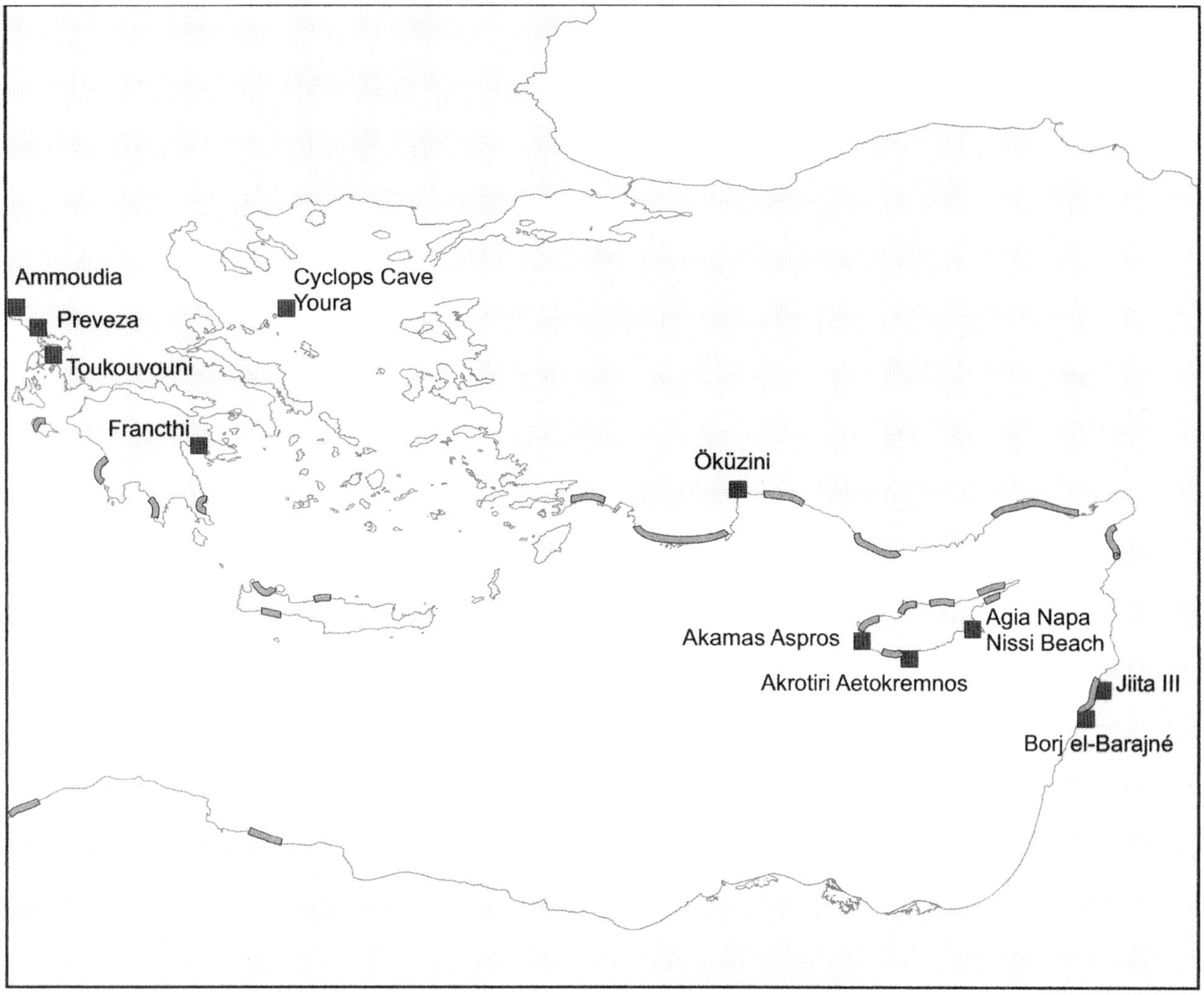

Fig. 13.2. Locations of (modern) sea-turtle nesting sites (shaded areas) and selected Mesolithic/Epipalaeolithic coastal sites (squares). (Based on Ammerman et al. *2006, 2, fig. 1; Runnels* et al. *2006, 260, fig. 1; Bar Yosef 2001, 135, fig. 2, Copeland 1991, 28, fig. 1; Camiñas 2006, fig. 1; Aureggi* et al. *2005, 9, fig. 2, Euroturtle, www.euroturtle.org).*

would have varied through time, even the low fish stocks of modern times would have supported small groups of early prehistoric foragers.

Faunal data from stratum 2 at Aetokremnos clearly demonstrate a familiarity with molluscs, sea birds and reptiles, suggesting that such species were endemic to the island in the 11th millennium BC (Simmons 1999, 157, table 7.1; Reese 2006). During the Cypro-PPNB limpets and fish from Mylouthkia and Akanthou and mollusca from Tenta provide evidence for the continued exploitation of sea resources, while deer, once introduced, thrived and remained a significant wild resource throughout early prehistory (Todd 1987; Croft 1991; Sevketoglu 2002; Peltenburg 2003). Pigs present an interesting case since, of the future domesticates, they appear to have been introduced first and are associated with early animal husbandry on the mainland (Ervynck *et al.* 2001). They were found in small numbers at Aetokremnos but dominate the Early Neolithic fauna at Asprokremnos, and a feral population of pigs persisted alongside domestic pigs at Shillourokambos (Simmons 1999, 164; McCartney *et al.* 2007; 2008, Guilaine and Briois 2006, 171). If a hunted pig population became established during the Late Pleistocene (certainly possible given a predator-free environment and high birth rate) then the occupants of Asprokremnos and other Early Neolithic sites would have arrived into a far less impoverished environment than generally assumed even by mainland standards. Alternatively, even if they stocked the island themselves, the occupants of Asprokremnos still benefited from the endemic birds and freshwater crab also consumed at the site (McCartney *et al.* 2007; 2008).

Both on the coast and inland the potential of wild plants for early foragers was significant; some 66 species are still utilized on the island (Savvides 2000; Wilcox 2003). Wilcox (2003, 235) suggests that barley, endemic to Cyprus, could have been cultivated independently on the island c. 8200 BC. Some 357 endemic and migratory bird species, over 120 species of snails (important sources

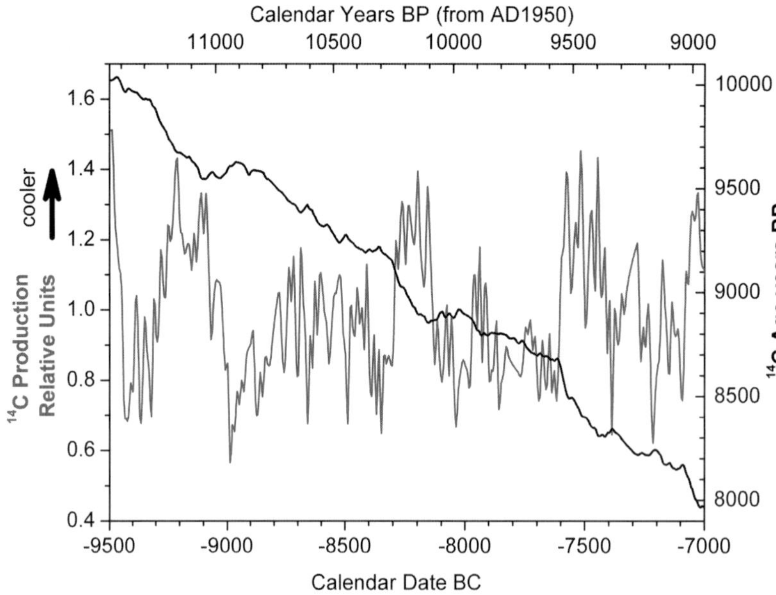

Fig. 13.3a. The radiocarbon calibration curve (IntCal04: Reimer et al. *2004) in black and an approximation of atmospheric radiocarbon (^{14}C) production in grey (calculated from the decadal tree-ring data of IntCal98, Stuiver* et al. *1998, using a standard box model of the carbon cycle, Siegenthaler* et al. *1980, with the assumption that oceanic ^{14}C uptake was constant during this time, as described by Usoskin and Kromer 2005, 32 – data courtesy of Bernd Kromer). The general pattern is one of steep slopes with more plateau-like periods in between. Steep slopes in the radiocarbon calibration curve correspond to periods when there is high radiocarbon (^{14}C) production. This occurs because more cosmic-ray-produced radiation reaches the Earth's stratosphere, which means that the sun is relatively quiet and providing less shielding; these high ^{14}C production periods correspond to periods of cooling on the Earth (such as the Little Ice Age: Eddy 1977 or the 9th–8th centuries BC: van Geel* et al. *1998) over the Holocene (Bond* et al. *2001). In reverse, plateaux and especially the reversals in the radiocarbon curve after major slopes indicate the reverse: periods of more active sun and generally warmer conditions – which in the context of the earlier Holocene also means wetter conditions (Rosen 2007) (on radiocarbon and climate, see further in Manning n.d.). The radiocarbon calibration curve is expressed in terms of ^{14}C years BP (Before Present = AD 1950 for radiocarbon) against calendar years BC (from known-age tree-ring series in the period shown) and the ^{14}C production curve is expressed in relative units (describing the production change to balance the observed $\Delta^{14}C$ change normalized to AD 1950).*

of fats and proteins), as well as reptiles, and amphibians combine to provide a wealthy larder. The ethnographic literature is replete with examples of modern foragers who currently (or in the recent past) make use of such resources in coastal or riverine economies (see contributions in Lee and Daly 1999). They demonstrate a range of behaviours little considered because a limited biomass for the island environment of Cyprus has been assumed, when in reality, though different from the mainland, Cyprus was certainly not bereft of sustenance. The controversial pygmy hippopotamus and dwarf elephant could also have been significant resources on Cyprus if human occupation reaches back into the Pleistocene (cf. Simmons 1999).

Mobility, climate and time

Predicated within current interpretations of the introduction of farming to Cyprus c. 8200 BC is the assumption that farming communities are by definition sedentary and thereby entail the first permanent settlements on the island (Guilaine *et al.* 2000; Peltenburg *et al.* 2001; Peltenburg 2003; 2004). A permanent settlement does not, however, necessarily imply year-round occupation, or indeed farming, and there is abundant ethnographic evidence that mobile foragers are often also cultivators, demonstrating the futility of strict oppositions (Halstead 2005; Milner 2005). Different groups can utilize the same place, permanence may be defined in terms of re-occupations, or only a part of a group may occupy a single place year-round while others remain mobile or periodically mobile (Halstead 2005; Milner 2005 with references). These alternatives can apply equally to foragers/collectors as to early cultivators/farmers, resulting in sites that shift or increase in size as palimpsests rather than growing as single entities developed through a unilinear progression from mobility to sedentism afforded by the production of surplus food.

In an environment characterized by seasonal resources mobility is likely to have been instrumental as sedentism has the effect of limiting choices in subsistence strategies (Halstead 2005, 38). The Cypriot prehistoric record, which is defined by shifting sites rather than tell formation, shows that mobility remained important to the islander's use of place. While evidence for temporary shelters, pits and

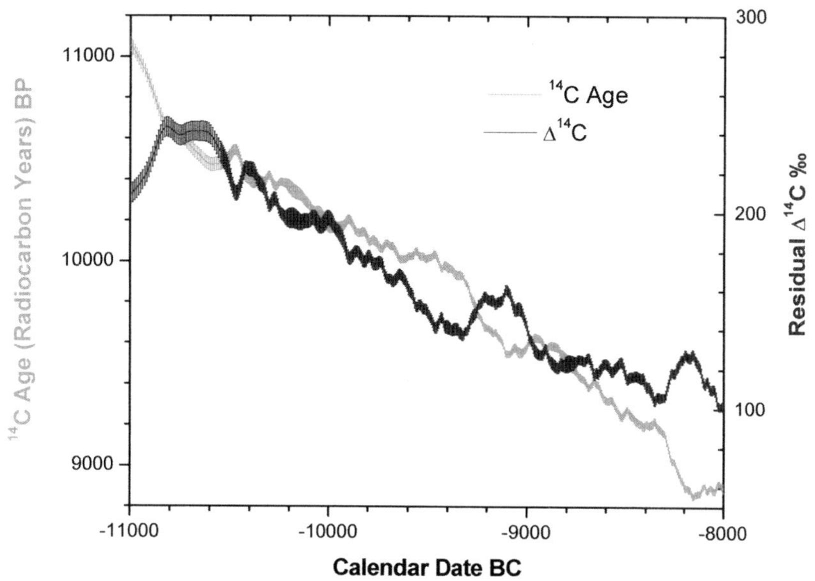

Fig. 13.3b. The light grey line (with 1σ errors indicated) shows the IntCal04 radiocarbon calibration curve for the period 11,000–8000 BC (Reimer et al. *2004). The Younger Dryas to Holocene transition, the change from the 10,000–9600 BP ^{14}C plateaux, is in the centre. The dark grey line (with 1σ errors indicated) shows the Δ^{14}C record (1000-year moving average) from the IntCal04 radiocarbon data-set (Reimer* et al. *2004) (changing ^{14}C levels show a picture similar to the production model in Fig. 13.3a, but without carbon cycle integration and small lag time applied). The two early Holocene 14C inversions (cooling periods) in the late 10th and late 9th millennia BC are again clear. As these inversions end, and subsequent plateaux follow, they might perhaps form the two early Holocene 'climatic context' moments for heightened interactivity between Cyprus and the mainland.*

even the location of wells demonstrates an incremental displacement across the landscape, only Tenta provides a real example of vertical stratigraphy (Todd 1987; Guilaine *et al.* 2000; Peltenburg *et al.* 2001; Peltenburg 2003). What is interesting is the absence of villages or indeed substantial stone architecture until the Cypro-M/LPPNB, when circular stone structures appear at Shillourokambos, then Tenta and Ais Yiorkis (Todd 1987; Simmons 2003; Guilaine and Briois 2006, 164–7). The same temporary structural features are apparent at the pre-farming site of Asprokremnos (McCartney *et al.* 2008). Thus while there is a material shift with the start of the Neolithic from the open campsites or rockshelters used in the Late Epipalaeolithic, structures remained simple and continued to be built with temporary building materials through the next millennium (or more).

If we cannot assume a unilinear development combining sedentism and farming practice, it seems worth considering other potential models of economic and social interaction between Cyprus and the mainland. One alternative perspective may be provided by consideration of climate history and radiocarbon chronology, with which we can speculate at a recurrent cycle of increased utilization/settlement on the island linked to global cooling and warming episodes (Bond *et al.* 2001, Manning n.d.) and the relatively stable periods that followed (Figs 13.3a and b). Examining Fig. 13.3a we see marked cooling episodes (steep slopes in the radiocarbon calibration curve = ^{14}C production peaks) in the periods c. 9300–9100 BC, 8300–8100 BC and 7600–7400 BC, and a more minor episode around 7000 BC. The first two peaks are followed by relatively stable multi-century phases (approximate plateaux on the radiocarbon calibration curve). One can observe that the current earliest Cypro-PPNB farming settlements on Cyprus appear around and especially following the second steep radiocarbon slope (the process is slightly circular since the steep slope in the radiocarbon calibration curve serves to attract dating probability to this area). In the Early Holocene climate context the plateaux following the steep slopes indicate probable warmer and wetter conditions operating on Cyprus (and the East Mediterranean region, Rosen 2007), and so Cyprus at these periods in particular may have been more attractive to early populations. We therefore speculate that the earlier production peak in the late 10th millennium BC and the following plateau might have provided a similar attractive period on Cyprus in the earlier 9th millennium BC, and that Early Neolithic presence may have followed through the intervening millennium prior to the Cypro-PPNB. This situation reversed the earlier Late Epipalaeolithic desiccation trend of the Younger Dryas: thus more favourable climate-related cycles for the Early Holocene perhaps encouraged prolonged use and increasingly permanent settlement on Cyprus over time during the Neolithic.

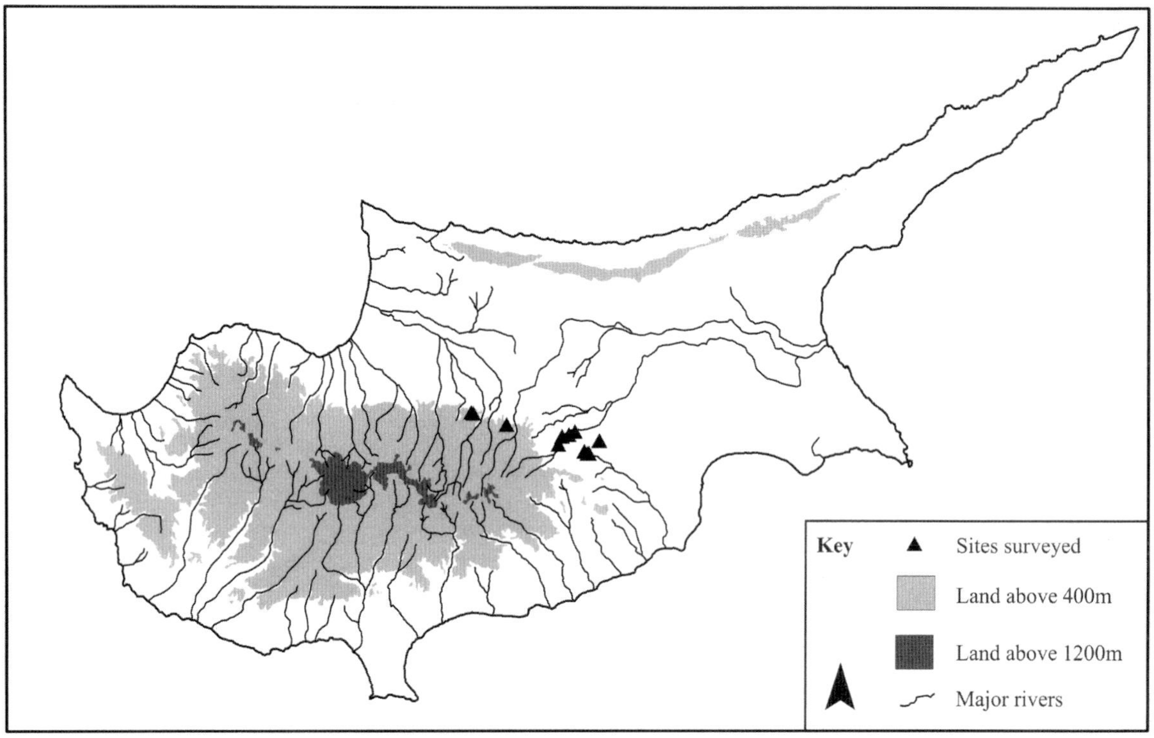

Fig. 13.4. The location of the EENC survey area.

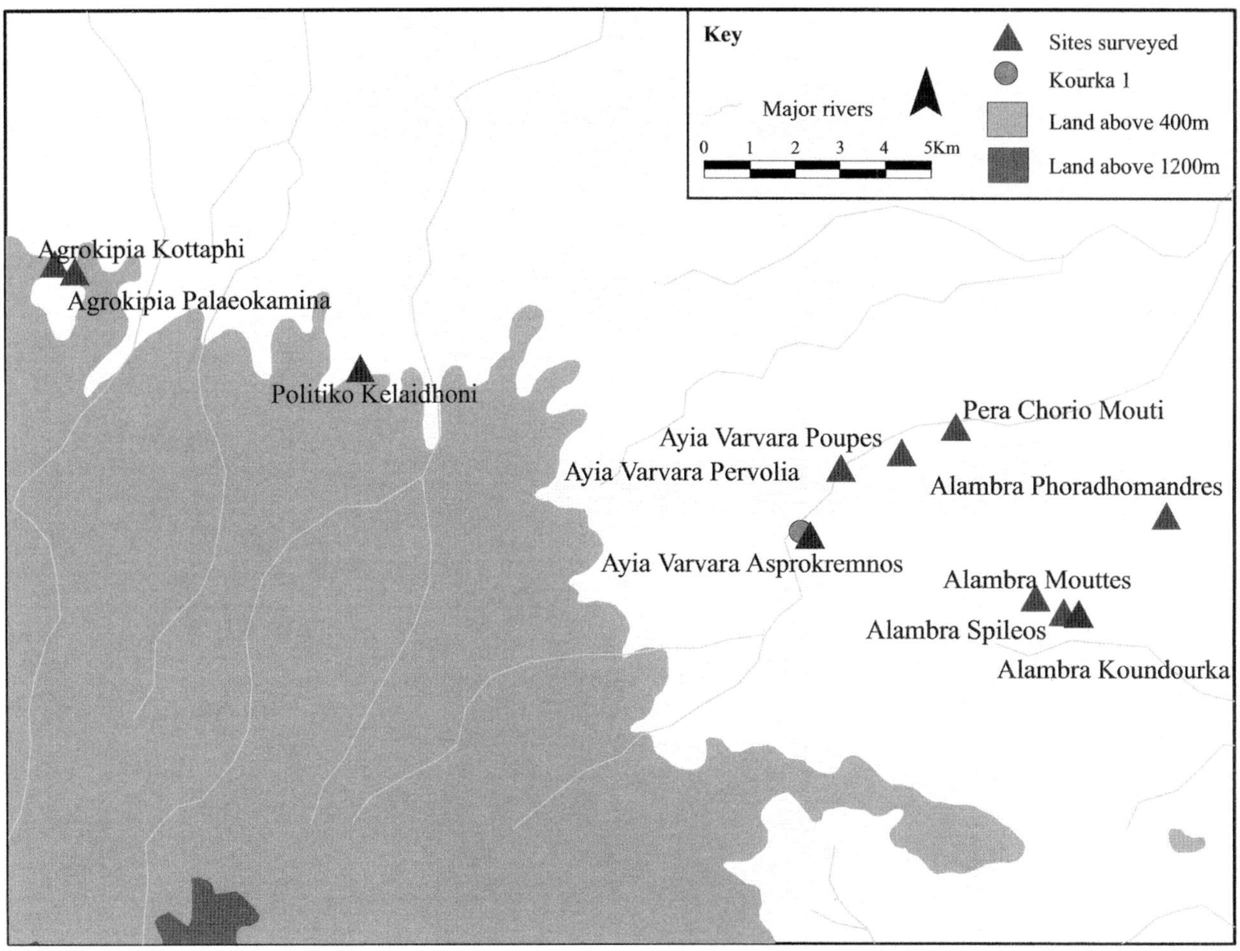

Fig. 13.5. Location of Asprokremnos in EENC survey landscape.

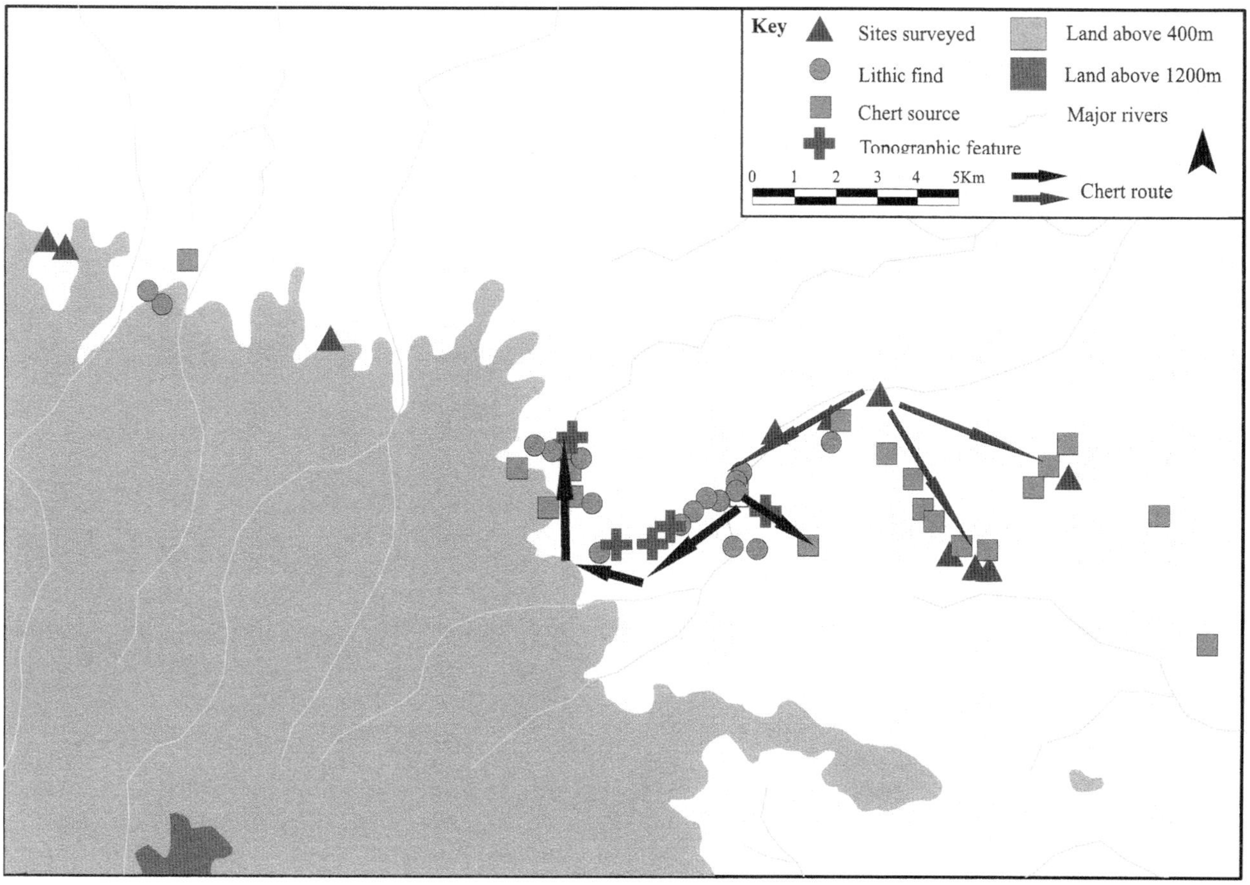

Fig. 13.6. Chert source map. Potential relationships between sites in the EENC landscape. The grey arrows indicate possible site relationships between flake-based assemblage sites along the Yialias River and from these sites to the Alambra chert sources. The black arrows indicate possible relationships between Asprokremnos and chert sources located further west.

Materiality in the Early Holocene landscape

The discussion above has focused on dispelling earlier ideas of island marginality once embedded in interpretations of early Cyprus, and on widening our perspective to view Cyprus within a wider Eastern Mediterranean landscape. We have considered how both foragers and farmers may have responded to such an environment through flexibility in both subsistence practice and settlement patterns. Cyprus is not a formidable sea-faring target, but it was able to support foragers and subsequently herder-hunters who cultivated, and these human groups continued to camp and live in temporary shelters from the difficult times of the Younger Dryas into the warmer wet climate of the Early Holocene. Ammerman (n.d.) has suggested that subsistence requirements were not the only resources sought by early coastal foragers, who may have also exploited sea salt. For Neolithic groups with a heightened materiality it seems unlikely that subsistence alone kept them returning to the sea or to the island of Cyprus in particular. To understand this choice we need to think beyond subsistence technology, since even farming, which provides a technological solution (particularly in verdant environments or climatic episodes), is not a sufficient catalyst on its own. The Neolithic was not simply a shift (and a gradual one at that) in subsistence technology. Instead, it was transformational because of the momentum gained in many areas, including more diverse material culture, ritual, social organization and perhaps cognitive ability (Bar Yosef 2001; Watkins 2004b). Because of this heightened materiality, Neolithic assemblages contain many more diagnostics than their early coastal forager counterparts and parallels may easily be drawn with the Levant (Guilaine *et al.* 2000; Peltenburg *et al.* 2001; McCartney *et al.* 2008).

What Neolithic groups were doing in the Cypriot landscape, however, has been little explored. Data collected between 2005 and 2008 by the Elaborating Early Neolithc Cyprus Project (EENC) project in central Cyprus have begun to provide new evidence of human activity during this transformational time (Figs 13.4, 13.5; McCartney *et al.* 2006; 2007; 2008). Though data analysis is still at a preliminary stage, a more complex and interactive landscape can already be illustrated for the Cypriot Neolithic. In an area some 26 km across between the villages of Lymbia and Agrokipia at the foot of the central Troodos massif, 22 aceramic surface scatter sites have been documented by targeted survey, illustrating a site

	INAA - EENC Chert Samples			
Sample #	Name	Location	Chert Type	Colour
1	Asprokremnos		Translucent	red/brown
2	Asprokremnos		Translucent	grey/green
3	Asprokremnos		Translucent	tawny
4	Kourka 1	LF2	Basal	pink
5	Kourka 2	LF3	Translucent	tawny
6	Kourka 3	LF4	Translucent	grey/green
7	Argakin ton Villourkon 1	LF7	Translucent	red/brown
8	Argakin ton Villourkon 2	LF8	Translucent	red/brown
9	Argakin ton Villourkon 4	LF10	Translucent	red/brown
10	Kourka 3	LF4	Basal	white
11	Koundourka	CS6	Translucent	tawny
12	Koundourka	CS6	Translucent	red/brown
13	Koundourka	CS6	Translucent	tawny
14	Mouttes	CS7	Lefkara T.	red
15	Mouttes	CS7	Translucent	tawny
16	Nostos	CS9	Lefkara T.	red
17	Laxia	CS8	Lefkara T.	grey
18	Platyvounia	CS12	Basal	pink
19	Alakathkia	CS16	Translucent	tawny
20	Analiondas Village	CS17	Translucent	red/brown
	LF=Lithic Findspot			
	CS=Chert Source			

Table 13.2. The results of a pilot study of chert source exploitation. Different bands indicate distinct chert types shown by Neutron Activation Analysis. LF = lithic findspot, CS = chert source (a full account of the analysis will be provided elsewhere).

density not previously shown by transect methodologies (e.g. Given and Knapp 2001, 265). Importantly, these lithic scatters did not exist in a vacuum. Survey investigation has also revealed some 42 chert sources, 78 isolated lithic finds, 6 major topographic features and a wealth of other resources, including volcanic and calcareous rock for the manufacture of ground-stone artefacts, and abundant ochre. The topographic features, prominent conical hills capped by andesitic pillow lava intrusives (Gass 1960, 72–3) may have provided features (in the absence of tells) for mapping the landscape or other communication purposes by foragers and early farmers (Cooney 1999; Rockman and Steele 2003, 25–43, 144–68, 222–41; Evans 2005, 112). While both isolated finds and topographic markers are likely to illustrate patterns of movement across the landscape, the abundant chert, ochre and other rock types represent a vast store of manufacturing resources. Additional island resources not available in the immediate study area include the local green stone (picrolite) and marine and fossil shell prized by Neolithic foragers and farmers alike, as well as by their Late Epipalaeolithic predecessors (Bar Yosef-Mayer and Porat 2008). Many of these resources were exploited at the Early Neolithic site of Asprokremnos, where more than 800 kg of high-quality chert was used to make Late PPNA-type arrowheads (among other tools) in numbers never before seen on the island; in addition, simple stone vessels were carved and even painted, showing at least one use of the significant quantities of red and yellow ochre processed at the site (McCartney *et al.* 2007, fig. 4; 2008, figs 2–3). No doubt there are many other material resources worth considering, but it is with chert that we may begin to document material utilization in relation to the landscape in greater detail.

The resource landscape: chert utilization

The utilization of chert is the feature of the landscape that can be perhaps most readily evaluated in order to consider the structure of landscape utilization (Fig. 13.6). Chert sources lie close to the pillow lava/sedimentary boundary that runs along the central Troodos massif (Stewart 2004). The quality, quantity and variety of cherts available consistently diminishes (in all directions) as one moves away from the major chert sources, namely Koundourka, Spileos and Mouttes, located on the Alambra ridge in the southeast part of the survey zone (McCartney *et al.* 2006). Towards the west, good-quality sources become increasingly scarce. Indeed, one consistent feature of the lithic assemblages already discovered in the western half of the survey zone is an increased reliance on the use of jasper. Considering the size and high quality of the Alambra cherts, and parallels in the dominant chaîne opératoire seen at both Koundourka and Asprokremnos, it was originally assumed that chert exploitation at Asprokremnos would relate to the exploitation of the Alambra cherts at Koundourka, given the visible qualities shown by the various chert types used at both sites.

Continued survey efforts have now demonstrated an effectively unbroken chain of chert-bearing ridges cross-cutting the Lefkara Formation from Ayia Varvara to Alambra, making this central area of Cyprus the largest single zone of chert currently known on the island. Quarry sites have now been found within a kilometre of the Asprokremnos site, but they failed to exhibit some of the chert types known from Koundourka, some 6 km distant. Given that knappers at Asprokremnos utilized at least 12 visibly different types of chert, the task of discovering which sources were used is daunting, particularly in a region so prolific in natural chert-bearing formations. The issue is further complicated because many 'quarry' site assemblages contain significant numbers of tools in addition to the perhaps unsurprising abundance of core reduction materials. Although site typologies have been founded upon such functional dichotomies (e.g. Shiffer 1995, 83–6), these distinctions fail more in-depth questioning since different activities appear to have been embedded in related movements across this landscape. Some sites, such as Koundourka and Asprokremnos, served a dual role as both camp and knapping station. Significant tool production and use at all quarry sites suggests that raw material extraction was not isolated from other forager activities, a view likely to demonstrate webs of inter-related material practice, or 'collectivities of action' (Whittle 2005, 65), that are critical to our understanding of the Early Holocene landscape in Cyprus.

Interestingly, results from a trial Neutron Activation Analysis (conducted by Ron Hancock, McMaster University, Canada) contradict the above intuitive associations and demonstrate unexpected geographic clusters (Table 13.2). Owing to its heterogeneous nature chert is notoriously difficult to source, and geochemical analyses can often yield wildly different results from a single sample, or, just as troubling, similar results from different sources and formations (Ron Hancock pers. comm.). In spite of this caveat, the results from the pilot study are worthy of note though additional tests will be necessary to substantiate the results. Intuitively the red-brown high-quality translucent chert utilized for blade manufacture at Asprokremnos appeared closely related to the visually similar material used to produce blades at the quarry site of Koundourka. The analysis results, however, suggest that the Asprokremnos knappers obtained their chert from the area to the west of the site. Thus despite its excellent quality, abundance and equal proximity (both sets of sources are c. 6 km from the site) the Koundourka cherts appear not to have been preferred at Asprokremnos and blade production at each site took place independently of the other.

Cherts analyzed from the Kourka 1 hill abutting Asprokremnos show a sample unlike any found to date at the latter site or elsewhere and thus may have been introduced to the area from elsewhere on Cyprus, or further afield. The hilltop material belongs to a flake-based assemblage with diagnostic elements recalling the early coastal assemblages noted above. It is tempting to suggest that this difference reflects a tendency of more mobile foragers to curate materials as they moved from one landscape to another (Rockman and Steele 2003, 19). Similarly, samples tested from the flake-based Kourka 3 assemblage show the use of both local and more distant sources, suggesting the use of a variety of sources as a part of risk-avoiding strategies (Stewart 2004; 2007). In contrast, blade manufacture at Asprokremnos and Koudourka represented a more significant investment in which raw material selection was intensified to such a degree that stricter raw material preferences may have been enacted. Here chert acquisition is likely to have been embedded in other activities, including the collection of ochre from the copper-bearing zones that begin only about a kilometre from the Asprokremnos site.

Conclusion

The dramatically altered Early Holocene landscape of Cyprus has paved the way for new perspectives on the early prehistory of the island. As an island, Cyprus was originally thought to represent a clear-cut case of Neolithic diffusion like Crete, which saw the arrival of fully developed farming practice (cf. Broodbank and Strasser 1991, though interpretations of Crete may need to be re-considered). But Cyprus belongs to the Near Eastern landscape, not the European one. Its geographic location made it an immediate part of the Levantine landscape once sea technology had been created. Maritime foraging began during the Late Epipalaeolithic and Neolithic foragers followed. One thing that distinguishes the Cypriot Early Neolithic foragers from their Late Epipalaeolithic coastal counterparts are the clear material culture parallels that can be demonstrated between Cyprus and the Northern Levant with the advent of the Neolithic, a reality of the increased materiality that expanded rapidly during the Neolithic period.

It is possible to argue that these groups were still merely 'utilizing' Cyprus, therefore not leaving evidence of 'permanent settlement', but this dichotomy seems increasingly artificial on the basis of growing evidence. The Neolithic of Cyprus did not begin with farming; this technology arrived later, perhaps gradually. The archaeological obsession with identifying 'the' colonization event (or 'earliest' horizons generally) has led us to look for populations in 'shock', struggling to find their way in a new unknown environment (Cooney 1999; Rockman and Steele 2003). In reality, the environment of Cyprus is not very different from that of the Levant. It is easily stocked with wild game, but periodic droughts and resource shortages are part of island life. The Cypriot landscape presented opportunities to early foragers, but no doubt demanded flexibility.

As an island context, Cyprus can be nothing but a coastal landscape, but Cypriots do not go *out* to sea; rather, they go *into* it, and they embrace it as part of everyday life. We have in the past forced our interpretations of Cyprus into a black box of insularity typically imbued with isolation (undoubtedly influenced by recent imperial landscapes) while failing to grasp the contingent circumstances that structured this island's history. Though we may have generated more questions than answers, it is necessary to alter our perceptions of Cyprus as something remote and uninhabitable. Human populations repeatedly embarked from the Levant seeking the island and no doubt more resources than land. Early Holocene Cyprus can only be understood as part of this Eastern Mediterranean landscape.

Acknowledgements

We thank Chris Hulin for work to produce the original version of Fig. 13.1. We thank Bernd Kromer for data and collaboration with regard to Fig. 13.3. The EENC project acknowledges with grateful thanks support from the Social Sciences and Humanities Research Council of Canada, and Cornell University. We thank the University of Cyprus for collaboration and support of the EENC. Finally, we thank the Department of Antiquities of Cyprus and especially its former Director, Dr Pavlos Flourentzos, for permission to conduct our EENC fieldwork and for support. We also thank George Demetriades for invaluable insights on natural resources of Cyprus, and salute the team members, without whom the past seasons of work would never have taken place.

Bibliography

Albrecht, G., Albrecht, B., Berke, H., Burger, D., Moser, J., Rähle, W., Schoch, W., Storch, G., Uerpmann, H. P. and Urban, B. (1992) Late Pleistocene and Early Holocene finds from Öküzini: a contribution to the settlement history of the Bay of Antalya, Turkey. *Paléorient* 18/2, 123–41.

Ammerman, A. (n.d.) The first argonauts: towards the study of the earliest seafaring in the Mediterranean. In A. Anderson (ed.), *Global Origins and the Development of Seafaring*. Cambridge: McDonald Institute for Archaeological Research.

Ammerman, A., Flourentzos, P., McCartney, C., Noller. J. and Sorabji, D. (2006) Two early sites on Cyprus. *Report of the Department of Antiquities, Cyprus*, 1–21.

Aureggi, M., Rizk, C. and Venizelos, L. (2005) Survey on sea turtle nesting activity in South Lebanon, 2004. http://www.medasset.org/PDF/Lebanon_report_2004.Pdf.

Bar Yosef, O. (2001) The world around Cyprus: from Epi-Palaeolithic foragers to the collapse of the PPNB civilization. Pp. 129–64 in S. Swiny (ed.), *The Earliest Prehistory of Cyprus*. CAARI Monograph Series 2. Boston: American Schools of Oriental Research.

Bar Yosef-Mayer, D. and Porat, N. (2008) Green stone beads at the dawn of agriculture. *Proceedings of the National Academy of Science*, 105/25, 8548–51

Bond, G., Kromer, B., Beer, J., Muscheler, R., Evans, M. N., Showers, W., Hoffmann, S., Lotti-Bond, R., Hajdas, I. and Bonani, G. (2001) Persistent solar influence on North Atlantic climate during the Holocene. *Science* 294, 2130–6.

Briois, F., Gratuze, B. and Guilaine, J. (1997) Obsidiennes du site Néolithique Précéramique de Shillourokambos (Chypre). *Paléorient* 23/1, 95–112.

Broodbank, C. (2000) *An Island Archaeology of the Early Cyclades*. Cambridge: Cambridge University Press.

Broodbank, C. (2006) The origins and early development of Mediterranean maritime activity. *Journal of Mediterranean Archaeology* 19/2, 199–230.

Broodbank, C. and Strasser, T. F. (1991) Migrant farmers and the Neolithic colonization of Crete. *Antiquity* 65, 233–45.

Carmiñas, J. (2006) Sea turtles of the Mediterranean Sea: population dynamics, sources of mortality and relative importance of fisheries impacts. Paper presented at the Expert Consultation on Interactions between Sea Turtles and Fisheries. Internet edition: http://www.fao.org/docrep/007/y5750e/y5750e03.htm.

Chapman, R. W. (1990) *Emerging Complexity*. Cambridge: Cambridge University Press.

Cherry, J. F. (1981) Pattern and process in the earliest colonisation of the Mediterranean islands. *Proceedings of the Prehistoric Society* 47, 41–68.

Cherry, J. F. (1990) The first colonization of the Mediterranean islands: a review of recent research. *Journal of Mediterranean Archaeology* 3, 145–221.

Cooney, G. (1999) Social landscapes in Irish prehistory. Pp. 46–64 in P. J. Ucko and K. Layton (eds), *The Archaeology and Anthropology of Landscape*. London: Routledge.

Copeland, L. (1991) Natufian sites in Lebanon. Pp. 27–42 in O. Bar Yosef and F. Valla (eds), *The Natufian Culture in the Levant*. Ann Arbor, MI: International Monographs in Prehistory.

Croft, P. (1991) Man and beast in Chalcolithic Cyprus. *Bulletin of the American Schools of Oriental Research* 282/283, 63–79.

Eddy, J. A. (1977) Climate and the changing sun. *Climate Change* 1, 173–90.

Ervynck, A., Dobney, K., Hongo, H. and Meadow R. (2001) Born free? New evidence for the status of Sus Scrofa at Neolithic Çayönü Tepesi (Southeastern Anatolia, Turkey). *Paléorient* 27/2, 47–73.

Euroturtle (2006) http://www.euroturtle.org/.

Evans, J. G. (2005) Memory and ordination: environmental archaeology in tells. Pp. 112–25 in D. Bailey, A. Whittle and V. Cummings (eds), *(Un)settling the Neolithic*. Oxford: Oxbow Books.

Finlayson, B. (2004) Island colonization, insularity or mainstream? Pp. 103–22 in E. Peltenburg and A. Wasse (eds), *Neolithic Revolution*. Levant Supplementary Series 1. Oxford: CBRL and Oxbow Books.

Galili, E. and Weinstein-Evron, M. (1985) Prehistory and paleoenvironments of submerged sites along the Carmel coast of Israel. *Paléorient* 11/1, 37–52.

Gass, I. G. (1960) *The Geology and Mineral Resources of the Dhali Area*. Memoir No. 4. Nicosia: Geological Survey Department, Cyprus.

Given, M. and Knapp, A. B. (2001) *The Sydney Cyprus Survey Project*. Monumenta Archaeological 21. Los Angeles, CA: The Cotsen Institute of Archaeology.

Gomez, B. and Pease, P. (1992) Early Holocene Cypriot coastal palaeogeography. *Report of the Department of Antiquities, Cyprus*, 1–8.

Guilaine, J. and Briois, F. (2006) Shillourokambos and the Neolithization of Cyprus: some reflections. *Eurasian Prehistory* 4/1–2, 159–75.

Guilaine, J. and Le Brun, A. (2003) *Le Néolithique de Chypre*. Bulletin de Correspondance Hellénique, Supplément 43. Athènes: École Française D' Athènes.

Guilaine, J., Briois, F., Vinge, J.-D. and Carrère, I. (2000) Découverte d'un Néolithique précérqmique ancien chypriote (fin 9e, début 8e millénaires cal. BC), apparenté au PPNB ancien/moyen du Levant nord. *Earth and Planetary Sciences* 330, 75–82.

Halstead, P. (2005) Resettling the Neolithic: faunal evidence for seasons of consumption and residence at Neolithic sites in Greece. Pp. 38–50 in D. Bailey, A. Whittle and V. Cummings (eds), *(Un)settling the Neolithic*. Oxford: Oxbow Books.

Hayden, B. (1981) Research and development in the Stone Age: technological transitions among hunter-gatherers. *Current Anthropology* 22, 519–48.

Held, S. (1989) Colonization cycles on Cyprus 1: the biogeographic and paleontological foundations of early prehistoric settlement. *Report of the Department of Antiquities, Cyprus*, 7–28.

Held, S. (1993) Insularity as a modifier of culture change: the case of prehistoric Cyprus. *Bulletin of the American Schools of Oriental Research* 292, 25–33.

Horden, P. and Purcell, N. (2000) *The Corrupting Sea*. Oxford: Basil Blackwell.

Horwitz, L., Tchernov, E. and Hongo, H. (2004) The domestic status of the early Neolithic fauna of Cyprus: a view from the mainland. Pp. 35–48 in E. Peltenburg and A. Wasse (eds), *Neolithic Revolution*. Levant Supplementary Series 1. Oxford: CBRL and Oxbow Books.

Irwin, G. (1992) *The Prehistoric Exploration and Colonization of the Pacific*. Cambridge: Cambridge University Press.

Knapp, B., Held, S. and Manning, S. (1994) The prehistory of Cyprus: problems and prospects. *Journal of World Prehistory* 8/4, 377–453.

Last, J. (1995) The nature of history. Pp. 141–57 in I. Hodder, M. Shanks, A. Alexandri, V. Buchli, J. Carman, J. Last and G. Lucos (eds), *Interpreting Archaeology*. London: Routledge.

Le Brun, A. (1986) Khirokitia: une Civilisation Originale? Pp. 1–11 in V. Karageorghis (ed.), *Cyprus Between the Orient and The Occident*. Nicosia: Department of Antiquities of Cyprus.

Lee, R. B. and Daly, R. (1999) *The Cambridge Encyclopedia of Hunters and Gatherers*. Cambridge: Cambridge University Press.

Manning, S. and Hulin, L. (2005) Maritime commerce and geographies of mobility in the Late Bronze Age of the Eastern Mediterranean: problematizations. Pp. 270–302 in E. Blake and A. B. Knapp (eds), *The Archaeology of Mediterranean Prehistory*. Oxford: Blackwell Studies in Global Archaeology.

Manning, S. W. (n.d.) Radiocarbon dating and climate change. In A. B. Mainwaring, R. Giegengack and C. Vita-Finzi (eds), *Climate Crises in History*. Philadelphia: The American Philosophical Society.

McCartney, C. (2004) Cypriot Neolithic chipped stone industries and the progress of regionalization. Pp. 103–22 in E. Peltenburg and A. Wasse (eds), *Neolithic Revolution*. Levant Supplementary Series 1. Oxford: CBRL and Oxbow Books.

McCartney, C. (2005) Preliminary report on the re-survey of three Early Neolithic sites in Cyprus. *Report of the Department of Antiquities, Cyprus*, 1–21.

McCartney, C. (in press) The lithic assemblage of Ayia Varvara Asprokremnos: a new perspective on the Early Neolithic of Cyprus. In E. Healy, S. Campbell and O. Maeda (eds), *Studies in Technology, Environment, Production and Society (S.T.E.P.S.)*. Berlin: *ex oriente*.

McCartney, C. and Gratuze, B. (2003) The chipped stone. Pp. 11–34 in E. Peltenburg (ed.), *Lemba Archaeological Project III.1, The Colonization and Settlement of Cyprus*. Studies in Mediterranean Archaeology 70:4. Sävedalen: Paul Åströms Förlag.

McCartney, C. and Todd, I. (2005) Tenta chipped stone report. Pp. 177–264 in I. Todd (ed.), *Vasilikos Vally Project 7, Excavations at Kalavasos-Tenta 2*, Studies in Mediterranean Archaeology 71:7. Sävedalen: Paul Åströms Förlag.

McCartney, C., Manning, S., Sewell, D. and Stewart, S. (2006) Elaborating Early Neolithic Cyprus (EENC): preliminary report on the 2005 field season. *Report of the Department of Antiquities, Cyprus*, 39–62.

McCartney, C., Manning, S., Sewell, D. and Stewart, S. (2007) The EENC 2006 field season: excavations at Agia Varvara-Asprokremnos and survey of the local early Holocene landscape. *Report of the Department of Antiquities, Cyprus*, 27–44.

McCartney, C., Manning, S., Rosendhal, S. and Stewart, S. (2008) Elaborating Early Neolithic Cyprus (EENC): preliminary report on the 2007 field season: excavations at Ayia Varvara-Asprokremnos and field survey. *Report of the Department of Antiquities, Cyprus*, 1–20.

Milner, N. (2005) Can seasonality studies be used to identify sedentism in the past? Pp. 32–7 in D. Bailey, A. Whittle and V. Cummings (eds), *(Un)settling the Neolithic*. Oxford: Oxbow Books.

Peltenburg, E. (2003) *Lemba Archaeological Project III.1, The Colonization and Settlement of Cyprus*. Studies in Mediterranean Archaeology 70:4. Sävedalen: Paul Åströms Förlag.

Peltenburg, E. (2004) Cyprus: a regional component of the Levantine PPN. *Neo-lithics* 1/04, 3–7.

Peltenburg, E. and Wasse, A. (2004) *Neolithic Revolution*. Levant Supplementary Series 1. Oxford: Oxbow Books.

Peltenburg, E., Colledge, S., Croft, P., Jackson, A., McCartney, C. and Murray, M. (2001) Neolithic dispersals from the Levantine corridor: a Mediterranean perspective. *Levant* 33, 35–64.

Perlès, C. (2001) *The Neolithic in Greece. The First Farming Communities in Europe*. Cambridge: Cambridge University Press.

Reese, D. (2006) Marine invertebrate exploitation at Akrotiri-Aetokremnos. *Report of the Department of Antiquities, Cyprus*, 23–37.

Reimer, P. J., Baillie, M. G. L., Bard, E., Bayliss, A., Beck, J. W., Bertrand, C. J. H., Blackwell, P. G., Buck, C. E., Burr, G. S., Cutler, K. B., Damon, P. E., Edwards, R. I., Fairbanks, R. G., Friedrich, M., Guilderson, T. P., Hogg, A. G., Hughen, K. A., Kromer, B., McCormac, G., Manning, S., Bronk Ramsey, C., Reimer, R. W., Remmele, S., Southon, J. R., Stuiver, M., Talamo, S., Taylor, F. W., van der Plicht, J. and Weyhenmeyer, C. E. (2004) IntCal04 terrestrial radiocarbon age calibration, 0–26 cal kyr BP. *Radiocarbon* 46/3, 1029–58.

Rockman, M. and Steele, J. (2003) *Colonization of Unfamiliar Landscapes*. London: Routledge.

Rosen, A. R. (2007) *Civilizing Climate: Social Responses to Climate Change in the Ancient Near East*. Lanham: AltaMira Press.

Runnels, C., Panagopoulou, E., Murray, P., Tsartsidou, G., Allen, S., Mullen, K. and Tourloukis, E. (2006) A Mesolithic landscape in Greece: testing a site-location model in the Argolid at Kandia. *Journal of Mediterranean Archaeology* 18/2, 259–85.

Sampson, A. and Katsarou, S. (2004) Cyprus, Aegean, and Near East during the PPN. *Neo-Lithics* 1/04, 13–15.

Savvides, L. (2000) *Edible Wild Plants of the Cyprus Flora*. Nicosia: Savvides.

Sevketoglu, M. (2002) Akanthou-Arkosyko (Tatlisu-Ciftlikdüzü) the anatolian connections in the 9th millennium BC. Pp. 98–106 in W. H. Waldren and J. A. Ensenyat (eds), *World Islands in Prehistory*. British Archaeological Reports, International Series 1095. Oxford: BAR.

Shiffer, M. (1995) *Behavioural Archaeology: First Principles*. Salt Lake City, UT: University of Utah Press.

Siegenthaler, U., Heimann, M. and Oeschger, H. (1980) ^{14}C variations caused by changes in the global carbon cycle. *Radiocarbon* 22, 177–91.

Simmons, A. (1999) *Faunal Extinctions in an Island Society*. New York: Kluwer Academic\Plenum Academic.

Simmons, A. (2003) Excavations at Kritou Marottou Ais Yiorkis, an Early Neolithic site in Western Cyprus: preliminary report. *Neo-Lithics* 2/03, 8–12.

Simmons, A. H. and Mandel, R. (2007) Not such a new light: a response to Ammerman and Noller. *World Archaeology* 39, 475–82.

Stanley-Price, N. P. (1977) Khirokitia and the initial settlement of Cyprus. *Levant* 10, 66–89.

Stewart, S. T. (2004) Hill and vale: understanding prehistoric lithic use in Cyprus. Pp. 123–32 in E. Peltenburg and A. Wasse (eds), *Neolithic Revolution*. Levant Supplementary Series 1. Oxford: Oxbow Books.

Stewart, S. T. (2007) Walking in Cyprus. Unpublished PhD thesis, University of Toronto.

Stuiver, M., Reimer, P. J., Bard, E., Beck, J. W., Burr, G. S., Hughen, K. A., Kromer, B., McCormac, G., van der Plicht, J. and Spurk, M. (1998) IntCal98 radiocarbon age calibration, 24,000–0 cal BP. *Radiocarbon* 40(3), 1041–83.

Swiny, S. (2001) *The Earliest Prehistory of Cyprus*. CAARI Monograph Series 2. Boston: American Schools of Oriental Research.

Todd, I. (1986) The foreign relations of Cyprus in the Neolithic and Chalcolithic periods: new evidence from the Vasilikos Vally. Pp. 13–28 in V. Karageorghis (ed.), *Cyprus Between the Orient and Occident*. Nicosia: Department of Antiquities.

Todd, I. (1987) *Vasilikos Valley Project 6, Excavations at Kalavasos-Tenta 1*. Studies in Mediterranean Archaeology 71:6. Göteborg: Paul Åströms Förlag.

Usoskin, I. G. and Kromer, B. (2005) Reconstruction of the 14C production rate from measured relative abundance. *Radiocarbon* 47, 31– 7.

Van Geel, B., van der Plicht, J., Kilian, M. R., Klaver, E. R., Kouwenberg, J. H. M., Renssen H., Reynaud-Farrera, I. and Waterbolk, H. T. (1998) The sharp rise of Delta C-14 ca. 800 cal BC: possible causes, related climatic teleconnections and the impact on human environments. *Radiocarbon* 40, 531–50.

Waldren, W. (2002) Links in the chain: evidence of sustained prehistoric contact and cultural interaction between the Balearic Islands and Continental Europe. Pp. 152–85 in W. H. Waldren and J. A. Ensenyat, *World Islands in Prehistory*. British Archaeological Reports, International Series 1095. Oxford: BAR.

Watkins, T. (1973) Some problems of the Neolithic and Chalcolithic periods in Cyprus. *Report of the Department of Antiquities, Cyprus*, 34–61.

Watkins, T. (2004a) Putting the colonization of Cyprus into context. Pp. 23–34 in E. Peltenburg and A. Wasse (eds), *Neolithic Revolution*. Levant Supplementary Series 1. Oxford: CBRL and Oxbow Books.

Watkins, T. (2004b) Architecture and 'theatres of memory' in the Neolithic of Southwest Asia. Pp. 97–106 in E. DeMarrias, C. Gosden and C. Renfrew (eds), *Rethinking Materiality: The Engagement of Mind with the Material World*. Cambridge: McDonald Institute for Archaeological Research.

Whittle, A. (2005) Lived experience in the Early Neolithic of the Great Hungarian Plain. Pp. 64–70 in D. Bailey, A. Whittle and V. Cummings (eds), *(Un)settling the Neolithic*. Oxford: Oxbow Books.

Wilcox, G. (2003) The origins of Cypriot farming. Pp. 231–8 in J. Guilaine and A. Le Brun (eds), *Le Néolithique de Chypre*. Bulletin de Correspondance Hellénique, Supplément 43. Athènes: École Française D'Athènes.

14. The Last of the Old: A Homogeneous Later Mesolithic Ireland?

Thomas Kador

This paper is not concerned with the adoption of agriculture per se. *It investigates the changes in material culture, and in particular lithic artefacts, associated with the Late Mesolithic and the Early Neolithic periods in Ireland. It is important to make this distinction, as the phrase Mesolithic–Neolithic transition in Northwest Europe has several, frequently conflated, meanings. Changes in material culture do not necessarily represent a shift in subsistence economy (from hunting and gathering to farming) and in settlement pattern (from highly mobile to more sedentary). In fact, finding a common denominator for these phenomena represents arguably the greatest challenge to early prehistoric research in Northwest Europe. Perhaps the most viable approach to recognizing the relation in which they stand to each other is to treat them as separate and unrelated occurrences, until proven otherwise. In this paper I initiate this by discussing Ireland's later Mesolithic assemblages, and in particular their perceived special status, by suggesting an 'isolated' but internally 'homogeneous' material culture.*

Isolating Ireland's later Mesolithic

There are two key assumptions that appear to lie at the heart of most studies concerned with the material dimensions of the Mesolithic–Neolithic transition in Ireland; both bring with them a number of corollaries. First, the known technologies of later Mesolithic Ireland were very distinct from those of Britain and its continental neighbours, suggesting to some (Woodman 2000; Woodman *et al.* 2006), but not others (Thomas 2004, 115), that 'the Irish Mesolithic … is insular with very little contact outside Ireland with the possible exception of the Isle of Man' (Woodman 1977, 195). Apart from this lack of 'evidence for contacts between insular hunter-gatherers and continental farmers', there is a perceived absence of signs for social complexity in the insular Late Mesolithic, suggesting that 'Ireland became isolated from Britain and Britain became isolated from the European mainland' (Bradley 2007, 35). Second, there was a significant break of traditions between (later) Mesolithic and Early Neolithic lifeways. Along with this second assumption goes a view that the material cultures of Britain and Ireland from earliest Neolithic times onwards, especially in terms of lithic artefacts but also regarding pottery styles, were very similar, suggesting a direct link between the two islands (Woodman 2000, 247–9; Sheridan 2000; 2004; Thomas 2004):

> One of the most overlooked aspects of this phase in prehistory [the Mesolithic–Neolithic transition] is the contrast between the diverse Mesolithic occupation of the British Isles and the emergence by 5000BP of a relatively uniform Neolithic material culture in many regions of these islands. (Woodman 2000, 247–8)

These 'similarities are clearest in material dated to after 3800 cal BC and transcend what appears to be very considerable regional diversity in settlement practices' (Cooney 2007, 548–9), while 'what was going on around 4000 cal BC remains stubbornly and frustratingly unclear and certainly varied' (Whittle 2003, 150).

To begin I will discuss the first of these assumptions, concerning Ireland's isolation in Mesolithic times, which tends to be paired with a view that technology within virtually the whole of Ireland was homogeneous throughout this period, showing little local or regional variation or change over time (cf. Woodman 1987; Kimball 2000; Cooney and Grogan 1999):

> Remarkably, the Irish Late Mesolithic flaked stone industry was *homogeneous throughout the island*, and quite distinct from any other European assemblage.(Thomas 2004, 115; my emphasis)

Considering that the later mesolithic period lasted in Ireland for at least two millennia, it seems remarkable indeed, if it was true that neither technological change occurred nor cultural exchange across a rather moderate sea crossing. Yet at the same time, cultural homogeneity is supposed to have reigned within the confines of the island, parts of which would have been much easier reached by water than over land.

While in recent years several researchers have attempted to address this perceived isolation and homogeneity, the conclusions they have arrived at are generally rather unsatisfactory and often quite ad hoc: for instance, the 'generalized stone toolkit' recognized as Irish Mesolithic may have occluded a much wider material culture (perhaps made of non-durable wooden objects) (Kimball 2000, 41); or 'the very narrow range of implement types and locations where they are found may be a product of our own typological criteria', as the 'distinctive "Larnian" technique and related implements [may] reflect only part of the range of tools used during the Later Mesolithic' (Woodman and Anderson 1990, 384). Others have endeavoured to explain away the significance of this 'homogeneity' and the incompatibility of material cultures encountered on either side of the Irish sea, drawing on studies such as Hodder's investigations among the San (Hodder 1982):

> There is no sense then, in which the existence of mutually exclusive assemblages of artefacts in different geographical areas can be taken as an index of the degree of contact between the human populations involved. (Thomas 2004, 115)

I would contend that if there is clear evidence for an isolated yet homogeneous and island-wide Irish later Mesolithic lithic artefact assemblage without parallels with contemporary material culture elsewhere we must not disregard such evidence, but aim to interpret and come to understand it. Clearly such a distinct cultural expression shared by people throughout later Mesolithic Ireland would have had significant implications for the nature of the transition to Neolithic technologies. Yet none of the above statements about Ireland's perceived material cultural isolation and homogeneity provide much supporting evidence for the conclusions at which the authors arrived. Hence we must ask on what evidence these assertions are actually based, and whether they stand up to basic scrutiny. I will test these notions, beginning with that of a homogeneous island-wide lithic assemblage.

Drawing up parameters

The *Oxford English Dictionary* defines 'homogeneous' as 'consisting of parts or elements all of the same kind' or being 'of uniform nature or character throughout' (OED online 2007). This sets out the parameters against which we must test the assemblages of later Mesolithic Ireland. If it is homogeneous, Irish later Mesolithic material must consist of elements all of the same kind and/or be uniform in its nature throughout the country. In principle this should be easily tested: all one needs do is compare the finds from Irish later Mesolithic contexts with one another and assess the extent of the similarity and difference between the assemblages. To start this examination we must decide on the sample of assemblages to be included. As our aim is to discuss later Mesolithic artefact typology, surface finds dated to the later Mesolithic period on typological grounds must be ruled out, because their inclusion would lead to a circular argument. The same can be said about excavated assemblages that have been dated entirely by typology. What we need are reliably (radiocarbon) dated contexts for lithic artefact material that point to activity between the 7th and the 4th millennia BC.

The number of stratified and dated Irish later Mesolithic lithic assemblages is extremely small. In a recent discussion of Mesolithic technology in Ireland Costa *et al.* (2005) suggested that while 'the number of radiocarbon dates associated with the Irish Mesolithic is approaching 100, nearly 50 per cent come from three sites: Ferriter's Cove, Co. Kerry (Woodman *et al.* 1999), Newferry, Co. Antrim (Woodman 1977), and Mount Sandel, Co. Derry (Woodman 1985)' (Costa *et al.* 2005, 21). According to Costa *et al.* (2005, 21), there are 17 sites dated to the period 6000–4000 BC, although their identities are not specified. Since Costa *et al.*'s study a number of new sites have been discovered and others have been re-dated, somewhat increasing the number of ^{14}C dates associated with Mesolithic artefacts. Yet the published literature and a small number of unpublished reports allowed me to identify only 14 locations with relatively reliable dates from this period associated with lithic artefacts. Of these only 10 were suitable for comparison with one another (see Table 14.3). Before we can review the evidence from these 10 locations we must briefly reflect on the currently accepted typological indicators of Irish later Mesolitic lithic technology.

Cores

One of the most typical indicators of Irish later Mesolithic artefact production is the presence of large, single-platform, uniplane, so-called 'Larnian' cores, generally using less than 50 per cent of the diameter of any given pebble. However, it has been demonstrated that this technology is restricted to the availability of large flint pebbles (e.g. Woodman *et al.* 1999, 33–4), which in turn are largely confined to the northeast of Ireland. Hence Larnian cores are not regularly found in areas outside the northeast and are rarely found on materials other than flint (Woodman 1987, 143–4). Consequently, the homogeneity of the Irish

		total	Retouched	Later Meso	% of total	elongated pebbles	stone axe heads	collection type	source
Bay Farm 1		c.30000	54	51	0.1	Not mentioned		Excavation	Woodman & Johnson 1996
Corralana	dates not directly associated with the artefacts	282	25	8	2.8	None	1	Surface collection	Warren *et al.* 2008
Dalkey, northern basal midden (Site V)		11-12,000		46	c.0.4	26	4	Excavation	Leon 2005, Liversage 1968
Ferriter's Cove		7166	34	25	0.3	42	14	Excavation	Woodman *et al.* 1999
Clonava / L. Derravaragh		111	3	2	1.8	Absent		Excavation	Mitchell 1972b
Moynagh Lough		167	49	8	4.8	Absent		Excavation	Bradley 1991
Newferry		2929	582	486	16.6	4	50	Excavation	Woodman 1977
Rockmarshall, midden 1		100	15	9	9.0	'large numbers'		Excavation	Mitchell 1947; 1949
Sutton 1 & 2		353	23	14	4.0	52	2	Excavation	Mitchell 1956; 1972a

Table 14.1. Locations dated to the later Mesolithic period that have also produced typical later Mesolithic artefact types.

later Mesolithic assemblages can clearly not be extended to core reduction technology and thus we can dismiss the latter as a safe indicator of later Mesolithic activity, in that the absence of Larnian cores from a location does not indicate the absence of later Mesolithic activity there. Moreover, the fact that Larnian core technology is largely confined to northeast Ireland could also be viewed as evidence for regional variation within the 'homogeneous' Irish later Mesolithic craft traditions.

Flake technology

Woodman suggested that 'the absence of the Larnian technique where chert or rhyolite was used shows that it was not so much the type of core that was important but rather the need to have a series of large blades' (Woodman 1987, 144). These generally large, hard-hammer-flaked blades (and flakes), whether produced through Larnian reduction or not, appear to represent the only definite diagnostic object for later Mesolithic activity at a location. The famous 'type fossil' here is the leaf-shaped 'Bann Flake' with distinct proximal trimming (see Woodman 1978). However, strictly speaking, the term 'Bann Flake' relates to only one very specific artefact type, which in turn forms part of a much broader object class of butt-trimmed and related forms (see Woodman 1978, 82–4). As we will see below, other types of objects occur at Irish later Mesolithic sites but the macrolithic, proximally trimmed industries appear to be the only relatively safe chronological indicators, as some of the other object types transcend the chronological boundaries of the later Mesolithic period. As with Larnian cores, much of our understanding of later Mesolithic flake typologies appears to be heavily influenced by expectations based on material evidence from the flint-rich northeast, the area most intensively studied and with the longest history of research (Woodman *et al.* 2006). But of course 'there is no reason why the range of artefacts from the North East should represent the norm against which everything else should be tested' (Woodman 2003, 10). This limitation of recognizable later Mesolithic artefact types to styles recognized only in the northeast invariably leads us to a problem of tautology. Almost by default, we can only discover artefacts assemblages and find locations that closely match those already known to be later Mesolithic. Therefore, without testing it against the evidence from other parts of the country, this perceived homogeneity might well just be a product of our northeastern research bias.

Contextualizing Irish later Mesolithic material

On reviewing the proportion of diagnostic later Mesolithic flakes and blades at the 10 later Mesolithic locations identified (Table 14.3), we can see that their percentage within the overall assemblages varies quite strongly (Table 14.1). Even so, with the exception of Newferry their proportion does not exceed 10 per cent. Two factors are responsible for the inflated proportion of diagnostics at this location. Firstly, at Newferry 'waste' flakes were weighed rather than counted during analysis and if these 6414 g were to be added to the total number of artefacts the proportion of diagnostic later Mesolithic material would be significantly lower. Secondly, Newferry very much represents the 'type site' for the Irish later Mesolithic artefact industry, setting the standard against which other assemblages are tested. Therefore, some of the objects

Site	mean length (mm)	Predominant raw material
Newferry 3, Zone 4, Co. Antrim	69	Flint
Curran Point, Larne, Co. Antrim	64	Flint
Rough Island, Co. Down	50	Flint
Sutton, Co. Dublin	54	Flint?
Dunpower, Co. Cork	50	Flint?
Clonava/Derravarragh, Co. Westmeath	71	Chert
Corralanna, Co. Westmeath	56	Chert
Ferriter's Cove, Co. Kerry	28	Flint
Ferriter's Cove (southern area)	47	Greenstone

Table 14.2. Average object size in comparable later Mesolithic assemblages (not exclusively from excavated contexts; based on Woodman et al. 1999, 72).

identified there as later Mesolithic would perhaps not be ascribed to this date in a different context. It is also possible that the proportions of later Mesolithic objects provided are too high for Rockmarshall and Sutton: here the reports do not make it clear whether the figures provided are those for all artefacts encountered or for diagnostic or known object types only.

Even accounting for a possible bias towards large and retouched objects Table 14.1 clearly illustrates that known diagnostic later Mesolithic forms make up only a small proportion of each collection. Thus it is possible that other object types might be present that have simply not been identified as being characteristic for the period. Woodman *et al.* (1999, 76) have pointed out that the 'diagnostic implements, such as "Bann Flakes" and/or "butt-trimmed forms", have been ascribed a "false significance" and that this can in some examples, as at Ferriter's Cove, be contrasted with an interesting and diverse range of other artefacts'. Furthermore, they also noted that at Ferriter's Cove only a 'small number of butt-trimmed forms [were] recovered and few of them could be considered classic examples' (Woodman *et al.* 1999, 76). In other words, despite notions of a 'definite template' (Woodman 1987, 144), it appears clear that the most diagnostic Irish later Mesolithic objects, the butt-trimmed forms, are only part of a much more varied pattern of lithic artefact technology throughout this period. Moreover, even butt-trimmed forms themselves are not uniform across Ireland. They can be made of several different raw materials and using various reduction strategies (i.e. with or without Larnian cores). In addition, there is also quite significant variation in object size among diagnostic later Mesolithic forms found in different assemblages from across Ireland (Table 14.2).

The size variation among later Mesolithic artefacts (see Table 14.2) may relate to factors such as location and the raw material used. Comparison of object sizes illustrates that the largest objects are not those made from flint from Co. Antrim but actually those made of midland chert. In addition, there is the possibility of a chronological dimension to the development of these artefacts. Woodman suggested a division based on the sequence from Newferry 'between the laminar earlier industries and the later flake-blade industries' (Woodman 1978, 209–10; 1977, 191). While we must acknowledge that there are shared morphological features recognizable on a number of objects within each of the assemblages reviewed, we should cautiously reconsider whether we can be confident that this degree of similarity represents homogeneity. It is not within the remit of this paper to further pursue the analysis of these collections in terms of identifiable patterns of variation; however, there is clearly much scope for further work on this. For instance, we could test to what extent the differences in size relate to the types of location, region or part of the landscape, whether they are primarily dependent on the raw materials used or if there is a chronological dimension to variation. Instead, while continuing to look for further evidence for similarities and variation in Irish later Mesolithic assemblages, I will also begin to ask some questions about Ireland's perceived isolation during the period.

The 'BET' factor

Away from the chipped stone artefacts there are a number of object types also frequently encountered in Irish later Mesolithic contexts, such as the elongated pebbles sometimes described as bevel-ended tools (BETs) or traditionally as limpet scoops and hammers (Warren 2006, 18–20). As Table 14.1 illustrates, 5 of the 10 tested locations have produced such objects. Some, especially the coastal shell-midden sites, have yielded larger quantities of the BETs than of the more typical later Mesolithic chipped stone artefacts. So why are these objects not also acknowledged as diagnostic Irish Mesolithic artefacts? It seems that elongated pebbles are one of the often forgotten implements of Mesolithic Ireland. While they clearly occur regularly and in significant numbers in later Mesolithic contexts on excavated sites, they are not a frequent feature of (early prehistoric) surface collections, possibly because of their rather non-distinctive appearance.

Surface collections, however, make up the vast bulk of the studied later Mesolithic artefact assemblages, leading in turn to an over-representation of flaked objects and an under-representation of pebble tools in our collections. Interestingly, elongated pebbles are also known from an Early Mesolithic context at Mount Sandel (Woodman 1978, 115) and thus appear to transcend the chronological divide between Early and Late Mesolithic in Ireland. Similar implements have been found at Mesolithic sites in Scotland, other coastal parts of Britain and Brittany (David and Walker 2004, 323–5; Pailler and Dupont 2007; Saville 2004, 191; Warren 2005, 100). If we are looking for a possible link between Ireland and other parts of Western Europe, contradicting the idea of isolation, these objects appear to present a real possibility. While such objects are known from all the above listed regions from secure (later) Mesolithic contexts, there are some chronological problems. Organic examples from Scotland – made of bone and antler – show 'a long currency of use; from the Mesolithic into the Bronze Age' (Warren 2005, 100), with 26 directly dated Scottish examples having produced 'an astonishing age range from c. 8470 to 3010 BP' (Saville 2004, 197). This casts some doubt on whether all the known stone examples from Scotland, Ireland and elsewhere can be ascribed a Mesolithic date. Clearly these objects were made and used in many parts of Western Europe during Mesolithic times, yet they are not as striking and shiny as flints and neither do they resemble arrowheads or other types of weapon that facilitate the 'boys and arrows' narratives of Mesolithic Europe (see Finlay 2000a and 2000b).

Axing isolationism

Other types of stone object found frequently in Irish Mesolithic contexts are ground and polished stone axe and adze heads. Though only half of the above reviewed locations have produced such objects in Mesolithic contexts, they are known in association with Mesolithic material from many parts of Ireland (Woodman 1978, 109–14; Woodman *et al.* 1999, 63–5). Thus the object formerly considered a 'type fossil' of Neolithic life is in fact 'one of the most consistently occurring aspects of the Irish Mesolithic rather than an occasional anomaly or chance intrusion' (Woodman *et al.* 1999, 79). Although comparable objects have not been encountered in pre-Neolithic contexts in Scotland or England (Saville 2004, 193) there are some examples from Wales (David and Walker 2004, 325–7). Polished stone axes are also known in Mesolithic contexts from other parts of Europe, such as Sweden (Welinder 1977).

The problem in Ireland would seem to be distinguishing between Mesolithic and Neolithic axe heads:

> A typical problem of identifying what is Later Mesolithic can be best seen from the use of polished mudstone axes. A small cache of axes found in the Ferriter's Cove excavation, near a group of cattle bones, are typical of those found in Later Mesolithic contexts, e.g. at Newferry, Co. Antrim, but there is no reason why these particular axes could not be Neolithic. (Woodman 1994, 216)

Having said that, Cooney (2007, 545–6), in part following Woodman (1977a, 189; Woodman *et al.* 1999, 78–9; Woodman *et al.* 2006, 299), suggests that there may be a distinction in both the raw materials used and the ways in which axes were produced in the Mesolithic and Neolithic periods respectively (see also Cooney 2000a, 210; Cooney and Mandal 1998). The example from the Danish Ertebølle, where Danubian shaft-hole axes have been encountered in strata dating to c. 6600 BP [c. 5620–5480 cal BC] (Fischer 2003, 410–11), is interesting for interpretations of the transition from Mesolithic to Neolithic ways of life. These axes are an oft-cited example for the existence of contacts between early farmers and hunter-gatherer groups in Northwest Europe. It is widely accepted that stone axe heads represented important items for trade, exchange and social contacts across large swathes of Europe in later parts of prehistory (e.g. Bradley and Edmonds 1993; Cooney 2000a), so is it so inconceivable that the origins of such networks in Ireland (or Britain) could lie in Mesolithic times? We know that other objects – flints, cherts, rhyolites, bloodstone and so on – were already moving over significant distances throughout this region (e.g. Rynne 1983/84; see also Kador 2007a; and forthcoming for Ireland; Cunliffe 2001, 128–9 for England; and Warren 2005, 139–41, for Scotland). In short, it seems clear that issues of continuity and change in relation to Irish early prehistoric lithic assemblages, as illustrated by ground and polished stone axe heads, have not been fully explored.

Apart from axe heads there are also other objects that some appear to have difficulty in accepting as later Mesolithic 'object types'. In particular, the large ground and chipped slate points found at Moynagh Lough, Ferriter's Cove, Toome (Woodman 2005, 132–3; Woodman *et al.* 1999, 82) and recently at Belderrig (Warren and Rice 2007), which Woodman (2005, 130–2) has termed 'Moynagh Points', are also encountered on Irish later Mesolithic sites. The number of Mesolithic associations for such objects is growing and there is no reason why – rather than considering them as peculiar oddities – we should not see them as expressions of 'interesting alternative ways of making tools here' (Woodman 2005, 133).

Normalizing Mesolithic evidence: issues of scale

The range of stone artefacts and the number of ways of making and using them in later Mesolithic Ireland is clearly much larger and more diverse than many of the recent synthetic reviews of the period would have us believe. The issue often appears to be primarily one of scale. That is, many of these reviews are mixing up a number of spatial scales in their reading of the material, confusing the

Site name	County	Site type	Dates bp	Date type	Substance dated	Laboratory	Laboratory number	Source
Bay Farm I	Antrim	Settlement	6940±65	Conventional	Unspecified	Belfast	UB-3552	Woodman & Johnson
Bay Farm I	Antrim	Settlement	7015±45	Conventional	Unspecified	Belfast	UB-3553	Woodman & Johnson
Belderrig	Mayo	Settlement	5380± 38	AMS	Hazelnut shell	Belfast	UB-7589	Rice & Warren 2007
Belderrig	Mayo	Settlement	5631±39	AMS	Hazelnut shell	Belfast	UB-6882	Rice & Warren 2007
Belderrig	Mayo	Settlement	5545±40	AMS	Hazelnut shell	Belfast	UB-7585	Rice & Warren 2007
Belderrig	Mayo	Settlement	5845±39	AMS	Hazelnut shell	Belfast	UB-7586	Rice & Warren 2007
Belderrig	Mayo	Settlement	5433±39	AMS	Hazelnut shell	Belfast	UB-7583	Rice & Warren 2007
Belderrig	Mayo	Settlement	5362±38	AMS	Hazelnut shell	Belfast	UB-7584	Rice & Warren 2007
Belderrig	Mayo	Settlement	5255±36	AMS	Hazelnut shell	Belfast	UB-7587	Rice & Warren 2007
Belderrig	Mayo	Settlement	5201±37	AMS	Hazelnut shell	Belfast	UB-7588	Rice & Warren 2007
Clonava (Lough Derravaragh)	Westmeath	Settlement	5360±110	Conventional	Charcoal	Teledyne Isotopes	I-4234	Mitchell 1972b
Clonava (Lough Derravaragh)	Westmeath	Settlement	5237±37	AMS	Hazelnut shell	Belfast	UBA-8105	Warren *et al.* 2007
Clonava (Lough Derravaragh)	Westmeath	Settlement	5706±37	AMS	Hazelnut shell	Belfast	UBA-8106	Warren *et al.* 2007
Corralana	Westmeath	Surface find	6418±44	AMS	Hazelnut shell	Belfast	UB-6349	Warren *et al.* 2007
Corralana	Westmeath	Surface find	6319±49	AMS	Hazelnut shell	Belfast	UBA-8109	Warren *et al.* 2007
Corralana	Westmeath	Surface find	6378±39	AMS	Hazelnut shell	Belfast	UBA-8110	Warren *et al.* 2007
Dalkey Island	Dublin	Midden	5300±170	Conventional	Charcoal	Dublin	D-38	McAulay & Watts 1961
Dalkey Island	Dublin	Midden	5600±80	AMS	Bone	Oxford	OxA-4570	Hedges *et al.* 1997
Dalkey Island	Dublin	Midden	6410±110	AMS	Bone	Oxford	OxA-4572	Hedges *et al.* 1997
Dalkey Island	Dublin	Midden	6870±90	AMS	Bone	Oxford	OxA-4568	Hedges *et al.* 1997
Dalkey Island	Dublin	Midden	7250±100	AMS	Bone	Oxford	OxA-4569	Hedges *et al.* 1997
Ferriter's Cove	Kerry	Midden	5245±55	Conventional	Charcoal	Cambridge	Q-2641	Woodman & O'Brien 1993
Ferriter's Cove	Kerry	Midden	5400±25	Conventional	Shell	Belfast	UB-3761	Woodman *et al.* 1999
Ferriter's Cove	Kerry	Midden	5400±220	Conventional	Charcoal	British Museum	BM-2227R	Bowman *et al.* 1990
Ferriter's Cove	Kerry	Midden	5420±150	Conventional	Charcoal	British Museum	BM-2227AR	Bowman *et al.* 1990
Ferriter's Cove	Kerry	Midden	5480±55	Conventional	Charcoal	Belfast	UB-3597	Woodman & O'Brien 1993
Ferriter's Cove	Kerry	Midden	5490±160	Conventional	Charcoal	British Museum	BM-2229R	Bowman *et al.* 1990
Ferriter's Cove	Kerry	Midden	5500±130	Conventional	Charcoal	British Museum	BM-2229AR	Bowman *et al.* 1990
Ferriter's Cove	Kerry	Midden	5505±45	Conventional	Charcoal	Belfast	UB-3599	Woodman & O'Brien 1993
Ferriter's Cove	Kerry	Midden	5510±70	AMS	Bone	Oxford	OxA-3869	Hedges *et al.* 1997
Ferriter's Cove	Kerry	Midden	5545±65	AMS	Human bone	Oxford	OxA-4918	Hedges *et al.* 1997

Ferriter's Cove	Kerry	Midden	5590±60	AMS	Human tooth	Oxford	OxA-5770	Hedges *et al*. 1997
Ferriter's Cove	Kerry	Midden	5620±80	Conventional	Hazelnut shell	Groningen	GrN-18771	Woodman & O'Brien 1993
Ferriter's Cove	Kerry	Midden	5620±130	Conventional	Hazelnut shell	Groningen	GrN-18770	Woodman & O'Brien 1993
Ferriter's Cove	Kerry	Midden	5680±70	Conventional	Marine shell	Cambridge	Q-2634	Woodman & O'Brien 1993
Ferriter's Cove	Kerry	Midden	5725±80	Conventional	Charcoal	Belfast	UB-3598	Woodman & O'Brien 1993
Ferriter's Cove	Kerry	Midden	5750±140	Conventional	Charcoal	British Museum	BM-2228R	Bowman *et al*. 1990
Ferriter's Cove	Kerry	Midden	5850±130	Conventional	Charcoal	British Museum	BM-2228AR	Bowman *et al*. 1990
Ferriter's Cove	Kerry	Midden	5900±110	Conventional	Charcoal	Groningen	GrN-18769	Woodman & O'Brien 1993
Ferriter's Cove	Kerry	Midden	6300±140	Conventional	Charcoal	Groningen	GrN-18772	Woodman & O'Brien 1993
Moynagh Lough	Meath	Settlement	5270±60	Conventional	Charcoal	Groningen	GrN-11443	Bradley 1991
Newferry	Antrim	Settlement	5290±170	Conventional	Charcoal	Dublin	D-36	McAulay & Watts 1961
Newferry	Antrim	Settlement	5310±85	Conventional	Charcoal	Belfast	UB-503	Smith *et al*. 1973
Newferry	Antrim	Settlement	5415±90	Conventional	Charcoal	Belfast	UB-489	Smith *et al*. 1973
Newferry	Antrim	Settlement	5705±90	Conventional	Charcoal	Belfast	UB-630	Woodman 1977
Newferry	Antrim	Settlement	5795±105	Conventional	Charcoal	Belfast	UB-508	Smith *et al*. 1973
Newferry	Antrim	Settlement	6215±100	Conventional	Charcoal	Belfast	UB-490	Smith *et al*. 1973
Newferry	Antrim	Settlement	6440±185	Conventional	Charcoal	Belfast	UB-653	Woodman 1977
Newferry	Antrim	Settlement	6590±60	Conventional	Charcoal	Belfast	UB-890	Woodman 1977
Newferry	Antrim	Settlement	6605±170	Conventional	Charcoal	Belfast	UB-505	Woodman 1977
Newferry	Antrim	Settlement	6915±60	Conventional	Charcoal	Belfast	UB-886	Woodman 1977
Newferry	Antrim	Settlement	6955±60	Conventional	Charcoal	Belfast	UB-516	Smith *et al*. 1973
Newferry	Antrim	Settlement	6980±115	Conventional	Charcoal	Belfast	UB-887	Woodman 1977
Newferry	Antrim	Settlement	7075±120	Conventional	Charcoal	Belfast	UB-885	Woodman 1977
Newferry	Antrim	Settlement	7175±105	Conventional	Charcoal	Belfast	UB-514	Smith *et al*. 1973
Newferry	Antrim	Settlement	7190±110	Conventional	Charcoal	Belfast	UB-517	Smith *et al*. 1973
Newferry	Antrim	Settlement	7485±115	Conventional	Charcoal	Belfast	UB-496	Smith *et al*. 1973
Newferry	Antrim	Settlement	7630±195	Conventional	Wood	Belfast	UB-641	Woodman 1977
Newferry	Antrim	Settlement	8175±145	Conventional	Organic mud	Belfast	UB-888	Woodman 1977
Newferry	Antrim	Settlement	8190±120	Conventional	Wood	Belfast	UB-487	Smith et al. 1973
Newferry	Antrim	Settlement	8895±125	Conventional	Charcoal	Belfast	UB-637	Woodman 1977
Rockmarshall	Louth	Midden	5470±110	Conventional	Charcoal	Teledyne Isotopes	I-5323	Woodman 1978
Rockmarshall	Louth	Midden	5705±110	AMS	Human bone	Oxford	OxA-4604	Hedges *et al*. 1997
Sutton	Dublin	Midden	5250±110	Conventional	Charcoal	Teledyne Isotopes	I-5067	Mitchell 1972a
Sutton	Dublin	Midden	6560±75	AMS	Bone	Oxford	OxA-3960	Hedges *et al*. 1997
Sutton	Dublin	Midden	6660±80	AMS	Bone	Oxford	OxA-3691	Hedges *et al*. 1997
Sutton	Dublin	Midden	7140±100	AMS	Bone	Oxford	OxA-4449	Hedges *et al*. 1997

Table 14.3. Dates and sources for Irish later Mesolithic locations reviewed in the paper.

evidence derived from local assemblages, which as we have seen can often be quite diverse across the country, with the bulk of 'diagnostic' later Mesolithic material filling up the drawers of our museums and collections. I have already drawn attention to the fact that most of this material has been collected as surface finds and thus we return to the circular argument that only those flakes and blades fitting the narrow typological criteria – largely drawn up following examples from the flint-rich northeast – were entered under the category of later Mesolithic objects, or indeed kept in the first place. Any other objects would have been placed in the pile of generic flint, chert and stone artefacts. It is possible that many of them are still somewhere in that pile. Hence Woodman (2005, 140) reminds us that we need to open the museum drawers as well as our minds 'and accept that we may have tried too hard to normalize Mesolithic typologies and technologies'.

Similar points can be made about the types of location at which Mesolithic material has been identified (see below), where we keep going back to the same types of place. In addition, we tend to approach them very much in a top-down fashion, looking at them on maps from a generic point of view, invariably identifying the course of a river or stream, a lake or the seashore in sufficiently close proximity to fit the bill of a Mesolithic 'close to water' location (Cooney and Grogan 1999, 13). Consequently, we fail to appreciate the more small-scale make-up of the individual locations and their surroundings, instead moving from the micro-scale of the artefact that marks out the spot as later Mesolithic straight to the generic scale of distribution maps, missing all that lies between these two dimensions, including the levels at which people in the past would have led their lives. This jumping between scales has allowed us to ignore the complexities of human actions and interactions and provided an opportunity to avoid having to explain aspects that cannot be easily rationalized within the existing models. For example, how do we account for the increasing number of sites that would appear to belong to a Mesolithic tradition but have not produced any diagnostic artefacts? (Woodman 2003, 15). Woodman's comment that 'important sites do not always have to be overloaded with artefacts' (2003, 15) serves as a useful reminder that there are clearly elements to the Irish artefact assemblages of the 5th and 6th millennia BC that we have not yet learned to recognize and it appears more and more that the only sites that we identify as later Mesolithic are those that confirm what we already believe to know about the period. This applies both to the parts of the landscape that these places occupy and to the material evidence they produce.

Discussions of later Mesolithic Ireland also frequently appear to incorporate a direct relationship between the distinct material culture, landscape use and a subsistence economy:

> The Irish later Mesolithic has distinctive characteristics such as the lithic assemblage and a focus on fishing and gathering which do not suggest much contact with either Britain or adjacent continental Europe. (Cooney 2000a, 13)

This perceived focus on 'fishing and gathering' is thought to be reflected in the types of location and environment in which Mesolithic material is encountered. Once again there is an element of bias here, as we tend to search for evidence of Mesolithic activity in locations where we expect it to be present – coastal areas, raised beaches, river valleys and lakeshore locations. This is very much reflected in the siting of large-scale research projects with the brief of investigating early prehistory in Ireland since the early 1980s (see Green and Zvelebil 1990; Zvelebil *et al.* 1996; Woodman 1983). It seems neither our patience nor our budgets are great enough to allow more objective but by definition generally less reward-rich strategies (see e.g. Rowley Conwy *et al.* 1987). While the archaeological programmes associated with large infrastructural projects that have succeeded in reshaping much of our perception of Neolithic settlement in Ireland (e.g. Cooney 2000b) have not had an equally dramatic impact on changing narratives of Mesolithic activity, they are clearly producing new and unexpected evidence for the period (e.g. McQuade and O'Donnell 2007; Fitzgerald 2007). There is an ever-growing body of Mesolithic material, some of which comes from parts of the country where it is not supposed to be found (e.g. Warren and Rice 2007; Driscoll 2006; Noel Dunne pers. comm.). Moreover, new approaches to those sites recently added as well as longer-known ones, especially those studies that emphasize people's activities across the landscape (e.g. Kador 2005; 2007a; 2007b; Little 2005; Driscoll 2006), have highlighted the fact that the location of Mesolithic material is not simply the result of opportunistic exploitation of the available natural resources. Many such locations highlight people's desires to communicate and stay in touch with one another and present concerns with working and depositing materials in appropriate ways. Thus the simple category of the Irish Mesolithic 'fisher-gatherer' and the explanation that both the uniform 'Larnian assemblages' and the locations of sites are the functional answers to this lifestyle appear to crumble under more detailed examination.

Irish homogeneity, what is left? Reversing the argument

To conclude this examination we can return to the original question and ask what remains of the 'remarkably homogeneous' Irish later Mesolithic assemblages. The first point that needs to be stressed is that the variety within Irish later Mesolithic artefact assemblages is much greater than commonly believed. Yet clearly there are certain object types – or more accurately objects made according to a certain set of templates – that seem to occur more frequently than others in Irish later Mesolithic contexts and, importantly, they appear to have few parallels outside the island. Having said that, even within this group of objects there are variations in dimensions, raw materials and reduction sequence. Elsewhere (Kador 2007a; 2007b; forthcoming) I have demonstrated that some of these artefacts have travelled considerable distances and it is just as likely that they are objects that not only served a functional purpose but also communicated affinity or distance between groups of people in various parts of the country. There is a clear indication, for example, through 'caches and other structured deposits of stone tools ... that the processes of stone-tool production were symbolically significant to the Mesolithic populations of Ireland' (Warren 2005, 138; see also Finlay 2003; Leon 2005). Thus it is conceivable that the manufacture of these particular objects was consciously maintained as distinct from technologies in other places, such as Britain and continental Europe. At the same time, other objects that were in use in Ireland, such as stone axe heads and elongated pebbles, have comparable types in other parts of Northwest Europe and there is no reason to rule out the idea that the knowledge of making and using these objects, or even some artefacts themselves, may have moved between various regions of Western Europe.

At the end of this review it seems that the only claim to homogeneity that can be sustained is that of the absence of microlithic material from Ireland after c. 6000 BC, or perhaps even before this date (see Costa *et al.* 2005). However, the fact that one element of artefact manufacture – even though an important one elsewhere – is not present does not make an assemblage homogeneous. Besides, we could also turn the argument on its head and suggest that, rather than Ireland and its lithic assemblage being out of touch with the rest of Europe during later Mesolithic times, perhaps it is Britain, and in particular the south, that represents the odd one out. This is reflected in the fact that 'in contrast to much of Europe and Ireland, the later Mesolithic of southern Britain is associated with the continued use of microliths' (Woodman 2000, 246). While many other parts of Northern and Western Europe already witness the arrival of/return to macrolithic chipped stone tool technologies. Examples for this can be found in Spain/Catalonia (Angelucci 2003), Portugal (Lubell *et al.* forthcoming; Araujo *et al.* 2009), Denmark and Northwest Germany (e.g. Bokelmann 1999). Additionally there is the apparent refusal by people in most parts of later Mesolithic southern Britain to integrate polished and ground stone axe heads into their stone artefact repertoire, as they were doing already in Ireland and many other parts of the continent.

Returning to transitions

In itself my contribution, focusing as it does almost exclusively on later Mesolithic material, may not appear to contribute greatly to our understanding of the landscapes of the Mesolithic–Neolithic transition. Towards the beginning I made the point that if Ireland's often assumed status as insular yet culturally homogeneous represents a true reflection of the situation in later Mesolithic times it would have had significant implications for the nature of the transition to Neolithic life. So where do my findings – that the Irish later Mesolithic assemblage does not appear to have been exceptionally homogeneous and that people may not have been quite as insular as has been proposed – leave us? It is clear that there are differences in the material cultures of later Mesolithic Britain and Ireland; however, there were equally significant differences in material culture within Britain and indeed, as I hope to have demonstrated here, Ireland. Yet often the scales at which we analyse the datasets do not allow us to identify these variations. Besides, people in Mesolithic times did not spend all of their time making and using stone tools; these were just one element in a world as rich and diverse in material culture as that of Neolithic times or indeed any other period. Therefore to suggest that there were cultural, or perhaps even ethnic, differences between people in two areas based on some different elements in their flaked stone tool assemblage could be compared to the paradigms of culture history, where 'every people had its pot'. Having said all that, people during the 6th and 5th millennia BC in Ireland, Britain and elsewhere clearly had reasons to design their stone tool kits in certain ways often distinct from those in other regions. There must have been equally strong reasons for the appearance of a relatively uniform material culture, in terms of lithic artefact assemblages and pottery styles, in many regions of Ireland and Britain from early in the 4th millennium (Woodman 1994, 217; 2000, 247; Thomas 2004; Edmonds 1995; but see Cooney 2007, 548–9). At the same time, regional differences were maintained or at least established rather quickly, with the diversification of pottery styles (e.g. Armit 1996; Sheridan 1995), the construction of substantial timber post-built structures and use of land enclosure in parts of Ireland, phenomena not recognized in Britain. It is aiming to achieve detailed understandings of some of these similarities and differences and the processes that underlie them, rather than propagating platitudes about generic material expression and landscape use, that will engender a wider understanding of the transition in Northwest Europe. This will require an adjustment of our scales of analysis and a refocusing of the archaeological lens to allow an in-depth engagement with the material evidence at a local and regional level, while not losing sight of the bigger picture.

Bibliography

Angelucci, D. E. (2003) Geoarchaeology and micromorphology of Abric de la Cativera (Catalonia, Spain). *Catena* 54, 573–601.

Araujo, A. C., Almeida, F. and Valente, M. J. (2009) Macrolithic industries of the Portuguese Mesolithic: a human adaptive response. Pp. 779–87 in S. McCartan, R. Schulting, G. M. Warren and Woodman, P. C. (eds), *Mesolithic Horizons: Proceedings of the seventh international Conference on the Mesolithic in Europe (Belfast 2005)*. Oxford: Oxbow Books.

Armit, I. (1996) The transition to agriculture: Scotland. Pp. 280–90 in T. Pollard and A. Morrison (eds), *The Early Prehistory of Scotland*. Edinburgh: Edinburgh University Press.

Bokelmann, K. (1999) Zum Beginn des Spätmesolitikums in

Südskandinavien. Geweihaxt, Dreieck und Trapez, 6100 cal BC. *Offa* 56, 183–97.

Bowman, S. G. E., Ambers, J. C. and Leese, M. N. (1990) Re-evaluation of British Museum radiocarbon dates issued between 1980 and 1984. *Radiocarbon* 32, 59–79.

Bradley, J. (1991) Excavations at Moynagh Lough, county Meath. *Journal of the Royal Society of Antiquaries of Ireland* 121, 5–26.

Bradley, R. (2007) *The Prehistory of Britain and Ireland.* Cambridge: Cambridge University Press.

Bradley, R. and Edmonds, M. (1993) *Interpreting the Axe Trade: Production and Exchange in Neolithic Britain.* Cambridge: Cambridge University Press.

Cooney, G. (2000a) *Landscapes of Neolithic Ireland.* London: Routledge.

Cooney, G. (2000b) Recognising regionality in the Irish Neolithic. Pp. 47–65 in G. J. A. Desmond, M. McCarthy, J. Sheehan and E. Shee Twohig (eds), *New Agendas in Irish Prehistory.* Dublin: Wordwell.

Cooney, G. (2007) Parallel worlds or multi-stranded identities? Considering the process of 'going over' in Ireland and the Irish Sea zone. Pp. 543–66 in A. W. R. Whittle and V. Cummings (eds), *Going Over: The Mesolithic–Neolithic Transition in North-West Europe Europe.* Oxford: British Academy and Oxford University Press.

Cooney, G. and Grogan, E. (1999) *Irish Prehistory: a Social Perspective.* Bray: Wordwell.

Cooney, G. and Mandal, S. (1998) *The Irish Stone Axe Project – Monograph 1.* Bray: Wordwell.

Costa, L. J., Sternke, F. and Woodman, P. C. (2005) Microlith to macrolith: the reasons behind the transformation of production in the Irish Mesolithic. *Antiquity* 79, 19–33.

Cunliffe, B. (2001) *Facing the Ocean: The Atlantic and its Peoples 8000 BC–AD 1500.* Oxford: Oxford University Press.

David, A. and Walker, E. A. (2004) Wales during the Mesolithic period. Pp. 299–337 in A. Saville (ed.), *Mesolithic Scotland and its Neighbours: The Early Holocene Prehistory of Scotland, its British and Irish Context and some Northern European Perspectives.* Edinburgh: Society of Antiquaries of Scotland.

Driscoll, K. (2006) The Early Prehistory in the West of Ireland: Investigations into the Social Archaeology of the Mesolithic, West of the Shannon, Ireland. Unpublished MLitt thesis, National University of Ireland, Galway.

Edmonds, M. (1995) *Stone Tools and Society: Working Stone in Neolithic and Bronze Age Britain.* London: Batsford.

Finlay, N. (2000a) Deer prudence. *Archaeological Review Cambridge* 17/2, 1–8.

Finlay, N. (2000b) Microliths in the making. Pp. 23–31 in R. Young (ed.), *Mesolithic Lifeways: Current Research in Britain and Ireland.* Leicester Archaeology Monographs 7. Leicester: University of Leicester, School of Archaeological Studies.

Finlay, N. (2003) Cache and carry: defining moments in the Irish later Mesolithic. Pp. 87–94 in L. Bevan and J. More (eds), *Peopling the Mesolithic in a Northern Environment.* British Archaeological Reports, International Series 1157. Oxford: BAR.

Fischer, A. (2003) Trapping up the rivers and trading across the sea: steps towards the neolithisation of Denmark. Pp. 405–13 in L. Kindgren, H. Knutsson, K. Loeffler, D. Larsson and A. Akerlund (eds), *Mesolithic on the move: Papers presented on the sixth international conference on the Mesolithic in Europe, Stockholm 2000.* Oxford: Oxbow.

Fitzgerald, M. (2007) Revolutionising our understanding of prehistoric basketry. *Seanda: The NRA Archaeology Magazine* 2, 49–51.

Green, S. W. and Zvelebil, M. (1990) The Mesolithic colonisation and agricultural transition of south east Ireland. *Proceedings of the Prehistoric Society* 56, 57–88.

Hedges, R. E. M., Housley, R. A., Ramsey, C. B. and van Klinken, G. J. (1997) Radiocarbon dates from the Oxford AMS system: Archaeometry datelist 24. *Archaeometry* 39, 445–71.

Hodder, I. (1982) *Symbols in Action.* Cambridge: Cambridge University Press.

Ingold, T. (2000) *The Perception of the Environment: Essays in Livelihood, Dwelling and Skill.* London: Routledge.

Kador, T. (2005) Reopening old trails – rethinking mobility: a study of the Mesolithic in northeast Ireland. Pp. 107–20 in D. Hoffmann, J. Mills and A. Cochrane (eds), *Elements of Being: mentalities, identities and movements.* British Archaeological Reports, International Series 1437. Oxford: Archaeopress.

Kador, T. (2007a) Stone Age motion pictures: an object's perspective from early prehistoric Ireland. Pp. 33–44 in R. Johnston and V. Cummings (eds), *Prehistoric Journeys.* Oxford: Oxbow Books.

Kador, T. (2007b) Marking out a New Path: People, Places and Movement in Early Prehistoric Ireland. Unpublished PhD thesis, University College Dublin

Kador, T. (forthcoming) Moving perceptions: movement, mobility and the material dimension. In S. McCartan, R. Schulting, G. M. Warren and P. C. Woodman (eds), *Proceedings of the MESO2005 Conference on the Mesolithic in Europe, Belfast 2005.* Oxford: Oxbow Books.

Kimball, M. J. (2000) Variation and context: ecology and social evolution in Ireland's later Mesolithic. Pp. 31–47 in G. J. A. Desmond, M. Mccarthy, J. Sheehan and E. Shee Twohig (eds), *New Agendas in Irish Prehistory: Papers in Commemoration of Liz Anderson.* Bray: Wordwell.

Leon, B. C. (2005) Mesolithic and Neolithic activity on Dalkey Island: a reassessment. *Journal of Irish Archaeology* 14, 1–21.

Little, A. P. (2005) Reconstructing the social topography of an Irish Mesolithic lakescape. Pp. 79–94 in H. L. Cobb, F. Coward, L. Grimshaw, and S. Price (eds), *Investigating Prehistoric Hunter-Gatherer Identities: Case Studies from Palaeolithic and Mesolithic Europe.* British Archaeological Reports, International Series 1411. Oxford: Archaeopress.

Liversage, G. D. (1968) Excavation at Dalkey Island, Co. Dublin, 1956–1959. *Proceedings of the Royal Irish Academy* 66, 53–233.

Lubell, D., Jackes, M., Sheppard, P. and Rowley-Conwy, P. (forthcoming) The Mesolithic–Neolithic in the Alentejo: archaeological investigations, 1984–1986. *Proceedings of the IV Congresso de Arqueologia Peninsular, Faro, September 2004.*

McAulay, I. R. and Watts, W. A. (1961) Dublin radiocarbon dates I. *Radiocarbon* 3, 26–38.

McQuade, M. and O'Donnell, L. (2007) Late Mesolithic fish traps from the Liffey estuary, Dublin, Ireland. *Antiquity* 81, 569–84.

Mitchell, G. F. (1947) An early kitchen-midden in County Louth. *Journal of the County Louth Archaeological Society* 11, 169–74.

Mitchell, G. F. (1949) Further early kitchen-middens in County Louth. *Journal of the County Louth Archaeological Society* 12, 14–20.

Mitchell, G. F. (1956) An early kitchen-midden at Sutton, Co. Dublin. *Journal of the Royal Society of Antiquaries of Ireland* 86, 1–26.

Mitchell, G. F. (1972a) Further excavation of the early kitchen-midden at Sutton, Co. Dublin. *Journal of the Royal Society of Antiquaries of Ireland* 102, 151–9.

Mitchell, G. F. (1972b) Some ultimate Larnian sites at Lake Derravaragh, Co. Westmeath. *Journal of the Royal Society of Antiquaries of Ireland* 102, 160–73.

Pailler, Y. and Dupont, C. (2007) Analyse fonctionnelle des galets biseautés du Mésolithique à la fin du Néolithiue dans l'Ouest de la France, la Grande-Bretagne et l'Irelande. *Bulletin de la Société préhistorique française* 104, 31–54.

Rowley-Conwy, P. A., Zvelebil, M. and Blankholm, H. P. (eds) (1987) *Mesolithic Northwest Europe: Recent Trends*. Sheffield: Department of Archaeology and Prehistory.

Rynne, E. (1983/84) An Antrim Bann flake from near Monasterevin. *Journal of the Kildare Archaeological Society* 16, 328–30.

Saville, A. (2004) The material culture of Mesolithic Scotland. Pp. 185–220 in A. Saville (ed.), *Mesolithic Scotland and its Neighbours: The Early Holocene Prehistory of Scotland, its British and Irish Context and some Northern European Perspectives*. Edinburgh: Society of Antiquaries of Scotland.

Sheridan, J. A. (1995) Irish Neolithic pottery: the story in 1995. Pp. 3–21 in I. Kinnes and G. Varndell (eds), *Unbaked Urns of Rudely Shape: Essays on British and Irish Pottery for Ian Longworth*. Oxbow Monograph 55. Oxford: Oxbow Books.

Sheridan, J. A. (2000) Achnacreebeag and its French connections: vive the 'Auld Alliance'. Pp. 1–15 in J. C. Henderson (ed.), *The Prehistory and Early History of Atlantic Europe*. British Archaeological Reports, International Series 861. Oxford: BAR.

Sheridan, J. A. (2004) Neolithic connections along and across the Irish Sea. Pp. 9–21 in V. Cummings and C. Fowler (eds), *The Neolithic of the Irish Sea: Materiality and traditions of practice*. Oxford: Oxbow Books.

Smith, A. G., Pearson, G. W. and Pilcher, J. R. (1973) Belfast radiocarbon dates VI. *Radiocarbon 15*, 599–610.

Thomas, J. (2004) Current debates on the Mesolithic–Neolithic transition in Britain and Ireland. *Documenta Praehistorica* 31, 113–30.

Warren, G. M. (2005) *Mesolithic Lives in Scotland*. Stroud: Tempus.

Warren, G. M. (2006) Technology. Pp. 13–34 in C. Conneller and G. M. Warren (eds), *Mesolithic Britain and Ireland: new approaches*. Stroud: Tempus.

Warren, G. M. and Rice, K. (2007) Excavations in 2007 at Belderrig, Co. Mayo. Stratigraphic Report. Unpublished excavation report, University College Dublin.

Welinder, S. (1977) *The Mesolithic Stone Age of Eastern Middle Sweden*. Stockholm: Antikvariskt Arkiv.

Whittle, A. (2003) *The archaeology of people: Dimensions of Neolithic Life*. London: Routledge.

Woodman, P. C. (1977) Recent excavations at Newferry, County Antrim. *Proceedings of the Prehistoric Society* 43, 155–200.

Woodman, P. C. (1978) *The Mesolithic in Ireland: Hunter-Gatherers in an Insular Environment*. British Archaeological Reports, British Series 58. Oxford: BAR.

Woodman, P. C. (1983) The Glencloy project in perspective. Pp. 25–34 in T. Reeves-Smyth and F. Hammond (eds), *Landscape Archaeology in Ireland*. British Archaeological Reports, British Series 116. Oxford: BAR.

Woodman, P. C. (1985) *Excavations at Mount Sandel 1973–1977*. Belfast: Northern Ireland Archaeological Monographs.

Woodman, P. C. (1987) The impact of resource availability on lithic industrial traditions in prehistoric Ireland. Pp. 138–49 in P. Rowley-Conwy, M. Zvelebil and H. P. Blankholm (eds), *Mesolithic Northwest Europe: Recent Trends*. Sheffield: Department of Archaeology and Prehistory.

Woodman, P. C. (1994) Towards a definition of Irish early neolithic lithic assemblages. Pp. 213–18 in N. Ashton and A. David (eds), *Stories in Stone*. Oxford: Lithic Studies Society Occasional Paper No. 4.

Woodman, P. C. (2000) Getting back to basics: transition to farming in Ireland and Britain. Pp. 219–59 in D. T. Price (ed.), *Europe's First Farmers*. Cambridge: Cambridge University Press.

Woodman, P. C. (2003) *Pushing Back the Boundaries for Early Human Settlement in Ireland*. John Jackson Memorial Lectures, Occasional Papers in Irish Science and Technology 27. Dublin: The Royal Dublin Society.

Woodman, P. C. (2005) It's too light to be right. Pp. 126–46 in N. Milner and P. C. Woodman (eds), *Mesolithic Studies at the Beginning of the 21st century*. Oxford: Oxbow Books.

Woodman, P. C. and Anderson, E. (1990) The Irish later Mesolithic: a partial picture. Pp. 377–87 in P. M. Vermeersch and P. Van Peer (eds), *Contributions to the Mesolithic in Europe*. Leuven: Leuven University Press.

Woodman, P. C. and Johnson, G. (1996) Excavations at Bay Farm 1, Carnlough, Co. Antrim, and the study of the 'Larnian' technology. *Proceedings of the Royal Irish Academy 96C*, 137–235.

Woodman, P. C. and O'Brien, M. (1993) Excavations at Ferriter's Cove, Co. Kerry: an interim statement. Pp. 25–34 in E. Shee Twohig and M. Ronayne (eds), *Past Perceptions: The Prehistoric Archaeology of South-West Ireland*. Cork: Cork University Press.

Woodman, P. C., Anderson, E. and Finlay, N. (1999) *Excavations at Ferriter's Cove 1983–95: Last Foragers, First Farmers in the Dingle Peninsula*. Bray: Wordwell.

Woodman, P. C., Finlay, N. and Anderson, E. (2006) *The Archaeology of a Collection: The Keiller-Knowles Collection of the National Museum of Ireland*. National Museum of Ireland Monograph Series 2. Bray: Wordwell.

Zvelebil, M., Macklin, M. G., Passmore, D. G. and Ramsden, P. (1996) Alluvial archaeology in the Barrow Valley, Southeast Ireland: the 'River Ford' Culture re-visited. *Journal of Irish Archaeology* 7, 13–40.

Part Three

Landscapes of Settlement

15. Farmers, Gatherers or Horticulturalists? Reconstructing Landscapes of Practice in the Early Neolithic

Eleni Asouti and Andrew S. Fairbairn

Research into subsistence practices focusing on the 'Origins of Agriculture' has been a staple of archaeobotanical investigations in the Early Neolithic of the Near East since the first formulation of the 'Neolithic Revolution' by Childe. Our paper offers a critical review of the historical background of research on Early Neolithic plant-based subsistence practices in this part of the world. It also outlines an alternative theoretical approach, one that takes issue with the dependence on ethnographic and ethnoarchaeological accounts of historically documented peasant communities and the uncritical use of traditional categories such as 'agriculture', 'cultivation', 'gathering', 'village' and their variants. We argue instead in favour of adopting approaches which aim at reconstructing the range and gradients of the different practices and social realities associated with Early Neolithic plant procurement and consumption. Drawing from examples of recent research, we suggest an agenda for archaeobotany and landscape archaeology that emphasizes the data-informed reconstruction of the dynamic ecological relationships between communities as socio-economic agents and targeted resources, and the reconstruction of the practices associated with food production and consumption.

Aims and objectives

The main aims of this paper are:

1. To provide an overview of the current state of archaeobotanical research into the Epipalaeolithic/Natufian and Pre-Pottery Neolithic A (PPNA) chronological horizons, which framed the transition to agriculture in the Near East, and review its theoretical basis.
2. To suggest an alternative theoretical framework that focuses on practices instead of categorical distinctions (i.e. social evolutionary stages) between 'hunter-gatherers' and 'cultivator-farmers'.
3. To demonstrate why and how a landscape perspective (as opposed to one that treats habitation sites as isolated entities) is perhaps the only way to break the theoretical straightjacket of established social evolutionary approaches such as the 'origins of agriculture' narrative.
4. To propose some practical guidelines for empirical research.

The archaeobotany of the Early Neolithic of Southwest Asia: current state of research

Research into plant-based Neolithic subsistence in Western Asia has focused on the investigation of models pertaining to the origins of agriculture, dealing primarily with plant biology and biogeography (biological domestication, distribution of wild progenitors, identification of early domesticates and so on) and the ecology (*sensu stricto*) of early cultivation practices. These developments follow a long tradition of research extending back to the work of Nikolai Vavilov (with the formulation of the concept of 'centres of origins'), the 'Neolithic Revolution' of V. Gordon Childe and the field explorations of Robert Braidwood, who sought to discover evidence for the earliest stages of food production in the 'hilly flanks' of Zagros (for summaries see Flannery 1973; Wright 1992; for history and a critique of more recent intellectual developments see McCorriston 2009). Research conducted by archaeobotanists in recent decades provides a great deal of detail about the context, chronology and regional mechanisms of plant exploitation and domestication. A survey of the recent archaeobotanical literature dealing with early sites from the Levant and southeast Anatolia indicates that they are dedicated to the investigation of agricultural origins and fall into three categories:

1. Archaeobotanical reports that describe individual site assemblages with emphasis on early cereal and/or pulse domesticates. Detailed descriptions of seeds and chaff and discussion of potential 'weed' floras predominate. Major foci for these papers are: a) the establishment of criteria for identifying biological traits of domestication in plant remains; b) the identification of cultivation and/or management of morphologically non-domestic seed plants, especially cereals; c) the identification of ancient cultivation and crop-processing practices; d) the identification of spatial distributions and chronological trajectories of plant cultivation and domestication (e.g. Hillman 1981; 1984; Harris 2002; Colledge 1998; 2001).
2. Synthetic review papers concerned with the origins and spread of agriculture. These compile data with emphasis on radiocarbon evidence in order to define 'centres of origin' for particular crop species and/or crop packages and trace their subsequent spread in the Near East and Europe (e.g. Colledge *et al.* 2004; 2005).
3. Articles focused on the genetic evidence for 'centres of origin' (e.g. Jones and Brown 2000; Salamini *et al.* 2002; Ozkan *et al.* 2005).

Epipalaeolithic and Early Neolithic archaeobotanical research has focused on the questions of where, when and why seed crop species were first domesticated, concentrating mainly on cereals, which have a range of morphological indicators, including chaff morphology, grain morphology and size, that are useful in identifying biological domestication. For the chronological horizon corresponding to the Levantine PPNA, the current consensus is that domestication was probably polycentric in origin (Willcox 2005). Domestication events probably occurred independently in different parts of the region, emerging from the various regional subsistence strategies adopted by pre-farming groups and a prolonged phase of pre-domestication cultivation and gathering of seed plants (Fuller 2007). Outside the specialist literature, archaeobotanical evidence and interpretations have been used extensively in a broad range of explanatory theories on the origins of agriculture in conjunction with climate change. It has been argued, for example, that sedentism in the Early Natufian was a response to the increased availability of storable plant resources resulting from climate warming during the Late Pleistocene. Climate deterioration associated with the Younger Dryas resulted in a 'food crisis' with the reduction of these high-ranked seed-food resources (Hillman *et al.* 2001; for an elaboration of the impacts of environmentally induced resource stress on plant-based subsistence strategies see Keeley 1995). Thus prehistoric communities that wished to maintain their sedentary lifeways took up cultivation as an adaptive response to climate-induced resource stress (Henry 1989; Bar-Yosef and Belfer-Cohen 2002). Following climatic amelioration, 'agriculture' (equated with the appearance of the first plant domesticates, the widespread adoption of plant cultivation practices and the emergence of 'founder crops') was the inevitable development on the evolutionary pathway to the establishment and spread of food-producing economies, in response to (among other factors) new environmental opportunities, population growth and increasing inter-group competition (see Sherratt 1997; Richerson *et al.* 2001; Byrd 2005).

Contrasting to the profusion of studies on Neolithic and, to a lesser degree, Epipalaeolithic plant exploitation is the limited amount of evidence pertaining to earlier periods. Few pre-Epipalaeolithic datasets have been collected, analysed and published for Southwest Asia, reflecting a combination of poor preservation of plant remains and inconsistent data collection. Rare Lower and Middle Palaeolithic finds show the use of a wide range of plant food species over a long period of prehistory: examples include the nut-rich finds at the Acheulian site of Gesher Benot Ya'aqov (Goren-Inbar *et al.* 2002) and the legume-rich finds in Kebara cave (c. 60–48 ka BP; Lev *et al.* 2005). Rich Upper Palaeolithic finds at Ohalo II (c. 23 ka BP) have provided an unparalleled window into plant use at that time, suggesting the use of many plant species for food, especially small-seeded grasses (Weiss *et al.* 2004). Somewhat later, Epipalaeolithic Abu Hureyra has also demonstrated a high dietary diversity (Hillman in Moore *et al.* 2000), although studies in Anatolia suggest a concentration of gathering effort on key oil-rich species, such as wild almond at Okuzini (Martinoli and Jacomet 2004) and a thistle species with edible parts (*Gundelia tournefortii*) at Hallan Çemi (Rosenberg *et al.* 1995). These sites demonstrate the great antiquity of plant food use in the Near East and have been used to argue that Flannery's notion of a 'broad spectrum revolution' (BSR) pre-dating agriculture occurred in the Upper Palaeolithic of the Levant by at least 23 ka BP, being preceded and followed by a narrower plant subsistence focus (Weiss *et al.* 2004; see Lev *et al.* 2005 for evidence that broad spectrum subsistence strategies were practised in the Levant from the Middle Palaeolithic). The BSR is thought to have provided a necessary pre-agricultural adaptation that induced communities into new ecological/subsistence relationships with a range of previously ignored plant species, including seed-producing plants, some of which gave rise later to domesticated crop species.

The contribution of archaeobotany to the development of theoretical models investigating the local and regional contexts of prehistoric plant-based subsistence in Southwest Asia is framed (sometimes implicitly) within the intellectual tradition of cultural ecology (Steward 1955), assuming that social organization is ultimately adaptive and optimizing, being primarily driven by the need to provide material provisions that are essential for survival. This has been the result of a distinct preference within Southwest Asian archaeology for ecological–environmental and/or demographic explanations for changes in subsistence and economy (Asouti 2006). In archaeobotany this has developed from the intellectual heritage of the pioneer archaeobotanists, drawn from botany and agricultural science, who established epistemological and methodological frameworks in a solidly positivist tradition. In this context, social and economic domains are often treated as exclusive, while the system of food production/procurement is of key importance for classifying the social evolutionary stage reached by a particular community. A comparable theoretical tendency is replicated in publications devoted to the exploration of the cultural and symbolic domains of Neolithic societies, where the focus is similarly set upon 'domestication' and its psycho-cultural correlates and effects on Early Neolithic communities (e.g. Cauvin 2000; Verhoeven 2004). In these archaeobotanical evidence is often treated as 'hard data', a phenomenon elsewhere referred to as 'black-boxing' (Jones 2002): it is brought forward to support theoretical agendas and debates that seek to distance themselves from purely ecological or subsistence-orientated approaches and look instead for explanations of the Neolithic transition in cultural, symbolic and ideological developments.

A major theoretical focus in the investigation of economic practices has been the development of appropriate 'middle-range' theory to assist with the interpretation of archaeobotanical assemblages. Models of Early Neolithic crop production have been constructed through the systematic collection and recording of tillage, sowing, weeding, manuring, watering and other management requirements of particular crop species, as well as their ecological associations with specific weed floras *and* socio-economic requirements (see Jones 2005; Bogaard and Isaakidou, this volume). However, such models may not be entirely appropriate for reconstructing and interpreting the various practices and ecological contexts of Early Neolithic plant-based economies in Southwest Asia. They may conflate modern ecological regimes (e.g. the use of modern weed species autecology and community ecology) with those of prehistoric plant habitats (the latter being characterized sometimes by now extinct varieties of plant cultivars and their wild progenitors). In addition, many 'weed' taxa in Southwest Asia display highly variable ecological/environmental tolerances and habitat preferences (including their natural occurrence in vegetation communities), while a significant portion of them could have been gathered as food or fodder (the latter being often incorporated in dung fuel) or both (see Colledge 2001; Fairbairn *et al.* 2002 and references therein). Bogaard (2004) has drawn attention to these issues and advocates instead the use of multivariate analysis of behavioural functional attributes (e.g. onset and length of flowering period, canopy size, leaf size and so on), rather than individual species properties and ecological preferences, in order to reconstruct prehistoric crop management practices. This approach, however, does not obviate the need to identify and isolate Early Neolithic 'weed floras' in the first place, a task that can prove problematic for the reasons outlined above.

As regards perceptions of Early Neolithic subsistence practices outside the specialist literature, the contribution of ethnography and ethnoarchaeology has not resulted in theoretical elaboration beyond the essential distinction drawn between the PPNA hunting-gathering/cultivator economies and the PPNB mixed agro-pastoral economies. The PPNB has been largely viewed as the period when domesticated cereal and pulse crops became regionally widespread and mixed agro-pastoral economies emerged, with important repercussions for cultural formations and the process of 'Neolithization' (for a useful summary see Harris 2002). At least one reason for this can probably be located in the conceptualization of Early Neolithic communities as permanently settled 'village' societies, potentially comparable to the peasant communities of historical times (for a general critique of such theoretical concepts see Finlayson and Warren, this volume, 'Time, Scale, Practice: Landscapes in Transition?'). The Neolithic plant-based subsistence economy has been considered on the whole as potentially comparable to 'traditional' agricultural village life, a perception very much affected by ethnoarchaeological work in Turkey, Iran and the Levant (e.g. Watson 1979; Kramer 1982; Ertuğ-Yaraş 1997; Palmer 2001). On the other hand, the PPNA has remained the near-exclusive domain of the 'origins' quest, and PPNA plant management practices have become obscured by the grand narratives of cereal domestication and agricultural origins.

The lack of plausible modern-day ethnographic analogues for modelling lifeways and subsistence patterns in Southwest Asia is even more marked for earlier periods than it is for the Early Neolithic: widely accepted interpretative models of Epipalaeolithic hunter-gatherer plant exploitation are almost always drawn from comparisons to New World and African modern-day ethnographic and ecological contexts (e.g. Keeley 1995). An exception has been the judicious application of hunter-gatherer ethnographic parallels to model the diversity of potentially exploitable plant resources. The best single example from Southwest Asia remains Gordon Hillman's work, which aimed to understand the human palaeoecology of Abu Hureyra (Hillman 1996; Hillman in Moore 2000). Even these accounts, however, are concerned more with the logistics and adaptive value of plant management practices, measured explicitly or implicitly as caloric returns, and with the impact of macro-environmental (climate) change on subsistence strategies (e.g. Hillman *et al.* 2001). As with most archaeobotanical studies of Epipalaeolithic, Natufian, PPNA and PPNB plant-based subsistence strategies, they are primarily concerned with issues of resource distribution and availability and less so with the socio-economic contexts of subsistence practices (see Asouti 2004).

Towards an alternative theoretical framework: rethinking 'origins' and 'transitions'

How could one classify the 'PPNA economy'? The usual conceptualization of the PPNA portrays communities engaged mainly in hunting and gathering, with plant cultivation (or, inasmuch as plant domesticates can be detected, 'agriculture') forming only a minor part of subsistence practices. This is usually contrasted with the PPNB, where the identification of biologically domesticated plant crops and the subsequent widespread adoption of animal domesticates are considered to clarify the picture. In effect, our perception of subsistence practices in the Neolithic of the Levant is largely framed in social evolutionary terms and categorical distinctions between hunter-gatherer, cultivator and farmer stages. Furthermore, as argued above, such perceptions of Neolithic subsistence economies have been formulated in direct relation to known ethnographic and ethnoarchaeological analogues – be they 'traditional' village-based economies, or general and geographically non-specific hunter-gatherer subsistence economies. In many cases production, processing and consumption activities and decisions are assumed on the presence of one or more archaeological indicators, such as domestic plant species, within an optimizing socio-economic system. The alternative theoretical viewpoint that we suggest here is to start from the presumption that Early Neolithic societies represent, in the terminology of ecological theory, '*non-analogue communities*'. Our own theoretical approach has been shaped by two main theoretical currents: a) historical ecology and b) recent debates on the concept of 'transition' as applied to the study of hunting-gathering and food-producing economies.

Historical ecology

Historical ecology arose within anthropology as a reaction to what was perceived as the 'neofunctional–ahistorical' bias of traditional social evolutionary approaches: that socio-economic practices are *primarily* adaptive and developmental in the long-term (Headland 1997). It pointed out the biases inherent in 'synchronic ethnographies' that are composed based on direct observation, and overlook the fact that human lifeways are produced by long and usually complex cultural and environmental histories. The shortcomings of its polemic notwithstanding, we have found the application of a historical ecological perspective to archaeological research questions such as the investigation of Epipalaeolithic and PPNA plant-based subsistence economies particularly useful. It offers a theoretical framework that allows us to focus on practices instead of broadly sketched 'adaptations' and categorical distinctions, thus accounting for the impact of long-term human–landscape interactions and the influence of local histories in human decision-making. Such a research agenda may also eschew the limitations of modern ethnography and neoecology that restrict archaeology's ability to infer past ecologies and behaviours for which there are no viable analogues in the present or the recent past (Trigger 1998, 34). Furthermore, it allows for the integration of the archaeobotanical evidence into the broader archaeological record, thus providing the historical context that enables us to examine how human practices and the economies they create are historically, socially, culturally and ecologically situated within specific local and regional contexts.

Theorizing the 'transition' from hunting-gathering to food production

Social evolutionary views of prehistoric subsistence offer classificatory schemes that separate 'hunting-gathering' from 'cultivation' and 'agriculture' stages. As a number of authors have observed, however, the critical question for understanding transitions is what happens within the border zones (see, for example, Ford 1985; Harris 1989; 1996; Zvelebil 1996; Smith 2001). What most of these analyses share is the conviction that biological domestication offers the most appropriate vantage point for assessing ancient societies occupying the 'middle ground' between agricultural and hunter-gatherer lifeways. Despite the fact that the boundary remains ambiguous, according to these views domestication offers a clear (and archaeologically detectable) marker that can be used as a 'solid point of reference in an otherwise dimly seen developmental landscape' (Smith 2001, 14). Societies occupying the 'middle ground' cannot be classified as hunter-gatherers or agriculturalists, despite the fact that domesticates contribute to their economies. Smith (2001) chooses to characterize such economies as 'low-level' food production economies, comprising variable and, in the long-term, successful socio-economic solutions that are fine-tuned to a wide range of cultural and environmental contexts.

Although we share Smith's view regarding the necessity of accepting this 'middle ground' for what it is and not as an evolutionary interlude between hunter-gatherer and farming societies, we find the applicability of such an approach to Southwest Asia limited; the appearance of traits of biological domestication is not solely a function of 'intervention in, and control of, the life cycle and reproductive process of the target species … specifically the control and management of their reproduction, which results in genetic and morphological changes that make them dependent on humans for their continued existence, from generation to generation' (Smith 2001, 16). This scenario for domestication places too much emphasis on *behavioural adaptation* (i.e. the adjustment of human tactics to fit and/or alter the characteristics of the plant species they have chosen to exploit). Furthermore, we question the universal applicability of the term 'low-level food production' for describing subsistence strategies that are not based exclusively on biologically domesticated species (see examples from other regions in Denham 2004 and Fairbairn 2005a), particularly in the light of prolonged periods of pre-domestication cultivation and management of edible plant species in Southwest Asia (see overview in Fuller 2007).

The alternative of mapping *landscape practices* aimed at plant exploitation via greater environmental preparation or manipulation (Terrell *et al.* 2003) has been rarely explored. As we will argue later on, in the course of the Late Pleistocene–Early Holocene transition in Southwest Asia processes of plant cultivation unfolded in ways that were context-dependent (the context in this case being the environment itself, both ecological and societal) rather than being solely attributable to human control and human-induced selective pressures *per se*.

Redefining domestication: why the 'earliest' or 'first' will remain forever elusive on the ground – cultivation as landscape practice

There is one question that is left unaddressed by an emphasis on biological domestication as the primary vantage point: in the light of long-term pre- or even never-domestication practices, what exactly distinguishes at the level of human practices, perceptions and decision-making the 'wild' from the 'non-wild', not only as biological categories but as socio-economic ones too?

Domestication cannot be considered as a static end-state, but should be seen rather as an ongoing process of human-induced transformation of plant genotype and phenotype, which is co-determined by cultivation/harvesting practices, plant behaviour and environmental circumstances. Cultivation/harvesting practices necessarily take place in specific places (plants being dependent on particular environmental conditions for successful growth), comprising environments that may be managed in order to improve food productivity for human communities (Terrell *et al.* 2003). Domestication can thus be sensed at a landscape scale, the landscape being the places in which plants are encountered. This approach opens a range of possibilities, allowing degrees of domestication to exist between the extremes of rarely utilized 'wild' plants in largely unmanaged landscapes and intensively managed 'cultivated' plants in managed landscapes. These degrees of domestication vary among different species of plants, as well as among varieties of the same species (Fuller 2007). Such a re-conceptualization of domestication also leaves open the possibility of movement in both directions, between greater and lesser degrees of biological domestication, depending on the intensity of particular practices, rather than representing a constant progressive development towards full control by humans of plant genotypes.

The logical conclusion to this line of argument is that viewing plant domestication as a linear process whereby the 'first' or 'earliest' cradle of origin, from which all subsequent developments emanated, can be detected becomes a theoretical exercise perhaps of some interest to plant biologists and botanists, but is one bearing little relevance for understanding the diversity and developmental trajectories of prehistoric economies. Concurring with Terrell *et al.* (2003, 329), we also think that archaeologists need to be able to study subsistence skills and resource exploitation practices without having to conclude first on the presence or absence of the biological traits of domestication and intentional human genetic modification of particular plant species, as well as the degree of human dependence on such products of biological evolution. In this sense it becomes imperative, in our opinion, to move from the study of the *essential* (biological domestication) to that of the *habitual* (landscape practices and associated subsistence strategies resulting in *socio-ecological* processes of 'domestication'). Such an approach renders any categorical distinction between 'farming' and 'gathering' as social evolutionary stages largely obsolete. In the remainder of this section we will outline a number of landscape practices that could become the focus of further investigations, while in the next section we will review the development of associated subsistence strategies in the archaeological context of Southwest Asia.

In the case of Southwest Asia, landscape practices associated with food production occurring in varying combinations and intensities may include, but are not limited to:

1. Deliberate disturbance of park woodlands and annual grasslands, including the use of fire (see Roberts 2002).
2. Modification of alluvial surfaces and moist steppe ecotonal habitats to assist cultivation (see Hillman in Moore *et al.* 2000; Colledge 2001). Specifically for the Levantine Epipalaeolithic it has also been suggested that minimal or no modification at all, such as tillage and planting seeds in the ground, would have been necessary for human management of cereal propagation (Anderson 1992). However, this has been questioned by Fuller (2007), who has argued that tillage and burial of cereal seeds in the ground might have been the ecological factors chiefly responsible for the early development of large seed size in pre-domestic cereals.
3. Dislocation and/or transplantation of targeted plant species (see Anderson 1992; Willcox *et al.* 2008).
4. Targeted procurement and processing for consumption of cereals, legumes and other plants, as reviewed in the next section.

When combined with other archaeological and palaeo-environmental research, archaeobotany provides the key archaeological technique for obtaining direct information about plant use practices in past landscapes. Plants do, after all, grow in a specific range of places determined by the autecology of each species and the attributes of the environment. Management, collection, cultivation, harvest, use and discard all require a series of practices to be enacted that may lead to the preservation of plant remains. With reference to both modern plant ecology and an understanding of environmental conditions surrounding a prehistoric habitation site, it is theoretically possible to spatially locate plant species found in archaeobotanical assemblages and hence directly identify ancient landscapes (see Fairbairn 2008). This type of approach is part of the standard archaeobotanical method and has been used to reconstruct the resources available to a given human community through methodologies developed under the umbrella of site catchment analysis (e.g. Vita-Finzi and Higgs 1970). More nuanced analyses have modelled the effects of temporal change in plant distributions caused by climate change, as in the case of Gordon Hillman's detailed work in northern Syria (see above). This type of information has the potential to go beyond the identification of resource availability and provide direct evidence of how specific economic practices involving plants were spatially distributed across the landscape, as demonstrated in the central Anatolian Neolithic (Asouti 2005a; Fairbairn *et al.* 2002; Fairbairn 2005a). The Anatolian example also demonstrates how consideration of the social as well as the environmental and economic contexts plays a key role in understanding the specific local character and trajectories of subsistence practices (Fairbairn 2005b). In turn such an approach offers a suitable setting for a critical rethinking of the positivist epistemology that underpins most Southwest Asian archaeobotanical research.

The evidence from Southwest Asia

From agricultural origins to landscape processes and practices

In Southwest Asia, a generalized scenario of 'domestication' accepts that human management of food plants (including various forms of cultivation) had been ongoing since at least the Late Pleistocene and more specifically the Late Natufian after c. 11,000 BC. This correlates with the Younger Dryas climatic oscillation, which led to a rapid decline in precipitation and the contraction of moister vegetation habitats. It remains possible that intensive plant management practices had been established long before but a number of factors, including the force and unpredictability of Late Pleistocene climatic oscillations that would have disrupted such processes, argue against this eventuality (Richerson *et al.* 2001). Very little can be deduced about landscape practices during this period, and it is frustrating that few Natufian sites (especially from the Southern Levant) have been investigated with such aims in mind. Ethnographic research and the limited archaeobotanical record suggest that a number of strategies might have been at work, including broadcast sowing of alluvial surfaces with minimal or no tillage and shifting cultivation (probably involving the seasonal burning of grasslands) alongside regionally varied gathering strategies (see Harris 2002).

Tree fruits are a constant food presence throughout the Epipalaeolithic and the same species (oak, terebinth, fig etc) continued in use into and through the Neolithic, although their archaeological traces decline after that. It was assumed that trees provided gathered resources from unmanaged woodland stands (Zohary and Hopf 2000) until the recent claim by Kislev *et al.* (2006) of fig 'domestication' at Gilgal by c. 9500 cal BC. Management of tree crop resources is possible without the need for domestication (Terrell *et al.* 2003) and can include tending, protection of seedlings and transplantation, at which point one can suggest effective cultivation. While we cannot yet prove the direct management of tree species in Southwest Asia at this time, woodland management practices and Neolithic impact on arboreal vegetation have been postulated on the basis of conjectural and/or indirect evidence. Roberts (2002) has suggested, for example, that the Early Holocene expansion of oak park-woodland in central Anatolia was not synchronous with rising temperatures and humidity, and might have been kept in check by human landscape management practices (such as small-scale clearance, vegetation burning, foddering, fuel collection and so on; for a much more detailed consideration of Early Neolithic woodland exploitation and ecology see Asouti and Hather 2001; Asouti 2005a).

A more interesting aspect, and one amenable to empirical investigation, is the correlation between social organization and patterns of food resource exploitation (see Bar-Yosef and Belfer-Cohen 2002; Harris 2002; Hayden 1995). Were the sites that were continuously occupied in the Epipalaeolithic already developing territorial systems that were radically different from those of more mobile adaptations to a deteriorating climate and resource base during the Pleistocene–Holocene transition? If so, did such developments create 'new' socio-economic environments presaging the successful implementation of cultivation practices across the region in the ensuing PPNA?

Broad-scale models such as that proposed by Sherratt (1997) have suggested that the adoption of cereal cultivation in response to climate change and the ensuing shifts in PPNA settlement location and distribution (with the preferential location of habitation sites on alluvial fans) were near-universal phenomena across the region. However, evaluating local patterns against such broad-scale models does not leave a clear-cut picture. The available archaeobotanical evidence points to significant variation between different sites and areas in Southwest Asia as regards plant management practices. The selection and management of crop staples suited to the particular needs and ecological conditions of specific habitation areas and environments suggests 'multiple domestications' (Willcox 2005; Weiss *et al.* 2006). Furthermore, several of these domestication events did not lead to long-term widespread adoption, but stand out as temporally isolated examples of crop development (Weiss *et al.* 2006). The available evidence speaks of diversity in plant use that is often subsumed under the grand 'origins' narrative. Research considering the use of plant food species in the

Epipalaeolithic and the PPNA has generally focused on the wild progenitors of cereal domesticates, and there has been much less serious discussion of how other seed-producing plants, beyond the ever present tree fruits/nuts, were used. Consideration of the management and exploitation of plant species beyond the founder crops has thus been neglected. There has been little interest in exploring insights that may arise from the realization that all plant use (but especially the large-scale extraction, management and production that we assume were required to support even the smaller sedentary or semi-sedentary sites) necessarily represents a form of landscape intervention that was embodied in specific ecological practices.

Reliance primarily on cereals ('wild' and/or 'proto-domesticated') was by no means a universal pattern; small-seeded grasses and legumes occurred in abundance on Upper Palaeolithic, Natufian and PPNA sites (Savard *et al.* 2006). On the basis of the available evidence it is clear that the founder seed crops (emmer wheat, einkorn wheat, barley, lentil, chickpea, pea and flax) formed only a small proportion of the plants used by Upper Palaeolithic and Epipalaeolithic/PPNA communities. While the developmental trajectories of cereals are of obvious interest, they represent only part of the story. Epipalaeolithic and PPNA sites show a large range of variability in the species utilized as foodstuffs and it is not until the PPNB that the founder crops become universally present. Some communities appear to have targeted specific high-ranking oil-rich seeds; others, such as Abu Hureyra I, utilized a range of starch-rich seeds (Hillman in Moore *et al.* 2000); while sites in northern Syria appear to have exploited the docks/knotweeds and various crucifers until they disappeared from the archaeobotanical record by c. 9000 cal. BC, thereafter dominated by founder crop progenitors (Willcox *et al.* 2008). Ninth-millennium cal BC sedentary communities in central Anatolia exploited a variety of arboreal species (Fairbairn, pers. comm.). A general decline in the diversity of the exploited plant food species appears to be a region-wide process, coming to a head between 10,000 and 9000 cal BC (Weiss *et al.* 2004), although central Anatolia seems to differ in this respect (albeit possibly as a result of preservation and/or retrieval biases, discussed in more detail later in this paper; see also Fairbairn *et al.* 2005).

The critical point for the development of the founder crop package comes precisely at this point in time, the Early–Middle PPNB juncture, as settlement fission and diffusion start to be attested in the regional settlement record. This is the time when new territories start to become colonized by Neolithic cultivators across Southwest Asia. The story of Cyprus is indicative, whereby the verified presence by the 9th millennium BC of plant domesticates has triggered an intense debate aiming at pinpointing the 'original' homeland of its PPNB settlers (cf. Peltenburg 2004; Asouti 2006). Much of this debate seems to have overlooked the speed with which biological domestication status could have been reached – within the space of a few generations (20–100 yrs) – in environments that differed radically from the native habitats of plant cultivars (i.e. soils lacking wild cereal seed banks), coupled with the regular use of sickle harvesting technology (Hillman and Davies 1990). It is very unlikely that the initial stages of such a process will ever be detectable in the archaeobotanical record. This situation can be contrasted with the ecology and the (much slower) rates of cereal biological domestication in areas (such as the northern Levant) set within or close to the natural distribution of the wild cereal populations (Willcox 2005).

In his recent overview of domestication processes and their ecological setting Fuller (2007) has suggested that the widespread occurrence of cereal domesticates and the founder crop suite in the PPNB was largely the outcome of technological changes (sickle harvesting replacing the 'archaic' hunter-gatherer practice of beating ears into baskets) that sped up the effects of long-term unconscious selective pressures for domestic mutants. We have already noted the critical role that settlement diffusion played (originally *within* Southwest Asia) in creating new windows of opportunity for biological domestication. There are, however, many more interesting implications in viewing Early Neolithic diffusion and the development of the founder crop package as fully integrated processes than merely postulating the creation of new ecological contexts for the occurrence of biological domestication. It seems increasingly likely that the process whereby PPNA diversity of plant-based subsistence practices gave way to the widespread adoption of the PPNB 'founder crop package', eventually integrated with animal herding, cannot be understood solely in terms of ecological, social evolutionary or deterministic cause and effect. The PPNB is the period that most researchers identify with 'real' domestication and the beginnings of mixed agro-pastoral economies (Bar-Yosef and Meadow 1995; Harris 2002). It is not coincidental that this chronological horizon is largely identified with increasing social inequality, territorial claims, factional tensions and attendant mechanisms for the levelling of social differences (cf. Bar-Yosef and Meadow 1995; Kuijt 2000; Asouti 2006). Very little debate has been directed to identifying and evaluating the contribution of subsistence strategies to these processes. Was such a contribution realised along the lines suggested by Hayden (1995) in his modelling of domestication driven by social competition imperatives? Although the available evidence would seem to preclude this interpretation (see Kuijt and Goring-Morris 2002), the potentially important role of subsistence economies in socio-cultural change during the 9th and 8th millennia cal BC has remained largely conjectural. Investigating Early Neolithic 'economies' or the 'strategic relationships' (*sensu* Wolf 1982) linking household and kinship structures, exchange, group identity, technology and subsistence practices has not been a high research priority; instead, ecological-functionalist and/or descriptive-integrative culture historical approaches (examples of the latter are the 'PPNB interaction sphere' and various core–periphery models) have largely prevailed (cf. Asouti 2006). In specialist research, the main obstacles have been the general reluctance of archaeobotanists to engage with the evidence beyond the basic consideration of Neolithic subsistence as adaptive, and the inadequacy of the 'traditional village societies' and 'complex hunter-gatherer' ethnographic models in reconstructions of the social dimensions of the Epipalaeolithic and Early Neolithic economies of Southwest Asia (see Fairbairn *et al.* 2005; Asouti 2006). In the next section we will explore how the available evidence can be deployed in ways that may allow us to take a step in this direction and demonstrate the potential of such an approach.

From crop processing to social relations

We know very little about the socio-economic correlates of plant exploitation (production as well as consumption) in the Epipaleolithic and the PPNA. The few available examples of well-preserved and well-sampled *in situ* archaeobotanical finds provide some insights. A number of communities appear to have invested significantly in communal facilities, including communal storage features at Jerf el Ahmar and Tell 'Abr, the latter containing *in situ* evidence of crop storage in the form of charred grain (Willcox *et al.* 2008). Recent studies have indicated that grain storage was widespread across the region, as suggested by the presence of similar facilities as far south as the PPNA site of Dhra' on the southern edge of the Dead Sea (Finlayson and Kuijt 2005) and the occurrence of rodent droppings at Jerf el Ahmar, Tell Qaramel and D'Jade (Willcox *et al.* 2008).

Jerf el Ahmar has also provided evidence of food processing/preparation facilities within buildings (Willcox 2002). Concentrations of *in situ* charred plant remains (systematic sampling was of paramount importance for their discovery as charred seeds were found in low quantities owing to the nature of the formation processes in an area that had not been used for refuse disposal) showed that the room had been used for grinding and perhaps soaking seeds of different cultivars and collected plants (all 'wild' in biological terms), including barley, einkorn wheat/rye, lentils, caper, almond, mustard seeds and grasses. The dominant species had distinct distributions and associations with artefacts and installations inside the room while, as in Tell Qaramel, processed seed 'cake' remains provided further evidence of elaborate culinary practices (Willcox *et al.* 2008). The rare evidence for the use of mustard seeds at Jerf serves to demonstrate the targeted procurement of plant resources beyond the usual range of high-calorie cereal staples. Later examples of non-staple seeds in unambiguous storage contexts, as at Neolithic Çatalhöyük (Fairbairn *et al.* 2005), indicate this to be a potentially widespread (if archaeologically rarely visible) behaviour.

At Jerf el Ahmar communal grain storage can be contrasted with the apparent occurrence of household-scale food processing. If we accept this provisional interpretation, storage is seen as the focus of cross-community social interaction and negotiation, while smaller-scale, kin or family group relations determine processing (and consumption?) activities. Storage and processing of plant foods thus embody somewhat different functions in the social construction of PPNA communities. This pattern can be contrasted with cases of household-scale storage *and* processing at settlements dated to the Late PPNB (LPPNB) chronological horizon, such as Çatalhöyük (Fairbairn *et al.* 2005).

How did communal storage come about and how was it articulated with production and consumption at Jerf el Ahmar and Tell 'Abr? We can speculate that communal storage reflected group-based production and procurement of grain crops and other exploited plant species, and might indicate community-wide attitudes to the ownership and use of the site territory and the broader landscape. Other forms of concentration of stored resources are of course possible, including communal or individual coercion and/or the operation of inter- and intra-community networks of social (kin-based?) obligation. The distribution among community members of stored foodstuffs for consumption is another potentially promising field of investigation. However, the obscurity of detail about consumption, caused in great part by the vagaries of the preservation and retrieval of archaeobotanical remains combined with the lack of integration of archaeobotanical remains and their contextual attributes and a limited understanding of the location of procurement activities in the landscape, renders further refinements non-feasible at present.

A similar lack of detailed evidence does not permit a data-informed evaluation of the Middle and LPPNB Levantine sites. Generally, the archaeological record seems to suggest that production *and* consumption were probably controlled by individual households (for an overview of other types of archaeological evidence besides plants, see Byrd 2000). However, a number of issues which might elucidate the nature and socio-economic contexts of these activities remain open. Most pertinent is the actual size of these habitation sites and their respective communities (particularly for the so-called 'mega-sites' such as 'Ain Ghazal, Basta, Wadi Shu'eib and others) Recent observations derived from the excavation of the broadly contemporary 'mega-site' of Çatalhöyük in central Anatolia have indicated that not all parts of the site were continuously occupied throughout its lifetime (Hodder 2005, 19). Additional evidence from the subsistence datasets has furthermore indicated the exploitation by its inhabitants of a very diverse range of resources, an observation which in turn is not necessarily compatible with the picture projected by ethnographically derived reconstructions of archetypical agrarian 'village' communities (Asouti and Fairbairn 2002; Fairbairn *et al.* 2002; 2005). This is most clearly evident in the uniquely rich and diverse wild 'non-food' flora manifested in the plant remains from Çatalhöyük, which is compatible with procurement activities and paths of movement across the landscape that cannot be easily accommodated with models of a static, predictable and optimizing (at least by traditionally accepted standards) 'agricultural' economy (Fairbairn *et al.* 2005, 200). Indeed, recent research has inferred the existence of a complex pattern of resource exploitation integrating the management of 'wild' and domesticated plant and animal resources in ways that challenge the expectations of widely invoked models of 'egalitarian', household-based 'village' economies (see Asouti 2005b; Fairbairn *et al.* 2005; Bogaard and Isaakidou, this volume). However, as we lack detailed reconstructions of settlement size alongside the intensive sampling for subsistence datasets and comparable levels of landscape sourcing for other sites of this time period, we cannot evaluate the applicability of these observations to different areas and Early Neolithic sites in Southwest Asia.

Suggestions for the development of appropriate research and field practices

Ultimately, we rely on empirical data to directly investigate ancient practices. Such data are produced by archaeological field practices, which in many cases are inadequate for providing archaeobotanical assemblages with the spatial and chronological resolution that is required in order to move beyond existing analytical priorities (see also Denham 2004; Fairbairn 2005a for comparative perspectives from a different geographical region). Poor and uneven preservation is widely recognized as a problem in obtaining quality data, particularly as regards Epipalaeolithic seasonal encampments and, more generally, sites without well-defined and well-maintained features such as buildings, fire installations and refuse disposal areas (Colledge 2001, 103–4). The outstanding examples which illuminate the minutiae of past practices are restricted to those few cases in which the conditions of large-scale preservation *and* the systematic retrieval of *in situ* plant remains are met. These include, for example, Ohalo II, Jerf El Ahmar and Netiv Hagdud (plant remains preserved by waterlogging and large-scale catastrophic burning) and Dhra' (systematic retrieval of plant remains and large-scale processing of sediments via machine-assisted water flotation), where detailed information about past plant-use practices has relied on extensive spatial sampling of individual structures and archaeological features. Yet these occurrences remain rare and it is clear from many sites, including key sites such as Abu Hureyra and Mureybit, that (albeit less well-resolved) information about practices and landscape use is also recoverable from generalized site fill deposits, dumps and other mixed rubbish deposits. Nuanced analytical and sampling strategies are required, focusing on:

1. A shift away from token sampling of sites aimed at a generalized understanding of the subsistence base and/or (especially for the PPNA chronological horizon) individual cases of 'agricultural origins', towards the collection of representative samples (normally amounting to at least 30–50 L of sediment per sample) processed on a large scale by machine-assisted flotation.
2. The combination of spatial with 'vertical' sampling across sites and features.
3. Intensive and *comprehensive* sampling of *in situ* preserved plant remains, for example in burnt structures, in order to provide direct evidence of procurement, processing and storage practices from contexts with minimal post-depositional disturbance. The value of this sampling strategy has been demonstrated at, for example, Netiv Hagdud, Jerf el Ahmar and Çatalhöyük.
4. Full integration of archaeobotanical datasets with their associated contextual information and the production of non-exclusive archaeological narratives; that is, adopting a contextual approach to archaeobotanical analysis and interpretation that moves beyond (but does not exclude) traditional analytical and epistemological concerns.
5. Integration of archaeobotanists within field teams and their active participation in designing and conducting fieldwork.
6. The implementation of 'off-site' research programmes with a scope extending beyond traditional survey aims (e.g. the recording of visible features, lithic scatters, etc.) to encompass integrated multidisciplinary landscape investigations (geoarchaeology, off-site excavation, environmental sampling and the latest prospection techniques).

Conclusion

A methodological as well as a theoretical shift, largely translated in the abandonment of categorical distinctions drawn between 'hunter-gatherer' and 'farming' economies for describing Early Neolithic socio-economics, is required in order to provide the richness of material datasets that will allow us to address new analytical and epistemological priorities in the investigation of Epipalaeolithic and Early Neolithic socio-economics. Seen from a modern perspective, plant cultivation, the evolutionary 'accidents' of plant domestication and eventually the emergence of 'agriculture' all form parts of an evolutionary continuum where food production is identified as the cornerstone of later human expansion and socio-cultural development across the Old World. Seen from the perspective of the very same societies that fostered them, however, they can also be viewed as culturally and environmentally mediated solutions to geographically and temporally specific problems of community survival and reproduction. As such, they are of obvious interest to the archaeologist and the social historian alike, because they can provide a rich source of information about historically constituted human attitudes, perspectives and decision-making that probably differed radically from those of later periods as well as modern ones. Methodologically, a widespread adjustment of both archaeobotanical and broader archaeological fieldwork and reporting practices is needed to maximize the potential of the hitherto largely elusive and at best fragmentary traces of plant resource management and landscape exploitation practices in the Epipalaeolithic/Natufian and the PPNA chronological horizons.

Bibliography

Anderson, P. C. (1992) Harvesting of wild cereals during the Natufian as seen from experimental cultivation and harvest of wild einkorn wheat and microwear analysis of stone tools. Pp. 521–56 in O. Bar-Yosef and F. Valla (eds), *The Natufian Culture in the Levant*. Ann Arbor, MI: International Monographs in Prehistory.

Asouti, E. (2004) The contribution of subsistence archaeology to generating supra-regional models for understanding Near Eastern Neolithisation. *Neo-lithics* 1/04, 23–4.

Asouti, E. (2005a) Woodland vegetation and the exploitation of fuel and timber at Neolithic Çatalhöyük: report on the wood charcoal macro-remains. Pp. 213–58 in I. Hodder (ed.), *Inhabiting Çatalhöyük: Reports from the 1995–99 Seasons*. McDonald Institute Monographs 38. Cambridge and Ankara: McDonald Institute for Archaeological Research and British Institute of Archaeology at Ankara.

Asouti, E. (2005b) Group identity and the politics of dwelling at Neolithic Çatalhöyük. Pp. 75–91 in I. Hodder (ed.), *Çatalhöyük Perspectives: Reports from the 1995–99 Seasons*. McDonald Institute Monographs 40. Cambridge and Ankara: McDonald Institute for Archaeological Research and British Institute of Archaeology at Ankara.

Asouti, E. (2006) Beyond the 'Pre-Pottery Neolithic B Interaction Sphere'. *Journal of World Prehistory* 20/2, 87–126.

Asouti, E. and Fairbairn, A. (2002) Subsistence economy in central Anatolia during the Neolithic – the archaeobotanical evidence. Pp. 181–92 in F. Gérard and L. Thissen (eds), *The Neolithic of Central Anatolia. Internal Developments and External Relations During the 9th–6th Millennia cal. BC*. Istanbul: Ege Yayınları.

Asouti, E. and Hather, J. (2001) Charcoal analysis and the reconstruction of ancient woodland vegetation in the Konya Basin, south-central Anatolia, Turkey: results from the Neolithic site of Çatalhöyük East. *Vegetation History and Archaeobotany* 10, 23–32.

Bar-Yosef, O. and Belfer-Cohen, A. (2002) Facing environmental crisis – societal and cultural changes at the transition from the Younger Dryas to the Holocene in the Levant. Pp. 55–67 in R. T. J. Cappers and S. Bottema (eds), *The Dawn of Farming in the Near East*. Berlin: *ex oriente*.

Bar-Yosef, O. and Meadow, R. (1995) The origins of agriculture in the Near East. Pp. 39–94 in T. D. Price and A. B. Gebauer (eds), *Last Hunters, First Farmers: New Perspectives on the Prehistoric Transition to Agriculture*. Santa Fe, NM: School of American Research Press.

Bogaard, A. (2004) *Neolithic Farming in Central Europe*. London: Routledge.

Byrd, B. F. (2000) Households in transition: Neolithic social organization within Southwest Asia. Pp. 63–98 in I. Kuijt (ed.), *Life in Neolithic Farming Communities. Social Organization, Identity and Differentiation*. London: Kluwer Academic/Plenum.

Byrd, B. F. (2005) Reassessing the emergence of village life in the Near East. *Journal of Archaeological Research* 13/3, 231–90.

Cauvin, J. (2000) *The Birth of the Gods and the Origins of Agriculture*. Cambridge: Cambridge University Press.

Colledge, S. (1998) Identifying pre-domestication cultivation using multivariate analysis. Pp. 121–31 in A. B. Damania, J. Valkoun, G. Willcox and C. O. Qualset (eds), *The Origins of Agriculture and Crop Domestication*. Aleppo, Syria: ICARDA.

Colledge, S. (2001) *Plant Exploitation on Epipaleolithic and Early Neolithic Sites in the Levant*. British Archaeological Reports, International Series 986. Oxford: Archaeopress.

Colledge, S., Conolly, J. and Shennan, S. (2004) Archaeobotanical evidence for the spread of farming in the Eastern Mediterranean. *Current Anthropology* 45(S), 35–58.

Colledge, S., Conolly, J. and Shennan, S. (2005) The evolution of Neolithic farming from SW Asian origins to NW European limits. *European Journal of Archaeology* 8, 137–56.

Denham, T. (2004) The roots of agriculture and arboriculture in New Guinea: looking beyond Austronesian expansion, Neolithic packages and indigenous origins. *World Archaeology* 36, 610–20.

Ertuğ-Yaraş, F. (1997) An Ethnoarchaeological Study of Subsistence and Plant Gathering in Central Anatolia. Unpublished PhD thesis, Washington University, Department of Anthropology, St Louis, MI.

Fairbairn, A. (2005a) An archaeobotanical perspective on Holocene plant use practices in lowland northern New Guinea. *World Archaeology* 37/4, 487–502.

Fairbairn, A. (2005b) A history of agriculture at Çatalhöyük East, Turkey. *World Archaeology* 37/2, 197–210.

Fairbairn, A. (2008) Beyond economy: seed analysis and landscape archaeology. Pp. 442–50 in B. David and J. Thomas (eds), *Handbook of Landscape Archaeology*. World Archaeological Congress Research Handbooks. Walnut Creek, CA: Left Coast Press.

Fairbairn, A., Asouti, E., Near, J. and Martinoli, D. (2002) Macrobotanical evidence for plant use at Neolithic Çatalhöyük, south-central Anatolia, Turkey. *Vegetation History and Archaeobotany* 11, 41–54.

Fairbairn, A., Near, J. and Martinoli, D. (2005) Macrobotanical investigations of the North, South and KOPAL areas at Çatalhöyük. Pp. 137–20 in I. Hodder (ed.), *Inhabiting Çatalhöyük: Reports from the 1995–1999 Seasons*. McDonald Institute Monographs 38. Cambridge and Ankara: McDonald Institute for Archaeological Research and British Institute of Archaeology at Ankara.

Finlayson, B. and Kuijt, I. (2005) Dhra' Excavation Project 2004. *CBRL 2005 Newsletter of the Council for British Research in the Levant*, 36–8.

Flannery, K. V. (1973) The origins of agriculture. *Annual Review of Anthropology* 2, 271–310.

Ford, R. I. (1985) The processes of plant food production in prehistoric North America. Pp. 1–18 in R. I. Ford (ed.), *Prehistoric Food Production in North America*. Anthropological Papers no. 75. Ann Arbor, MI: Museum of Anthropology, University of Michigan.

Fuller, D. Q. (2007) Contrasting patterns in crop domestication and domestication rates: recent archaeobotanical insights from the Old World. *Annals of Botany* 100/5, 903–24.

Goren-Inbar, N., Gonen, S., Melamed, Y. and Kislev, M. (2002)

Nuts, nut cracking, and pitted stones at Gesher Benot Ya`aqov, Israel. *Proceedings of the National Academy of Sciences* 99/4, 2455–60.

Harris, D. (1989) An evolutionary continuum of people–plant interaction. Pp. 11–26 in D. Harris and G. Hillman (eds), *Foraging and Farming: The Evolution of Plant Exploitation*. London: Unwin Hyman.

Harris, D. (1996) Domesticatory relations of people, plants and animals. Pp. 437–63 in R. Ellen and K. Fukui (eds), *Redefining Nature*. Oxford: Berg.

Harris, D. R. (2002) Development of the agropastoral economy in the Fertile Crescent during the Pre-Pottery Neolithic period. Pp. 67–83 in R. T. J. Cappers and S. Bottema (eds), *The Dawn of Farming in the Near East*. Berlin: *ex oriente*.

Hayden, B. (1995) A new overview of domestication. Pp. 273–300 in T. D. Price and A. B. Gebauer (eds), *Last Hunters, First Farmers: New Perspectives on the Prehistoric Transition to Agriculture*. Santa Fe, NM: School of American Research Press.

Headland, T. N. (1997) Revisionism in ecological anthropology. *Current Anthropology* 38, 605–30.

Henry, D. O. (1989) *From Foraging to Agriculture. The Levant at the End of the Ice Age*. Philadelphia: University of Pennsylvania Press.

Hillman, G. C. (1981) Reconstructing crop husbandry practices from charred remains of crops. Pp. 123–62 in R. J. Mercer (ed.), *Farming Practice in British Prehistory*. Edinburgh: Edinburgh University Press.

Hillman, G. C. (1984) Interpretation of archaeological plant remains: the application of ethnographic models from Turkey. Pp. 1–42 in W. van Zeist and W. A. Casparie (eds), *Plants and Ancient Man*. Rotterdam: Balkema.

Hillman, G. (1996) Late-Pleistocene changes in wild plant-foods available to hunter-gatherers of the northern Fertile Crescent: possible preludes to cereal cultivation. Pp. 159–203 in D. Harris (ed.), *The Origins and Spread of Agriculture and Pastoralism in Eurasia*. London: UCL Press.

Hillman, G. and Davies, D. (1990) Measured domestication rates in wild wheats and barley under primitive cultivation and their archaeological implications. *Journal of World Prehistory* 42, 157–219.

Hillman, G. C., Hedges, R., Moore, A. M. T., Colledge, S. M. and Pettitt, P. (2001) New evidence for Late Glacial cereal cultivation at Abu Hureyra on the Euphrates. *The Holocene* 11, 383–93.

Hodder, I. (2005) Changing entanglements and temporalities. Pp. 221–83 in I. Hodder (ed.), *Changing Materialities at Çatalhöyük: Reports from the 1995–1999 Seasons*. McDonald Institute Monographs 39. Cambridge and Ankara: McDonald Institute for Archaeological Research and British Institute of Archaeology at Ankara.

Jones, A. (2002) *Archaeological Theory and Scientific Practice*. Cambridge: Cambridge University Press.

Jones, G. (2005) Garden cultivation of staple crops and its implications for settlement location and continuity. *World Archaeology* 37/2, 164–76.

Jones, M. and Brown, T. (2000) Agricultural origins: the evidence of modern and ancient DNA. *The Holocene* 10/6, 769–76.

Keeley, L. (1995) Protoagricultural practices among hunter-gatherers: a cross cultural survey. Pp. 243–72 in T. D. Price and A. Gebauer (eds), *Last Hunters-First Farmers*. Santa Fe, NM: School of American Research Press.

Kislev, M. E., Hartmann, A. and Bar-Yosef, O. (2006) Early domesticated fig in the Jordan valley. *Science* 312, 1372–4.

Kramer, C. (1982) *Village Ethnoarchaeology. Rural Iran in Archaeological Perspective*. New York: Academic Press.

Kuijt, I. (ed.), (2000) *Life in Neolithic Farming Communities. Social Organization, Identity and Differentiation*. London: Kluwer Academic/Plenum.

Kuijt, I. and Goring-Morris, N. (2002) Foraging, farming, and social complexity in the Pre-Pottery Neolithic of the southern Levant: a review and synthesis. *Journal of World Prehistory* 16, 361–440.

Lev, E., Kislev, M. E. and Bar-Yosef, O. (2005) Mousterian vegetal food in Kebara Cave, Mt. Carmel. *Journal of Archaeological Science* 32, 475–84.

Martinoli, D. and Jacomet, S. (2004) Identifying endocarp remains and exploring their use at Epipalaeolithic Öküzini in southwest Anatolia, Turkey. *Vegetation History and Archaeobotany* 13, 45–54.

McCorriston, J. (2009) Domestication and the dialectic: Archaeobotany and the future of the Neolithic Revolution in the Near East. Pp. 27–36 in A. S. Fairbairn and E. Weiss (eds), *From Foragers to Farmers; papers in honour of Gordon C. Hillman*. Oxford: Oxbow Books.

Moore, A. M. T., Hillman, G. C. and Legge, A. J. (2000) *Village on the Euphrates: From Foraging to Farming at Abu Hureyra*. Oxford: Oxford University Press.

Ozkan, H., Brandolini, A., Pozzi, C., Effgen, S., Wunder, J. and Salamini, F. (2005) A reconsideration of the domestication geography of tetraploid wheats. *Theoretical and Applied Genetics* 110, 1052–60.

Palmer, C. (2001) Traditional agriculture. Pp. 621–9 in R. Adams, P. Bienkowski and B. MacDonald (eds), *The Archaeology of Jordan*. Sheffield: Sheffield Academic Press.

Peltenburg, E. (2004) Social space in early sedentary communities of Southwest Asia and Cyprus. Pp. 71–90 in E. Peltenburg and A. Wasse (eds), *Neolithic Revolution. New Perspectives on Southwest Asia in Light of Recent Discoveries in Cyprus*. Levant Supplementary Series 1. Oxford: CBRL and Oxbow Books.

Richerson, P. J., Boyd, R. and Bettinger, R. L. (2001) Was agriculture impossible during the Pleistocene but mandatory during the Holocene? A climate change hypothesis. *American Antiquity* 66/3, 387–411.

Roberts, N. (2002) Did prehistoric landscape management retard the post-glacial spread of woodland in Southwest Asia? *Antiquity* 76, 1002–10.

Rosenberg, M., Nesbitt, R. M., Redding, R. W. and Strasser, T. R. (1995) Hallan Cemi Tepesi: some preliminary observations concerning early Neolithic subsistence behaviours in eastern Anatolia. *Anatolica* 21, 1–12.

Salamini, F., Ozkan, H., Brandolini, A., Schafer-Pregl, R. and Martin, W. (2002) Genetics and geography of wild cereal domestication in the Near East. *Nature Reviews Genetics* 3/6, 429–41.

Savard, M., Nesbitt, M. and Jones, M. K. (2006) The role of wild grasses in subsistence and sedentism: new evidence from the northern Fertile Crescent. *World Archaeology* 38/2, 179–96.

Sherratt, A. (1997) Climatic cycles and behavioural revolutions: the emergence of modern humans and the beginning of farming. *Antiquity* 71, 271–87.

Smith, B. D. (2001) Low-level food production. *Journal of Archaeological Research* 9/1, 1–43.

Steward, J. H. (1955) *Theory of Culture Change*. Urbana, IL: University of Illinois Press.

Terrell, J. E., Hart, J. P., Barut, S., Cellinese, N., Curet, A., Denham, T., Kusimba C. M., Latinis, K., Oka, R., Palka, J., Pohl, M. E. D., Pope, K. O., Williams, P. R., Haines, H. and Staller, J. E. (2003) Domesticated landscapes: the subsistence ecology of plant and animal domestication. *Journal of Archaeological Method and Theory* 10/4, 323–68.

Trigger, B. (1998) Archaeology and epistemology: dialoguing across the Darwinian chasm. *American Journal of Archaeology* 102/1, 1–34.

Verhoeven, M. (2004) Beyond boundaries: nature, culture and a holistic approach to domestication in the Levant. *Journal of World Prehistory* 18/3, 179–282.

Vita-Finzi, C. and Higgs, E. S. (1970) Prehistoric economy in the Mount Carmel area of Palestine: site catchment analysis. *Proceedings of the Prehistoric Society* 36, 1–37.

Watson, P. J. (1979) *Archaeological Ethnography in Western Iran*. Viking Fund Publications in Anthropology no. 57. Tucson, AZ: University of Arizona Press.

Weiss, E., Wetterstrom, W., Nadel, D. and Bar-Yosef, O. (2004) The broad spectrum revisited: evidence from plant remains. *Proceedings of the National Academy of Sciences* 101, 9551–5.

Weiss, E., Kislev, M. E. and Hartmann, A. (2006) Autonomous cultivation before domestication. *Science* 312, 1608–10.

Willcox, G. (2002) Charred plant remains from a 10th millennium kitchen at Jerf el Ahmar, Syria. *Vegetation History and Archaeobotany* 11, 55–60.

Willcox, G. (2005) The distribution, natural habitats and availability of wild cereals in relation to their domestication in the Near East: multiple events, multiple centres. *Vegetation History and Archaeobotany* 14, 534–41.

Willcox, G., Fornite, S. and Herveux, L. (2008) Early Holocene cultivation before domestication in northern Syria. *Vegetation History and Archaeobotany* 17, 313–25.

Wolf, E. R. (1982) *Europe and the People without History*. Berkeley, CA: University of California Press.

Wright, G. A. (1992) Origins of food production in Southwestern Asia: a survey of ideas. *Current Anthropology* 33(S), 109–39.

Zohary, D. and Hopf, M. (2000) *Domestication of Plants in the Old World*. Oxford: Oxford University Press.

Zvelebil, M. (1996) The agricultural frontier and the transition to farming in the circum-Baltic region. Pp. 323–45 in D. Harris (ed.), *The Origins and Spread of Agriculture and Pastoralism in Eurasia*. London: UCL Press.

16. Modelling the Agricultural Impacts of the Earliest Large Villages at the Pre-Pottery Neolithic–Pottery Neolithic Transition

Dana Campbell

One predominant explanation for the transition between the Pre-Pottery and the Pottery Neolithic (PPN and PN respectively) periods in the southern Levant asserts that the abandonment of many large Middle and Late PPNB villages, and the apparent concurrent shift towards a more dispersed settlement pattern, might be attributed to localized ecological degradation. These phenomena are often thought to have resulted from the impacts of ecological mismanagement or an unsustainable form of Early Neolithic mixed farming. Advances in agro-pastoral and ecological research dictate that it is now appropriate to critically re-examine these assumptions. 'Ain Ghazal, the site of the proposal's first articulation, presents an ideal testing ground for re-evaluating the evidence of localized anthropogenically induced ecological degradation, for considering the true nature of Neolithic agronomy, and for evaluating the actual environmental impact of prolonged mixed farming within the village's immediate landscape. I argue that the ecological impact of Neolithic farmers should be sought in the results of analogous agronomic experimentation, and question the degree to which early agricultural practices altered the physical appearance, and perhaps perceptions, of local landscapes. Finally, I argue that settlement abandonment at or near the PPN–PN transition would appear to have had little to do with the environmental impact of early mixed farming systems.

Introduction

The reasons why the transformations of material culture and settlement patterning that define the so-called Pre-Pottery Neolithic B (PPNB) 'collapse' occurred between the PPN and Pottery Neolithic (PN) in the Levant have long been debated (e.g. de Vaux 1970; Kenyon 1957; Kirkbride 1968; Moore 1973; 1985; Perrot 1968). Climatic fluctuation (e.g. Simmons 2000), the effects of social stresses associated with life in (unprecedentedly) large communities (e.g. Kuijt 1995) or the increasing importance of mixed farming (e.g. Banning and Siggers 1997) have been highlighted as means to explain, at least partially, these cultural transformations. The predominant explanatory model, however, asserts that the abandonment of many large Middle and (in particular) Late PPNB communities (MPPNB and LPPNB), and the seemingly concurrent shift towards more dispersed settlement patterns across the region, might be attributed to localized ecological degradation (Köhler-Rollefson 1988; Köhler-Rollefson and Rollefson 1990; Rollefson and Köhler-Rollefson 1989; Rollefson *et al.* 1992; Rollefson and Köhler-Rollefson 1993). It suggests that gradual and cumulative localized ecological degradation was caused by the combined impacts of increasing community size over the course of the PPNB, agricultural mismanagement or ecologically insensitive practice and the allegedly unsustainable character of Neolithic mixed farming.

In light of recent advances in agro-pastoral and environmental research, it is appropriate to re-examine critically concepts of agricultural sustainability and societal collapse at (or near) the end of the PPNB. Specifically, the ecological degradation argument's concept of sustainability does not reflect current agronomic research. This, when combined with the economic and environmental data from key Neolithic settlements, permits a re-evaluation of the evidence for agricultural over-exploitation within the landscapes of large PPN villages and allows a better definition of the potential ecological impacts of those means of agricultural production that were likely to have been adopted by PPN communities (Campbell 2008). This paper summarizes one such re-evaluation of the ecological degradation argument by re-considering the evidence from LPPNB 'Ain Ghazal. This is the best known of the so-called 'mega-sites' and first prompted Rollefson and Köhler-Rollefson to develop the degradation model (Köhler-Rollefson and Rollefson 1990). I, however, suggest that 'Ain Ghazal was, in fact, set within a landscape easily capable of sustaining even the largest proposed population levels of the LPPNB settlement indefinitely.

Attempting to explain settlement transformations at the end of the PPNB

The ecological degradation argument states that the subsistence economy of these large villages represented the first sustained experiment in large-scale sedentism and mixed farming, and the agricultural practices of these groups were responsible for destabilizing local ecologies. It goes on to suggest that these negative ecological impacts ultimately led to ineffective agricultural production and the abandonment of large communities in favour of smaller settlements that were more dispersed throughout the landscape (Köhler-Rollefson and Rollefson 1990). Given the scale of the communities involved, it was suggested that the cultivation of select species at the expense of natural biological heterogeneity would have compromised ecological stability and, over time, caused soil fertility to decline to levels from which it could not easily recover. Additionally, the spatial and practical needs of cultivation and herding were suggested to have been inherently contradictory: post-harvest, cultivated fields would have been bare of vegetation and not useful as pasture (Köhler-Rollefson 1988, 89), and the feeding behaviour of goats would have prevented pasture and particularly woodland regeneration (Rollefson and Köhler-Rollefson 1989, 76), compounding the purported negative impact of cultivation and causing greater land pressures as herds are presumed to have been kept largely separated from cultivated fields (Köhler-Rollefson and Rollefson 1990). Ultimately, these combined pressures are thought to have denuded the accessible landscape to such a degree that this large community could no longer sustain itself without significant adaptation.

The evidence indicates that a number of long-established, evidently successful communities were vacated throughout the MPPNB (e.g. Jericho, Beidha, Ghwair), LPPNB (Ba'ja, Munhatta, Abu Gosh) and beyond ('Ain Ghazal, Wadi Shu'eib). Whether these settlement transformations are symptomatic of a so-called societal 'collapse' or not is not entirely relevant to the question at hand, which is whether it is likely that these large communities practised sustainable agriculture. As the timing of these events coincides with broad economic trends involving the widespread incorporation of domestic herds into the agricultural regimes of most sedentary populations, the relationships between these populations and the landscapes they inhabited, as mediated by subsistence economic pursuits, are worth careful consideration.

A 'mega-site' case in point: 'Ain Ghazal

Settled near the beginning of the MPPNB (8320–8090 cal BC, GrN-12966; 8259–7979 cal BC, AA-1164; Rollefson 1998a, table 1), the village at 'Ain Ghazal appears to have expanded considerably throughout its subsequent phases, and its built environment is estimated to have covered an area of approximately 8 ha by the later LPPNB, or between 7500/7200–6800 cal BC (Rollefson 1998a). As this site is included in the select category of LPPNB villages described as 'mega-sites,' the scale of settlement is key. That the size of these communities, also including Wadi Shu'eib (Simmons *et al.* 2001), 'Ain Jammam (Waheeb and Fino 1997), and Basta (Nissen *et al.* 2004), had exceeded anything yet known in the Levant is not in dispute. The implications for social organization of such large concentrations of people, and the putative stresses experienced by populations that were ideologically equipped for life in much smaller groups, has been the focus of much current discussion (e.g. Kuijt 2000). Such discussion serves to highlight alternative suggestions for transformations at the end of the PPNB.

The main surface scatter relates largely to the final phase of Neolithic (PN) occupation at the site, and was reported to have covered nearly 9.4 ha; an additional 10 per cent is thought to have been destroyed by the 1974 construction of the Amman–Zarqa highway (Rollefson *et al.* 1984, 140). The evidence for MPPNB occupation suggests that the community's built environment was not likely to have exceeded 3.2 ha (Fig. 16.1), contrary to the higher estimates described by preliminary excavation reports (e.g. Rollefson *et al.* 1992; Rollefson and Köhler-Rollefson 1993; Rollefson 1998b). To clarify, both the Central and North Fields contain MPPNB remains, but neither pre-LPPNB deposits nor sterile sediments were reached in nearby test square 3300. An undifferentiated PPNB deposit and architecture characterized as 'probably PPNC [Pre-Pottery Neolithic C]' were found in squares 6891 and 5493 respectively (Rollefson and Köhler-Rollefson 1993, 108). Radiocarbon and AMS assays confirm that build-up of the South and East Fields did not pre-date, at the earliest, a late moment in the MPPNB (Rollefson 1998a, 9, fig. 1). Drawing a boundary halfway between where MPPNB remains were and where they were not results in an estimate of 3.01 ha, an extent that includes some of the space affected by road construction and assumes that the built environment was restricted to the west and south banks of Wadis al-Zarqa and al-Fakhit respectively. If this suggested boundary stretched further towards each river and to the inner edge of where MPPNB remains were not found, an estimate of 3.7 ha results, still smaller than the

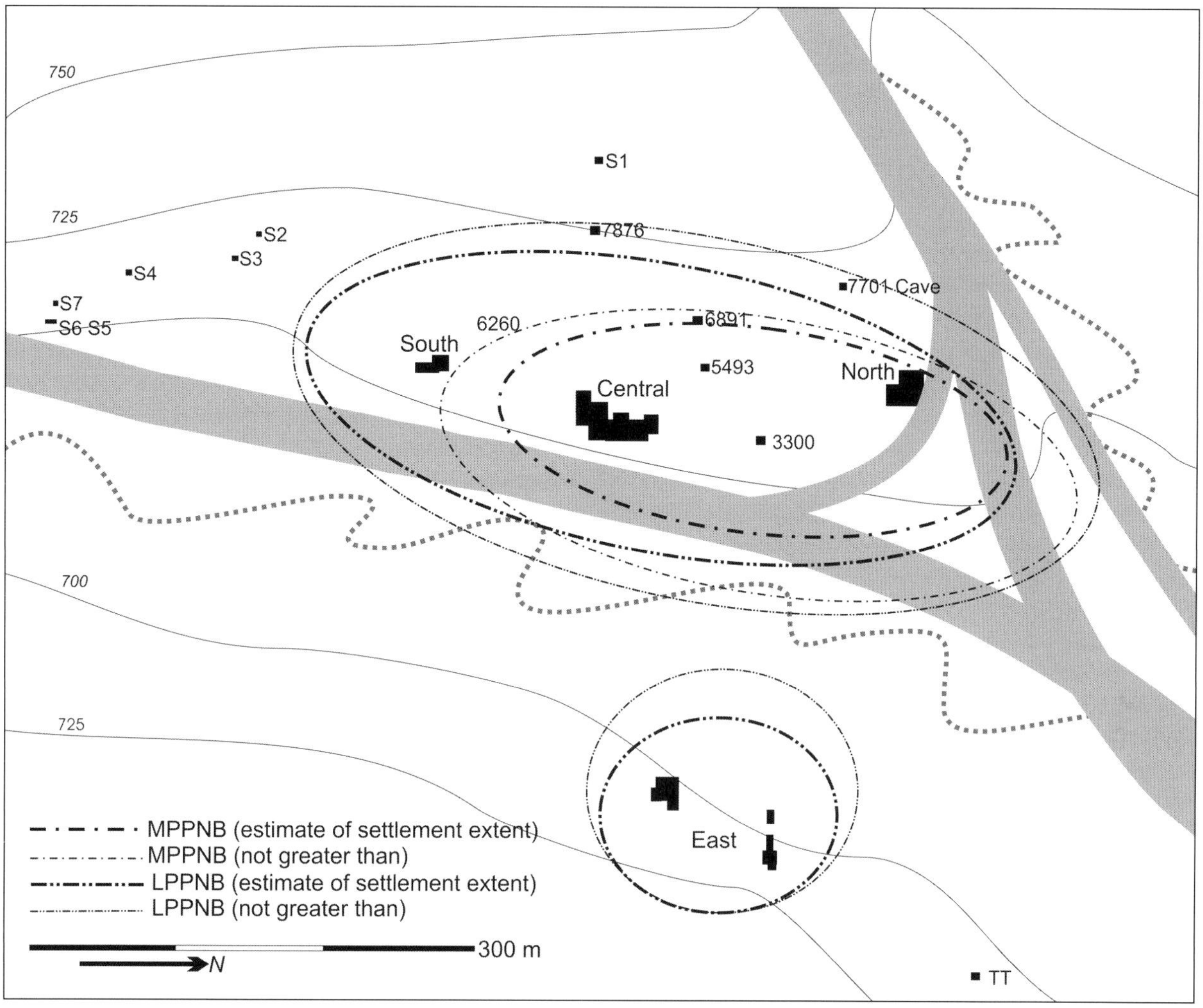

Fig. 16.1. Settlement extent estimates for MPPNB and LPPNB 'Ain Ghazal. Re-drawn after Rollefson 1983, fig. 1; Rollefson 1997, fig. 1; Rollefson et al. 1992, fig. 2.

4–5 ha that has been suggested previously (e.g. Rollefson *et al.* 1992, 446).

The evidence suggests that expansion had occurred with the earliest LPPNB deposits. Architecture in square 6260 was described as reminiscent of the LPPNB 'Terraced House' in the North Field, yet no Neolithic material was present in squares 7876 or 7704 (the cave), immediately to the west (Rollefson and Köhler-Rollefson 1993, 108). Although excavation in the walled compound was not possible, cobble layers in units S2 and S4 were comparable to PPNC–PN strata in the main excavation areas. If the built environment did not include areas closest to the rivers, or existed in two parts, one on either side of the wadi, the extent of LPPNB settlement would have therefore encompassed 7.8 ha (Figure 16.1). If the settlement did, in fact, encroach on this space, the revised estimate rises to a maximum area of 10.6 ha (Campbell 2008; note that the excavators' preliminary estimates of the LPPNB extent has been variously reported as between 9 and 12–13 ha; Rollefson *et al.* 1992; Rollefson and Köhler-Rollefson 1993; Rollefson 1998b).

PPNC material does not seem to be present in the East Field (Rollefson 1998a, table 1), nor does this area appear to have had a strictly residential use during the PN (Rollefson 1998a, 9). Undifferentiated Late Neolithic material was found in S2 and S4 and, in test-squares immediately south, nothing pre-dating the Early Bronze Age was recovered (Figure 16.1). PPNC modification and re-use of LPPNB buildings characterizes the North Field and, with the 'rare' exception of potentially intrusive finds, no PN deposits are thought to exist in this area (Rollefson *et al.* 1992). That this pattern may relate to preservation issues, cannot, of course, be discounted. By the same standards that estimated the extent of the MPPNB and LPPNB villages, the whole of the PPNC community may have covered 9.3 ha, but it could have contracted to 7.5 ha during the early PN (Fig. 16.1). The later village includes a Yarmukian pottery scatter of about 50 × 50 m (since destroyed) on the northern side of Wadi al-Fakhit (Rollefson and Kafafi 1994, fig. 1). Suggestions that PPNC and PN architectural densities decreased in comparison to those of the earlier villages are intriguing (e.g. Rollefson *et al.* 1992, 446;

	MPPNB			*LPPNB*			*PPNC*		
	If extent was…								
Estimated population density	3 ha	3.9 ha	4.7 ha	7.8 ha	9.2 ha	10.6 ha	9.73 ha	11.6 ha	13.5 ha
Low range[a]	259	333	406	671	793	914	837	997	1158
Mid range[b]	419	539	657	1089	1284	1481	1355	1616	1876
High range[c]	860	1106	1349	2234	2634	3037	2780	3314	3848

a Based on a total site are of 116.3 m^2 per person; analogous to Tell-i Nun, southwest Iran (Jacobs, 1979).
b Based on 71.8 m^2 per person; analogous to Shahabad, central Iran (Kramer, 1979).
c Based on 35 m^2 per person; analogous to Marib, north Yemen (van Beek, 1982).

Table 16.1. Estimated population levels for MPPNB, LPPNB and PPNC 'Ain Ghazal, based on various settlement extents (see Figure 16.1) and ethnographic analogy (see notes).

Köhler-Rollefson and Rollefson 1990, 7) but difficult to substantiate in light of the fact that 'Ain Ghazal's uppermost strata have been significantly impacted by both modern construction and agricultural activities.

Thus the evidence demonstrates a substantial and expanding settlement at 'Ain Ghazal from the MPPNB into the LPPNB. Table 16.1 outlines estimated population levels for a range of settlement sizes and population densities for the LPPNB that form the basis of a revised assessment of the ecological impact of 'Ain Ghazal's subsistence needs (Campbell 2008).

The environmental and economic context of PPN 'Ain Ghazal

'Ain Ghazal is situated near the boundary of two major vegetational zones, with typically Mediterranean vegetation to the west and Irano-Turanian elements towards the east, and so its inhabitants were within convenient reach of several ecological niches. Moisture- and mineral-rich micro-environments were present throughout the bisected topography near the confluence of the Wadis al-Fakhit and al-Mudawwa'ib, two major tributaries to the Wadi al-Zarqa, the longest drainage system in modern Jordan. The Zarqa survived as an active, reliable resource until its former sinusoidal course was altered to a more controlled flow managed by recent construction and water treatment facilities (Kafafi and Rollefson 1997, 242). Although dramatically affected by Amman's sprawl, its banks have been used for extensive garden cultivation until very recently (Rollefson 1983, 1).

The 400 mm isohyet, understood as the boundary representing the minimum rainfall level required for strictly Mediterranean plant communities, is currently just over 4 km west of 'Ain Ghazal, and recent meteorological records from Amman airport, 2 km to the south, suggest that the site receives an average of 320 mm of annual rainfall (Anon. 1984). Present hydrological conditions, therefore, present no serious limitations to the rain-fed cultivation of those plants described as the 'founder crops' of the Neolithic Levant, many of which continued to grow on the broad, flat hilltops and valleys east of 'Ain Ghazal when preliminary surveys of the sites surroundings were conducted (Rollefson 1984; Simmons and Kafafi 1988). Few would dispute, however, that PPN rainfall levels were, in fact, higher than those of today (e.g. Robinson *et al.* 2006; Rossignol-Strick 1995; Wilkinson 2003), and should the marked negative shift in stable oxygen and carbon isotope ratios in the Soreq cave profile (Bar-Matthews and Avalon 1997, 155; Bar-Matthews *et al.* 1998; Bar-Matthews *et al.* 2003) have applicability to this particular area, average rainfall levels at the site may have been as high as 425–475 mm per year. Reliable rain-fed crop production surrounding Neolithic 'Ain Ghazal would have been, therefore, theoretically assured.

Geomorphological surveys of areas up to 5 km away from 'Ain Ghazal are published in a preliminary manner (Simmons and Kafafi 1988; Mandel and Simmons 1988), and the landscape seems typical of the lower east Jordan highland–plateau interface. Gently undulating towards the northeast, extensive arable fields have only recently fallen into disuse. It has been reported that the degree of slope surrounding the site may have been severe enough to render the area productive only for pasture (e.g. Rollefson 1997); however, very little of the landscape (less than 2 per cent within 5 km) constitutes slopes of 50 per cent (27°) or greater, and only slightly more (3.5 per cent of the area within 5 km) consists of slopes greater than 30 per cent (17°) (Campbell 2008). This is significant: reliable cultivation is not limited by the local topography. Ethnographic summaries of 'traditional' practices in the area tend to assert that farmers do not cultivate slopes greater than 30 per cent (see Hillman 1973; Palmer 1998; Turkowski 1969). However, the 'traditional' practices under observation are usually accomplished by means of mechanized ploughs; a 30 per cent slope may represent the limit under which one can safely operate heavy machinery, but does not impose a serious restriction on the cultivation of most cereals and pulses (Zohary 1973; Zohary and Hopf 2000). With the exception of a relatively large expanse of chalky marls found nearly 12 km south-east of 'Ain Ghazal (Bender 1968; Mandel and Simmons 1988), the surface geology and corresponding major soil associations surrounding the site also seem to present no major limiting factor to the cultivation of any species in the typical Neolithic crop assemblage.

The surrounding ecological diversity ensured that a wide range of subsistence-economic choices were readily available to 'Ain Ghazal's PPN community, and in addition would have allowed a degree of flexibility to be built into the community's economic regimes. Although botanical and faunal assemblages document significant economic trends over the duration of occupation (the increasing importance of controlled or domestic animals over hunted fauna, for example), they clearly demonstrate that resources from a variety of habitats were used throughout the village's long history (Donaldson 1985; Rollefson and Köhler-Rollefson 1993; Köhler-Rollefson *et al.* 1988; Neef 1997; von den Driesch and Wodtke 1997; Wasse 1997; 2002). From the earliest MPPNB settlement, up to and including the early PN (Yarmukian) phase, 'Ain Ghazal's inhabitants were acquiring resources endemic to Mediterranean woodland: arboreal species include *Pistacia* sp. (terebinth) and *Ficus* sp. (fig); representative faunal species include *Meles meles* (Eurasian badger), *Mellivora capensis* (Ratel or honey badger), *Caracal caracal* (Caracal lynx), *Felis sylvestrus* (wild cat), *Aquila chrysaetos* (golden eagle)), steppic Irano-Turanian species including *hemionus* (Asiatic wild ass), *Hemichinus aurtus* (long-eared hedgehog), *Coturnix coturnix* (common quail), *Alectoris chukar* (chukar partridge), *Otis tarda* (great bustard)), and resources more typical of the desert such as *Vulpes rueppelli* (Ruepelli's, or sand fox). The presence of cf. *Anas* (unspecified marsh duck) and cf. *Aythya* (unspecified diving duck or mallard) suggest that wetlands were also within reach of the village, at least during earlier phases (Köhler-Rollefson and Rollefson 1990, 4).

It has been suggested that 'Ain Ghazal's faunal assemblage includes evidence for decreasing ecological diversity through time, and that this pattern may be attributed to over-exploitation (Köhler-Rollefson 1988; Köhler-Rollefson *et al.* 1988, 96). The number of identified species decreases from 'over 50' in the MPPNB (Köhler-Rollefson *et al.* 1988) to a lesser number during the LPPNB. This purported difference, however, is due to 'small carnivores' for that part of the assemblage analysed before 1993 (Köhler-Rollefson *et al.* 1993, table 1). According to the 1993 analysis, the 'small carnivore' category accounts for 7.6 per cent of the MPPNB number of identified specimens, but only one or two unspecified carnivorous elements were recovered from LPPNB and early PN contexts, and none from PPNC deposits. All of these fur-bearing 'small carnivores' (which included *Martes foina* (beech marten), *Meles meles* (Eurasian badger) and *Herpestes ichneumon* (Egyptian mongoose)) were recovered from one context inside an MPPNB building (square 3082), and may suggest some economic specialization rather than ecological events occurring nearby; hunting for fur may have become redundant with domestic animals (Köhler-Rollefson 1988; Köhler-Rollefson *et al.* 1993, table 1; Köhler-Rollefson *et al.* 1988, tables 1 and 2; Von den Driesch and Wodtke 1997, 534). Furthermore, a number of these species tend to inhabit their landscapes in low densities, and decreases in their number may therefore be expected with an increase in domestic carnivores (Douglas Baird pers. comm.).

It is suggested that by the PPNC there is evidence for depletion of local arboreal resources caused by the over-exploitation of woodland for fuel and construction material, and by the interference of mixed farming in woodland regeneration. MPPNB houses seem to have incorporated timber structural supports of relatively large diameter, and internal walls or partitions replace posts during later Neolithic phases (Köhler-Rollefson 1988, 90; Rollefson and Köhler-Rollefson 1989, 77–8). To credit architectural developments over time with local ecological degradation is to disregard contemporary regional trends and to underestimate the importance of concurrent transformations in social spheres to architectural developments (see Banning and Byrd 1984; Byrd 2000; Kuijt and Goring-Morris 2002). The suggestion also assumes that wood would have been used as a primary or important source of fuel, with relatively high amounts needed for quick-lime production (Rollefson *et al.* 1992; Rollefson and Köhler-Rollefson 1989, 78; Simmons 2000, 219). However, the potential for the use of deadwood as fuel is rarely discussed (e.g. Rollefson and Köhler-Rollefson 1989, 78). In any thriving riparian habitat excess density tends to kill parts of individual trees, and occasionally seasonal flooding may fell whole stands (Asouti 2005, 252; note that that *Populus* sp. (poplar) and *Tamarix* cf. *jordanis* (tamarix) comprise the most frequently reported charcoal at 'Ain Ghazal: Kafafi and Rollefson 1997, 116). At Neolithic Çatalhöyük, for example, nearly half of *Quercus* sp. (oak) charcoal, and that from a number of riparian and fruit species from selected deposits, was partially decayed or deadwood at the time of use (Asouti 2005, table 10.3). Additionally, the potential that any PPN community practised formal woodland management strategies (i.e. coppicing) has not yet been adequately explored; it is notable that poplar is the most extensively cultivated tree throughout Southwest Asia (see Halstead *et al.* 1998; Rackham 1982; Willcox 1974; 1992).

Within this broad ecological setting, the LPPNB inhabitants of 'Ain Ghazal cultivated domestic species of wheat (*Triticum dicoccum* or emmer and *T. monococcum* or einkorn wheat), barley (*Hordeum distichum*), lentil (*Lens culinaris*, or *L. orientalis*, its wild equivalent) and field pea (*Pisum* cf. *arvense*), but also had access to several other edible plants including chickpea (*Cicer reticulatum* or *C. arietinum*), vetch (possibly *Vicia faba* – broad bean or *V. erivilia* – bitter vetch) (Donaldson 1985, 100; Kafafi and Rollefson 1997; Neef 1997). The remains of sheep (*Ovis orientalis*) and goat (*Capra aegagrus* or the domestic *C. hircus)* are ubiquitous throughout LPPNB deposits and together represent the highest proportion of the identified specimens (Wasse 2002). Cattle (*Bos primigenius*), boar (*Sus scrofa*) and gazelle (likely *Gazella gazella*) are among the other significant faunal resources, but would not have

Crop	*Suitability level*	*Area (in dunum)*	*Proportion (of 1.25 hours)*
Barley	Ideal	17,566	31%
	Moderate	30,974	55%
Emmer wheat	Ideal	17,550	31%
	Moderate	30,986	55%
Field pea	Ideal	-	-
	Moderate	18,163	32%
Lentil	Ideal	-	-
	Moderate	44,651	79%

Table 16.2. Estimated areas (in dunum, 1000 m²) of land suitable for the reliable cultivation of barley, emmer wheat, field pea and lentil within a walk of 1.25 hours' duration from 'Ain Ghazal, under a precipitation regime where rainfall amounts are 50 mm higher than current levels; and percentage of the total amount of land within that same distance.

been herded during the LPPNB (Wasse 2002, 61; Von den Dreisch and Wodtke 1997, 535).

Farming and 'Ain Ghazal's PPN landscape

While ever-increasing amounts of data concerning which plant and animal species were being managed during the PPN in Southwest Asia are readily available (e.g. Colledge *et al.* 2004; Garrard 1999; Horwitz *et al.* 1999; Wasse 2002), we have very little information on how farmers achieved cultivation and pastoralism. One might assume, however, that the 'worst case scenario' of agricultural management would involve a cultivation regime where crops are grown in extensive field systems (as opposed to more intensive strategies), and without the incorporation of techniques to ensure the preservation of soil organic matter (SOM), nutrient supplying capacity or the maintenance of soil aggregation (e.g. Dalal and Mayer 1986; Masri and Ryan 2005). The most ecologically insensitive (and spatially demanding) strategies would also view cultivation and herding as mutually exclusive activities, with animals kept at densities too high to allow for the preservation of self-regenerating pasture.

Consider, then, if this form of agricultural management had been in effect at LPPNB 'Ain Ghazal. In this scenario, the highest estimated population of LPPNB 'Ain Ghazal, just over 2600 people, is assumed to have received 50 per cent of its total caloric requirements from cultivated cereals, and 30 per cent from cultivated pulses. This scenario is just one of a series that were undertaken as part of more detailed site catchment analyses discussed elsewhere (Campbell 2008). It is assumed that each person requires an average of 2500 kcal per day, and that each kilogram of cereal grain contains 3300 kcal, while each kilogram of legume grain contains 3250 kcal (USDA 1960/61). It is also assumed that rainfall at this time was 50 mm higher than today's average values, a relatively conservative approximation, and a difference far less than the 175–300 mm range suggested by the Soreq cave data (Bar-Matthews *et al.* 2003). The 'arable' landscape surrounding 'Ain Ghazal is defined as those areas where major soil associations, depth, precipitation levels and topographic circumstances (slope) conform to either the ideal or to the preferred habitat that each of the relevant plant species (domestic einkorn or emmer wheat, barley, field pea or lentil) require. Those areas of the landscape that present even minor limiting factors – *T. dicoccum* (emmer), for example, is known not to thrive in soil associations that include marls (Zohary 1973; Zohary and Hopf 2000) – are classed as unsuitable for cultivation. 'Ideally' suitable land (that which reflects the ideal ecological preferences of each crop) is assumed to produce 54 kg of cereal or legume grain per dunum (1000m²), while 'moderately' suitable land could produce 36 kg; these are conservative yield estimates based on the low range value that organic dry-farming tends to produce in the region today (e.g. Diaz-Ambrona and Minguez 2001; Hillman 1973; Jones and Singh 2000; Lopez-Bellido *et al.* 2004). This scenario imposes further limitations by excluding all potential flood plains, or areas where the survival of young plants that may be emerging in time with spring floods would be at risk, from 'arable' land. This exclusion reflects a somewhat wasteful land management strategy, one that is out of step with current thought regarding the practicalities of Neolithic cultivation. Such alluvium- and moisture-rich contexts are ideal for particular cultivation strategies (e.g. localized fixed-plot horticulture (Sherratt 1980), or intensive garden horticulture (Bogaard 2005; Jones 2005)) and are not likely to have been overlooked by 'Ain Ghazal's Neolithic farmers.

To provide 50 per cent of this large LPPNB village's caloric requirements, 6744 dunum of 'ideally' suitable land would have needed to have been under cereal cultivation at any given time. This represents 39 per cent of the 'ideal' land within just over one hour's walk from the settlement (Table 16.2). Land ascribed to this class represents only 31 per cent (17,550 dunum) of the total area within an hour and a quarter's walk of 'Ain Ghazal, and *excludes* an additional 30,986 dunum considered only 'moderately' suitable for cultivation. If an additional 30 per cent of the village's total caloric requirements were derived from cultivated

Scenario: If 50% of caloric requirements came from cultivated cereals				
Estimated population level	2634 (high-range estimate for LPPNB)		1284 (mid-range estimate for LPPNB)	
	Area (in dunum)	Proportion of total area *within* suitability class	Area (in dunum)	Proportion of total area *within* suitability class
'Ideally' suitable land required	6744	39%	3288	19%
'Moderately' suitable land required	10 116	33%	4931	16%

Scenario: If 30% of caloric requirements came from cultivated legumes				
Estimated population level	2634 (high-range estimate for LPPNB)		1284 (mid-range estimate for LPPNB)	
Lentil	Area (in dunum)	Proportion of total area *within* suitability class	Area (in dunum)	Proportion of total area *within* suitability class
'Moderately' suitable land required	6163	14%	3004	7%
Field pea	Area (in dunum)	Proportion of total area *within* suitability class	Area (in dunum)	Proportion of total area *within* suitability class
'Moderately' suitable land required	6163	34%	3004	17%

Table 16.3a (top) and 16.3b (bottom): Areas (in dunum, 1000 m²) and percentage of land available in each suitability class (see Table 16.2) within a walk of 1.25 hours' duration from 'Ain Ghazal, to provide 50% of the inhabitants' total caloric requirements in cultivated cereals (Table 16.3a), and 30% in cultivated legumes (Table 16.3b).

pulses, 6163 dunum would have needed to have been under cultivation at any given time; this area represents 14 per cent of land considered 'moderately' suitable for lentil cultivation near 'Ain Ghazal, or 34 per cent of land classed as suitable for growing field pea.

To suggest that 80 per cent of LPPNB 'Ain Ghazal's caloric requirements came from cultivated plants probably overestimates the role of domesticated plants in Neolithic diet. The point, however, is that in this scenario, where caloric requirements and the amount of land needed to meet them are deliberately inflated, spatial requirements remain relatively minimal. To bring less than half of land classed as 'ideally' suited to cultivation under cultivation (under low-yielding and extensive field systems), this representing less than one-third of the total land available, would still allow large areas of the surrounding landscape to remain untouched and unexploited. Even marginal improvements to this scenario could ensure that very little of 'Ain Ghazal's landscape would have been impacted by the needs of cultivation. The exploitation of alluvial plots along Wadi al-Zarqa, for example, capable of producing far more than 54 kg per dunum (cultivation trials suggest that such contexts can produce as much as 100–150 kg per dunum: e.g. Willcox 1999, 488), has not yet been explored, nor has the potential of higher average annual rainfall levels. Also recall that this scenario is based on the highest estimated LPPNB population figure. If the village had instead been home to just over 1200 people, then only 3288 dunum (or 19 per cent, as opposed to 39 per cent) of the most suitable arable land would have needed to have been under cultivation to provide a 50 per cent contribution from cultivated cereals (Table 16.3a), and only 7–17 per cent of land well suited to lentil or field pea cultivation could provide an additional 30 per cent caloric contribution to the inhabitants' diet (Table 16.3b).

Estimating the spatial requirements of herding is far more difficult. Resources that are 'on the hoof' are not fixed in space, and may not have been regularly kept in close proximity to the village (e.g. Halstead 1996, 301–4; Köhler-Rollefson and Rollefson 1990). Cultivation and herding are not, in fact, mutually exclusive activities (e.g. Harris 1999), although to assess the 'worst case scenario' of agricultural management, they are here considered to have been. For domestic herds of sheep and goat to provide 20 per cent of the annual caloric requirements of just over 2600 people, I estimate that no more than 41,496 dunum would have been required as year-round pasture (Campbell 2008). The living herd size is estimated at three times the number of adult sheep or goats needed for annual consumption, and assumes that each adult animal produces 25 kg of usable meat valued at 3500 kcal per kg (USDA 1960/61). This spatial requirement figure assumes that herd densities reflect the best sustainable practice, and do not exceed 4 animals ha^{-1}. Range experiments undertaken in the Negev (Tadmor *et al.* 1974) and the Galilee (Gutman and Seligman 1979) demonstrate that stocking rates of 5 sheep or goats ha^{-1}, or 3.35 ha^{-1} respectively, had a negligible ecological impact on pasture regeneration, and that such animal densities could be sustained indefinitely. Just over 41,000 dunum represents 43 per cent of all land classed as 'not ideally' suitable for cultivation within a 2 hours' walk from 'Ain Ghazal, and therefore excludes an additional 46,625 dunum of highly suitable farmland.

The ecological degradation argument suggests that the feeding behaviour of goats would have prevented pasture and, particularly, woodland regeneration (e.g. Rollefson and Köhler-Rollefson 1989, 76). The potential impact of browsing goats on woodland is, indeed, more negative than their impact on 'typical' pastures that tend to be a mixture of Mediterranean grassland and legume swards. In my view, however, the important question is 'at what density do herds have a negative ecological impact on woodland habitats?' Open park-woodland range experiments in Tunisia demonstrated that the optimum stocking rate in such habitats is 1–2 goats ha^{-1}, and that with densities exceeding 9 animals ha^{-1} seedling establishment was compromised and the growth of new wood impaired (Meurer and Steinbach 1986). Under an alternate herding scenario, then, where

LPPNB 'Ain Ghazal's goats were kept at lower stocking rates (in this case 1.6 animals ha^{-1}), a total of 103,714 dunum could support their year-round maintenance. In this case, all land within a 2-hour walk of the village (excluding, however, all 'ideally' arable land, or an additional 46,625 dunum), would have been grazed (Campbell 2008). If, however, herds were permitted to graze within 3 hours of 'Ain Ghazal, these spatial requirements would have been met comfortably. An area of this size would also incorporate a hypothetical 'herd exclusion zone,' representing *all* land within 5 km of the settlement, and so rules out the pasture potential of harvested plots, fallow land, and all other resources within this boundary (Campbell 2008). In other words, 'Ain Ghazal's herders clearly had the potential opportunity to maintain their flocks within a reasonable distance of the community without adversely affecting the local ecology or infringing on space that may have been allocated to the needs of cultivation.

Finally, these figures reflect an economic strategy where *all* of a villager's diet is based on locally produced, domestic food resources (50 per cent cereals, 30 per cent pulses, 20 per cent sheep/goat). This is clearly not an accurate reflection of 'Ain Ghazal's LPPNB economy, and the macro-botanical evidence from the site indicates that a wide range of wild plant foods from a variety of habitats was also regularly exploited (Donaldson 1985; Rollefson 1984).

The impact of Neolithic agricultural practices

Modern ecologically insensitive agricultural management practices can profoundly affect soil fertility and plant diversity, and may ultimately render vast areas of the landscape susceptible to erosion (see Chang 1950; Dalal and Mayer 1986). Although several factors are responsible for this predicament, the most significant seems to be the extensive use of deep-impact, often mechanized, tillage practices. The impacts of what modern farmers describe as 'conservation' tillage practices (using only seed-drills, digging sticks or hoes, or their mechanized equivalents), on the other hand, are far less severe. Most, if not all, of the negative ecological consequences associated with modern or 'conventional' cultivation techniques can be effectively negated under alternate management regimes, and the incorporation of these techniques may ensure the sustainability (defined as reliable yields consistently achieved without adversely affecting the long-term fertility of arable soils) of cultivation (e.g. Hemmat and Eskandari 2005; Lopez-Bellido *et al.* 1996; Masri and Ryan 2005). SOM (as a convenient indicator of soil structural stability and nutrient supplying capacity) may remain largely unaffected over the long-term (e.g. Bauer and Black 1981).

Shallow-impact tillage tends to preserve soil aggregation and maintain SOM content, which in turn preserves the microbial populations of soils, which are essential for the breakdown and processing of organic residues and for nutrient supplying capacity (e.g. Doran 1980a; 1980b). Water storage capacity tends to be increased largely because surface residues, even if these are only 'standing stubble,' minimize the potential for soil surface sealing, and therefore promote water infiltration (e.g. O'Leary and Connor 1997). The associated erosion risks are reduced and a higher water storage capacity tends to translate into improved crop yields. Improved moisture retention capacity is obviously of particular importance to semi-arid agriculture and several agronomic trials demonstrate that this advantage decreases proportionally in areas of higher rainfall (e.g. Hemmat and Eskandari 2005, 75). Finally, microsite-scale plant heterogeneity, common and essential to the ecological stability of arid to semi-arid environments, tends to survive on land that has been cultivated only by conservation-based, rather than conventional, means (Robles and Burke 1997, 354).

The position here is two-fold: that analogues to the ecological impact of Neolithic farming should be sought in cultivation trials conducted according to the principles of conservation-based management as such practices are likely to have been more in step with how PPN farmers achieved cultivation, and that particular assumptions about the ecological impact of cultivation are rooted in impacts associated with modern, 'conventional' farming.

The ecological degradation argument implies that soil fertility may not have recovered once it had been compromised; it suggests that the negative ecological impacts of farming were both cumulative and long-lasting, and that '... by the middle of the PPNC, the farmers had to travel at least 3 km from the settlement to reach the edge of most of the arable fields' (Köhler-Rollefson and Rollefson 1990, 9). Cultivation trials aimed at quantifying the process of recovery, however, demonstrate that the level of SOM content, in particular, may be restored to near-original levels following periods as short as 20 years (e.g. Chang 1950), or an estimated 50–60 years (e.g. Burke *et al.* 1995). To suggest that large expanses of arable land would have needed to be removed entirely from a Neolithic agricultural system for perhaps hundreds of years is not an accurate reflection of the true impacts of cultivation, nor of the resilience of semi-arid Southwest Asian ecosystems.

Summary

My argument is based on two complementary positions. First, that the spatial requirements of cultivation and herding for even the largest LPPNB community at 'Ain Ghazal would not have been so great as to preclude the preservation of unaffected or 'natural' areas within the village's surrounding landscape. Second, that although we know very little of how Neolithic people achieved cultivation or pastoralism (in practical and logistical terms), it seems likely that these groups were technologically limited to conventions that a modern farmer would describe as conservation-based agricultural management, and so the potential ecological impacts of Neolithic agriculture should fall within the parameters set by those practices. In my opinion, the purported negative impacts of Neolithic cultivation and herding are unlikely, and settlement patterning transformations or community abandonments and the seemingly concurrent shift towards more dispersed settlement patterning across the region would appear to have had little to do with the environmental impact of early mixed farming economies.

Bibliography

Anon. (1984) *National Atlas of Jordan, Part I, Climate and Agroclimatology*. Amman: Jordan National Geographic Centre.

Asouti, E. (2005) Woodland vegetation and the exploitation of fuel and timber at Neolithic Catalhöyük: report on the wood charcoal macro-remains. Pp. 213–58 in I. Hodder (ed.), *Inhabiting Çatalhöyük: Reports from the 1995–99 Seasons*. McDonald Institute Monographs 38. Cambridge and Ankara: McDonald Institute for Archaeological Research and British Institute of Archaeology at Ankara.

Banning, E. B. and Byrd, B. F. (1984) The architecture of PPNB 'Ain Ghazal, Jordan. *Bulletin of the American School of Oriental Research* 255, 15–20.

Banning, E. B. and Siggers, J. (1997) Technological strategies at a Late Neolithic farmstead in Wadi Ziqlab, Jordan. Pp. 319–32 in H. G. K. Gebel, Z. Kafafi and G. O. Rollefson (eds), *The Prehistory of Jordan II. Perspectives from 1997. Studies in Early Near Eastern Production, Subsistence, and Environment 4*. Berlin: *ex oriente*.

Bar-Matthews, M. and Avalon, A. (1997) Late Quaternary palaeoclimate in the Eastern Mediterranean region from stable isotope analysis of speleothems at Soreq Cave, Israel. *Quaternary Research* 47, 155–68.

Bar-Matthews, M., Avalon, A. and Kuafman, A. (1998) Middle to Late Holocene (6500 year period) palaeoclimate in the Eastern Mediterranean Region from stable isotopic composition of speleothems from Soreq Cave, Israel. Pp. 203–14 in A. Issar and N. Brown (eds), *Water, Environment and Society in Times of Climatic Change*. Dordrecht: Kluwer.

Bar-Matthews, M., Ayalon, A., Gilmore, M., Matthews, A. and Hawkesworth, C. J. (2003) Sea-land oxygen isotopic relationships from planktonic foraminifera and speleothems in the Eastern Mediterranean region and their implication for paleorainfall during interglacial intervals. *Geochimica et Cosmochimica Acta* 67, 3181–99.

Bauer, A. and Black, A. L. (1981) Soil carbon, nitrogen, and bulk density comparisons in two cropland tillage systems after 25 years and in virgin grassland. *Soil Science Society of America Journal* 45, 1166–70.

Bender, F. (1968) *Geologie von Jordanien*. Berlin: Bornträger.

Bogaard, A. (2005) 'Garden agriculture' and the nature of early farming in Europe and the Near East. *World Archaeology* 37/2, 177–96.

Burke, I. C., Laurentoth, W. K. and Coffin, D. P. (1995) Soil organic matter recovery in semiarid grasslands: implications for the conservation reserve program. *Ecological Applications* 5/3, 793–801.

Byrd, B. F. (2000) Households in transition: Neolithic social organization within Southwest Asia. Pp. 63–98 in I. Kuijt (ed.), *Life in Neolithic Farming Communities, Social Organization, Identity, and Differentiation*. New York: Kluwer Academic and Plenum.

Campbell, D. (2008) Sustainable Assumptions: Modelling the Ecological Impacts of Pre-Pottery Neolithic Farming Communities in the Levant. Unpublished PhD thesis, University of Liverpool.

Chang, C. W. (1950) The effect of long-time cropping on soil properties in Northeastern New Mexico. *Soil Science* 69, 359–68.

Colledge, S., Conolly, J. and Shennan, S. (2004) Archaeobotanical evidence for the spread of farming in the Eastern Mediterranean. *Current Anthropology* 45(S), 35–58.

Dalal, R. C. and Mayer, R. J. (1986) Long-term trends in fertility of soils under continuous cultivation and cereal cropping in southern Queensland. I: overall changes in soil properties and trends in winter cereals. *Australian Journal of Soil Research* 24, 265–79.

De Vaux, R. (1970) Palestine during the Neolithic and Chalcolithic Periods. Pp. 499–538 in I. E. S. Edwards, C. J. Gadd, and N. G. L. Hammond (eds), *The Cambridge Ancient History*, Vol. 1/1. Cambridge: Cambridge University Press.

Diaz-Ambrona, C. H. and Minguez, M. I. (2001) Cereal–legume rotations in a Mediterranean environment: biomass and yield production. *Field Crops Research* 70, 139–51.

Donaldson, M. (1985) The plant remains. In G. O. Rollefson, A. H. Simmons, M. Donaldson, W. Gillespie, Z. Kafafi, I. Köhler-Rollefson, E. McAdam and S. L. Rolston (eds), Excavation at the Pre-Pottery Neolithic B village of 'Ain Ghazal (Jordan), 1983. *Mitteilungen der Deutschen Orient Gesellschaft* 117, 96–104.

Doran, J. W. (1980a) Microbial changes associated with residue management with reduced tillage. *Soil Science Society of America Journal* 44, 518–24.

Doran, J. W. (1980b) Soil microbial and biochemical changes associated with reduced tillage. *Soil Science Society of America Journal* 44, 765–71.

Garrard, A. (1999) Charting the emergence of cereal and pulse domestication in South-west Asia. *Environmental Archaeology* 4, 67–86.

Gutman, M. and Seligman, N. G. (1979) Grazing management of Mediterranean foothill range in the Upper Jordan river valley. *Journal of Range Management* 32/2, 86–92.

Halstead, P. (1996) The development of agriculture and pastoralism in Greece: when, how, who and what? Pp. 296–309 in D. R. Harris (ed.), *The Origins and Spread of Agriculture and Pastoralism in Eurasia*. London: UCL Press.

Halstead, P., Tierney, J., Butler, S. and Mulder, Y. (1998) Leafy hay: an ethnoarchaeological study in Northwest Greece. *Environmental Archaeology* 1, 71–80.

Harris, F. (1999) Nutrient management strategies of small-hold farmers in a short-fallow farming system in Northeast Nigeria. *The Geographic Journal* 165/3, 275–85.

Hemmat, A. and Eskandarie, I. (2005) Dryland winter wheat response to conservation tillage in a continuous cropping system in Northwestern Iran. *Soil and Tillage Research* 86/1, 99–109.

Hillman, G. (1973) Agricultural resources and settlement in the Aşvan Region. *Anatolian Studies* 23, 217–40.

Horwitz, L. K., Tchernov, E., Ducos, P., Becker, C., Von den Driesch, A., Martin, L. and Garrard, A. (1999) Animal domestication in the Southern Levant. *Paléorient* 25/2, 63–80.

Jacobs, L. (1979) Tell-I Nun: archaeological implications of a village in transition. Pp. 175–91 in C. Kramer (ed.), *Ethnoarchaeology: Implications of Ethnography for Archaeology*. New York: Columbia University Press.

Jones, G. (2005) Garden cultivation of staple crops and its implication for settlement location and continuity. *World Archaeology* 37/2, 164–76.

Jones, M. J. and Singh, M. (2000) Time trends in crop yields in long-term trials. *Experimental Agriculture* 36, 165–79.

Kafafi, Z. and Rollefson, G. O. (1997) Excavations at Neolithic 'Ayn Ghazal, 1993–1994. Pp. 235–40 in G. Bisheh, M. Zaghloul and I. Kehrberg (eds), *Studies in the History and Archaeology of Jordan VI*. Amman: Jordan Press Foundation.

Kenyon, K. (1957) *Digging Up Jericho*. London: Benn.

Kirkbride, D. (1968) Beidha: Early Neolithic village south of the Dead Sea. *Antiquity* 42, 263–74.

Köhler-Rollefson, I. (1988) The aftermath of the Levantine Neolithic revolution in the light of ecological and ethnographic evidence. *Paléorient* 14/1, 87–93.

Köhler-Rollefson, I. and Rollefson, G. O. (1990) The impact of Neolithic subsistence strategies on the environment: the case of 'Ain Ghazal, Jordan. Pp. 3–14 in S. Bottema, G. Entjes-Nieborg and W. van Zeist (eds), *Man's Role in Shaping the Eastern Mediterranean Environment*. Rotterdam: A. A. Balkema/Brookfield.

Köhler-Rollefson, I., Gillespie, W. and Metzer, M. (1988) The Fauna from Neolithic 'Ain Ghazal. Pp. 423–32 in A. N. Garrard and H. G. Gebel (eds), *The Prehistory of Jordan: The State of Research in 1986*. British Archaeological Reports, International Series 396. Oxford: BAR.

Köhler-Rollefson, I., Quintero, L. and Rollefson, G. O. (1993) A brief note of the fauna from Neolithic 'Ain Ghazal. *Paléorient* 19/2, 95–7.

Kramer, C. (1979) An archaeological view of a contemporary Kurdish village: domestic architecture, household size, and wealth. Pp. 139–63 in C. Kramer (ed.), *Ethnoarchaeology: Implications of Ethnography for Archaeology*. New York: Columbia University Press.

Kuijt, I. (1995) New Perspectives on Old Territories: Ritual Practices and the Emergence of Social Complexity in the Levantine Neolithic. Unpublished PhD thesis, Harvard University.

Kuijt, I. (2000) People and space in early agricultural villages: exploring daily lives, community size, and architecture in the Late Pre-Pottery Neolithic. *Journal of Anthropological Archaeology* 19, 75–102.

Kuijt, I. and Goring-Morris, N. (2002) Foraging, farming and social complexity in the Pre-Pottery Neolithic of the Southern Levant: a review and synthesis. *Journal of World Prehistory* 16/4, 361–440.

Lopez-Bellido, L., Fuentes, M., Castillo, J. E., Lopez-Garrido, F. J. and Fernandez, E. J. (1996) Long-term tillage, crop rotation, and nitrogen fertilizer effects on wheat yield under rainfed Mediterranean conditions. *Agronomy Journal* 88, 783–91.

Lopez-Bellido, L., Lopez-Bellido R. J., Castillo, J. E. and Lopez-Bellido F. J. (2004) Chickpea response to tillage and soil residual nitrogen in a continuous rotation with wheat: I. biomass and seed yield. *Field Crops Research* 88/2–3, 191–200.

Mandel, R. D. and Simmons, A. H. (1988) A preliminary assessment of the geomorphology of 'Ain Ghazal. Pp. 431–6 in A. N. Garrard and H. G. Gebel (eds), *The Prehistory of Jordan: The State of Research in 1986*. British Archaeological Reports, International Series 396. Oxford: BAR.

Masri, Z. and Ryan, J. (2005) Soil organic matter and related physical properties in a Mediterranean wheat-based rotation trial. *Soil and Tillage Research* 87/2, 146–54.

Meurer, M. and Steinbach, J. (1986) Effects of goat browsing on some floristic components of the Maquis and Garrigue in Northern Tunisia. Pp. 132–3 in P. J. Joss, P. W. Lynch and O. B. Williams (eds), *Rangelands: A Resource Under Siege*. Cambridge: Cambridge University Press.

Moore, A. M. T. (1973) The Late Neolithic in Palestine. *Levant* 5, 36–68.

Moore, A. M. T. (1985) The Development of Neolithic Societies in the Near East. *Advances in World Archaeology* 4, 1–69.

Neef, R. (1997) Status and perspectives of archaeobotanical research in Jordan. Pp. 601–9 in H. G. Gebel, G. O. Rollefson and Z. Kafafi (eds), *The Prehistory of Jordan II, Perspectives from 1997*. Studies in Near Eastern Production, Subsistence, and Environment 4. Berlin: *ex oriente*.

Nissen, H. J., Muheisen, M. and Gebel, H. G. (with contributions by M. T. C. Affonso, C. Becker, H. G. Gebel, A. Hauptmann, B. D. Hermansen, U. Kamp, W. Karasneh, M. Muheisen, R. Neef, H. J. Nissen, E. Pernicka and N. Qadi) (2004) *Basta I: The Human Ecology*. Biblioteca Neolithica Asiae Meridionalis et Occidentalis and Yarmouk University, Monograph of the Faculty of Archaeology and Anthropology 4. Berlin: *ex oriente*.

O'Leary, G. J. and Connor, D. J. (1997) Stubble retention and tillage in a semiarid environment: I. soil water accumulation during fallow. *Field Crops Research* 52, 209–19.

Palmer, C. (1998) 'Following the plough': the agricultural environment of Northern Jordan. *Levant* 30, 129–65.

Perrot, J. (1968) *Prehistoire Palestinienne*. Supplement au Dictionaire de la Bible 8. Paris: Letougey et Ane.

Rackham, O. (1982) The growing and transport of timber and underwood. Pp. 199–217 in S. McGrail (ed.), *Woodworking Techniques Before A.D. 1500*. British Archaeological Reports, 129. Oxford: BAR.

Robinson, S. A., Black, S., Sellwood, B. W. and Valdes, P. H. (2006) A review of palaeoclimates and palaeoenvironments in the Levant and Eastern Mediterranean from 25,000 to 5000 BP: setting the environmental background for the evolution of human civilization. *Quaternary Science Reviews* 25, 1517–41.

Robles, M. D. and Burke, I. C. (1997) Legume, grass, and conservation reserve program effects on soil organic matter recovery. *Ecological Applications* 7/2, 345–57.

Rollefson, G. O. (1983) The 1982 excavations at 'Ain Ghazal: preliminary report. *Annual of the Department of Antiquities of Jordan* 27, 1–15.

Rollefson, G. O. (1984) 'Ain Ghazal: an Early Neolithic community in Highland Jordan, near Amman. *Bulletin of the American School of Oriental Research* 255, 3–13.

Rollefson, G. O. (1997) Neolithic 'Ayn Ghazal in its landscape. Pp. 241–4 in G. Bisheh, M. Zaghloul and I. Kehrberg (eds), *Studies in the History and Archaeology of Jordan VI*. Amman: Jordan Press Foundation.

Rollefson, G. O. (1998a) Expanded radiocarbon chronology from 'Ain Ghazal. *Neo-lithics* 2/98, 8–10.

Rollefson, G. O. (1998b) The aceramic Neolithic of Jordan. Pp.

102–26 in D. O. Henry (ed.), *The Prehistoric Archaeology of Jordan*. British Archaeological Reports, International Series 705. Oxford: BAR.

Rollefson, G. O., Kafafi, Z. (with contribution by Hisahiko Wada) (1994) The 1993 Season at 'Ain Ghazal Preliminary Report. *Annual of the Department of Antiquities Jordan* 38: 11–32.

Rollefson, G. O. and Köhler-Rollefson, I. (1989) The collapse of Early Neolithic settlements in the Southern Levant. Pp. 73–89 in I. Herskovish, B. Arensburg, O. Bar-Yosef and P. Smith (eds), *People and Culture in Change: Proceedings of the Second Symposium on Upper Paleolithic, Mesolithic and Neolithic Populations of Europe and the Mediterranean Basin*. British Archaeological Reports, International Series 508(i). Oxford: BAR.

Rollefson, G. O. and Köhler-Rollefson, I. (1993) PPNC adaptations in the first half of the 6th millennium B.C. *Paléorient* 19/1, 33–42.

Rollefson, G. O., Banning, E. B., Byrd, B. F., Kafafi, Z., Köhler, I., Petocz, D., Rolston, S. and Villers, L. (1984) The Pre-Pottery Neolithic B village of 'Ain Ghazal (Jordan) –preliminary report of the 1982 excavation season. *Mitteilungen der Deutschen Orient Gesellschaft* 116, 139–83.

Rollefson, G. O., Simmons, A. H. and Kafafi, Z. (1992) Neolithic cultures at 'Ain Ghazal, Jordan. *Journal of Field Archaeology* 19, 443–70.

Rossignol-Strick, M. (1995) Sea-land correlation of pollen records in the Eastern Mediterranean for the glacial–interglacial transition: biostratigraphy versus radiometric time-scale. *Quaternary Science Reviews* 14, 893–915.

Sherratt, A. (1980) Water, soil and seasonality in early cereal cultivation. *World Archaeology* 11/3, 313–30.

Simmons, A. H. (2000) Villages on the edge, regional settlement change and the end of the Levantine Pre-Pottery Neolithic. Pp. 211–30 in I. Kuijt (ed.), *Life in Neolithic Farming Communities: Social Organization, Identity, and Differentiation*. New York: Kluwer Academic/Plenum Publishers.

Simmons, A. H. and Kafafi, Z. (1988) Preliminary report on the 'Ain Ghazal archaeological survey, 1987. *Annual of the Department of Antiquities of Jordan* 32, 27–39.

Simmons, A. H., Cooper, J., Mandel, R., Rollefson, G., Al-Nahar, M., Kafafi, Z., Köhler-Rollefson, I. and Roler-Durand, K. (2001) Wadi Shu'eib, final report of test excavations. *Bulletin of the American School of Oriental Research* 321, 2–38.

Tadmor, N. H., Eyal, E. and Benjamin, R. W. (1974) Plant and sheep production on semiarid annual grassland in Israel. *Journal of Range Management* 27/6, 427–32.

Turkowski, L. (1969) Peasant agriculture in the Judean Hills. *Palestine Exploration Quarterly* 101, 101–12.

USDA (United States Department of Agriculture) (1960/61) *Food Balances in Foreign Countries*. USDA FAS-M 108, II.

Van Beek, G. W. (1982) A population estimate for Marib: a contemporary tell village in North Yemen. *Bulletin of the American Schools of Oriental Research* 248, 61–7.

Von den Driesch, A. and Wodtke, U. (1997) The fauna of 'Ain Ghazal, a major PPN and Early PN settlement in Central Jordan. Pp. 511–56 in H. G. Gebel, G. O. Rollefson and Z. Kafafi (eds), *The Prehistory of Jordan II. Perspectives from 1997*. Studies in Early Near Eastern Production, Subsistence, and Environment 4. Berlin: *ex oriente*.

Waheeb, M. and Fino, N. (1997) 'Ayn el-Jammam: a Neolithic site near Ras en-Naqb, Southern Jordan. Pp. 215–19 in H. G. Gebel, G. O. Rollefson and Z. Kafafi (eds), *The Prehistory of Jordan II. Perspectives from 1997*. Studies in Early Near Eastern Production, Subsistence, and Environment 4. Berlin: *ex oriente*.

Wasse, A. (1997) Preliminary results of an analysis of the sheep and goat bones from 'Ain Ghazal, Jordan. Pp. 575–92 in H. G. Gebel, G. O. Rollefson, and Z. Kafafi (eds), *The Prehistory of Jordan II. Perspectives from 1997*. Studies in Early Near Eastern Production, Subsistence, and Environment 4. Berlin: *ex oriente*.

Wasse, A. (2002) Final results of an analysis of the sheep and goat bones from 'Ain Ghazal, Jordan. *Levant* 34, 59–82.

Wilkinson, T. J. (2003) Environmental context. Pp. 15–32 in T. J. Wilkinson (ed.), *Archaeological Landscapes of the Near East*. Tuscon, AZ: University of Arizona Press.

Willcox, G. (1974) A history of deforestation as indicated by charcoal analysis of four sites in Eastern Anatolia. *Anatolian Studies* 24, 117–33.

Willcox, G. (1992) Timber and trees: ancient exploitation in the Middle East. Evidence from plant remains. *Bulletin of Sumerian Agriculture* 6, 1–31.

Willcox, G. (1999) Agrarian change and the beginnings of cultivation in the Near East: evidence from wild progenitors, experimental cultivation and archaeobotanical data. Pp. 478–500 in C. Gosden and J. Hather (eds), *The Prehistory of Food, Appetites for Change*. One World Archaeology 32. London: Routledge.

Zohary, M. (1973) *Geobotanical Foundations of the Middle East 1–2*. Geobotanica Selecta 3. Stuttgart: Gustav Fischer.

Zohary, M. and Hopf, M. (2000) Domestication of Plants in the Old World: The Origin and Spread of Cultivated Plants in West Asia, Europe, and the Nile Valley. Oxford: Clarendon.

17. Taskscapes and the Transition

Chantal Conneller

This paper provides a critical review of Ingold's (1993) concept of the taskscape and its use within Mesolithic studies. I will argue that a focus on the taskscape has helped to overcome some of the traditional problems that Mesolithic archaeologists have had in conceptualizing landscape. However, the approach has also generated certain problems which are particularly relevant for thinking about the Mesolithic–Neolithic transition. In particular, a highly romanticized, apolitical vision of the Mesolithic landscape contrasts with the political, monumental landscapes of the Neolithic, perpetuating current stereotypes. I argue that we need to think about the divergent and changing ways particular Mesolithic landscapes were inhabited, both before and across the transition.

Introduction

Ingold's concept of the 'taskscape' has been influential in Mesolithic studies since its introduction in 1993 (Conneller 2000; Finlay 2004; Hind 2004; Warren 2000). This paper will explore the place of the 'taskscape' in British Mesolithic studies and its implications for thinking about the Mesolithic–Neolithic transition. It has provided a way to move beyond some of the problems Mesolithic archaeology has traditionally had with spatial scales and the nature of its evidence – in particular the relationship between lithic scatters and landscapes. Lithic scatters have been equated with bounded 'living areas' (Clark 1954) or seen as marking economically productive areas of the landscape (Mithen 2000; Barton *et al.* 1995), or even as referencing particular natural features (Tilley 1994). Ingold's idea of the landscape as the 'congealed taskscape' has also been attractive in that it sees Mesolithic people as actively producing the landscape, rather than simply responding to a set of environmental conditions (cf. Rowley-Conwy 1984; 2004). This lessens the previous contrasts between the 'natural' landscapes of the Mesolithic and the 'cultural' landscapes of the Neolithic.

The taskscape has seen less use in the Neolithic with a few notable exceptions, such as Edmonds' work on the transition (Edmonds 1997). In this paper I would like to describe the ways in which Ingold's work has enhanced our understanding of Mesolithic landscapes and explore the different ways it has been used by Mesolithic and Neolithic archaeologists. However, I would also like to highlight some aspects that I find problematic, and in particular explore the implications of the various ways it has been used for thinking about the Mesolithic–Neolithic transition.

The taskscape

The taskscape has been an extremely productive concept for thinking about the Mesolithic. The concept taskscape – particularly employed in conjunction with a knowledge of the *châine opératoire* – has been important in understanding the artefactual material usually encountered on Mesolithic sites (i.e. stone tools and their manufacturing debris) as the product of dynamic activities that produce space, rather than simply as demarcating a 'site'. This latter view has been prevalent in Mesolithic archaeology from Clark's definition of the living area of Star Carr by the number of flints per square yard and subsequent estimations of the number of people that could have lived in this space. That this is an inappropriate way of understanding archaeological material has previously been noted (Dunnell 1992); however, newer 'landscape' approaches have tended to follow suit in seeing lithic scatters as marking a particular (usually homogenized) type of space. So Barton *et al.*'s (1995) 'persistent places' mark good hunting grounds, and the sites discovered by the Southern Hebrides Mesolithic Project on Colonsay and Islay were located with reference to good views, good hunting grounds or raw material sources (Mithen 2000). The same criticism has also been made by McFadyen (2006, 125) of Tilley's work. She points out that, in contrast to his approach to Neolithic monuments, he treats Mesolithic sites as findspots, thus reducing them to something that marks particular landscape features. Despite the phenomenological basis of both Tilley's and Ingold's approaches, it is only the concept of the taskscape that permits a thorough engagement with the specificity of the Mesolithic artefactual evidence.

Rather than seeing sites as generated in response to particular resources or landscape features, Ingold sees activity episodes as producing the landscape. A focus on the taskscape forces a qualitative assessment of a place through a focus on the tasks undertaken at a particular locale and their relationship to other tasks in the broader landscape. This produces a busy, peopled space, rather than the relatively abstract or homogenized spaces of previous work. Ingold's idea of places as nodes in broader networks of activity means that sites are not demarcated by the presence or absence of finds, rather activity and movement extend into the landscape. The understanding of the landscape as the congealed taskscape has also been important in forcing an understanding of how hunter-gatherers produce the landscape through assembling a diverse range of things from different parts of the landscape in the pursuit of their tasks. In Ingold's account, meanings of a place are drawn up from the events and activities that took place there; the Mesolithic landscape thus becomes one of meaning, rather than simply function, and the landscape is therefore vital in considering the production of the identities of those who inhabited it.

The taskscape in action

The usefulness of the concept over previous approaches to understanding the occupation of a particular landscape can be better illustrated through a case study. The example comes from the Vale of Pickering, North Yorkshire (Figure 17.1). Here a series of Mesolithic sites dating between c. 9200 and 8400 BC clustered around an ancient lake, Lake Flixton. Many interpretations of how hunter-gatherers occupied this and adjacent areas have been offered as part of attempts to elucidate the part that Star Carr (located on the western shore of Lake Flixton) played in the seasonal round of its inhabitants.

Clark's original interpretation of Star Carr's place in the landscape saw it as a lowland base camp occupied in the winter months by people who followed red deer, their major prey, into adjacent uplands areas, generating microlith-dominated hunting camps in the process (Clark 1972). Subsequent studies followed suit, positing a limited number of site types occupied at specific seasons of the year. So Rowley-Conwy (1994) suggested that Star Carr was a spring hunting camp which was visited each year along with winter base camps in the Vale of Pickering, summer base camps on the coast and summer hunting camps in the uplands of the North York Moors. In a similar vein, Simmons (1996) suggested winter base camps on the coast and spring and summer hunting camps in the Vale of Pickering and North York Moors respectively. Recently Donahue and Lovis (2006) have suggested autumn/spring residential camps and logistical camps in the Vale of Pickering, autumn/winter upland hunting/kill sites and spring/summer coastal residential/logistical camps.

These understandings of hunter-gatherer use of the landscape are problematic at a number of levels (see Spikins 1999; Conneller 2005; or Wickham-Jones 2005 for a more detailed discussion of these issues). First the identification of site type emphasizes a single – usually economic – function for a site chosen from an extremely small list of possibilities (base camp/hunting camp/kill site), which tends to be based on an uncritical, functionalist reading of lithic typology. This is particularly problematic for large, complex palimpsests such as Star Carr. The second problem is the uncritical linking of a small number of sites, whose dating spans several hundred years or even millennia, into a single atemporal, unchanging seasonal round. Jochim, in a sophisticated critique of such approaches, uses ethnographic data to suggest that a normative seasonal round does not exist: though people routinely make varying decisions about group, sub-group and family mobility, these are not stable and thus cannot be discerned in the archaeological record (Jochim 1991).

A focus on the taskscape permits archaeologists to explore the occupation of the landscape in both a realistic and a meaningful way. Examination of even a single site – here Seamer C on the northern shore of Lake Flixton (Figure 17.2) – allows a focus on relationships in the broader landscape. At Seamer C the following relationships were produced through the undertaking of routine tasks (Conneller and Schadla-Hall 2003): People came to the edge of the lake, which was surrounded by an open landscape of grasses, tall herbs and shrubs – sorrel, meadowsweet, crowberry, dwarf birch. They carried with them an already partially worked core of high-quality brown till flint that derived from a source on the now-submerged North Sea Plain and also a number of large blades and a burin deriving from a variety of different flint nodules. This core and these blades would recall the place on the North Sea Plain where the raw material was procured, the place or places where the core was previously worked and where the blades and burin were produced and also the different individuals involved in these tasks. Here the core was further reduced and a number of fine

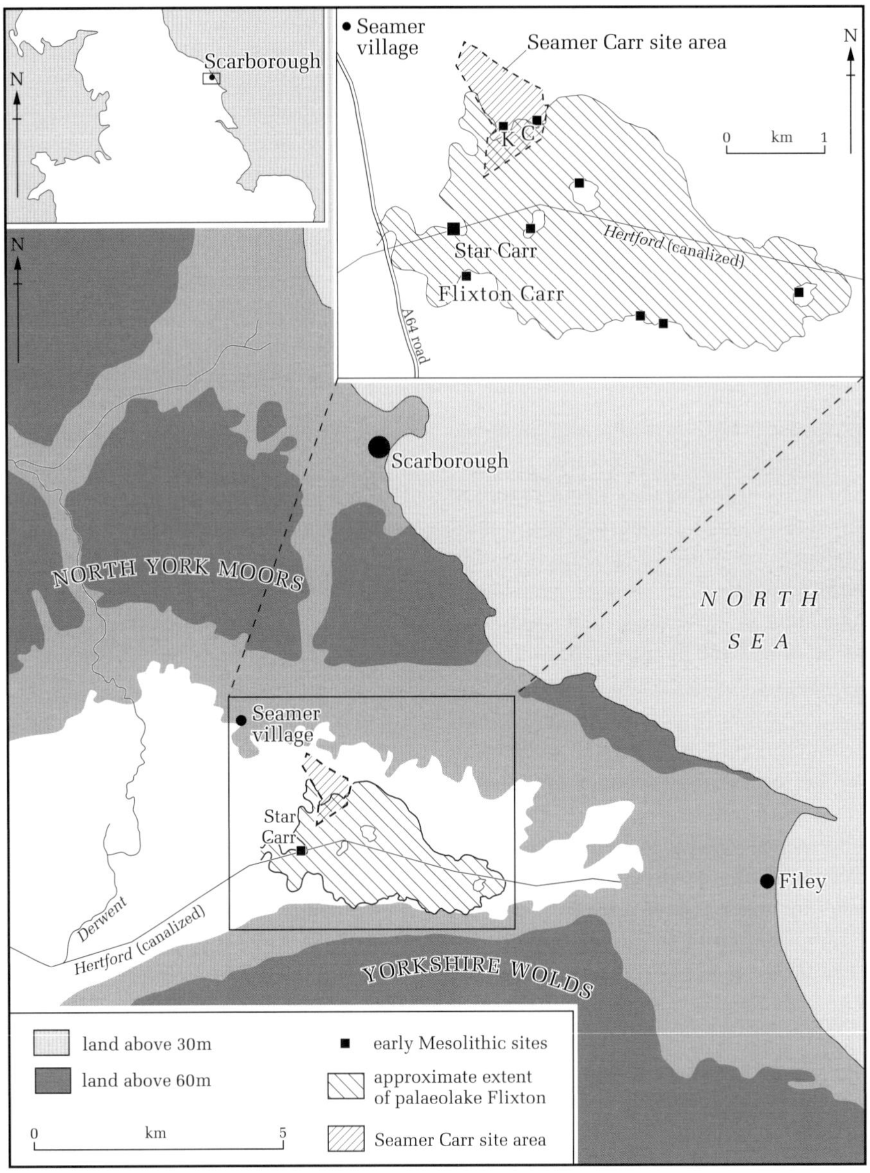

Fig. 17.1. Location of the Vale of Pickering. Many thanks to Archaeology International.

blades were produced. Many of these blades, and also the imported blades, were used; many were broken. Some blades, however, were collected up and taken on to be used elsewhere as the need arose.

On another visit, people brought to the edge of the lake, from a source on the North Sea Plain, five cores (three large and pre-formed, the other two unmodified tabular blocks) and several burins. Two horses were butchered here and the flint cores were worked. The three pre-formed cores were extensively reduced until they were no longer deemed workable. A vast number of blades were produced, some of which were used. Blades were also removed from the tabular flint blocks and a number of burins were made and used on the site, while one of the imported burins was used and resharpened seven times. But these two tabular blocks were still large and usable when work was finished and so were removed when their knappers left, as were a large number of blades that had been manufactured from the three pre-formed cores, though the cores themselves were abandoned.

Several hundred years later people returned, coming from the beach in the summer, through a landscape of birch and pine woodland, to the lake edge fringed with dense reedswamp. They brought with them a lot of small and medium-sized beach pebbles. Here they undertook a variety of activities, involving the processing of the remains of red deer, roe deer and elk and the manufacturing of a number of different tools. However, most of the activities were focused on hide processing; 141 scrapers were manufactured from the small pebbles and used before people moved on.

Other sites in the Vale can, similarly, be regarded from a taskscape perspective as nodes in networks of movement across the landscape. Ready-made blades, tools, cores and

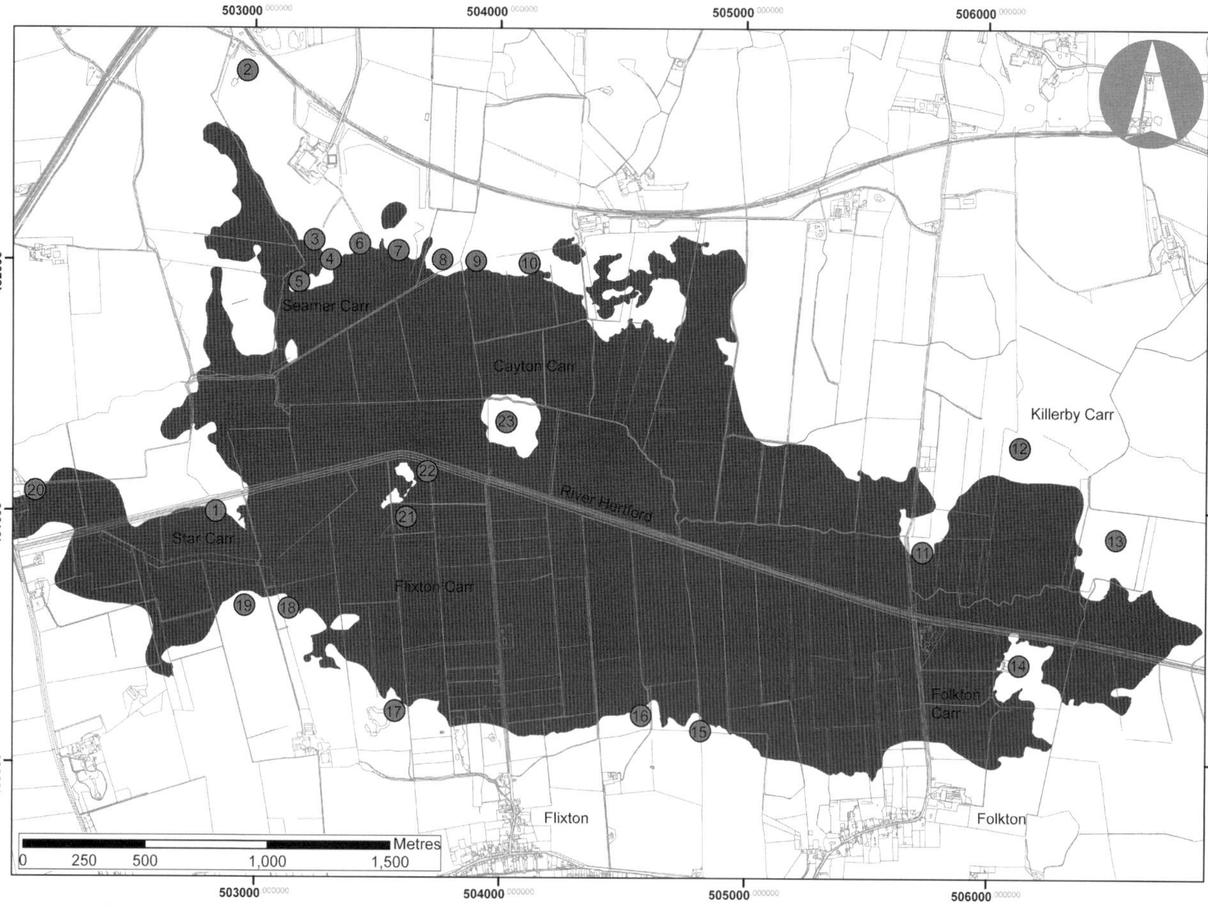

Fig. 17.2. Archaeological sites around lake Flixton. Seamer C is site 7. Many thanks to Barry Taylor.

unworked nodules were carried to other places in the Vale, such as sites Seamer L, VPD, No Name Hill and Seamer K. Here they were worked, and newly made cores, blades and tools were carried away. For example, at Barry's Island, people, or perhaps a single person, carried 16 nodules to the reedbeds on the lake margins. Here they carefully removed the cortex of these nodules and prepared the cores. When the cores were prepared ready for working, they were collected up and taken away to be used as the need arose. At Scatter 30 of Seamer K people brought with them a partially reduced core which they worked to produce flakes and blades to make microliths. In total 11 microliths were manufactured, as demonstrated by the deposition of 11 microburins. The microliths were carried away to be used elsewhere. We see also different rhythms and intensities of working. In contrast to this intensive production of 11 microliths from a single core, at VPD a complete refit sequence contains only one microlith. At a number of locations in the Vale of Pickering, particularly those at the water's edge, in the reedbeds, people repeatedly came with tools and blades which had been manufactured at knapping stations. Here they undertook tasks which tended to involve cutting, perhaps the cutting of reeds or the processing of meat or plant resources. When they were finished they abandoned the tools and moved on.

Other objects established relationships between people and places through their continued presence at particular places. The knowledge of objects left at a place sustained it in memory. At Seamer K a dump of nodules was left at the site; this consisted of 8 completely exhausted cores, another 8 cores that were abandoned because of flaws in the raw material and 6 unmodified or minimally tested nodules, which appear unknappable. None of these cores appears to have been worked on the site and they seem to have been brought here especially to be abandoned. Similarly, at the site of Flixton school, 12 extremely large nodules were deposited in a tight cluster – these must have been especially selected for their size. Most of them are of very poor-quality material, possibly too poor for knapping.

A focus on taskscapes – importantly in conjunction with a study of the *châine opératoire* – helps us to see how the landscape of the Vale of Pickering was produced by Mesolithic people through networks of connections between people, places and things. Even a single act of microlith manufacture would create relationships between the people and places involved in the procurement of raw material, in the manufacture of containers to carry the raw material, and in the hunting or gathering activities that produced the antler soft hammer to work the raw material. But a single task cannot be considered in isolation; it is the interlocking of a whole range of tasks that produces the landscape in its entirety, relations between people and the intricate temporalities of social life. The Vale of Pickering would be produced as a familiar landscape, then, through the movement of these tasks, and as a meaningful landscape through the activities and events that produced it.

Problems with the taskscape

It is clear that a focus on the taskscape provides a more nuanced view of the varied and complex ways that people occupied the Vale of Pickering during the Early Mesolithic. However, such an approach is not without its problems and I highlight the limitations of this approach in the remainder of this paper. In particular two weaknesses of the taskscape – the romanticization of the landscape and the focus on the routine – have implications for our understanding of the Mesolithic–Neolithic transition.

Romanticized nature

The vision of the landscape in Ingold's taskscape is a highly romantic one. This is evident in his analysis of Breugel's painting *The Harvesters*, where the politics of the depiction of peasants in a bucolic idyll commissioned by a wealthy Antwerp merchant is dismissed in a couple of sentences (Ingold 1993, 165). What the painting shows is a landscape that was owned by particular people and to which access was controlled. The church in Ingold's account is a place where 'people's ancestors are buried in the graveyard … The church, too, resonates to the cycles of human life and subsistence' (Ingold 1993, 169). The fact that this would also have been a highly politicized space, where social distinctions were played out and reinforced both in life and death, is ignored. Access to parts of the church – and sacred objects – would have been highly restricted, while the social hierarchy would also have been evident through the prominent commemoration of certain individuals.

The romantic vision of the landscape that the taskscape can engender becomes a problem because of the different research histories of Mesolithic and Neolithic archaeology. With a few notable exceptions (e.g. Edmonds 1997), the taskscape has been employed in Mesolithic rather than Neolithic archaeology (Conneller 2000; Warren 2000). This produces Mesolithic landscapes as apolitical and free from inequality or trauma, while the landscapes of the Neolithic, with those monuments that guide bodies in processions and restrict access to certain people (e.g. Tilley 1984), are constructed as inherently political. As Mesolithic archaeologists, we should be aware of the political aspects of tasks and technologies, what they enable people to do, and in what ways certain practices are restricted to certain sorts of people. As technologies of the body, tasks are things that need to be done in particular ways (Mauss 1979) and techniques can also be coercive (Foucault 1977). These kinds of practices can create landscapes where spaces are restricted to particular individuals and groups.

The romanticized view of the landscape feeds back into longstanding ideas of hunter-gatherers as free from the iniquities of western civilization, as 'at one with nature' (Rousseau 1964), ideas that have been reiterated as a result of more recent technological and ecological concerns (e.g. Shepard 1998). Various ethnographic accounts indicate that hunter-gatherers do in fact inhabit highly politicized spaces and more general arguments against these romantic ideas have been made, in particular by Sandall (2001). Furthermore, as Headland and Reid (1989, 50) point out, such stereotypes have played their part in the recent and historical repression of hunter-gatherers.

Temporality and change

One of the most positive aspects of the taskscape is that this approach enables a detailed consideration of temporality. To a certain extent this is not new: the historic emphasis on the seasonal round in Mesolithic archaeology has generated some relatively sophisticated attempts to understand seasonal rhythms (e.g. Simmons 1996). However, thinking about tasks forces us to think about interconnecting rhythms at a range of different temporal scales and, indeed, there have been some extremely productive attempts to discuss the temporalities of social life (see Warren 2006; Edmonds 1997).

However, the focus on tasks can also be problematic in terms of the types of temporality it generates. In particular, the emphasis on quotidian tasks – effectively flintknapping, hunting animals and processing their meat, or collecting shellfish or plant foods – produces a British Mesolithic that seems to alter little throughout the period, with the possible exception of the change between Early and Late Mesolithic around 9000–8700 BP. This transition (in England and Wales at least) manifests itself technologically through a change from broad obliquely blunted points, isosceles triangles and trapezes to a range of smaller geometric microliths, which itself has been equated to a change in hunting strategies as a result of changing ecological conditions (Myers 1989).

The atemporality of the Mesolithic is a problem that exists independently of work on Mesolithic taskscapes. It is a problem that has been produced by the emphasis on the Mesolithic as characterized by particular economic adaptations (Early Mesolithic terrestrial hunting versus Late Mesolithic marine resources or Early Mesolithic intercept versus Late Mesolithic encounter hunting) to particular sets of environmental circumstances (Early Mesolithic open forests versus Late Mesolithic closed canopy forests). It has been exacerbated by the palaeoeconomists' emphasis on an unchanging seasonal round and by the reliance on Star Carr as 'standing for' the Early Mesolithic and the Oronsay middens for the Late Mesolithic. Newer approaches to the British Mesolithic have attempted to move beyond economic considerations and provide a more nuanced understanding of the period; however, the focus on daily tasks – particularly those that are archaeologically visible – panders to previous stereotypes. Tasks such as flintworking and hunting and gathering are simply too generic in the way that they are currently understood to provide anything other than a generalized stereotype of the period.

The trouble with viewing the Mesolithic as homogenous is that, when we come to the Neolithic, there is the sudden transformation of a 'stable way of life'. The point that a homogenized Mesolithic and Neolithic makes the transition difficult to examine has been made before (e.g. Tilley 1994; Edmonds 1997). Edmonds, in particular, attempts to circumnavigate these problems through a discussion of taskscapes across the transition. This is an excellent discussion which highlights some continuity across the two periods, but also how the introduction of new animals and material culture transformed certain routine tasks and produced different landscapes. However, just as an examination of change between Mesolithic and Neolithic creates the Mesolithic as an entity, so the emphasis on continuity produces an unchanging Late Mesolithic. Thus Edmonds suggests: 'People were exploiting a wide variety of resources and modifying the land through limited woodland clearance during the fifth millennium, *much as they had done for generations*' (Edmonds 1997, 100, my emphasis).

Continuity is argued on the basis of routine tasks and re-occupation of particular places. However, as argued above, generic descriptions of tasks (lacking an understanding of their political, social and ideological significance) produce sameness. Re-occupation of similar parts of the landscape is also problematic as an indicator of continuity. Lesley McFadyen's work on long barrows and long cairns – where Mesolithic material is often also discovered – makes it clear that coincidence of place cannot be taken as evidence of continuity (McFadyen 2003). At sites such as Hazleton North, for example, there are millennia between the two occupations (Saville 1990). It seems that what is significant on all these sites is the Neolithic people's encounter with Mesolithic material, rather than continuity in practices. The Mesolithic material was recognized by people as something significant (though whether taken as evidence of past occupation of ancestors or some contemporary 'other' is not clear) and on this basis often incorporated into the fabric of monuments themselves.

The problem with the taskscape for thinking about the transition (and change more generally) is that it produces an archaeology which is at once too general (generic tasks) and too specific (particular sites). If we take the middle ground and look at regional patterns we can see that there were a variety of Mesolithic ways of life at the end of the 5th millennium BC, some of which seem to have started and stopped abruptly around the time of the transition, while others seem to have started just before it and then continued. In the Pennines, for example, we have a way of life dated to around 4200–3700 BC that involved small ephemeral camps around stone-built hearths in very high upland areas, where people manufactured rod microliths (Spikins 1999; 2002; Chatterton 2006). This way of life appears to have been short-lived and vanished fairly abruptly. In western Scotland a new understanding of death arose involving the patterned arrangement of human bones in shell middens, which seems to date at Cnoc Coig to 4200–3800 BC (Richards and Mellars 1998; Meikeljohn *et al.* 2005), but thereafter continues at sites where the human bone produces 'Neolithic' dates, such as Raschoille (Wickham-Jones 1994; Ashmore 2004).

The point is not so much that there were continuities and disjunction over the transition; rather that the Mesolithic was not one homogenized way of being and that there were both continuities and radical change over short temporal scales within the Mesolithic. It is important for Mesolithic archaeologists to stress this, otherwise the Neolithic becomes another instance of the 'surprise of origin' (Wagner 2001): the reification of our analytical categories 'Mesolithic' and 'Neolithic', then our 'discovery' of change. It is vital to recognize a variety of broad and small-scale transformations throughout the Mesolithic. Some of these were across a broad geographical scale. There are nuanced changes in the shape of microliths, for example: the change from Star Carr to Deepcar to Horsham types in the early Mesolithic (Reynier 1998), the appearance of microscalene triangles around 6000 BC (Barton and Roberts 2004) and the seeming reliance on rod microliths at the very end of the Mesolithic (Jacobi *et al.* 1976; Spikins 2002; Chatterton 2006). Significant changes can also be seen in burial practices, with the longstanding tradition in western Britain of depositing human bone in caves ceasing around 6000 BC (Chamberlain 1996; Conneller 2006).

At a local scale we can also see significant changes. Returning to the Vale of Pickering, there are major discontinuities in the evidence: caches were forgotten; relations between people and things were severed; familiar objects abandoned. Places in the Vale were re-occupied, but rarely for a similar purpose. Seamer D, for example, was successively a place where microliths were made and later where a cache was established (Conneller and Schadla-Hall 2003). Star Carr, where our understanding of the temporality of occupation is more sophisticated, shows both continuity and disjunction. Petra Dark's work on the macrocharcoal reveals sporadic burning in the area of the site for a period of 80 years, then a hiatus during which visits were rare for 100 years, then a second phase of intermittent burning lasting 130 years (Mellars and Dark 1998). It has previously been noted that activities based on the mass deposition of rare and unusual artefacts, such as barbed points, antler frontlets, beads, axes and animal bone, took place at Star Carr (Chatterton 2003; Conneller and Schadla-Hall 2003; Conneller 2004), and Chatterton (2003) has suggested that Star Carr was a ritual site. Radiocarbon dating of the artefacts from the site is unfortunately rather limited, since the material recovered by Clark was impregnated with organic preservative. However, recent dating links antler artefacts such as barbed points and antler working debris with both of these phases of occupation (Dark *et al.* 2006). This long-lasting deposition is unusual in the context of the Vale of Pickering and also raises the question of why the site appears to have been more or less abandoned for a century in between these two phases. In addition, the recommencement of antler deposition may have had a very different meaning in the second phase, given the period of abandonment and the long history of the site. Finally, these practices ceased at the site once more, demonstrating a complex landscape history of continuity, disjunction and reinterpretation.

It is important that we do not forget the dynamism of the Mesolithic occupation of the landscape. The evidence for the Vale of Pickering demonstrates the complexity of the Mesolithic occupation, with contingent decisions and events which led people to return to places or to abandon them for years or even generations. We are not seeing a timeless, unyielding seasonal round in the Vale of Pickering. It is only Star Carr that demonstrates any fidelity of purpose, but even here we see a major disjunction and a potential shift in the understanding of the place. The change in mortuary treatment at c. 6000 BC described above also suggests a radical reconceptualization of landscape in which long-standing traditions of placing disarticulated human remains in caves abandoned. Martin Bell's work on the Late Mesolithic of the Severn Estuary also demonstrates this dynamism – but in this case, that Mesolithic people's understanding of the landscape itself was as inherently dynamic (Bell 2007). Here people seem to have sought out dynamic landscapes and activity is often particularly concentrated at times of dramatic change, arguably because these landscapes offered opportunities and unimagined possibilities for living (McFadyen 2006). So, here at least, change was actively part of society and people's understandings of the landscape.

Conclusions

I have argued in this paper that the concept of the taskscape offers a productive way to think about landscapes. However, it is not without its problems: in particular, the focus on routine tasks produces a sameness in our accounts and also makes it difficult to think about the complex ways in which understandings were generated and inequalities incorporated into these landscapes. It is particularly problematic when thinking about the transition, in part owing to the very different research traditions of the Mesolithic and Neolithic periods. Instead, we need to bring out the divergent and changing ways in which Mesolithic landscapes are occupied; only then will we begin to have an idea of the extent to which change at the Mesolithic–Neolithic transition differs in frequency and intensity from that occurring throughout the Mesolithic.

Bibliography

Ashmore, P. (2004) A date list (to October 2002) for early foragers in Scotland. Pp. 95–157 in A. Saville (ed.), *Mesolithic Scotland and its Neighbours: The Early Holocene Prehistory of Scotland, its British and Irish Context and some Northern European Perspectives*. Edinburgh: Society of Antiquaries of Scotland.

Barton, R. N. E. and Roberts, A. J. (2004) The Mesolithic period in England: an overview of the current position. Pp. 339–58 in A. Saville (ed.), *Mesolithic Scotland and its Neighbours: The Early Holocene Prehistory of Scotland, its British and Irish Context and some Northern European Perspectives*. Edinburgh: Society of Antiquaries of Scotland.

Barton, R. N. E., Berridge, P. J., Walker, M. J. C. and Bevins, R. E. (1995) Persistent places in the Mesolithic landscape: an example from the Black Mountain uplands of South Wales. *Proceedings of the Prehistoric Society* 61, 81–116.

Bell, M. (2007) *Prehistoric Coastal Communities: The Mesolithic of Western Britain*. Council for British Archaeology Research Report 149. York: CBA.

Chamberlain, A. T. (1996) More dating evidence for human remains in British caves. *Antiquity* 70, 950–3.

Chatterton, R. (2003) Star Carr reanalysed. Pp. 69–80 in L. Bevan and J. Moore (eds), *Peopling the Mesolithic in a Northern Environment*. British Archaeological Reports, International series 1157. Oxford: BAR.

Chatterton, R. (2006) Transition and Persistence: Material Culture in the Mesolithic Landscape of North Yorkshire. Unpublished PhD thesis, University of Manchester.

Clark, J. G. D. (1954) *Excavations at Star Carr*. Cambridge: Cambridge University Press.

Clark, J. G. D. (1972) *Star Carr: A Case Study in Bioarchaeology*. Reading, MA: Addison-Wesley Modular Publications, Module 10.

Conneller, C. J. (2000) Fragmented space? Hunter-gatherer landscapes of the Vale of Pickering. *Archaeological Review from Cambridge* 17/1, 139–50.

Conneller, C. J. (2004) Becoming deer: corporeal transformations at Star Carr. *Archaeological Dialogues* 11/1, 37–56.

Conneller, C. J. (2005) Moving beyond sites: Mesolithic technology in the landscape. Pp. 42–55 in N. J. Milner and P. C. Woodman (eds), *Mesolithic Studies at the Beginning of the 21st Century*. Oxford: Oxbow Books.

Conneller, C. J. (2006) Death. Pp. 139–64 in C. Conneller and G. Warren (eds), *Mesolithic Britain and Ireland: New Approaches*. Stroud: Tempus.

Conneller, C. J. and Schadla-Hall, T. (2003) Beyond Star Carr: the Vale of Pickering in the 10th millennium BP. *Proceedings of the Prehistoric Society* 69, 85–106.

Dark, P., Higham, T., Jacobi, R. and Lord, T. (2006) New radiocarbon accelerator dates on artefacts from the Early Mesolithic site of Star Carr, North Yorkshire. *Archaeometry* 48/1, 185–200.

Donahue, R. E. and Lovis, W. (2006). Regional settlement systems in Mesolithic Northern England: scalar issues in mobility and territoriality. *Journal of Anthropological Archaeology* 25/2, 248–58.

Dunnell, R. (1992). The notion 'site'. Pp. 21–42 in J. Rossignol and L. Wandsnider (eds), *Space, Time and Archaeological Landscapes*. New York: Plenum Press.

Edmonds, M. R. (1997) Taskscape, technology and tradition. *Analecta Praehistorica Leidensia* 29, 99–110.

Finlay, N. (2004) E-scapes and E-motion: other ways of writing the Mesolithic. *Before Farming* 2004/1, article 4.

Foucault, M. (1977) *Discipline and Punish: The Birth of the Prison*. New York: Pantheon Books.

Headland, T., and Reid, L. (1989) Hunter-gatherers and their neighbours from prehistory to the present. *Current Anthropology* 30, 43–66.

Hind, D. (2004) Where many paths meet: towards an integrated theory of landscape and technology. Pp. 35–51 in A. M. Chadwick (ed.), *Stories from the Landscape: Archaeologies of Inhabitation*. British Archaeological Reports, International Series S1238. Oxford: BAR.

Ingold, T. (1993) The temporality of the landscape. *World Archaeology* 25, 152–74.

Jacobi, R. M., Tallis, J. H. and Mellars, P. A. (1976) The southern Pennine Mesolithic and the ecological record. *Journal of Archaeological Science* 3, 307–20.

Jochim, M. A. (1991) Archaeology as long-term ethnography. *American Anthropologist* 93, 309–19.

McFadyen, L. (2003) A Revision of the Materiality of Architecture: The Significance of Neolithic Long Mound and Chambered Monument Building Practice, with Particular Reference to the Cotswold–Severn Group. Unpublished PhD thesis, University of Wales, Newport.

McFadyen, L. (2006) Landscape. Pp. 121–38 in C. Conneller and G. Warren (eds), *Mesolithic Britain and Ireland*. Stroud: Tempus.

Mauss, M. (1979) Body techniques. Pp. 95–123 in M. Mauss, *Sociology and Psychology: Essays*. London: Routledge and Kegan Paul.

Meiklejohn, C., Merrett, D. C., Nolan, R. W., Richards, M. P. and Mellars, P. A. (2005) Spatial relationships, dating and taphonomy of the human bone from the Mesolithic site of Cnoc Coig, Oronsay, Argyll, Scotland. *Proceedings of the Prehistoric Society* 71, 85–106.

Mellars, P. A. and Dark, P. (1998) *Star Carr in Context*. Cambridge: McDonald Institute for Archaeological Research.

Mithen, S. J. (2000) *Hunter-gatherer Landscape Archaeology: The Southern Hebrides Mesolithic Project 1988–1998*. Cambridge: McDonald Institute for Archaeological Research.

Myers, A. M. (1989) Reliable and maintainable technological strategies. Pp. 78–91 in R. Torrence (ed.), *Time, Energy and Stone Tools*. Cambridge: Cambridge University Press.

Reynier, M. J. (1998) Early Mesolithic settlement in England and Wales: some preliminary observations. Pp. 174–84 in N. Ashton, F. Healy and P. Pettitt (eds), *Stone Age Archaeology: Essays in Honour of John Wymer*. Lithic Studies Society Occasional Papers 6. Oxbow Monograph 102. Oxford: Oxbow Books.

Richards, M. P. and Mellars, P. A. (1998) Stable isotopes and the seasonality of the Oronsay middens. *Antiquity* 72, 178–84.

Rousseau, J.-J. (1964) *The First and Second Discourses*. New York: Saint Martins.

Rowley-Conwy, P. (1984) The laziness of the short-distance hunter: origins of agriculture in Western Denmark. *Journal of Anthropological Archaeology* 3, 300–24.

Rowley-Conwy, P. (1994) Mesolithic settlement patterns: New zooarchaeological evidence from the Vale of Pickering, Yorkshire. *Archaeological Reports (University of Durham and Newcastle-upon-Tyne)* 1994, 1–7.

Rowley-Conwy, P. (2004) How the West was lost: a re-appraisal of agricultural origins in Britain, Ireland and southern Scandinavia. *Current Anthropology* 45 (Supplement), 83–111.

Sandall, R. (2001) *The Culture Cult: Designer Tribalism and Other Essays*. Boulder, CO: Westview Press.

Saville, A. (1990) *Hazleton North, Gloucestershire, 1979–82: The Excavation of a Neolithic Long Cairn of the Cotswold–Severn Group*. English Heritage Archaeological Report no. 13. London: HBMC(E).

Shepard, P. (1998) *The Tender Carnivore and the Sacred Game*. Athens: University of Georgia Press.

Simmons, I. (1996) *The Environmental Impact of Later Mesolithic Cultures*. Edinburgh: Edinburgh University Press.

Spikins, P. A. (1999) *Mesolithic Northern England. Environment, Population and Settlement*. British Archaeological Reports, British Series 283. Oxford: BAR.

Spikins, P. A. (2002) *Prehistoric People of the Pennines: Reconstructing the Lifestyles of Mesolithic Hunter-gatherers on Marsden Moor*. Leeds: West Yorkshire Archaeology Service.

Tilley, C. Y. (1984) Ideology and the legitimation of power in the Middle Neolithic of southern Sweden. Pp. 111–46 in D. Miller and C. Tilley (eds), *Ideology, Power and Prehistory*. Cambridge: Cambridge University Press.

Tilley, C. Y. (1994) *A Phenomenology of Landscape*. Oxford: Berg.

Wagner, R. (2001) *The Anthropology of the Subject*. Berkeley, CA: University of California Press.

Warren, G. M. (2000) Seascapes: peoples, boats and inhabiting the later Mesolithic in western Scotland. Pp. 97–104 in R. Young (ed.), *Mesolithic Lifeways: Current Research from Britain and Ireland*. Leicester Archaeology Monographs 7. Leicester: University of Leicester, School of Archaeological Studies.

Warren, G. M. (2006) Technology. Pp. 13–34 in C. Conneller and G. Warren (eds), *Mesolithic Britain and Ireland: New Approaches*. Stroud: Tempus.

Wickham-Jones, C. (1994) *Scotland's First Settlers*. London: Batsford.

Wickham-Jones, C. (2005) Summer walkers. Pp. 30–41 in N. J. Milner and P. C. Woodman (eds), *Mesolithic Studies at the Beginning of the 21st Century*. Oxford: Oxbow Books.

18. From Mega-Sites to Farmsteads: Community Size, Ideology and the Nature of Early Farming Landscapes in Western Asia and Europe

Amy Bogaard and Valasia Isaakidou

This paper considers the social and ecological nature of early farming landscapes in Western Asia and Europe from the perspective of community (settlement) size. The dispersal/agglomeration of households has important implications for the intensity of farming, on the one hand, and of social interaction, on the other. 'Mega-sites' emerged in the late PPNB while the package of Neolithic crops and livestock was in the process of forming. These sites posed significant challenges given available farming techniques and represent the culmination of a trend towards large-scale aggregation rooted in local complex hunter-gatherer traditions. Neolithic communities in Greece and Central Europe were shaped by a fully formed package of crops, livestock and intensive mixed farming practices. Nevertheless, variation through time and between regions reflects continuing negotiation between local traditions and outside influences, the farming household and the wider community.

Introduction

Early farming landscapes, as taskscapes (Ingold 1993), harbour fundamental insights into the nature and significance of the agricultural transition. Near Eastern 'mega-sites', Greek 'villages' and Central European 'farmsteads' present obvious contrasts in settlement form, despite common agricultural ingredients. In this paper we bring together bioarchaeological and settlement data, interpreted in the light of ethnoarchaeological analogy, to infer the nature of early farming landscapes in different regions and to consider their social and ideological implications.

The 'farming package' and potential for intensive mixed farming

The combined management of crops and ovicaprids had crystallized by the late Early PPNB in the upper Tigris–Euphrates basin (Nesbitt 2002; Peters *et al.* 2005) (Table 18.1). The process of experimentation with animals is becoming clearer with more systematic speciation of ovicaprid remains (Wasse 1997; 2002) and early finds from Cyprus (Cypro-EPPNB/MPPNB), where all four of the final main farmyard animals (sheep, goats, cattle and pigs) are present in an insular context – albeit accompanied by fallow deer (Peltenberg *et al.* 2001). It appears that a distinct 'package' of plants, animals and practices was in the process of developing (Bogaard 2005).

The *full* package (including cattle and pigs) had emerged by the end of the PPN in Western Asia and established itself in Crete and mainland Greece as early as 7000 cal BC (Table 18.1). From the end of the 7th millennium BC, farming settlements appeared in the Balkans and Western Mediterranean, and thereafter across the European interior. Remarkably, the remains of domestic crops and livestock over this vast region and time period are routinely found together.

It has been argued elsewhere that the combined emergence of cultivation and herding in Western Asia (Bogaard 2005; cf. Peters *et al.* 2005), and the subsequent spread of this complex across Europe (Halstead 1989a; Bogaard 2004b; 2005) do not merely reflect their dietary complementarity, since hunted game can provide cultivators with additional protein and fat. Rather, it is possible to infer a deeper *functional* interdependence in the system of production – an 'intensive' form of mixed

Years cal BC (approx.)	Levant	Central Anatolia	Crete (Knossos)	Greek Mainland	Central Europe LBK
4500					
			Late Neolithic		
5000	Chalcolithic			Late Neolithic	Jüngere
5500			Middle Neolithic	Middle Neolithic	Älteste
		ECA IV (Çatalhöyük West)			
6000	Pottery Neolithic				
		ECA IIIB (Çatalhöyük East Levels VIA-0)	Early Neolithic	Early Neolithic	
6500	PPNC				
			Aceramic/Initial Neolithic	Aceramic/Initial Neolithic	
7000	LPPNB	ECA IIIA (Çatalhöyük East Levels XII/IX – VIB)			
7500					
	MPPNB				
8000			Mesolithic		
	EPPNB	ECA II (Çatalhöyük East Levels Pre XIID- XII/ IX)			
8500					
	PPNA				
9000					

Table 18.1. Chronological table. Levant and Central Anatolia based on Cessford et al. *(2005); Kuijt and Goring-Morris (2002); Özbaşaran and Buitenhuis (2002); Greek Neolithic based on Andreou* et al. *1996, 260, table 1; Crete based on Tomkins (2008); for the LBK see Stäuble (1995), Strien and Gronenborn (2005).*

farming that involved high inputs of labour per unit area or animal. In this regime, cultivation areas provide grazing and fodder for livestock (cf. Flannery 1969), which in turn regulate crop growth and provide manure (Halstead 2006a). This small-scale form of cultivation exploits the potential of manure and midden material for creating a rich 'garden soil' over a restricted zone that can be tilled and weeded manually by a household; without fertilizer, yields of staple grain crops are lower and so must be grown on a larger scale, using animal traction and other labour-saving husbandry methods ('extensive cultivation') (cf. Halstead 1995b). In this perspective, it was the emergence of intensive mixed farming – and not cultivation or herding *per se* – that constituted the *decisive* transition, providing a diversified means of household production and highly 'buffered' habitats for domestic plants and animals that were ultimately reproduced over a range of climatic zones (Bogaard 2004a, 170; 2004b; 2005).

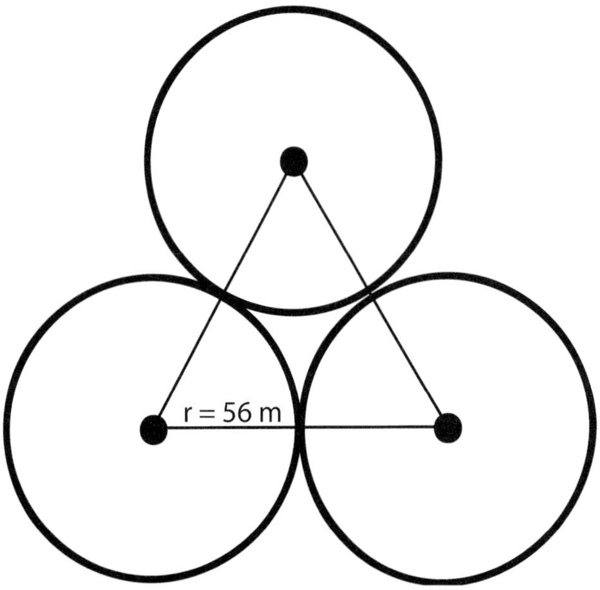

Fig. 18.1. A hypothetical household distribution in which each unit is surrounded by a cultivated area of 1 ha.

Intensity, productivity and settlement size

Close proximity of cultivation plots to the homestead enhances the feasibility of high inputs of manual labour, nutrients and, where necessary, water. In recent traditional agriculture, manuring occurs either by transfer of stall litter and midden material or by 'folding' animals on fields. Hauling of stall manure is feasible only over short distances (e.g. ≤1000 m, even with the aid of animal traction) (Halstead field notes). Animals can be penned overnight on cultivated plots, but only if fenced or guarded to prevent escape, and on distant fields they are normally vulnerable to predators. A study of present-day non-mechanized cultivation in Evvia, Greece (Jones *et al.* 1999) showed that there was a positive relationship between the proximity of plots to the village and the intensity of manuring, weeding and watering.

For intensive farming, therefore, a dispersed settlement pattern of isolated farmsteads, each surrounded by cultivation plots, would be optimal. For example, if each household used c. 1 ha for cultivation (assuming high area yields – 1500 kg/ha – and the need for c. 250 kg per person per year in households of six people), a cultivation radius of c. 50–60 m is implied and hence a minimum distance of just over 100 m between houses (Figure 18.1). In this scenario, manuring would happen almost as a matter of course, as refuse was removed from the immediate vicinity of the house.

The isolated household, however, is vulnerable to losses in human labour, livestock and crops owing to disease,

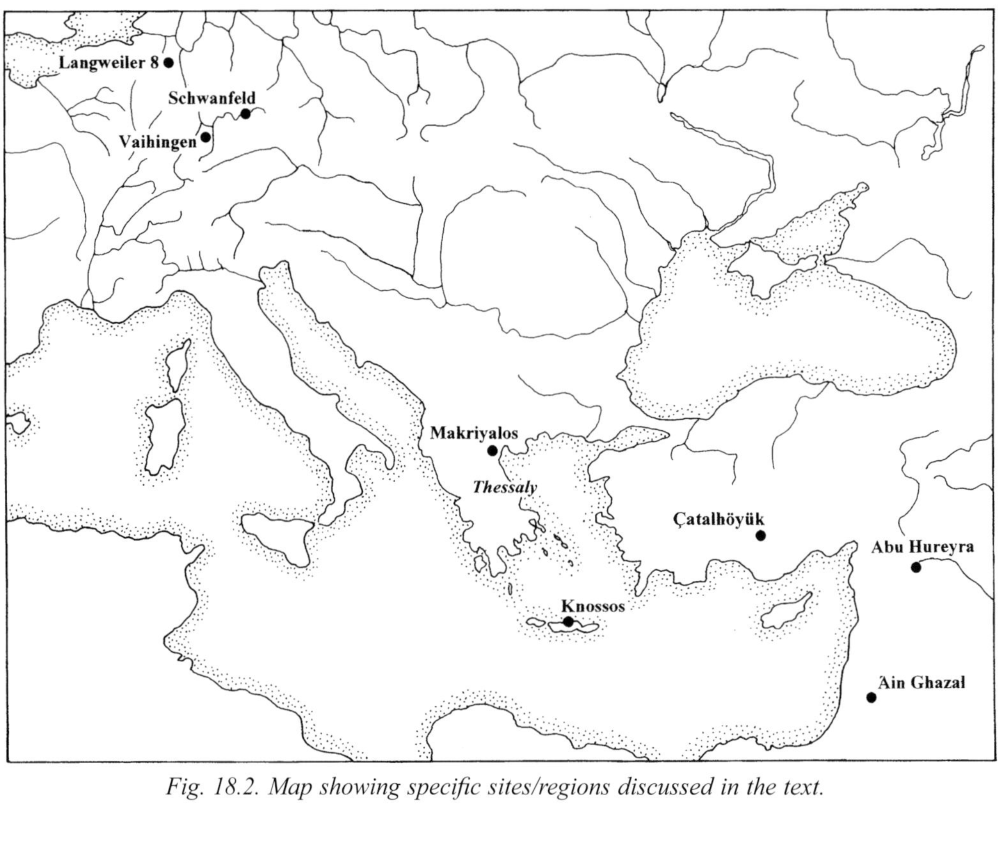

Fig. 18.2. Map showing specific sites/regions discussed in the text.

predation, bad weather and so on (Forbes 1982; Halstead 1989b). Households can to some extent insulate themselves by producing a 'normal surplus', but this is limited in abundance and temporal scope unless combined with reciprocity beyond the household (Sahlins 1972). While sharing is mediated by social norms of hospitality, material culture and cooking practices, it is encouraged and even enforced by close spatial proximity between households (Whitelaw 1983). In practice, therefore, there is a trade-off between dispersion of households, enhancing proximity to cultivation plots, and aggregation of households, enhancing opportunities for sharing and communal security. An ideal balance would entail enough aggregation to allow ample scope for sharing, while also permitting reasonable proximity to cultivation plots.

Interplay between these factors arguably contributed to a variety of settlement patterns and farming landscapes from Western Asia to Northwest Europe. To explore this variety, we consider early farming landscapes associated with the 'indigenous' transition in Western Asia before turning to two European case studies: the Neolithic of Crete and mainland Greece and the early Neolithic *Linearbandkeramik* of Central Europe (Figure 18.2; Table 18.1).

'Mega-sites' and early farming landscapes in Western Asia

The roots of intensive mixed farming may be traced back to the Near Eastern PPNB(-C), when ovicaprid herding (and, by the end of the PPN, domestic pig and cattle keeping) emerged alongside a standard suite of crops. The PPNB(-C) is a long and complex period with marked variation in subsistence practices and residential patterns, but here we focus on the 'mega-sites' of 10+ ha that emerged in the Levant and continued in the Pottery Neolithic at Çatalhöyük in central Anatolia and at Sha'ar Hagolan in the Jordan valley. The mega-sites represent the upper end of the site size range (Kuijt 2000b), but their formation was localized (e.g. eastern edge of the Rift Valley in the southern Levant – Gebel 2004) and smaller farming settlements occurred across the PPNB zone (Hole 2000).

The inhabitants of mega-sites practised farming and herding alongside hunting and gathering. Although agriculture raises carrying capacity, peak population estimates of c. 3000–5000 or more (Köhler-Rollefson and Rollefson 1990; Kuijt 2000a; 2000b; Moore 2000, 273–5; Cessford 2005) represent aggregation beyond any plausible compromise between household production and collective subsistence security. Caution is needed (cf. Whitelaw 2004) in applying ethnographically based habitation densities (van Beek 1982); estimates of cultivated area, likewise, involve a range of assumptions, but distances to furthest fields of >1 km appear inevitable for mega-sites (Table 18.2). Such distances would be problematic for intensive farming (see above) and, in Table 18.2, even the lowest estimate for a modest 5-ha settlement implies a distance to furthest plots of c. 500 m, which would perhaps be prohibitive for hauling manure *without* the assistance of draught animals (and, on present evidence, *domestic* cattle do not appear until the end of PPN, a very late stage in the development of mega-sites). Consequently, lower area yields should be assumed, implying even greater distances to furthest plots.

Far from being an inevitable result of combined agriculture and herding, mega-sites reflect – more plausibly – a triumph of ideological motives over ecological feasibility. Unless we accept that they represent gathering/feasting/burial places that were not permanently and/or fully occupied (Hole 2000), the issue of their development and longevity is at least as problematic as that of their eventual collapse.

Previous approaches to mega-site economy

The model of mega-site farming developed by Köhler-Rollefson and Rollefson, drawing on faunal evidence from PPNB–PN 'Ain Ghazal, argued for herding principally of goats and hence (assuming these to be obligatory browsers) for enforced spatial segregation of crop and animal

	Radius (m) from centre of site to most distant cultivated land				
Site size (ha)	100 people/ha	150 people/ha	200 people/ha	250 people/ha	300 people/ha
1	237	288	331	369	403
5	530	643	739	824	901
6	581	705	810	903	987
7	628	761	875	975	1066
8	671	814	936	1043	1140
9	712	863	992	1106	1209
10	750	909	1046	1166	1274
15	919	1114	1281	1428	1561

Table 18.2. The distance to the furthest cultivation plots in settlements of differing size and for a range of population densities, assuming high area yields (1500 kg per ha) and a requirement of 250 kg per person per year (cereals). Distances are estimated from the centre of the site to the outer edge of cultivated land.

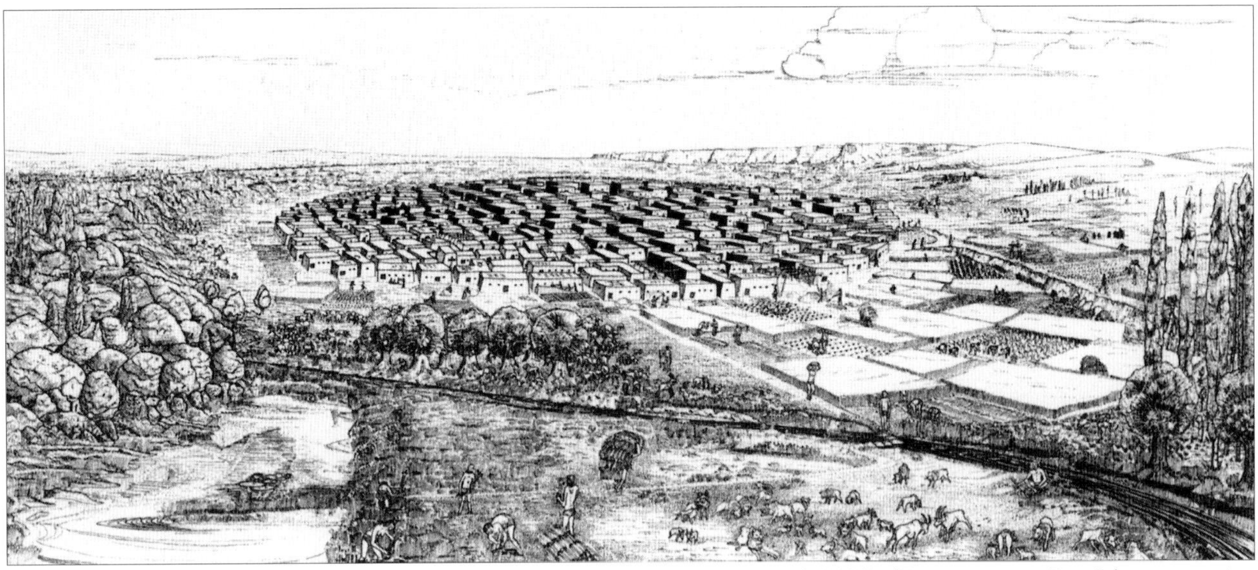

Fig. 18.3. Reconstruction of PPNB Abu Hureyra and its landscape (from Moore et al. *2000). (Frontispiece to Part I (reconstruction of the archaeological site of Abu Hureyra) from* Village on the Euphrates *© 1999 by Oxford University. By permission of Oxford University Press. 6001682.)*

husbandry. Increasing human and thus goat numbers led to the development of specialized nomadic pastoralism (reflected in the small, supposedly pastoral sites of the Jordanian steppe), environmental degradation (as heavy browsing and cutting of firewood for lime production prevented regeneration of woodland) and eventual mega-site collapse (Rollefson and Köhler-Rollefson 1989; Köhler-Rollefson and Rollefson 1990; Rollefson 2004).

Goats are not *obligatory* browsers, however, undermining the argument for segregation of plant and animal husbandry, but the model also overestimates the importance of goats in the 'Ain Ghazal ovicaprid assemblage (Wasse 1997, 578; Martin 2000, 96). Speciation of the ovicaprid remains from this mega-site has shown a dramatic decline of goat from MPPNB onwards and a corresponding increase of sheep, which dominate the assemblage from LPPNB/C onwards (Wasse 2002, 61, table 2), a pattern common in Near Eastern Neolithic sites (Martin 2000, 98). Furthermore, as Martin has demonstrated, inhabitants of steppic sites (the 'specialized pastoral component' of the Köhler-Rollefson model) were essentially hunter-gatherers during the PPN, with domesticates (principally sheep) making their appearance in the early LN (Martin 2000, 92, table 4). The importance of the predominance of sheep among ovicaprids in faunal assemblages is further discussed below.

Also potentially relevant is Sherratt's (1980; 2006; also Flannery 1969) model of flood-plain cultivation based on the location of many early farming sites, including Çatalhöyük, on or near alluvial soils offering highly productive conditions for spring-sown crops with minimal labour. A spring-only sowing regime is implausible (Hillman and Davies 1992, 134), however, and the claim that flooded alluvium is 'practically self-cultivating' (Sherratt 1980, 318) is also problematic. Clustering of early farming settlements near water sources may reflect the importance of water as a limiting factor on crop growth, but opportunistic cultivation of flooded alluvium is unreliable and, depending on local hydrology, attempts to stabilize production by the use of irrigation or drainage systems may have required significant labour input.

While previous models of Neolithic mega-site farming are problematic, neither intensive cultivation nor 'extensive' field agriculture provides an adequate alternative. Crucially, these sites developed when the farming 'package' was still forming. Abu Hureyra already covered 8 ha when gazelle hunting gave way to ovicaprid herding, and the husbandry of pigs and cattle followed in subsequent phases (Moore *et al.* 2000, 493, 497). 'Ain Ghazal had grown to 9.5 ha before domestic pigs and cattle appeared in the PPNC (Köhler-Rollefson and Rollefson 1990). Domestic pigs and cattle were apparently lacking throughout the earlier Neolithic sequence at Çatalhöyük (Russell and Martin 2005). Thus the roles of pigs (local disturbance and manure) and cattle (bulk manure production and potentially transport) in intensive household farming were lacking from mega-site farming systems, or came in only at a late stage, when the sites were reaching their maximum extent. Moreover, the late addition of domestic cattle largely precludes 'traditional' extensive agriculture based on animal traction (Halstead 1987), despite growing evidence (Isaakidou 2006; cf. Moore *et al.* 2000, 498; Helmer and Gourichon 2004) for *draught* use of cattle much earlier than has been assumed.

Mega-site farming and herding: towards a revised model

Faunal and archaeobotanical data from mega-sites such as

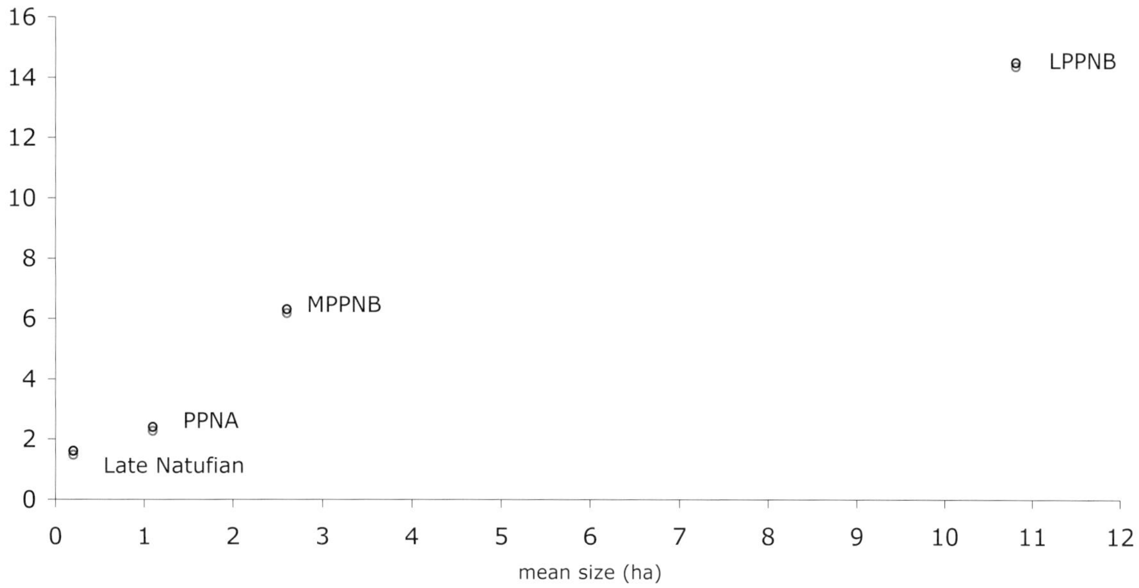

Fig. 18.4. Mean site size versus mean architectural compartmentalization of Late Natufian–LPPNB settlements in the Mediterranean zone of the south-central Levant (data from Kuijt 2000b, table 2).

Abu Hureyra (Figure 18.3), 'Ain Ghazal and Çatalhöyük suggest cultivation of grain crops and herding of domestic ovicaprids. Available ovicaprid mortality data (Legge 1996; Legge and Rowley-Conwy 2000; Wasse 2002) are compatible with a generalized 'meat' culling pattern (Martin 2000; Bogaard 2005) and not with the specialized dairying typical of recent pastoralists (Dahl and Hjort 1976). Contrary to previous arguments for a divorce between crop and animal husbandry, grazing and perhaps folded ovicaprids may have manured, and so maintained the fertility and tractability of, cultivated land around large settlements (possibly combined with middening). Sheep are particularly suited to close integration with arable farming (Halstead 2006a). The correlation of settlement growth to 9.5 ha at PPNC 'Ain Ghazal with the highest frequency of sheep (Wasse 2002), and the increase in sheep through the PPNB/C elsewhere, including at Abu Hureyra, may reflect this relationship (Bogaard 2005) rather than specialized pastoralism. Grazing animals could provide a mobile source of manure for distant cultivation plots. Concentration of folding and penning (Matthews 2005; Russell and Martin 2005) and perhaps middening near the settlement, however, coupled with greater opportunities for intensive cultivation, is likely to have made the inner 'garden belt' the most productive and valuable land (cf. Rollefson 2004). A sharp contrast in productivity between infield and outfield plots would surely have threatened community cohesion, unless cultivated land was in some form of communal 'ownership', with redistribution of plots among households/lineages. The collective ideology underpinning such a system may be reflected in the remarkable uniformity in household architecture and equipment that is so typical of mega-sites and of the PPNB in general (Banning 1998; Kuijt 2000a; 2000b).

Hunting and gathering also played a role in provisioning mega-sites, and wild resources coming in 'big packages' may have been particularly linked with an ethic of sharing (cf. Halstead 1999; Flannery 2002 and see below). A recently excavated burned building at Çatalhöyük contained *in situ* stores of crops and wild plants in a small, secluded room, while bucrania and horn cores of cattle – a species associated with feasting – were found in the larger, more public room (Twiss *et al.* 2009). A decrease in hunting at sites such as 'Ain Ghazal and Abu Hureyra during the later PPNB (Wasse 2002; Legge and Rowley-Conwy 2000) may reflect social changes associated with the rise of the household (see below) and increasing labour demands in relation to agriculture as the final elements of the 'package' (cattle and pigs) were incorporated. The emphasis on wild animals in Neolithic art (Russell and Meece 2005) may reflect the tenacity of an ideological outlook based in hunter-gatherer lifeways and a nostalgic attitude to hunting and food sharing (Benz 2000, 140).

Farming and the development and collapse of mega-sites

Various social or economic factors have been linked with aggregation at mega-sites: ritual authority (Kuijt 2000a; 2000b; Gebel 2004), control of dispersed resources through endogamy (Baird 2005), 'material entanglement' (Hodder 2006) and nodal location within trade networks (Sherratt 2006). Specific local circumstances are also relevant, such as population dislocation from the Mediterranean coast and Jordan Valley to the Jordanian Highlands (Gebel 2004; Rollefson 2004). Whichever combination of factors is most relevant, mega-sites represent the last stage in a long-term trend of increasing community size (Figure 18.4)

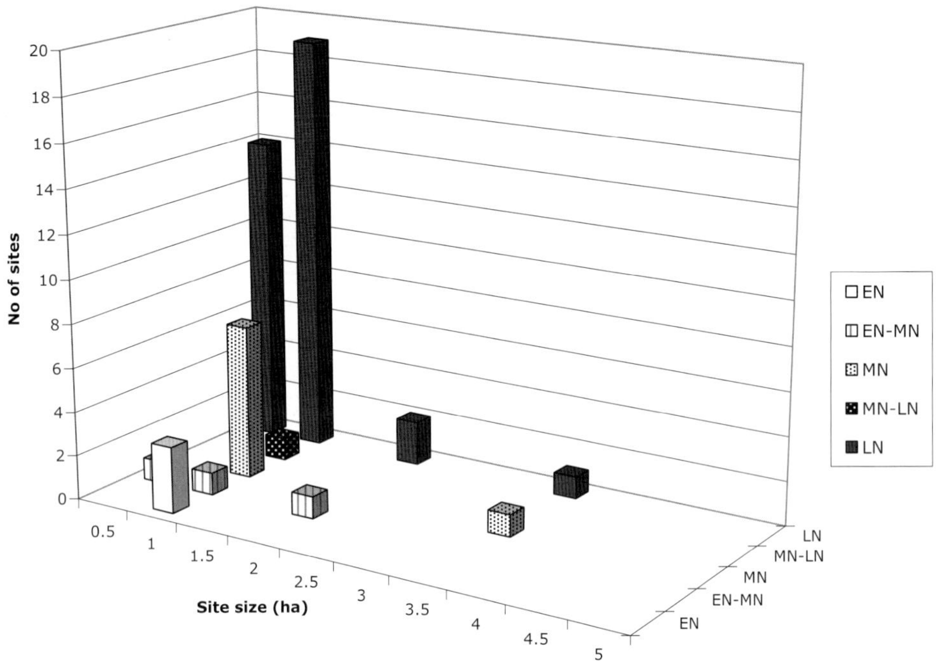

Fig. 18.5. Size of Thessalian sites during the Neolithic (data from Halstead 1984, fig. 6.21).

(Kuijt 2000a; 2000b; Byrd 2005), beginning with complex Epipalaeolithic hunter-gatherers and accelerating with ('pre-domestication') cultivation in the PPNA (Colledge 1998; Sherratt 2006). The development of crop and animal husbandry surely increased carrying capacity and facilitated the latter stages in the growth of mega-sites, but long-term aggregations of 3000–5000 or more persons would have presented a major challenge to early farming, whether this took the form of specialized pastoralism, flood-plain cultivation or the integrated alternative sketched out here.

A plausible response to this challenge may have been an intensification of labour inputs and hence of productive outputs. Such intensification is arguably more achievable if farming labour is primarily organized at a domestic (Flannery 1972) rather than collective (Sahlins 1972) level. Small-scale residential units can be identified throughout the extended sequence from sedentism to combined farming and herding in the southern Levant (Byrd 2000; Banning 2003), but the household asserted itself as a social and economic unit only gradually (Wright 2000). A marked continuity in ritual practices and symbolism from the Natufian or even earlier through to the PPNB and PN (Bar-Yosef 1998; Kuijt 2000a; Goring-Morris and Belfer-Cohen 2002) helps to explain how any 'centrifugal' impulse towards household productivity and self-sufficiency was long overwhelmed by stronger impulses towards social aggregation. The eventual decline of mega-sites could thus have more to do with the full emergence of intensive household farming and its associated 'centrifugal forces' (cf. Banning 1998, 214) than with the supposed development of specialized pastoralism resulting from resource exhaustion (see above).

Farming, household and community in the Greek Neolithic

The earliest Neolithic on the Greek mainland and Crete (Table 18.1) is rather poorly known owing to limited horizontal exposure of deposits in a few multi-period sites. Discussion here thus draws on evidence from later periods as well. From the Early Neolithic onwards, arguably *permanent* 'villages' (Halstead 2005) spread in the fertile lowlands of eastern mainland Greece, especially Thessaly, with the longest-lived and presumably most successful forming tells of c. 1–2 ha, spaced c. 2–3 km apart and with populations estimated at around 200–250 inhabitants (Figure 18.5) (Halstead 1984). Apparently shorter-lived are 'flat-extended' sites, where habitation drifts laterally over time, making estimates of community size extremely difficult. The earliest dwellings are flimsy pit-houses but, from later in the Early Neolithic, free-standing wattle-and-daub or mud-brick rectangular buildings, arguably housing families, become the norm on tell sites (Kotsakis 1999; Halstead 2006b). On long-lived tells rebuilding of individual houses on the same spot perhaps implies 'domestic' land ownership, as well as continuity of habitation and ancestral claims (Halstead 1984, 161). Conversely, on flat-extended sites, flimsier and spatially unstable huts perhaps betray more collective ownership of land (Halstead 2006b).

That subsistence strategies focused on domesticated crops and animals is demonstrated by the ubiquitous presence of domestic cereals (mostly emmer) and pulses and of the four main farmyard animals, dominated by sheep;

remains of wild animals and plants are rare, although the latter are subject to serious preservation biases (Halstead 1996). Claims for *local* domestication of some crops and animals are unsubstantiated to date. The simultaneous appearance of domestic crops and animals arguably as a 'package', ethnographic studies of 'traditional' husbandry and palynological evidence for limited clearance formed the basis of Halstead's (1981) model of an *integrated* 'horticultural' farming regime during the Neolithic.

This model, subsequently elaborated in light of ethnography (e.g. Halstead 1987; 1989b; 1995b; 2000; 2006a), is supported by recent bioarchaeological studies. Archaeobotanical evidence suggests balanced reliance on cereals and pulses more compatible with intensive horticulture than extensive agriculture (Halstead 1981; 1996). Microwear analysis of sheep and goat teeth from Late Neolithic Makriyalos implies similar diets for the two taxa, while high abrasion is compatible with 'pasturage in a confined space' (Mainland and Halstead 2005, 111), arguably on small plots of arable land. The biometric and perhaps isotopic separation of wild and domestic cattle and pigs further suggests that livestock were controlled sufficiently closely to restrict interbreeding with wild relatives, implying relatively small-scale animal husbandry (Halstead 2000). Stalling of some animals within settlements is suggested by the presence of foetal/newborn bones (Halstead 2005) and evidence for use of dung as fuel (Valamoti 2004, 130). Unfortunately, despite recent rigorous sampling, weed assemblages of the quality and quantity available from Central Europe (see below) are not yet available (e.g. Valamoti 2004, 130).

Present evidence thus suggests reliance since the Early Neolithic on a horticultural regime and settlement in small villages that, in contrast to the Near Eastern mega-sites, would have facilitated intensive land use. Early farmers will have had to balance household independence and control of resources with the need for collaboration within and beyond the 'village', since individual households and even individual villages were not demographically viable in the long term; distant social relations were also perhaps necessary for the acquisition of raw materials (evident, for example, in the chipped stone assemblage from Early Neolithic Argissa: Perlès 1990) and assistance in bad years. Regional affinities in material culture imply regular contact and shared technologies and ideologies, and thus potential exchange of marriage partners (Halstead 1989a). A concern with the strengthening of communal ties is implied by the dispersal of *adult* human remains in habitation contexts (Triantaphyllou 1999; 2008) and by large-scale earthworks, especially in flat-extended sites (Kotsakis 1999; Pappa and Besios 1999). Large-scale commensal event(s) identified at Late Neolithic Makriyalos also placed emphasis on collective cohesion (standardized serving vessels), as well as perhaps catering for a regional gathering (Pappa *et al.* 2004; Urem-Kotsou and Kotsakis 2007). The spatial and temporal distribution of food preparation facilities during the Neolithic suggests their gradual 'removal' from more 'public' to more 'private' areas and a progressive erosion of collective commensality in favour of domestic hoarding (Hourmouziadis 1979; Halstead 1995a).

Knossos, on Crete, is one of the earliest farming sites in Europe (Table 18.1), but is little known owing to limited publication and an obscure ceramic chronology (Tomkins 2007). Its establishment is roughly contemporary with that of the earliest Thessalian sites and shares similar architecture (a mud-brick wall in the earliest deposits suggests rectangular structures: Evans 1971, 104 and plate II) and subsistence (cereals and pulses: Evans 1968, 269; the four common farmyard animals with a clear preponderance of sheep: Isaakidou 2008). At Knossos, bone pathologies suggest precocious use from the Early Neolithic onwards (Table 18.1) of breeding *female* cattle for small-scale traction (Isaakidou 2006). This novel practice is as yet only poorly attested in the Neolithic of mainland Greece and so may initially have been a response to particular problems of the southern Aegean, where the early onset of summer drought places a premium on early sowing, which is often made difficult by the late onset of autumn rains (Halstead field notes). Use of cows for ploughing would have alleviated the acute time stress inevitable with manual tillage at sowing time (Isaakidou 2008). At any rate, use of draught cattle began when the Knossos settlement covered only 0.25–0.35 ha, perhaps representing a population of only 50–70, and so arguably served to facilitate intensive rather than extensive cultivation (Isaakidou 2008, table 1). Interestingly, the use of draught cattle is associated with only modest and slow growth of the settlement, which is estimated to have covered only c. 1.75–2.5 ha by the Late and Final Neolithic (Table 18.1) (Tomkins 2008). This sharp contrast with the early formation of Near Eastern mega-sites suggests that community size at Knossos (and elsewhere in the Neolithic of Greece) was under strong social constraints that perhaps prevented growth beyond levels observed among egalitarian societies (<500: Forge 1972).

By the time farming reached Greece it had apparently crystallized into an integrated system practised irrespective of differences in architecture, site type (tell or flat-extended) (e.g. Valamoti 2004, 131) and region. The abundant settlement data from lowland Thessaly suggests a dense network of small villages (≤1 ha) that provided opportunities for social interaction and mutual support without making intensive farming impracticable, while the larger aggregation at later Neolithic Knossos may have been enabled by the early use of cows for traction. As Halstead points out, however, architectural fluidity ('house' plans, distribution of cooking facilities) and material culture evidence (house models, animal and human figurines, burials) imply a continuing tension between household and community throughout the Greek Neolithic, thus undermining arguments for the existence and hence adoption in the Early Neolithic of an established Neolithic ('domus') ideology (e.g. Hodder 1990; also Childe 1957) alongside the 'subsistence package' (Halstead 1996; 2006b).

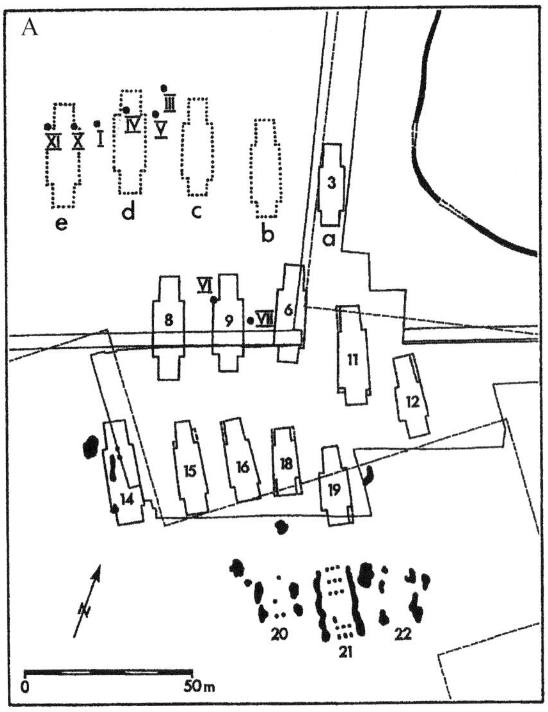

The LBK of Central Europe: household-centred farming landscapes?

The Early Neolithic of central Europe (*Linearbandkeramik* or LBK) developed from its earliest phase around the household as the primary social and farming unit. The earliest (*älteste*) LBK shows certain continuities in lithic technology and typology with the Late Mesolithic (Gronenborn 1999), and significant levels of indigenous acculturation appear plausible. Neolithization was radical, however, with daily life centred on a new form of dwelling – the longhouse – and constrained by the routines of cultivation and herding (Bogaard 2004a, 162; Lüning 2006). One of the earliest known LBK sites is Schwanfeld in Franconia, founded c. 5500 cal BC (Lüning 2005a; 2005b; 2006). Here, excavation of four longhouse sequences (*Hofplätze*) revealed a small 'hamlet' with three contemporary structures (numbers 3, 6 and 19 in Figure 18.6a) in its initial phase, shown reconstructed in Figure 18.6b. Some households used distinct lithic sources, perhaps linking them with distant points of origin. Though the archaeobotany of Schwanfeld is not yet published, earliest LBK data from elsewhere suggest that weed assemblages

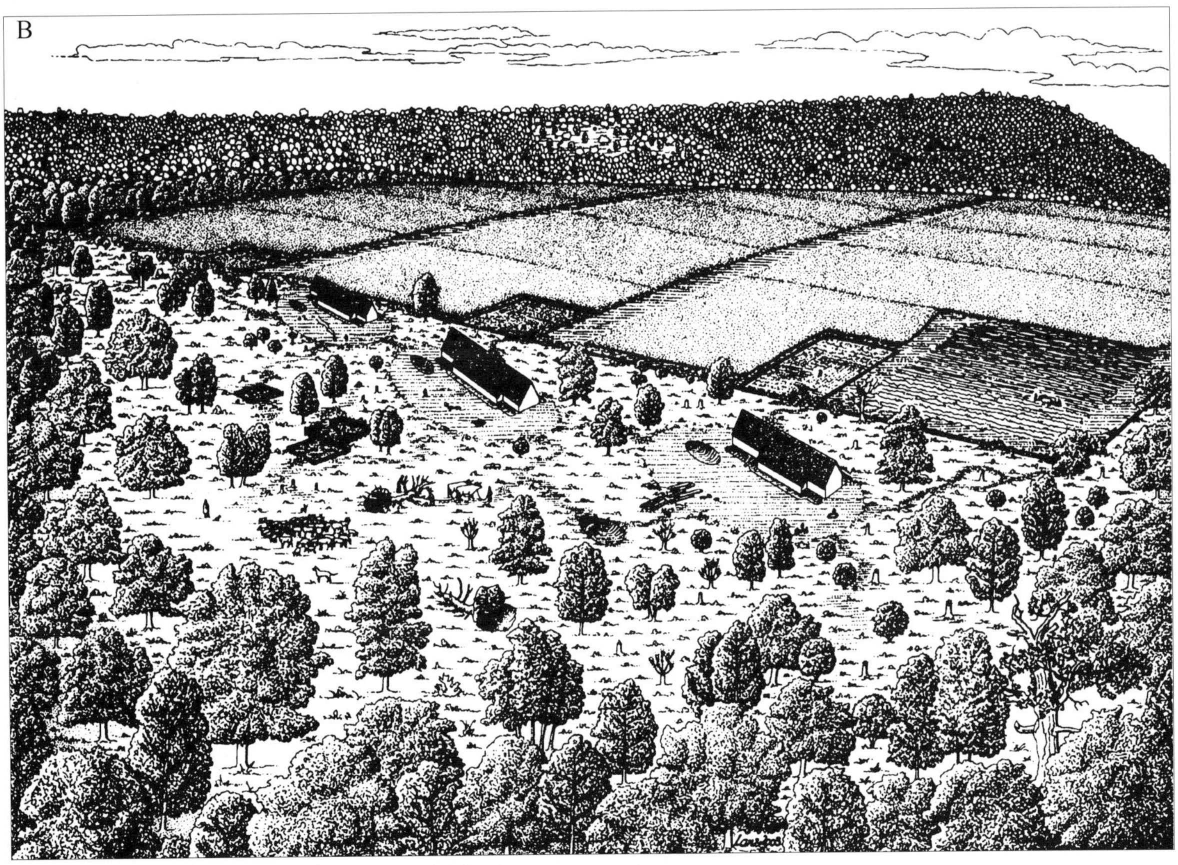

Fig. 18.6. A) reconstructed settlement plan of Schwanfeld, Franconia, showing four longhouse sequences (rows); B) reconstructed view of Schwanfeld (houses 3, 6 and 19) and its agricultural landscape (after Lüning 2005a, figs 2, 4).

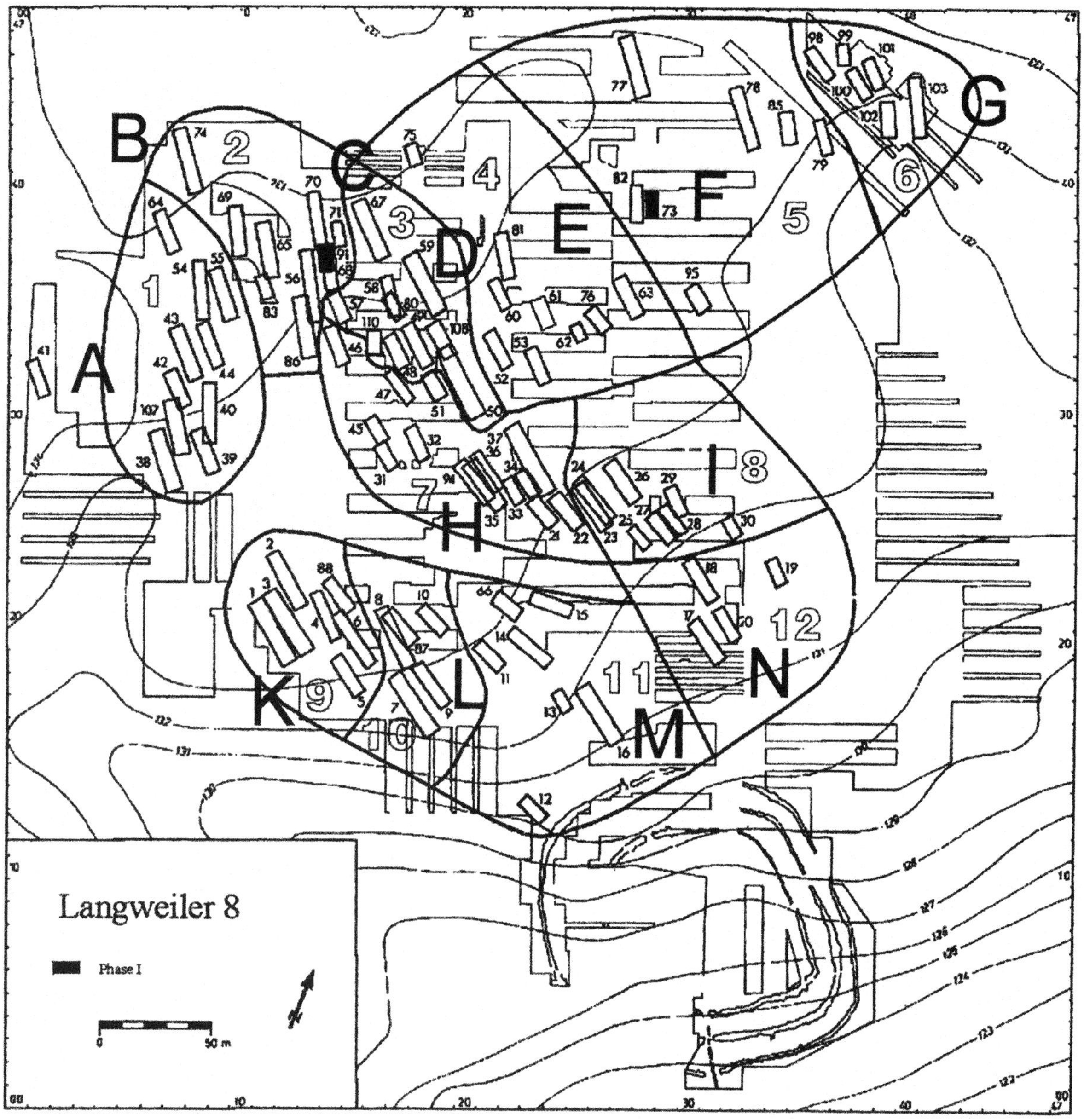

*Fig. 18.7. Plan of Langweiler 8, Aldenhovener Plateau, Rhineland, showing original (1–12) and revised (A–N) spatial divisions into groups of related longhouses (*Wohnplätze*) (from Lüning 2005a, fig. 16).*

and crop husbandry practices were not radically different to the later LBK and featured fixed plots that were more or less intensively managed (Bogaard 2004a, in prep; Kreuz *et al.* 2005, table 7). Faunal remains from the earliest LBK indicate a varying mixture of cattle/pig/sheep husbandry and hunting (Lüning 2000, 111–14). Mitochondrial DNA evidence from archaeological LBK cattle bone (Bollongino *et al.* 2006) supports metrical indications (Benecke 1994, 48–55; Döhle 1997; Lüning 2000, 105) that cattle arrived in a fully domesticated state.

The nature of farming in the later (*jüngere*) LBK has been investigated by comparing weed assemblages, in terms of their functional ecology, with modern weed floras developed under known conditions (Bogaard 2004a, in prep). The most relevant model of farming appears to be intensive cultivation, featuring high soil fertility through middening/manuring and high disturbance through careful tillage and weeding. Animal husbandry arguably involved relatively intensive and small-scale management of cattle and pigs, given their metrical distinctiveness from native wild populations (Benecke 1994, 48–55), and of ovicaprids. This diverse and integrated farming regime fits well with the household-centred nature of LBK society and its dispersed settlement pattern. The generally large size of LBK longhouses (50 m^2+) suggests that they may have housed two or three generations together (Lüning 2000, 15) and were capable of juggling diverse activities. Households to some extent developed their own 'sub-culture', with distinctive elements of ceramic decoration, lithic sources and so on (Fridrich 1994; Strien 2005).

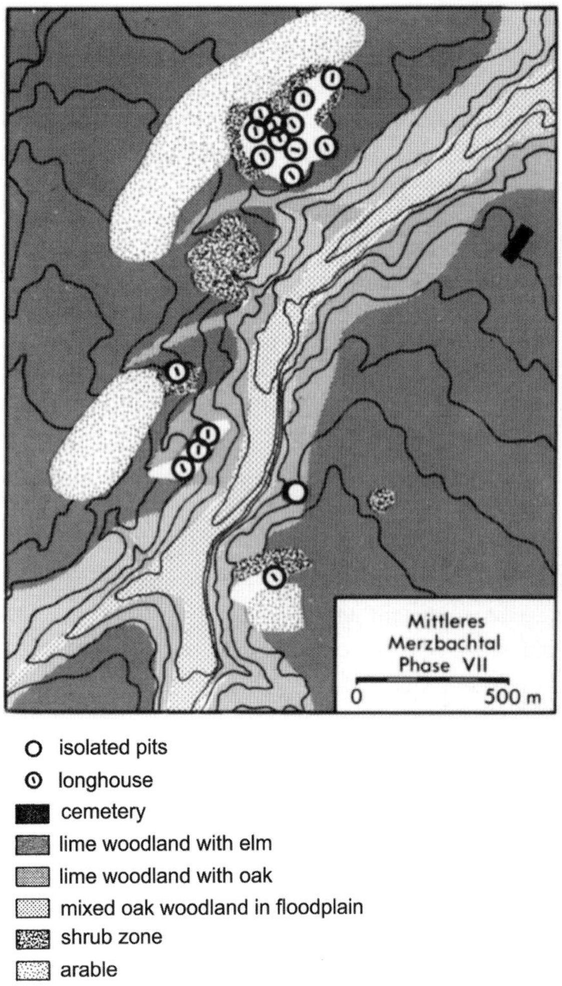

Fig. 18.8. Reconstructed landscape of the middle Merzbach valley, phase VII (from Stehli 1989, figs 9, 11). Langweiler 8 is the northernmost site.

Proximal replacement of longhouses may reflect descent-based claims to zones of the settlement (Figure 18.7) and perhaps to associated cultivation plots (Bogaard 2004a, 166; Bogaard *et al.* 2008).

Farming techniques and weed assemblages did vary somewhat within the later LBK, and in part this variation relates to settlement pattern. A dispersed settlement (*Streusiedlung*) model is well known from rescue excavations in the Rhineland, including an entire LBK landscape in the Merzbach valley of the Aldenhovener Plateau near Cologne (Lüning and Stehli 1994). Here, 'settlements' range from single longhouses to up to 11 contemporary longhouses dispersed across 7 ha (Langweiler 8) (Figure 18.8). The Merzbach valley sites approach the 'ideal' model of dispersed households (Figure 18.1), though contemporary houses in 'clusters' are often less than 100 m apart, suggesting that proximity to neighbours was more important than minimal distance to cultivation plots. Lüning (2005a) has recently postulated a more specific inflection on the dispersed household 'ideal': a pattern of 'central' founder settlements surrounded by isolated farmsteads and hamlets that first emerged in the Flomborn period (from c. 5300 cal BC).

A highly nucleated settlement type has recently been documented at Vaihingen in southwest Germany (Figure 18.9), where Strien (2005) estimates as many as 50 households across c. 5 ha during the later Flomborn phases. Comparison of weed assemblages suggests that relatively high nucleation at Vaihingen was associated with greater variability in cultivation intensity, while the Rhineland farmers achieved more consistently intensive growing conditions, perhaps because of their more dispersed household distribution (Bogaard 2004a, 145–6, in prep). Clearly, the scale and density of household nucleation – and by extension the intensity and configuration of the farming landscape – varied among sites and regions within the LBK.

The importance of communal identity at Vaihingen is underlined by the settlement ditch (Figure 18.9), which was open for a relatively short time in the Flomborn period before being filled in and used as a burial area. Together with the demonstrable impact of nucleation on the farming landscape, the ditch and its burials illustrate active negotiation in the LBK (as in the Neolithic of Greece) between the individual, household and community. The community was also intimately linked with wider society: Vaihingen females are morphologically indistinguishable from women buried 60 km away in the LBK cemetery at Schwetzingen (Meyer and Alt 2006), while variable strontium signals in human and livestock teeth appear to suggest various forms and scales of mobility, perhaps through intermarriage and exchanges of livestock (Bentley *et al.* 2003; Bentley and Knipper 2005). If it is correct to link women (and children) with cultivation near the home, emerging patterns of regional similarity in the ecological profile of weed assemblages may reflect similar cultivation practices spread by women circulating in a patrilocal settlement pattern (cf. Eisenhauer 2003; Bogaard 2004a, 147–8, 167).

Conclusions

Available data from Near Eastern mega-sites suggest distinctive farming landscapes that integrated sheep, goat and crop husbandry and placed great ideological emphasis on wild resources and on sharing, extending perhaps to the redistribution of cultivation areas. The nature and scale of occupation at these settlements, however, remains a critical and open question, and further bioarchaeological investigation is needed to determine the precise form and intensity of crop and herd management. We have argued that these extraordinary sites reflect the culmination of ideology/cult practices rooted in sedentary hunter-gatherer lifeways and presented major challenges for the emerging farming package. As Benz (2000) indicates, foragers-turned-farmers may resist or subvert the potential constraints of farming if their ideological outlook remains rooted in hunter-gatherer traditions.

Fig. 18.9. Vaihingen an der Enz, with the settlement ditch highlighted (R. Krause unpublished).

Farming landscapes from Greece to Central Europe (and beyond) were in various ways shaped by the functioning of a fully formed package of crops, livestock and intensive management practices, including the early use of secondary animal products (e.g. manuring, traction). At the same time, the diverse forms of settlement that emerged in different parts of Europe reflect much more than farming practice *per se*; indeed, the essential nature of farming appears to have been rather similar in Greece and Central Europe, despite obvious contrasts in residential patterns.

We conclude that community size in the Neolithic is expressive first and foremost of the *ideological context* of farming practice. It is only by bringing together direct evidence of farming and herding practices with archaeological evidence for settlement, burial and ritual behaviour that the full meaning of farming landscapes can be deconstructed.

Acknowledgements

This paper was written while Valasia Isaakidou held a Leverhulme-funded postdoctoral fellowship at the Department of Archaeology, University of Nottingham. We would like to thank David Taylor for drawing Figure 18.2, and Andrew Moore/Oxford University Press, Jens Lüning and Rüdiger Krause/Landesdenkmalamt Baden-Württemberg for permission to reproduce Figures 18.3, 18.6–18.8 and 18.9. This paper benefited greatly from discussion with Paul Halstead and Katheryn Twiss, and we wish to acknowledge their useful suggestions and bibliographic advice. We would also like to thank Paul Halstead for allowing us to cite unpublished work. The paper's shortcomings are entirely our responsibility.

Bibliography

Andreou, S., Fotiadis, M. and Kotsakis, K. (1996) Review of Aegean Prehistory V: the Neolithic and Bronze Age of Northern Greece. *American Journal of Archaeology* 100, 537–97.

Baird, D. (2005) The history of settlement and social landscapes in the Early Holocene in the Çatalhöyük area. Pp. 55–74 in I. Hodder (ed.), *Çatalhöyük Perspectives: Reports from the 1995–99 Seasons*. McDonald Institute Monographs 40. Cambridge and Ankara: McDonald Institute for Archaeological Research and British Institute of Archaeology at Ankara.

Banning, E. B. (1998) The Neolithic period: triumphs of

architecture, agriculture and art. *Near Eastern Archaeology* 61, 188–237.

Banning, E. B. (2003) Housing Neolithic farmers. *Near Eastern Archaeology* 66, 4–21.

Bar-Yosef, O. (1998) The Natufian culture in the Levant, threshold to the origins of agriculture. *Evolutionary Anthropology* 6/5, 159–77.

Benecke, N. (1994) *Archäozoologische Studien zur Entwicklung der Haustierhaltung in Mitteleuropa und Südskandinavien von den Anfängen bis zum ausgehenden Mittelalter*. Berlin: Akademie Verlag.

Bentley, R. A. and Knipper, C. (2005) Transhumance at the early Neolithic settlement at Vaihingen (Germany). *Antiquity* 79, Online Project Gallery.

Bentley, R. A., Krause, R., Price, T. D. and Kaufmann, B. (2003) Human mobility at the early Neolithic settlement of Vaihingen, Germany: evidence from strontium isotope analysis. *Archaeometry* 45, 471–86.

Benz, M. (2000) *Die Neolithisierung im Vorderen Orient: Theorien, Archäologische Daten und ein Ethnologisches Modell*. Berlin: *ex oriente*.

Bogaard, A. (2004a) *Neolithic Farming in Central Europe*. London: Routledge.

Bogaard, A. (2004b) The nature of early farming in central and south-east Europe. *Documenta Praehistorica* 31, 49–58.

Bogaard, A. (2005) 'Garden agriculture' and the nature of early farming in Europe and the Near East. *World Archaeology* 37/2, 177–96.

Bogaard, A. (in prep) *Plant use and crop husbandry in an early Neolithic village: Vaihingen an der Enz, Baden-Württemberg*. Frankfurter Archäologische Forschungen.

Bogaard, A., Bending, J. and Jones, G. (2008) Crop husbandry and its social significance in the Körös and LBK periods. Pp. 131–8 in D. Bailey and A. Whittle (eds), *Living Well Together: Settlement, Materiality and Subsistence in Neolithic and Copper Age South-east Europe*. Oxford: Oxbow Books.

Bollongino, R., Edwards, C. J., Alt, K. W., Burger, J. and Bradley, D. G. (2006) Early history of European domestic cattle as revealed by ancient DNA. *Biology Letters* 2, 155–9.

Byrd, B. F. (2000) Households in transition: Neolithic social organisation within Southwest Asia. Pp. 63–98 in I. Kujit (ed.), *Life in Neolithic Farming Communities: Social Organisation, Identity and Differentiation*. New York: Academic/Plenum Publishers.

Byrd, B. F. (2005) Reassessing the emergence of village life in the Near East. *Journal of Archaeological Research* 13, 231–90.

Cessford, C. (2005) Estimating the Neolithic population of Çatalhöyük. Pp. 323–6 in I. Hodder (ed.), *Inhabiting Çatalhöyük: Reports from the 1995–99 Seasons*. McDonald Institute Monographs 38. Cambridge and Ankara: McDonald Institute for Archaeological Research and British Institute of Archaeology at Ankara.

Cessford, C., Blumbach, M., Goze Akoglu, H., Higham, T., Kuniholm, P. I., Manning, S. W., Newton, M. W., Ozbakan, M. and Ozer, A. M. (2005) Absolute dating at Çatalhöyük. Pp. 65–99 in I. Hodder (ed.), *Changing Materialities at Çatalhöyük: Reports from the 1995–99 Seasons*. McDonald Institute Monographs 39. Cambridge and Ankara: McDonald Institute for Archaeological Research and British Institute of Archaeology at Ankara.

Childe, V. G. (1957) *The Dawn of European Civilization*. London: Routledge.

Colledge, S. (1998) Identifying pre-domestication cultivation using multivariate analysis. Pp. 121–31 in A. B. Damania, J. Valkoun, G. Willcox and C. O. Qualset (eds), *The Origins of Agriculture and Crop Domestication*. Aleppo, Syria: ICARDA.

Dahl, G. and Hjort, A. (1976) *Having Herds: Pastoral Herd Growth and Household Economy*. Stockholm: University of Stockholm.

Döhle, H.-J. (1997) Zum Stand der Untersuchungen an Neolithischen Tierknochen aus Mitteldeutschland. *Jahresschrift für mitteldeutsche Vorgeschichte* 79, 111–47.

Eisenhauer, U. (2003) Jüngerbandkeramische Residenzregeln: Patrilokalität in Talheim. Pp. 561–73 in J. Eckert, U. Eisenhauer and A. Zimmerman (eds), *Archäologische Perspektiven. Analysen und Interpretationen im Wandel (Festschrift Lüning)*. Rahden, Germany: Marie Leidorf.

Evans, J. D. (1968) Knossos Neolithic, part II: summary and conclusions. *Annual of the British School at Athens* 63, 267–76.

Evans, J. D. (1971) Neolithic Knossos: the growth of a settlement. *Proceedings of the Prehistoric Society* 37, 95–117.

Flannery, K. V. (1969) Origins and ecological effects of early domestication in Iran and the Near East. Pp. 73–100 in P. J. Ucko and G. W. Dimbleby (eds), *The Domestication and Exploitation of Plants and Animals*. London: Duckworth.

Flannery, K. V. (1972) The origins of the village as a settlement type in Mesoamerica and the Near East: a comparative study. Pp. 23–53 in P. J. Ucko, R. Tringham and G. W. Dimbleby (eds), *Man, Settlement and Urbanism*. London: Duckworth.

Flannery, K. V. (2002) The origins of the village revisited: from nuclear to extended households. *American Antiquity* 67, 417–33.

Forbes, H. (1982) Strategies and soils: technology, production and environment in the peninsula of Methana, Greece. Unpublished PhD thesis, University of Pennsylvania. Ann Arbor, MI: University Microfilms International.

Forge, A. (1972) Normative factors in the settlement size of Neolithic cultivators (New Guinea). Pp. 363–76 in P. J. Ucko, R. Tringham and G. W. Dimbleby (eds), *Man, Settlement and Urbanism*. London: Duckworth.

Fridrich, C. (1994) Kulturgeschichtliche Betrachtungen zur Bandkeramik im Merzbachtal. Pp. 207–393 in J. Lüning and P. Stehli (eds), *Die Bandkeramik im Merzbachtal auf der Aldenhovener Platte*. Cologne: Rheinland-Verlag GmbH.

Gebel, H. G. K. (2004) Central to what? The centrality issue of the LPPNB mega-site phenomenon in Jordan. Pp. 1–19 in H. D. Bienert, H. G. K. Gebel and R. Neef (eds), *Central Settlements in Neolithic Jordan*. Studies in Early Near Eastern Production, Subsistence, and Environment 5. Berlin: *ex oriente*.

Goring-Morris, N. and Belfer-Cohen, A. (2002) Symbolic behaviour from the Epipalaeolithic and Early Neolithic of the Near East: preliminary observations on continuity and change. Pp. 67–79 in H. G. K. Gebel, B. D. Hermansen and C. H. Jensen (eds), *Magic Practices and Ritual in the Near Eastern Neolithic*. Studies in Early Near Eastern Production, Subsistence, and Environment 8. Berlin: *ex oriente*.

Gronenborn, D. (1999) A variation on a basic theme: the transition to farming in southern central Europe. *Journal of World Prehistory* 13, 123–210.

Halstead, P. (1981) Counting sheep in Neolithic and Bronze Age Greece. Pp. 307–39 in I. Hodder, G. Isaac and N. Hammond (eds), *Pattern of the Past: Studies in Honour of David Clarke*. Cambridge: Cambridge University Press.

Halstead, P. (1984) Strategies for Survival: An Ecological Approach to Social and Economic Change in the Early Farming Communities of Thessaly, N. Greece. Unpublished PhD thesis, University of Cambridge.

Halstead, P. (1987) Traditional and ancient rural economy in Mediterranean Europe: plus ça change? *Journal of Hellenic Studies* 107, 77–87.

Halstead, P. (1989a) Like rising damp? An ecological approach to the spread of farming in southeast and central Europe. Pp. 23–53 in A. Milles, D. Williams, and N. Gardner (eds), *The Beginnings of Agriculture*. British Archaeological Reports, International Series 496. Oxford: Archaeopress.

Halstead, P. (1989b) The economy has a normal surplus: economic stability and social change among early farming communities of Thessaly, Greece. Pp. 68–80 in P. Halstead and J. O'Shea (eds), *Bad Year Economics: Cultural Responses to Risk and Uncertainty*. Cambridge: Cambridge University Press.

Halstead, P. (1995a) From sharing to hoarding: the Neolithic foundations of Aegean Bronze Age society? Pp. 11–20 in R. Laffineur and W.-D. Niemeier (eds), *POLITEIA: Society and State in the Aegean Bronze Age*. Liège: University of Liège.

Halstead, P. (1995b) Plough and power: the economic and social significance of cultivation with the ox-drawn ard in the Mediterranean. *Bulletin on Sumerian Agriculture* 8, 11–22.

Halstead, P. (1996) The development of agriculture and pastoralism in Greece: when, how, who and what? Pp. 296–309 in D. R. Harris (ed.), *The Origins and Spread of Agriculture and Pastoralism in Eurasia*. London: UCL Press.

Halstead, P. (1999) Neighbours from hell? The household in Neolithic Greece. Pp. 77–95 in P. Halstead (ed.), *Neolithic Society in Greece*. Sheffield Studies in Aegean Archaeology 2. Sheffield: Sheffield Academic Press.

Halstead, P. (2000) Land use in postglacial Greece: cultural causes and environmental effects. Pp. 110–28 in P. Halstead and C. Frederick (eds), *Landscape and Land Use in Postglacial Greece*. Sheffield Studies in Aegean Archaeology 3. Sheffield: Sheffield Academic Press.

Halstead, P. (2005) Resettling the Neolithic: faunal evidence for seasons of consumption and residence at Neolithic sites in Greece. Pp. 38–50 in D. Bailey, A. Whittle and V. Cummings (eds), *(Un)settling the Neolithic*. Oxford: Oxbow Books.

Halstead, P. (2006a) Sheep in the garden: the integration of crop and livestock husbandry in early farming regimes of Greece and southern Europe. Pp. 42–55 in D. Serjeantson and D. Field (eds), *Animals in the Neolithic of Britain and Europe*. Oxford: Oxbow Books.

Halstead, P. (2006b) *What's Ours is Mine? Village and Household in Early Farming Society in Greece*. Amsterdam: Achtentwintigste Kroon-Voordracht, Stichting Nederlands Museum voor Anthropologie en Praehistorie.

Helmer, D. and Gourichon, L. (2008) Premières données sur les modalités de subsistance à Tell Aswad (Syrie, PPNB moyen en récent, néolithique céramique ancien). Pp. 119–51 in E. Vila, L. Gourichon, A. M. Choyke, and H. Buitenhuis (eds), *Archaeozoology of the Near East VIII*, vol. 1. Lyon: Maison de l'Orient et de la Méditerranée.

Hillman, G. C. and Davies, M. S. (1992) Domestication rates in wild wheats and barley under primitive cultivation: preliminary results and archaeological implications of field measurements of selection coefficient. Pp. 113–58 in P. C. Anderson (ed.), *Préhistoire de l'Agriculture: Nouvelles Approaches Expérimentales et Ethnographiques*. Paris: Editions du Centre National de la Recherche Scientifique.

Hodder, I. (1990) *The Domestication of Europe*. Oxford: Blackwell.

Hodder, I. (2006) *Çatalhöyük: The Leopard's Tale*. London: Thames and Hudson.

Hole, F. (2000) Is size important? Function and hierarchy in Neolithic settlements. Pp. 191–209 in I. Kuijt (ed.), *Life in Neolithic Farming Communities: Social Organisation, Identity and Differentiation*. New York: Academic/Plenum Publishers.

Hourmouziadis, G. (1979) *To Neolithiko Dimini*. Volos: Society for Thessalian Studies.

Ingold, T. (1993) The temporality of the landscape. *World Archaeology* 25, 152–74.

Isaakidou, V. (2006) Ploughing with cows: Knossos and the 'secondary products revolution'. Pp. 95–112 in D. Field and D. Sergeantson (eds), *Animals in the Neolithic*. Oxford: Oxbow Books.

Isaakidou, V. (2008) 'The fauna and economy of Neolithic Knossos' revisited. Pp. 90–114 in V. Isaakidou and P. Tomkins (eds), *Escaping the Labyrinth: The Cretan Neolithic in Context*. Sheffield Studies in Aegean Archaeology 8. Oxford: Oxbow Books.

Jones, G., Bogaard, A., Halstead, P., Charles, M. and Smith, H. (1999) Identifying the intensity of crop husbandry practices on the basis of weed floras. *Annual of the British School at Athens* 94, 167–89.

Köhler-Rollefson, I. and Rollefson, G. O. (1990) The impact of Neolithic subsistence strategies on the environment: the case of 'Ain Ghazal, Jordan. Pp. 3–14 in S. Bottema, Entjes-Nieborg and W. van Zeist (eds), *Man's Role in the Shaping of the Eastern Mediterranean Landscape*. Rotterdam: Balkema.

Kotsakis, K. (1999) What tells can tell: social space and settlement in the Greek Neolithic. Pp. 66–76 in P. Halstead (ed.), *Neolithic Society in Greece*. Sheffield Studies in Aegean Archaeology 2. Sheffield: Sheffield Academic Press.

Kreuz, A., Marinova, E. M., Schäfer, E. and Wiethold J. (2005) A comparison of early Neolithic crop and weed assemblages from the Linearbandkeramik and the Bulgarian Neolithic cultures: differences and similarities. *Vegetation History and Archaeobotany* 14, 237–58.

Kuijt, I. (2000a) Keeping the peace: ritual, skull caching and community integration in the Levantine Neolithic. Pp. 137–64 in I. Kuijt (ed.), *Life in Neolithic Farming Communities: Social Organisation, Identity and Differentiation*. New York: Academic/Plenum Publishers.

Kuijt, I. (2000b) People and space in early agricultural villages: exploring daily lives, community size, and architecture in the late Pre-Pottery Neolithic. *Journal of Anthropological Archaeology* 19, 75–102.

Kuijt, I. and Goring-Morris, N. (2002) Foraging, farming and social complexity in the Pre-Pottery Neolithic of the southern Levant: a review and synthesis. *Journal of World Prehistory* 16, 361–440.

Legge, A. (1996) The beginning of caprine domestication in southwest Asia. Pp. 238–62 in D. R. Harris (ed.), *The Origins and Spread of Agriculture and Pastoralism in Eurasia*. London: UCL Press.

Legge, A. and Rowley-Conwy, P. (2000) The exploitation of animals. Pp. 423–71 in A. Moore, G. Hillman and A. Legge (eds), *Abu Hureyra: a Village on the Euphrates*. Oxford: Oxford University Press.

Lüning, J. (2000) *Steinzeitliche Bauern in Deutschland – die Landwirtschaft im Neolithikum*. Bonn: Dr. Rudolf Habelt GmbH.

Lüning, J. (2005a) Bandkeramische Hofplätze und absolute Chronologie der Bandkeramik. Pp. 49–74 in J. Lüning, C. Fridrich and A. Zimmerman (eds), *Die Bandkeramik im 21. Jahrhundert: Symposium in der Abtei Brauweiler bei Köln vom 16.9.–19.9.2002*. Rahden, Germany: Verlag Marie Leidorf.

Lüning, J. (2005b) Vorwort. Pp. 7–9 in H. Stäuble, *Häuser und absolute Datierung der Ältesten Bandkeramik*. Universitätsforschungen zur prähistorischen Archäologie 117. Bonn: Verlag Dr. Rudolf Habelt GMBH.

Lüning, J. (2006) Missionare aus dem Western bekehren und belehren. *Archäologie in Deutschland* 2006, 28–31.

Lüning, J. and Stehli, P. (eds) (1994) *Die Bandkeramik im Merzbachtal auf der Aldenhovener Platte*. Cologne: Rheinland-Verlag GMBH.

Mainland, I. and Halstead, P. (2005) The diet and management of domestic sheep and goats at Neolithic Makriyalos. Pp. 104–12 in J. Davies, M. Fabis, I. Mainland, M. Richards and R. Thomas (eds), *Diet and Health in Past Animal Populations: Current Research and Future Directions*. Oxford: Oxbow Books.

Martin, L. (2000) Mammal remains from the eastern Jordanian Neolithic, and the nature of caprine herding in the steppe. *Paléorient* 25, 87–104.

Matthews, W. (2005) Micromorphological and microstratigaphic traces of uses and concepts of space. Pp. 355–98 in I. Hodder (ed.), *Inhabiting Çatalhöyük: Reports from the 1995–99 Seasons*. McDonald Institute Monographs 38. Cambridge and Ankara: McDonald Institute for Archaeological Research and British Institute of Archaeology at Ankara.

Meyer, C. and Alt, K. W. (2006) Kulturwandel und Bevölkerungswandel? *Archäologie in Deutschland* 2006, 34–5.

Moore, A. M. T. (2000) The buildings and layout of Abu Hureyra 2. Pp. 261–75 in A. M. T. Moore, G. C. Hillman and A. J. Legge (eds), *Village on the Euphrates*. Oxford: Oxford University Press.

Moore, A. M. T., Hillman, G. C. and Legge, A. J. (eds) (2000) *Village on the Euphrates*. Oxford: Oxford University Press.

Nesbitt, M. (2002) Where and when did domesticated cereals first occur in southwest Asia? Pp. 113–32 in R. T. J. Cappers and S. Bottema (eds), *The Dawn of Farming in the Near East*. Berlin: *ex oriente*.

Özbaşaran, M. and Buitenhuis, H. (2002) Proposal for a regional terminology for central Anatolia. Pp. 67–77 in F. Gérard and L. Thissen (eds), *The Neolithic of Central Anatolia. Internal Developments and External Relations During the 9th–6th Millennia cal BC. Proceedings of the International CANeW Table Ronde Istanbul, 23–24 November 2001*. Istanbul: Ege Yayinlari.

Pappa, M. and Besios, M. (1999) The Neolithic settlement at Makriyalos, Northern Greece: preliminary report on the 1993–1995 excavations. *Journal of Field Archaeology* 26, 177–95.

Pappa, M., Halstead, P., Kotsakis, K. and Urem-Kotsou, D. (2004) Evidence for large-scale feasting at Late Neolithic Makriyalos, Northern Greece. Pp. 16–44 in P. Halstead and J. Barrett (eds), *Food, Cuisine and Society in Prehistoric Greece*. Sheffield Studies in Aegean Archaeology 5. Oxford: Oxbow Books.

Peltenberg, E., Colledge, S., Croft, P., Jackson, A., McCartney, C. and Murray, M. A. (2001) Neolithic dispersals from the Levantine corridor: a Mediterranean perspective. *Levant* 33, 35–64.

Perlès, C. (1990) L'outillage de pierre taillée néolithique en Grèce: approvisionnement et exploitation des matières premières. *Bulletin de Correspondance Hellénique* 114, 1–42.

Peters, J., von den Driesch, A. and Helmer, D. (2005) The upper Euphrates–Tigris basin: cradle of agro-pastoralism? Pp. 96–124 in J.-D. Vigne, J. Peters and D. Helmer (eds), *The First Steps of Animal Domestication*. Oxford: Oxbow Books.

Rollefson, G. O. (2004) The character of LPPNB social organization. Pp. 145–55 in H. D. Bienert, H. G. K. Gebel and R. Neef (eds), *Central Settlements in Neolithic Jordan*. Studies in Early Near Eastern Production, Subsistence, and Environment 5. Berlin: *ex oriente*.

Rollefson, G. O. and Köhler-Rollefson, I. (1989) The collapse of early Neolithic settlements in the southern Levant. Pp. 73–89 in I. Hershkovitz (ed.), *People and Culture in Change*. British Archaeological Reports, International Series 508. Oxford: Archaeopress.

Russell, N. and Martin, L. (2005) The Çatalhöyük mammal remains. Pp. 33–98 in I. Hodder (ed.), *Inhabiting Çatalhöyük: Reports from the 1995–99 Seasons*. McDonald Institute Monographs 38. Cambridge and Ankara: McDonald Institute for Archaeological Research and British Institute of Archaeology at Ankara.

Russell, N. and Meece, S. (2005) Animal representations and animal remains at Çatalhöyük. Pp. 209–30 in I. Hodder (ed.), *Çatalhöyük Perspectives: Reports from the 1995–99 Seasons*. McDonald Institute Monographs 40. Cambridge and Ankara: McDonald Institute for Archaeological Research and British Institute of Archaeology at Ankara.

Sahlins, M. (1972) *Stone Age Economics*. New York: Aldine de Gruyter.

Sherratt, A. (1980) Water, soil and seasonality in early cereal cultivation. *World Archaeology* 2, 313–30.

Sherratt, A. (2006) The origins of farming in south-west Asia. *ArchAtlas* October 2006, 2nd edition. Internet edition http://www.archatlas.org/OriginsFarming/Farming.php (accessed 19 August 2007).

Stäuble, H. (1995) Radiocarbon dates of the earliest Neolithic in central Europe. *Radiocarbon* 37, 227–37.

Stehli, P. (1989) Merzbachtal—Umwelt und Geschichte einer bandkeramischen Siedlungskammer. *Germania* 67, 51–76.

Strien, H.-C. (2005) Familientraditionen in der bandkeramischen Siedlung bei Vaihingen/Enz. Pp. 189–97 in J. Lüning, C. Fridrich and A. Zimmermann (eds), *Bandkeramik in the 21st Century: Symposium in the Brauweiler Abbey near Cologne, 16th–19th of September 2002*. Rahden, Germany: Verlag Marie Leidorf.

Strien, H.-C. and Gronenborn, D. (2005) Klima- und Kulturwandel während des mitteleuropäischen Altneolithikums (58./57.–51./50. Jh. v. Chr.). Pp. 131–49 in D. Gronenborn (ed.), *Klimaveränderung und Kulturwandel in neolithischen Gesellschaften Mitteleuropas, 6700–2200 v. Chr*. Mainz: Verlag des Römisch-Germanischen Zentralmuseums.

Tomkins, P. (2007) Neolithic: Strata IX–VIII, VII–VIB, VIA–V, IV, IIIB, IIIA, IIB, IIA and IC Groups. Pp. 9–48 in N. Momigliano (ed.), *Knossos Pottery Handbook: Neolithic and Bronze Age (Minoan)*. London: British School at Athens.

Tomkins, P. (2008) Time, space and the reinvention of the Cretan

Neolithic. Pp. 21–48 in V. Isaakidou and P. Tomkins (eds), *Escaping the Labyrinth: The Cretan Neolithic in Context*. Sheffield Studies in Aegean Archaeology 8. Oxford: Oxbow Books.

Triantaphyllou, S. (1999) Prehistoric Makriyalos: a story from the fragments. Pp. 128–35 in P. Halstead (ed.), *Neolithic Society in Greece*. Sheffield Studies in Aegean Archaeology 2. Sheffield: Sheffield Academic Press.

Triantaphyllou, S. (2008) Living with the dead: a reconsideration of mortuary practices in the Greek Neolithic. Pp. 136–54 in V. Isaakidou and P. Tomkins (eds), *Escaping the Labyrinth: The Cretan Neolithic in Context*. Sheffield Studies in Aegean Archaeology 8. Oxford: Oxbow Books.

Twiss, K. C., Bogaard, A., Charles, M. P., Henecke, J., Russell, N., Martin, L. and Jones, G. (2009) Plants and animals together: interpreting organic remains from Building 52 at Çatalhöyük. *Current Anthropology* 50, 885–95.

Urem-Kotsou, D. and Kotsakis, K. (2007) Pottery, cuisine and community in the Neolithic of north Greece. Pp. 225–46 in C. Mee and J. Renard (eds), *Cooking Up the Past: Food and Culinary Practices in the Neolithic and Bronze Age Aegean*. Oxford: Oxbow Books.

Valamoti, S. (2004) *Plants and People in Late Neolithic and Early Bronze Age Northern Greece: An Archaeobotanical Investigation*. British Archaeological Reports, International Series 1258. Oxford: Archaeopress.

Van Beek, G. (1982) A population estimate for Marib: a contemporary tell village in north Yeman. *Bulletin of the American School of Oriental Research* 248, 61–7.

Wasse, A. (1997) Preliminary results of an analysis of the sheep and goat bones from 'Ain Ghazal, Jordan. Pp. 575–92 in H. G. K. Gebel, Z. Kafafi and G. O. Rollefson (eds), *The Prehistory of Jordan II. Perspectives from 1997*. Studies in Early Near Eastern Production, Subsistence, and Environment 4. Berlin: *ex oriente*.

Wasse, A. (2002) Final results of an analysis of the sheep and goat bones from 'Ain Ghazal, Jordan. *Levant* 34, 59–82.

Whitelaw, T. (1983) People and space in hunter-gatherer camps: a generalising approach in ethnoarchaeology. *Archaeological Review from Cambridge* 2, 48–66.

Whitelaw, T. (2004) Estimating the population of Neopalatial Knossos. Pp. 147–58 in G. Cadogan, E. Hatzaki and A. Vasilakis (eds), *Knossos: Palace, City, State*. London: British School at Athens.

Wright, K. (2000) The social origins of cooking and dining in early villages of western Asia. *Proceedings of the Prehistoric Society* 66, 89–121.

19. The Temporality of Materials: Occupation Practices in Eastern England During the 5th and 4th Millennia BC

Duncan Garrow

'The Mesolithic' and 'the Neolithic' have often been conceptualized as being different in terms of the temporality of occupation within each period. This paper focuses on the relationship between three main elements of the Neolithic 'package' – settlement, cereals and pottery – in order to understand the temporality of landscape occupation in Britain c. 5000–3000 BC. It suggests that recent debates for and against 'mobile' or 'permanent' occupation within the Neolithic have been unnecessarily polarized. Rather than trying to understand the temporality of occupation in these black and white terms, it argues that we ought to consider the degrees and qualities of sedentism in evidence at that time.

Introduction

Discussions of 'the Mesolithic' and 'the Neolithic' have long stressed differences between them in terms of the temporality of occupation. This has usually been in order to bolster the barrier of difference erected between the two sides. The very idea of a Neolithic 'revolution' or 'package' assumed a series of close relationships between different 'Neolithic' traits which could be distinguished clearly from their equivalents in the preceding period. Thus substantial settlements, along with the permanent occupation it was assumed they were associated with, were seen to have been directly linked to the uptake of cereal crops. All three elements came into force together, distinguishing 'the Neolithic' from the mobile hunting and gathering way of life which came before.

Across large parts of Europe, the association between archaeologically visible houses, relatively permanent occupation and cereal cultivation is fairly unproblematic. In the LBK-dominated Early Neolithic of Central Europe, it is possible to link the three elements with reasonable certainty (e.g. Bogaard 2004; but see Whittle 1996a). Interestingly, it is at either 'end' of the European Neolithic – the places on which this volume is focused – that the link becomes rather more problematic. In Southwest Asia, for example, there are substantial settlements in which people did not use pottery, while cereal domestication appears to have been initiated by people who might otherwise be considered hunter-gatherers. In Britain, it has been argued – if by no means universally accepted – that while elements of the Neolithic package were taken on board, people remained relatively mobile until well into the Bronze Age.

In this paper, I focus on the relationship between three main elements of the Neolithic 'package' – settlement, cereals and pottery – in order to understand the temporality of landscape occupation in Britain c. 5000–3000 BC. Faunal evidence is not considered as the acidic sandy soils typical of the sites under consideration ensured an almost complete absence of evidence. The character of the archaeological evidence – or at least that which has received most of our attention – leads me to focus primarily on the Neolithic, or the period immediately after the transition. However, the results of this focus nevertheless allow us to think harder about the similarities and differences between both 'periods', and thus also about the reality of this 'transition'. In recent years, within Mesolithic archaeology there have been calls for interpretations which move beyond small-scale, detailed, empirical studies towards an archaeology which addresses broader interpretive issues (e.g. Conneller and Warren 2006). The opposite might be said for the Neolithic: discussions of the temporality of occupation on that side of 'the transition' require rather fewer studies conducted at a broad interpretive scale, and rather more detailed, nuanced and specific considerations of the evidence.

The temporality of occupation: the story so far

The notion of a Neolithic in Britain where people did not live in permanent settlements first gained currency in the late 1980s and early 1990s, primarily as a result of Thomas' work (Thomas 1988; 1991). His important, and perhaps at times intentionally provocative, argument was based on a series of related points: that the construction of monuments does not require sedentary communities; that small-scale cereal growth does not require permanent occupation; and that the general absence of archaeological evidence for permanent dwellings was meaningful in terms of Neolithic practices, rather than post-depositional processes or excavation strategies in the present:

> The population of Neolithic Britain did not live in major timber-framed buildings, quite probably did not reside in the same place all year round, did not go out to labour in great walled fields of waving corn, were not smitten by over-population or soil decline, and much of their day-to-day food may have been provided by wild crops (Thomas 1991, 28).

Thomas' arguments in relation to the temporality of occupation had a substantial effect on Neolithic archaeology in Britain, and over the course of the subsequent decade there were many and varied accounts of an essentially mobile British Neolithic (e.g. Edmonds 1997; Whittle 1997; Pollard 1999). These writers joined Thomas in placing key emphasis on the almost total absence of archaeologically visible structures across much of southern Britain.

This account of a Neolithic in which most people lived most of the time in impermanent settlements has by no means met with universal approval. Over the past decade there has been substantial criticism of the 'mobility orthodoxy'. Barclay and Cooney, for example, have pointed out that there are substantial numbers of archaeologically visible houses in certain areas, especially Ireland and Scotland (Cooney 1997; 2001; Barclay 2000). More controversially, and entirely ignoring such regional variations, Rowley-Conwy has argued for a total return to an old-style Neolithic – complete with permanent houses, substantial cereal agriculture and waves of immigration from continental Europe – across the whole of Britain, Ireland and Scandinavia (Rowley-Conwy 2004).

In relation to these discussions of the temporality of Neolithic occupation, I make three suggestions. Firstly, that far too much emphasis has been placed on the presence/absence of houses in the Neolithic (see also Gibson 2003). Secondly, that debate has become unnecessarily polarized into oppositions between mobile and sedentary occupation, and wild and domestic plant foods, which has the unhelpful effect of creating further dualisms to run alongside the division between 'Mesolithic' and 'Neolithic'. Thirdly, that discussions have often been conducted at too wide a scale, compressing a broad range of evidence into a single narrative rather than exploring the full range of variability across space and time (see also Bradley 2003; 2007).

In the rest of this paper, in order to discuss the character of occupation between c. 5000 and 3000 BC in a way which escapes some of the generalizations of previous arguments, I will focus on one region of England: East Anglia (the counties of Norfolk, Suffolk and Cambridgeshire). I challenge two of the central orthodoxies which, to different degrees, have pervaded *both* sides of the debate so far. The first is that an absence of archaeologically visible structures necessarily implies settlement mobility. The second is that the relative quantities of cereals or wild plants within 'the Neolithic economy' can be used to understand 'the permanence of occupation'. Approached at this scale, I argue that both concepts are too generalized to be relevant in terms of the lives that people led. I suggest that, if we are to understand 'the transition' fully in any region of Britain, we need to move away from narratives which come down on either side of the mobile/permanent, wild/domestic fence which has been erected. In fact, to extend the horticultural metaphor further, the most effective course of action would be to knock down the fence altogether, view the gardens on either side of it as one, and then assess whether it needs re-erecting or not. As Marshall puts it, 'sedentism should be measured in degrees and qualities, not by presence or absence' (2006, 158).

The effect of this polarization of debate is particularly visible once past discussions of settlement 'mobility' – itself an extremely flexible concept (Kelly 1992; Jones 2000) – are scrutinized in more detail. The majority of writers have in fact described a variable and fluid scenario, in which 'mobility' can mean moving on at intervals of a few days or of several decades. Thus, even among those who would be characterized as generally favouring a mobile Neolithic over a sedentary one, we find arguments for 'short-term sedentism' (Whittle 1997, 21), and for settlements lasting 'a few years or so' (Pollard 2000, 364) or even 'a generation or more' (Edmonds 1997, 104) – periods of time not all that different from the lower estimates suggested on the other side of the fence (e.g. Rowley-Conwy 2003, 15). Furthermore, those arguing for the predominance of either wild or domesticated plant foods have usually done so while admitting that both were used (e.g. Richmond 1999; Thomas 1999; Rowley-Conwy 2000).

It is also important that many of these arguments have been conducted in relation to either negative circumstantial evidence (i.e. an absence of houses) or analogy with ethnographically documented situations very different to those of 5th/4th-millennium BC Northern Europe (discussed in more detail below). In the case study here my intention is not necessarily to support either camp, but simply to investigate positive evidence relating to the temporality of occupation within the materials generated as a consequence of 4th-millennium lives. In an attempt to shed the baggage which accompanies any discussion of 'the Neolithic' or 'the Mesolithic' (see Thomas 1988), in my own discussion of the evidence I use the date ranges instead: essentially, '5th millennium' is equivalent to 'Mesolithic' and '4th millennium' to 'Neolithic'.

Landscape and place in East Anglia

The landscape of East Anglia is varied ecologically, if not topographically (much of the region lies below 30m OD). The area which has seen the most archaeological work is the fenland, an area of marshland leading to open water which is now drained but which would have provided a wealth of wetland and dryland resources during the 5th and 4th millennia BC (Hall and Coles 1994). Other major ecological zones include the clay 'uplands', which rarely rise above 100m OD, a chalk ridge running north–south up to the coast, and the Breckland chalky sands towards the centre of the region.

Artefactual evidence recovered through incidental surface collection suggests that activity was spread widely across East Anglia during the 5th millennium (Jacobi 1984; Sturt, this volume). However, there have been very few detailed archaeological investigations in terms of intensive fieldwalking or excavations. Consequently, our understanding of occupation practices in specific places comes primarily from environmental rather than artefactual evidence. As a result of the substantial number of palynological studies carried out over the past fifty years, a series of separate artificial clearings – some apparently covering a sizeable area – have been documented (Clark and Godwin 1962; Sims 1983; Smith *et al.* 1989; French and Pryor 1993; Wiltshire 2007). Mellars has outlined why it would have been beneficial to create such clearings; reasons include the encouragement of a broader spectrum of plants and animals that could then be gathered and hunted (Mellars 1976). A degree of movement between the different ecological zones of the fenland basin or coast and the upland forested clayey areas has generally been suggested (e.g. Tilley 1979), although the frequency at which this mobility cycle occurred is debatable (Smith *et al.* 1989).

If we could complement this level of environmental information with a parallel understanding of the 5th millennium artefactual evidence we would be able to produce a much more satisfying account. However, the kind of work required to understand flint scatters of this date has not been carried out at a broad scale in the region (although see Sturt 2006). The fact that 4th-millennium archaeology is much more readily visible has led to a very uneven narrative. We need to think harder about working with, and get better at understanding, the 5th-millennium evidence. Nevertheless, on the basis of the limited evidence recovered, we can argue that the temporalities of occupation were comparable with those elsewhere in Britain. We may well be looking at a picture in which certain places were occupied persistently if not permanently (e.g. Barton *et al.* 1995), while others were visited for certain specific reasons many times over, and others still became archaeologically visible as a result of single, small-scale visits (e.g. Conneller and Schadla-Hall 2003; McFadyen 2006).

In the 4th millennium BC the environmental picture is not vastly different from the earlier one. Cereals certainly appear in the pollen sequences around this time (Clark and Godwin 1962; Smith *et al.* 1989; Wiltshire 2007), but there is little evidence of woodland clearance on any substantial scale. While it is feasible that greater numbers of clearances were made, these do not seem to have been any larger in size than those cleared previously. In some cases, there is even evidence that forest was left to regenerate during the early part of the 4th millennium (Smith *et al.* 1989).

As mentioned above, the archaeological evidence for the 4th millennium is better understood than that of the preceding millennium. While East Anglia has fewer upstanding monuments than many other regions of Britain, both causewayed enclosures (Pryor 1998; Evans and Hodder 2006; Evans *et al.* 2006) and long barrows (Hogg 1940; Ellis 2004; Evans and Hodder 2006) have been excavated, and many more possible examples identified through aerial photography (Garrow 2006). The 4th-millennium flint scatters of East Anglia are among the best studied in Britain (e.g. Hall and Coles 1994; Healy 1996; Edmonds *et al.* 1999), and suggest a similar picture to that of the previous millennium: some persistent places visited for a wide range of reasons, other places occupied for quite specific tasks, and others visited perhaps only once (see, for example, Edmonds *et al.* 1999; Garrow 2006; Sturt 2006).

While flint scatters certainly suggest occupation of a sort, it is not clear whether they represent what we might want to term 'settlement' in the more general sense of the word (see Bradley 1987): flint scatters can be created by a wide range of activities. If we were to search for settlement sites by looking for 4th-millennium buildings, we would not get very far: there are in total only three structures potentially of that date, at Padholme Road, Fengate (Pryor 1974), Co-op, Fengate (Gibson 1998) and Spong Hill (Healy 1988). Considerable doubt about the date of all three structures remains, and as they all had very little material culture associated with them it is also difficult to ascertain their function. All three might be considered too small to have acted effectively as 'houses'. The number of structures found has remained extremely low despite the fact that numerous massive-scale development-led excavations, undertaken by skilled and experienced archaeologists, have taken place in the region over the past two decades. Greater numbers of Late Bronze Age and Iron Age structures are found regularly on single archaeological sites. We can be very confident that the absence of 4th-millennium buildings in this region is real.

In contrast to these structures, East Anglia does have a substantial body of evidence for another form of site which many have suggested would have related to 'settlement' – clusters of pits (Fig. 19.1). Most writers agree that these

Fig. 19.1. One of a cluster of 4th-millennium BC pits being excavated by Luis Huscroft at Kilverstone, Norfolk. The finds from the pit are visible in the foreground.

pits – which have been found close to monuments and flint scatters or on their own far from any other contemporary features, entirely in isolation or in groups of up to 200 or more – were related to occupation of some sort. They are found almost exclusively in places which might be considered especially suitable for 'settlement' (close to rivers, on light soils, on low-lying ground: see Garrow 2007b), and their fills, which usually include charcoal and other burnt material, broken pottery and flint flakes, are redolent of what might be termed 'domestic' activity.

In total, 646 pits dated securely to the 4th millennium BC had been found in East Anglia by the summer of 2002, on 33 different sites (Garrow 2006). It is the evidence from this rich source of artefactual information that I will concentrate on in the next section of this paper.

Pits in East Anglia

Pits have been increasingly discussed in recent accounts of the Neolithic in Britain. This change has arguably occurred for three main reasons: a substantial increase in the number of known sites following the introduction of developer-funded archaeology; a recognition that deposition in a variety of contexts was an important aspect of people's lives; and renewed interest in defining the character of 'settlement'. While most writers have seen them as being related to settlement occupation (e.g. Thomas 1999; Pollard 2000), they have not been treated as interpretatively equivalent to post-built structures. At times, their presence has been considered important primarily as a positive indicator of the desired but consistently absent house plans (e.g. Bradley 2007; C. Evans pers. comm.). Generally, the more ephemeral archaeological evidence that pits alone are seen to represent has been viewed as signifying more ephemeral 'settlement' (e.g. Edmonds 1999, 19; Whittle 1999, 64).

The association made between pits and ephemeral settlement is at least partly understandable. It is hard to view a couple of shallow scoops in the ground containing a few scraps of pottery and flint flakes as the archaeological signature of anything more than a short visit to that place. However, a number of sites have been found, particularly in East Anglia, which signify occupation of a rather more substantial kind: at Kilverstone, 226 pits were recovered (Garrow *et al.* 2006); at Hurst Fen there were at least 200 (Clark *et al.* 1960); at Broom Heath there were 67 (Wainwright 1972); at Spong Hill there were 56 (Healy 1988) and at Barleycroft Paddocks there were 27 (Evans and Knight 1997): while one possible post-hole structure was found at Spong Hill, it was associated only with one small cluster of pits and its date remains uncertain. The earthwork enclosure at Broom Heath appears to post-date the pits (Wainwright 1972, 12).

It is very difficult to link the number of pits on a site directly with the duration of occupation; if they had wanted to, it would have been possible for people to dig large numbers of pits very quickly, especially as most of these sites are on soft sandy soils. However, two of the material types found within these pits afford us further insight into the 'amounts' of occupation we are dealing with. In the following section, I will consider what the quantities of pottery found within them suggest about the temporality of occupation, before moving on to the temporal implications of their charred plant remains.

Pits, pottery and the temporality of occupation

On all of the sites in East Anglia which produced large numbers of pits, large numbers of vessels were found. All of these vessels were broken, and the vast majority were only partially represented (sometimes by very small numbers of sherds); substantial amounts of material never made it into the pits. At Hurst Fen, it was estimated that 283–317 different pots were represented, at Broom Heath 272, at Kilverstone there were at least 150, and at Spong Hill 113 (Clark *et al.* 1960, 228; Wainwright 1972, Appendix 3; Knight in Garrow *et al.* 2006; Healy 1988, chapter 3). Even on some of the sites with far fewer pits, the number of vessels was impressive. At Eaton Heath, for example, pieces of 26 different pots were found in just 12 pits, and at Hall Farm Reservoir a single pit contained the remnants of 12 vessels (Wainwright 1973; Birks 2000) (all vessel quantities are those given in the original site reports).

We must be cautious in reading the duration of occupation at these sites from the number of vessels represented. Nevertheless, as there was no clear evidence that pots had been deliberately smashed or brought specially to the sites for deposition, it does seem reasonable to assume that the material within the pits was 'created' (i.e. pots broken, flint working waste produced, etc.) during the course of occupation at these sites. I am usually extremely reluctant to employ ethnographic analogy – not least because, in making a comparison between practices of the archaeological past and practices observed in the present, it is usually necessary to ignore the fact that we are dealing with entirely different contexts. However, remaining mindful that we must be careful not to make *direct* comparisons across such vast distances in time, space and context, I would nevertheless like to explore the rates at which pots are broken in contemporary (or near contemporary) situations in an experimental attempt to escape the vagueness of previous estimates as to the length of 4th-millennium occupations.

As part of his study of deposition on Iron Age settlement sites, Hill carried out a wide-ranging survey of ethnographic and ethnoarchaeological studies which had investigated the rates at which pots are broken (Hill 1995, 129–31). Having surveyed the literature, he arrived at an average figure for the number of pots broken per 'household' per year of 5.4 (overall, numbers varied between 2.4 and 22). It is important to stress that the application of these figures to the past is a *heuristic* methodology. In reality, during the 4th millennium, there may well have been significant variability either side of the figures used in terms of pot breakage rates (as reflected by the different bars in Fig. 19.4). Similarly, and perhaps more importantly, it is impossible to know what proportion of the pots that were originally broken at these sites actually made it into the pits. As a result of these considerations, the numbers below must simply be viewed as *a way of thinking through* the vessel quantities recovered on 4th-millennium sites.

If we take Hill's average figure of 5.4 vessels per year and transfer it across space and time to pit sites in East Anglia, we might thus suggest that the c. 300 vessels at Hurst Fen may have been broken by the equivalent of one 'household' occupying that place over the course of 55 years (or five 'households' over the course of a decade). If the complete range of breakage rates stated by Hill were used, this estimate would vary between 14 years (using the figure of 22 pots) and 125 years (using the figure of 2.4 pots). It is also important to be aware that these figures are based on the vessels which actually made it into the pits at all; the original numbers may well have been significantly higher. Similarly, the material at Kilverstone (151 pots) equates to one 'household' occupying that place for 25 years (this figure could vary between 7 and 63 years), or five 'households' for five years. Even the single pit at Hall Farm Reservoir, with 12 vessels, could be viewed as being equivalent to two years' occupation (or between six months and five years).

On this basis it can certainly be argued that the link between pits and ephemeral and highly mobile occupation has been overplayed. Whatever the numbers set out above actually do imply in a quantitative sense, it is clear that many of these sites – all but one of which produced no evidence whatsoever for any structures – did witness substantial occupation(s). In order to think through in more detail the temporality of this occupation, I would like to turn now to another form of evidence within the pits – the charred plant remains.

Pits, plant remains and the temporality of occupation

Within accounts of the 4th millennium in Britain, it is generally agreed that both cultivated cereals and collected wild foods contributed to people's diets (e.g. Thomas 1999; Jones 2000; Rowley-Conwy 2000). However, as mentioned above, there has been a tendency to accentuate one body of evidence at the expense of the other, depending on which side of the mobile/sedentary divide an author has placed themselves. On one side it has been argued that, while a mixture of wild foods and cereals was undeniably used, the former predominated throughout the Neolithic, with the latter playing a more symbolic role (e.g. Richmond 1999; Thomas 1999). On the other side it has been suggested that wild foods, and particularly hazelnuts, are simply more visible due to biases of preservation, and that the Neolithic economy was actually based almost entirely on cereals (e.g. Rowley-Conwy 2000). Each camp has used the same evidence, but simply looked at it in an entirely different way. Consequently, it might well be argued, as Thomas himself put it, that 'a proper understanding of agricultural ... practices in Neolithic Britain is far from complete' (1999, 32).

The archaeobotanical evidence recovered from pit sites in East Anglia suggests that it is unwise to come down on either side of the wild/cultivated fence. Of 33 Early Neolithic pit sites known in the region, 11 had been flotation sampled for charred plant remains. Of these sites, 9 produced wild plant remains, with hazelnut shells predominant but apple also in evidence, while 7 produced charred cereal remains (dominated by emmer and to a lesser extent barley). A further 2 sites, excavated before environmental flotation samples were taken routinely, produced secondary evidence for cereals in the form of seed impressions on pottery (Clark *et al.* 1960, 211; Wainwright 1972, 21; see Fig. 19.2). Of the four sites where cereals were not found, in three cases only a single feature was sampled, and in the fourth sandy soils led to extremely poor preservation conditions even for much more recent contexts. Broken down into individual feature-specific samples, the picture is slightly less well-balanced but

Fig. 19.2. Cereal impressions in pottery from Hurst Fen (left) and Broome Heath (centre and right) (Clark et al. *1960, plate XXV; Wainwright 1972, plate X).*

nevertheless comparable: of the 61 pits sampled in total, 52 (85 per cent) produced hazelnuts and 35 (57 per cent) cereal grains.

Considering that every one of these sites was located on sand or gravel subsoils (leading to extremely poor preservation conditions, particularly for charred seeds), and that for a variety of different reasons hazelnut shells are much more likely to be preserved in the first place (Jones 2000), it can be argued that both wild and cultivated plant foods are actually well represented. Given the problems and biases of initial preservation and post-depositional survival, it is almost impossible to assess effectively the relative quantities of wild and cultivated resources used on these sites. What seems most important is that on 64 per cent of all sites sampled, and on 100 per cent of those sites where more than one sample was taken, both cereals and hazelnuts were found. On the basis of this evidence, they *both* formed an important part of diets.

In order to consider the implications that this archaeobotanical picture has for understanding the temporality of occupation, it is helpful to review previous discussions of the relationship between wild foods/cereals and mobility/permanence. As a result of the way in which the Neolithic has been perceived historically, it has been much easier to argue *for* a link between the cultivation of domesticated plants and permanent settlement rather than against. Consequently, those wishing to argue for substantial settlement mobility have felt it necessary to draw on ethnographic analogies to bolster their case, focusing on a small number of communities around the world who do grow crops while maintaining a degree of mobility (e.g. Whittle 1996b, 17; Thomas 1999, 25). However, these ethnographic examples deal with the Americas rather than Europe, and actually relate to communities growing different crops with requirements different from the wheats and barleys grown in Britain during the 4th millennium. Bogaard has pointed out that while some members of the Penobscot of Maine (the example cited by Whittle 1996b, 17) do live partly mobile lives, the crops they grow are actually located adjacent to permanently occupied villages (Bogaard 2004, 51). Similarly, the Siriono of Bolivia (one of the examples cited by Thomas 1999, 25) – who, it should be noted, were described by their ethnographer as only 'semi-nomadic' (Speck 1940, 29) – tended to plant their crops mainly in the rainy season, during which they actually lived in one place for six months at a time (Speck 1940).

Among the archaeobotanical community there appears to be consensus that in Neolithic Britain it would have been very risky, to say the least, and probably disastrous, to leave small fields or garden plots of cereals entirely unattended (M. Jones pers. comm.). It is impossible to find experimental work which exactly parallels the situation proposed by some for 4th-millennium Britain, with crops being sown and then left to grow untended until people returned to the spot several months later. However, other experiments have shown that – even in well-tended fields – the seed, then the seedlings and then the crops themselves can be eaten or destroyed by a wide variety of animals, including mice, beetles, birds and larger mammals (Willcox 1999; Dark and Gent 2001); these problems would have been severely exacerbated if the plots were abandoned. Relatively newly cleared plots are also much more effective if weeded throughout the growing season (Bogaard 2004, 42). Among those who have studied plant remains as a specialism there is widespread agreement that if cereals were to have formed any useful component of anyone's diet in the Neolithic they would have needed looking after throughout the growing season: there would have to have been 'a certain degree of sedentism, by at least part of the community' (Jones 2000, 83).

Given the large number of sites with cereal grains in East Anglia, it therefore seems reasonable to argue – in accordance with the pottery evidence discussed in the previous section – that, despite the superficially ephemeral character of the archaeology represented by pits, the occupation they were associated with was not necessarily short-term at all.

Landscape and place in transition in Eastern England

This discussion of occupation practices is strengthened by the fact that artefactual and archaeobotanical evidence have been considered together. The pottery from pit sites

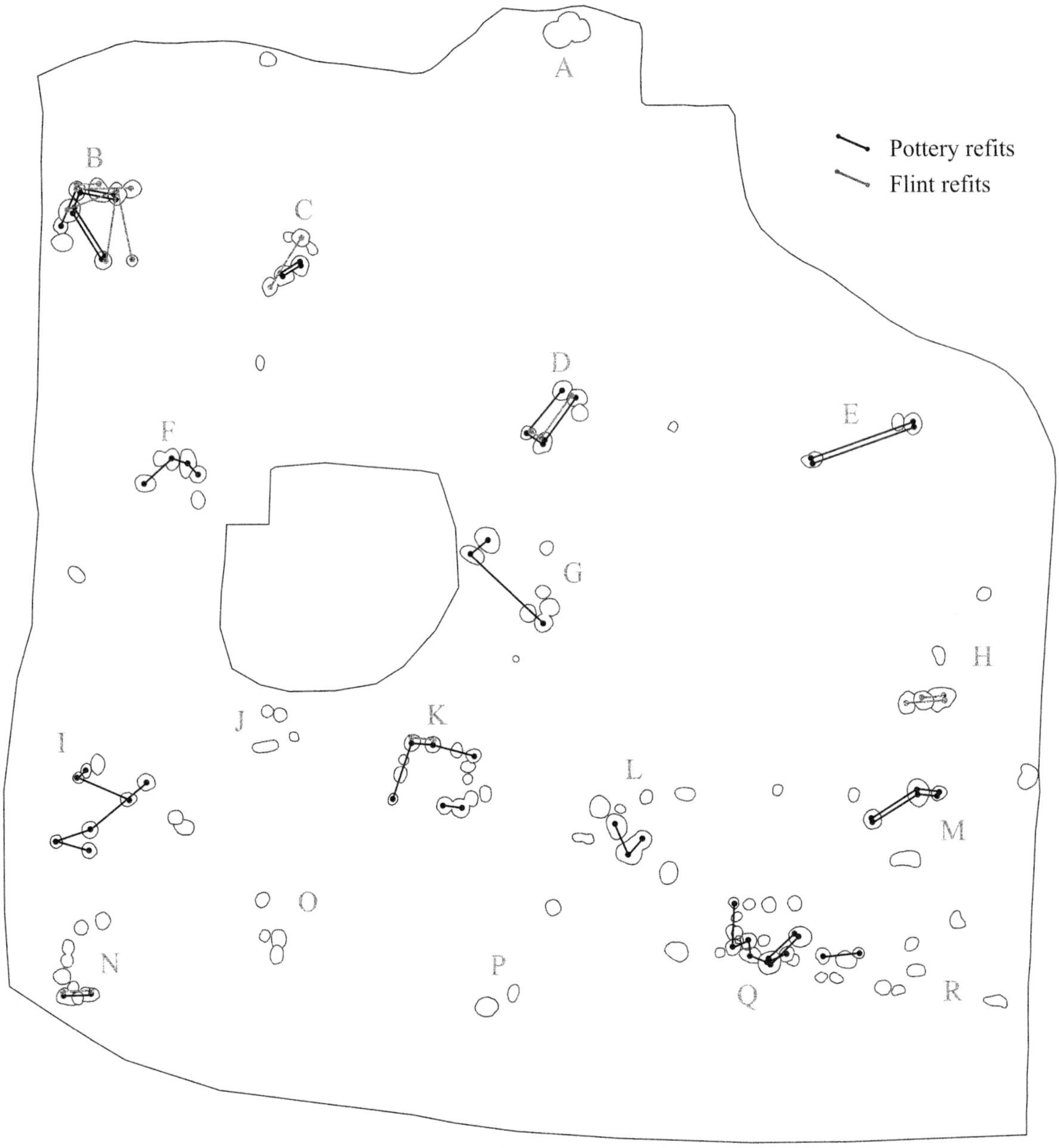

Fig. 19.3. Kilverstone Area E, showing refitting pottery and flint and named pit clusters.

suggests that the 'amounts' of occupation with which some of them were associated were very substantial indeed. This was the case despite the fact that, other than in a single (and by no means conclusive) case, no archaeologically visible structures were identified on them. The link between relatively long-term 'settlement' and archaeologically visible buildings is not nearly as direct or straightforward as it has been made to seem in the past. The archaeobotanical evidence indicates that people ate a combination of wild plant foods and cultivated cereals. The implications of the fact that cultivated resources were grown are highly significant in terms of our understanding of the temporality of occupation, as they would have been tended throughout the growing season. Although there is no direct evidence for the link, it therefore seems very likely that, if pit sites were occupied for substantial periods of time, the crops would have been grown at or close to them.

In order to explore the character of occupation in one of these places in more detail, I will turn briefly to the evidence from Kilverstone in Norfolk. The site has been discussed in substantial detail elsewhere (Garrow *et al.* 2005; 2006) so here I will focus only on the pottery and the charred seeds. Within Area E at Kilverstone, 138 pits were found within an excavation area of c. 50 × 40m (Fig. 19.3). Those pits contained the remains of 106 different vessels. Of the 22 pits sampled for environmental remains, all 22 produced charred hazelnuts, while 18 (82 per cent) produced charred seeds.

The pits within Area E fell into fairly distinct, spatially discrete clusters. These clusters proved to be materially

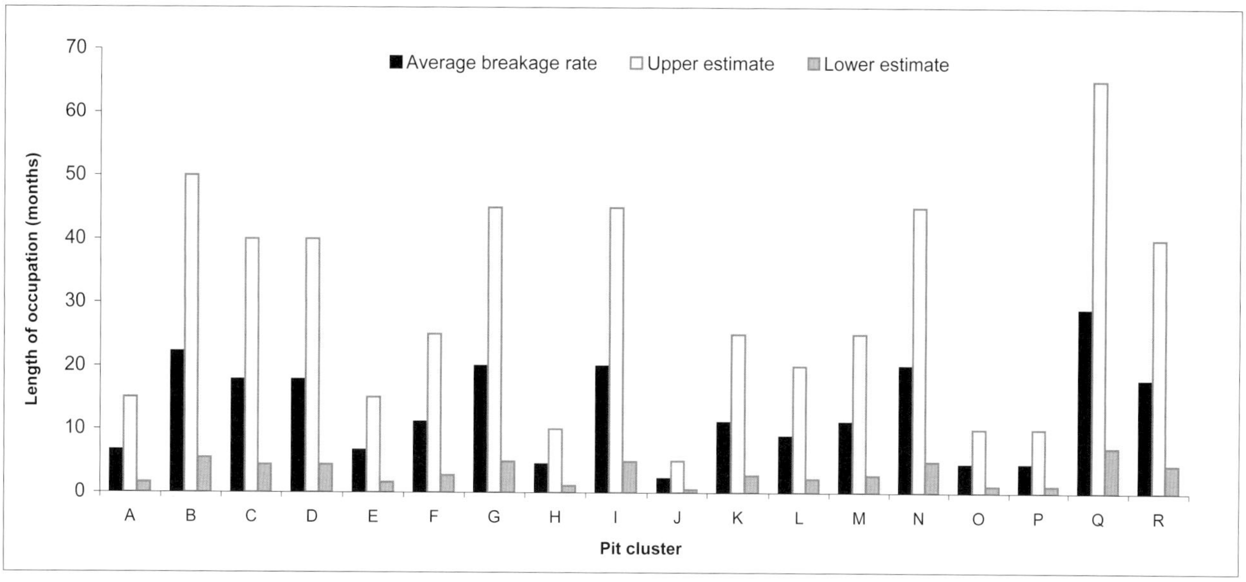

Fig. 19.4. The temporality of occupation at Kilverstone. Three different estimates (based on Hill's average, lowest and highest figures for pot breakage rates) are given as to the length of time with which each of the pit clusters may have been associated, given the number of different vessels each contained.

distinct as well: while numerous flint and pottery refits were made between pits in the same cluster, none could be established between pits in different clusters. I have argued, with colleagues, elsewhere that the material separation between the clusters at Kilverstone – and on several of the other sites discussed in this paper – could reflect a temporal separation, each being created as a result of separate occupations (Garrow *et al.* 2005; Garrow 2007a). In order to establish an even finer-grained account of settlement, I will consider the lengths of the various 'pulses' of occupation represented by each of the clusters at Kilverstone, again using Hill's estimates of pot breakage rates.

Figure 19.4 details the equivalent length of occupation (for one 'household') that each cluster may have represented, given the number of broken pots it contained (using Hill's lowest, highest and average figures for pot breakage rates). I am not suggesting that Kilverstone was necessarily occupied by one 'household'. Similarly, it is worth reiterating that, during the Neolithic, there may well have been significant variability either side of the recent/contemporary figures used in terms of pot breakage rates. As discussed above, I am simply trying to provide a way of *thinking through* the temporal implications of the evidence. According to the pottery breakage rates, the different clusters might therefore represent periods of occupation varying between 2 and 29 months using Hill's average figure, and between 2 weeks and 65 months using his upper and lower values. If there were periods of time when the site was not occupied, the average timescales would nevertheless fit the archaeobotanical record very comfortably, with all but four of the clusters lasting at least six months.

The broad spectrum of material culture recovered on site provides insight into the different activities that went on there: flint was knapped, pots used and broken, fires lit; some fishing, and probably hunting as well, took place, and we might assume that domestic animals were kept; cereals were grown and wild resources gathered. Even during periods of time when the site was occupied it is likely that some, and possibly most, people would have left the site for a period of days, weeks or even months, to take their domesticated animals to summer pasture, to exploit other food sources at other specific times, or to acquire things (flints, pots) from other places. In many ways, this might be viewed as a typical 4th-millennium place.

Conclusion: landscapes in transition

Owing to the very different character of the evidence in East Anglia, as things stand it is difficult to compare occupation practices on either side of 'the transition'. Within Mesolithic archaeology generally classic notions such as 'the seasonal round' and permanent 'homebases' have recently been called into question (Jochim 1991; Spikins 1999; Conneller 2005). In their place, more fluid, flexible and contextually specific understandings of occupation have been suggested. Again, Neolithic archaeology could learn from this shift away from rigid narratives (which focus on *permanent* settlements or *domesticated* plants). While the description of Kilverstone above does indeed sound like a 'typical' Neolithic place, it is important to realize that it would have been one place within a network of other, different, but equally 'typical' places. These were occupied in quite different ways, and thus left entirely different archaeological signatures: places which witnessed only one or two shorter-term visits, leaving a few pits containing relatively small amounts of material culture; places used persistently for specific flint-working tasks, leaving surface scatters heavily biased towards particular tools but with no cut features; places where small patches of forest were cleared but never settled, perhaps leaving a broken axe or two but little else. Just as sedentism should be 'measured in degrees and qualities, not by presence or absence' (Marshall 2006, 158), so too should landscape occupation. During the 4th millennium, the picture in East Anglia is not one of neat villages with fields around post-built houses, but nor is it one of people constantly on the move, never settling down. According to the evidence considered here, settlement was neither completely permanent nor completely mobile.

As far as it is possible to tell, the picture is not entirely dissimilar during the 5th millennium. While small-scale surface scatters suggest relatively fleeting or insubstantial occupations, the large and sustained clearings documented in the pollen record indicate places that were occupied and maintained persistently for long periods of time. It is quite possible that, during the 4th millennium, the choice to grow cereal crops (which necessitated more sustained occupation in certain places) and a slightly altered attitude to 'place' (suggested by cut features such as causewayed enclosures and pits) developed hand-in-hand, resulting in subtly different temporalities of landscape occupation.

This paper developed out of a frustration with previous discussions of the character of landscape occupation either side of the 'transition'. In the past, debate has been polemical, and arguments have come to be constructed on the basis either of negative evidence (an absence of houses), or often rather vague considerations of the archaeological evidence. I hope to have shown that it is possible to take discussion beyond this position. Landscape, place and time are perhaps best understood *in* transition, not either side of *a* transition. In focusing on one particular region, I have been careful to avoid the pitfalls of discussing occupation practices at too large a scale. Consequently, it was not necessary to force a variety of different and often contradictory evidence into one, uniform, all-encompassing narrative. Similarly, I have focused not on things that we do not have (archaeologically visible houses) but on the *positive* material residues of occupation. I have also considered both archaeobotanical and artefactual evidence together, rather than in isolation; each material was able to shed a subtly different light on occupation. The material generated as a result of 4th-millennium lives does have something to tell us, and we should be confident in using it to gain a full understanding of the temporality of occupation at that time.

Acknowledgements

I would like to thank Bill Finlayson and Graeme Warren for organizing the 'Landscapes in Transition' workshop; all of the delegates for their comments on my paper at the time and since, and for making it such an enjoyable and productive event; Julie Gardiner and the Prehistoric Society for permission to reproduce the images in Fig. 19.2; Emma Beadsmoore and Mark Knight for analysing the material dynamics of the site at Kilverstone with me (our interpretation was entirely a joint project); Rachel Ballantyne, Amy Bogaard and Martin Jones for their archaeobotanical advice (in this case, any faults within my subsequent translations of what they said really are my own); Chantal Conneller and Fraser Sturt for Mesolithic advice; and Anwen Cooper, Fraser Sturt, Lesley McFadyen and Marie Louise Stig Sørensen for their insightful comments on earlier written versions of this paper.

Bibliography

Barclay, G. (2000) Between Orkney and Wessex: the search for the regional Neolithics of Britain. Pp. 275–85 in A. Ritchie (ed.), *Neolithic Orkney in its European Context*. Cambridge: McDonald Institute for Archaeological Research.

Barton, N., Berridge, P., Walker, M. and Bevins, R. (1995) Persistent places in the Mesolithic landscape: an example from the Black Mountain uplands of south Wales. *Proceedings of the Prehistoric Society* 61, 81–116.

Birks, C. (2000) *Report on an Archaeological Watching Brief at Hall Farm Reservoir, Croxton*. Unpublished Norfolk Archaeological Unit Report 518.

Bogaard, A. (2004) *Neolithic Farming in Central Europe: An Archaeobotanical Study of Crop Husbandry Practices*. London: Routledge.

Bradley, R. (1987) Flint technology and the character of Neolithic settlement. Pp. 181–6 in A. Brown and M. Edmonds (eds), *Lithic Analysis and Later British Prehistory*. British Archaeological Reports, British Series 162. Oxford: BAR.

Bradley, R. (2003) Neolithic expectations. Pp. 218–22 in I. Armit, E. Murphy, E. Nelis and D. Simpson (eds), *Neolithic Settlement in Ireland and Western Britain*. Oxford: Oxbow Books.

Bradley, R. (2007) *The Prehistory of Britain and Ireland*. Cambridge: Cambridge University Press.

Clark, J. G. D. and Godwin, H. (1962) The Neolithic in the Cambridgeshire Fens. *Antiquity* 36, 10–23.

Clark, J. G. D., Higgs, E. and Longworth, I. (1960) Excavations at the Neolithic site at Hurst Fen, Mildenhall, Suffolk (1954, 1957 and 1958). *Proceedings of the Prehistoric Society* 26, 202–45.

Conneller, C. (2005) Moving beyond sites: Mesolithic technology in the landscape. Pp. 42–55 in N. Milner and P. Woodman (eds), *Mesolithic Studies at the Beginning of the 21st Century*. Oxford: Oxbow Books.

Conneller, C. and Schadla-Hall, T. (2003) Beyond Star Carr: the Vale of Pickering in the tenth millennium BP. *Proceedings of the Prehistoric Society* 69, 85–105.

Conneller, C. and Warren, G. (2006) *Mesolithic Britain and Ireland: New Approaches*. Stroud: Tempus.

Cooney, G. (1997) Images of settlement and the landscape in the Neolithic. Pp. 23–31 in P. Topping (ed.), *Neolithic Landscapes*. Oxbow Monograph 86. Oxford: Oxbow Books.

Cooney, G. (2001) Bringing contemporary baggage to Neolithic landscapes. Pp. 165–80 in B. Bender and M. Winer (eds), *Contested Landscapes: Movement, Exile and Place*. Oxford: Berg.

Dark, P. and Gent, H. (2001) Pests and diseases of prehistoric crops: a yield 'honeymoon' for early grain crops in Europe? *Oxford Journal of Archaeology* 20/1, 59–78.

Edmonds, M. (1997) Taskscape, technology and tradition. *Analecta Praehistorica Leidensia* 29, 99–110.

Edmonds, M. (1999) *Ancestral Geographies of the Neolithic: Landscapes, Monuments and Memory*. London: Routledge.

Edmonds, M., Evans, C. and Gibson, D. (1999) Assembly and collection – lithic complexes in the Cambridgeshire fenlands. *Proceedings of the Prehistoric Society* 65, 47–82.

Ellis, C. (2004) *A Prehistoric Ritual Complex at Eynesbury, Cambridgeshire*. East Anglian Archaeology Occasional Paper 17. Salisbury: Trust for Wessex Archaeology.

Evans, C. and Hodder, I. (2006) *A Woodland Archaeology: Neolithic Sites at Haddenham, Volume 1*. Cambridge: McDonald Institute for Archaeological Research.

Evans, C. and Knight, M. (1997) *The Barleycroft Paddocks Excavations, Cambridgeshire*. Unpublished Cambridge Archaeological Unit Report 218.

Evans, C., Edmonds, M. and Boreham, S. (2006) 'Total archaeology' and model landscapes: excavation of the Great Wilbraham causewayed enclosure, Cambridgeshire 1975–6. *Proceedings of the Prehistoric Society* 72, 113–62.

French, C. and Pryor, F. (1993) *The South-west Fen Dyke Survey Project, 1982–86*. East Anglian Archaeology 59. Peterborough: Fenland Archaeological Trust.

Garrow, D. (2006) *Pits, Settlement and Deposition during the Neolithic and Early Bronze Age in East Anglia*. British Archaeological Reports, British Series 414. Oxford: BAR.

Garrow, D. (2007a) It's 17km as the crow flies ... Neolithic journeys seen through the material at either end. Pp. 45–53 in V. Cummings and R. Johnston (eds), *Prehistoric Journeys*. Oxford: Oxbow Books.

Garrow, D. (2007b) Placing pits: landscape occupation and depositional practice during the Neolithic in East Anglia. *Proceedings of the Prehistoric Society* 73, 1–24.

Garrow, D., Beadsmoore, E. and Knight, M. (2005) Pit clusters and the temporality of occupation: an earlier Neolithic site at Kilverstone, Thetford, Norfolk. *Proceedings of the Prehistoric Society* 71, 139–57.

Garrow, D., Lucy, S. and Gibson, D. (2006) *Excavations at Kilverstone, Norfolk 2000–02: An Episodic Landscape History*. East Anglian Archaeology 113. Cambridge: Cambridge Archaeological Unit.

Gibson, A. (2003) What do we mean by Neolithic settlement? Some approaches, 10 years on. Pp. 136–45 in I. Armit, E. Murphy, E. Nelis and D. Simpson (eds), *Neolithic Settlement in Ireland and Western Britain*. Oxford: Oxbow Books.

Gibson, D. (1998) *Archaeological Excavations at the Co-op site, Fengate*. Unpublished Cambridge Archaeological Unit Report 264.

Hall, D. and Coles, J. (1994) *Fenland Survey. An Essay in Landscape and Persistance*. English Heritage Archaeological Report 1. London: English Heritage.

Healy, F. (1988) *The Anglo-Saxon Cemetery at Spong Hill, North Elmham, Part VI: Occupation during the Seventh to Second Millennia BC*. East Anglian Archaeology 39. Dereham: Norfolk Archaeological Unit.

Healy, F. (1996) *The Fenland Project, No. 11: The Wissey Embayment: Evidence for Pre-Iron Age Occupation Accumulated Prior to the Fenland Project*. East Anglian Archaeology 78. Dereham: Fenland Project Committee.

Hill, J. D. (1995) *Ritual and Rubbish in the Iron Age of Wessex*. British Archaeological Reports, British Series 242. Oxford: BAR.

Hogg, A. (1940) A long barrow at West Rudham, Norfolk. *Norfolk Archaeology* 27, 315–31.

Jacobi, R. (1984) The Mesolithic of northern East Anglia and contemporary territories. Pp. 43–76 in J. Barringer (ed.), *Aspects of East Anglian Prehistory*. Norwich: Geo Books.

Jochim, M. (1991) Archaeology as long-term ethnography. *American Anthropologist* 93, 309–19.

Jones, G. (2000) Evaluating the importance of cultivation and collecting in Neolithic Britain. Pp. 79–84 in A. Fairbairn (ed.), *Plants in Neolithic Britain and Beyond*. Neolithic Studies Group Seminar Papers 5. Oxford: Oxbow Books.

Kelly, R. (1992) Mobility/sedentism: concepts, archaeological measures, and effects. *Annual Review of Anthropology* 21, 43–66.

McFadyen, L. (2006) Landscape. Pp. 121–38 in C. Conneller and G. Warren (eds), *Mesolithic Britain and Ireland: New Approaches*. Stroud: Tempus.

Marshall, Y. (2006) Introduction: adopting a sedentary lifeway. *World Archaeology* 38/2, 153–63.

Mellars, P. (1976) Fire ecology, animal populations and man: a study of some ecological relationships in prehistory. *Proceedings of the Prehistoric Society* 42, 15–45.

Pollard, J. (1999) 'These places have their moments': thoughts on settlement practices in the British Neolithic. Pp. 76–93 in J. Bruck and M. Goodman (eds), *Making Places in the Prehistoric World: Themes in Settlement Archaeology*. London: UCL Press.

Pollard, J. (2000) Neolithic occupation practices and social ecologies from Rinyo to Clacton. Pp. 363–9 in A. Ritchie (ed.), *Neolithic Orkney in its European Context*. Cambridge: McDonald Institute for Archaeological Research.

Pryor, F. (1974) *Excavation at Fengate, Peterborough, England: The First Report*. Royal Ontario Museum Archaeology Monograph 3. Toronto: Royal Ontario Museum.

Pryor, F. (1998) *Etton – Excavations of a Neolithic Causewayed Enclosure near Maxey, Cambridgeshire 1982–7*. English Heritage Archaeological Report 18. London: English Heritage.

Richmond, A. (1999) *Preferred Economies. The Nature of the Subsistence Base throughout Mainland Britain during Prehistory*. British Archaeological Reports, British Series 290. Oxford: BAR.

Rowley-Conwy, P. (2000) Through a taphonomic glass, darkly: the importance of cereal cultivation in prehistoric Britain. Pp. 43–53 in J. Huntley and S. Stallibrass (eds), *Taphonomy and Interpretation*. Oxford: Oxbow Books.

Rowley-Conwy, P. (2003) No fixed abode? Nomadism in the northwest European Neolithic. Pp. 115–44 in G. Burenhult and S. Westergaard (eds), *Stones and Bones: Formal Disposal of the Dead in Atlantic Europe during the Mesolithic–Neolithic Interface 6000–3000 BC*. British Archaeological Reports, International Series 1201. Oxford: BAR.

Rowley-Conwy, P. (2004) How the West was lost: a re-appraisal of agricultural origins in Britain, Ireland and Southern Scandinavia. *Current Anthropology* 45 (Supplement), 83–113.

Sims, R. (1983) The anthropogenic factor in East Anglian vegetational history: an approach using APF techniques. Pp. 223–36 in H. Birks and R. West (eds), *Quaternary Plant Ecology*. Oxford: Blackwell.

Smith, A., Whittle, A., Cloutman, E. and Morgan, L. (1989) Mesolithic and Neolithic activity and environmental impact on the south-east fen-edge in Cambridgeshire. *Proceedings of the Prehistoric Society* 55, 207–49.

Speck, F. (1940) *Penobscot Man: The Life History of a Forest Tribe in Maine*. London: Oxford University Press.

Spikins, P. (1999) *Mesolithic Northern England. Environment, Population and Settlement*. British Archaeological Reports, British Series 283. Oxford: BAR.

Sturt, F. (2006) Landscape and Land-use Change during the Mesolithic and Neolithic of the Fen Edge. Unpublished PhD thesis, University of Cambridge.

Thomas, J. (1988) Neolithic explanations revisited: the Mesolithic–Neolithic transition in Britain and south Scandinavia. *Proceedings of the Prehistoric Society* 54, 59–66.

Thomas, J. (1991) *Rethinking the Neolithic*. Cambridge: Cambridge University Press.

Thomas, J. (1999) *Understanding the Neolithic*. London: Routledge.

Tilley, C. (1979) *Post-Glacial Communities in the Cambridge Region: Some Theoretical Approaches to Settlement and Subsistence*. British Archaeological Reports, British Series 66. Oxford: BAR.

Wainwright, G. (1972) The excavation of a Neolithic settlement on Broome Heath, Ditchingham, Norfolk. *Proceedings of the Prehistoric Society* 38, 1–97.

Wainwright, G. (1973) The excavation of prehistoric and Romano-British settlements at Eaton Heath, Norwich. *Archaeological Journal* 130, 1–43.

Whittle, A. (1996a) *Europe in the Neolithic: The Creation of New Worlds*. Cambridge: Cambridge University Press.

Whittle, A. (1996b) Houses in context: buildings as process. Pp. 13–26 in T. Darvill and J. Thomas (eds), *Neolithic Houses in Northwest Europe and Beyond.* Oxford: Oxbow Books.

Whittle, A. (1997) Moving on and moving around: Neolithic settlement mobility. Pp. 15–22 in P. Topping (ed.), *Neolithic Landscapes*. Oxbow Monograph 86/Neolithic Studies Group Seminar Papers 2. Oxford: Oxbow Books.

Whittle, A. (1999) The Neolithic period. Pp. 58–76 in J. Hunter and I. Ralston (eds), *The Archaeology of Britain: An Introduction from the Upper Palaeolithic to the Industrial Revolution*. London: Routledge.

Willcox, G. (1999) Archaeobotanical significance of growing Near Eastern progenitors of domestic plants at Jalès, France. Pp. 103–17 in P. Anderson (ed.), *Prehistory of Agriculture: New Experimental and Ethnographic Approaches*. Los Angeles, CA: University of California.

Wiltshire, P. (2007) Palynological analysis of paleochannel sediments. Pp. 62–77 in K. Gdaniec, M. Edmonds and P. Wiltshire (eds), *A Line across Land: Fieldwork on the Isleham–Ely Pipeline*. East Anglian Archaeology 121. Cambridge: CambridgeArchaeological Unit.

Part Four

Conclusion

20. Time, Scale, Practice: Landscapes in Transition?

Bill Finlayson and Graeme Warren

This final contribution provides an overview of key themes arising from the papers presented. Four interconnected areas were the focus for much discussion: time scales; scales of analysis and narrative; landscapes of practice and the connections between places; and finally the use of analogy. Using these focal points suggests new ways of understanding landscapes in transition.

Introduction

During the process of inviting people to participate in this workshop we were aware that we could only ask a small number of the many scholars working in the field to participate. The process of selection was not designed to be comprehensive, but to help us explore particular themes, and in so doing contrast approaches followed in Northwest Europe (mainly the British Isles) with those used in Southwest Asia, to provide a lens to expose our inbuilt assumptions and identify new ways forward. The choice of these two regions reflects the immediate research interests of the organizers; further comparative exercises including different regions would be of great interest. Many of the participants working in Southwest Asia were based in, or had strong connections with, British archaeological traditions: contrasts between their approaches and those working within Britain and Ireland were therefore particularly interesting. Within British and Irish archaeology a number of stereotypes have become embedded: for example, that archaeologists working in the Neolithic are interested in theory, those working in the Mesolithic are not. This is also restated (contradictorily of course) as those working in Britain being interested in theory with those working in Southwest Asia not. These binary oppositions are demonstrably false; we would suggest that our tendency to shorthand and simplify complex traditions of research by comparison and contrast is an interesting reflection of the way we divide the past, illustrating how powerful this way of interpreting the world into them and us can be.

An advantage of having a small group of people participating in a residential meeting is always the extended discussions that go on between the formal presentations. The papers presented in this volume reflect these discussions as much as the formal question sessions that followed presentations. In some cases the final papers have drawn much closer to one another than the initial presentations. It is, however, interesting to note that, with the exception of a small number of papers (e.g. Bogaard and Isaakidou, this volume), there remains very much a focus on one region or the other. This should not occasion surprise: synthesis of the two regions is rare in the literature generally and demonstrable differences in intellectual traditions are clearly apparent in the papers presented here. Many specific areas of comparison are present, such as the processes by which islands are integrated into new forms of material practice in the different regions. We leave the reader to draw their own conclusions about the worth of such direct parallels and focus here on broader scale concepts. Four key themes emerged during the course of the debates, and were discussed further in a closing session of the workshop:

1. **The significance of time scales:** in particular the need to move between long and short scales in understanding trajectories of change.
2. **Scales of analysis:** although clearly related to issues of temporality, this theme focuses more on how to integrate grand-scale narratives with intimate richly detailed local studies.
3. **'landscapes of practice':** understanding how varied

bodies of evidence, especially archaeobotanical, carry implications for the organization of the landscape more widely and how these, in turn, impact upon social relations and our understandings of them. Related to this are issues about understanding specific places and the connections between them.

4. **Analogies and familiarity:** in particular, discussion highlighted the difficulty and insidiousness of several key concepts, all based firmly, if often implicitly, on analogical reasoning as we move from the 'other' of hunter-gatherer sites to the assumed familiarity of the village.

The significance of time scales

Archaeology is about time and change, but we often appear to be very bad at expressing this. The problems in using stadial terminologies such as Epipalaeolithic, Mesolithic, Neolithic and dichotomous terminologies such as hunter-gatherer and farmer have been addressed many times; they emphasize the static and make it hard for us to consider change, unless it be revolutionary. Gamble (2007) highlights the key role of revolutions in structuring archaeological thought about time, arguing that the concept of revolution is itself a fiction associated with nation states and the idea that through the interventions of individuals one can control the flow of history (cf. Berman 1983). In particular, Gamble focuses on the Neolithic Revolution as expressed by Childe, suggesting that it 'acts as a pair of scissors to cut the conceptual network, taking the continuum of archaeological time and snipping it into a big, but manageable problem. Then our archaeological imagination can get to work' (Gamble 2007, 19). Many of the papers presented here suggest more fluid processes, not to deny the scale of the changes that occurred, but to show how they arose from within existing societies, often showing long-term continuity matched by faster rhythms of change.

The variations between different time scales are significant. At one extreme we have the long sweep, where the Neolithic is seen by some (including Watkins in this volume) as the moment when anatomically modern people finally come into their own and become cognitively modern, a sweep of history that encompasses tens of thousands of years (cf. Barker 2006). At the other, we have archaeological evidence from the single contexts we excavate; time formed by individual actions. All too rarely do we build our temporalities from these latter contexts, a bottom-up process deeply based on our evidence. Garrow, focusing on Eastern England, shows how important this is for our interpretation of different forms of archaeological evidence.

Goring-Morris and Belfer-Cohen, and Maher, argue for the importance of long-term cultural perspectives in Southwest Asia, with the former paper providing a detailed synthesis of the southern Levantine data and the latter taking the specific feature of mortuary data as its focal point. Benz also considers long-term processes, citing ethnographic data and suggesting that once people make critical decisions regarding how to respond to environmental change they can become locked into certain courses: parallels here with Tipping's paper, or wider discussions of the incompatibility of hunting and gathering and agricultural practices, are important (e.g. Sahlins 1968; Rowley-Conwy 1983; Bender 1985; Hodder 1990; Thomas 1999; Price and Gebauer 1995; Cauvin 2000; Zvelebil 2002; see also 'Landscapes of Practice' below).

A different dimension of the problem with time scale is discussed by Robinson *et al.*, where they take a long-term climatic perspective, using a number of different proxy sources, and then try to integrate this with archaeological data. As with the different archaeological timescales, it is hard to link overall patterns to specific events. Robinson *et al.* make the specific point that although the common language between climate reconstruction and archaeology is time, there are many practical difficulties in combining different datasets – including the need to calibrate radiocarbon dates before comparing them with other radiometric series, the need to include error bars, and the need to remember what is actually dated and what, as is common in geological or palaeoenvironmental sequences, is simply interpolated between known dates (see varied discussions in Telford *et al.* 2004; Parnell *et al.* 2008). Related problems exist with adequately relating local, regional and global data for climate and environmental change. Goring-Morris and Belfer-Cohen remind us that in the Levant, while we are aware that the landscape is highly locally variable, and the climate changing with specific local effects, we are as yet unable to really put these changes together in the detailed manner that we would like. The acceptance of clear and straightforward climate stories is in part no doubt produced by a desire for simple causes (see the discussion of Big Beat, below), but an important factor is the 'black box' effect, where the debate is hard to follow for the non-specialist. The difficulty of combining different forms of evidence is significant, and we should be wary of taking the line of least resistance and uncritically accepting an oversimplified environmental narrative.

The time scale of long-term processes, whatever their ultimate cause, may well have rendered people unaware of the cumulative effects of change as they occurred, but it seems likely that certain events encapsulated these changes (possibly marking revolutions where transitions were occurring). For example, if we assume that small-scale societies conceive time to be comprised of repeated and long standing rhythms then developing patterns of change present a challenge to established cosmologies founded upon a circular, repetitive temporality (see also Zvelebil 2008). Depending on the rate of change, the more 'linear dimension' of historical temporality may be largely imperceptible, possibly only to be recognized by older people who can recall how things once were; or, occasionally, change may occur so fast that an entire society can perceive it. There are moments when it appears almost certain that change was not only perceptible, but intentionally created – such as with the building of purpose-made granaries for storing harvests, which create a real and fixed investment in the future – but, of course, the final outcome of these changes could not have been known.

These problems, of course, raise issues of resolution and many papers stress the need for greater accuracy and precision in relating local data and dates with broader explanatory models. As Whittle and colleagues have argued, 'we need far better chronologies than the ones to which we have become accustomed and to which we have, through complacency or lack of imagination, resigned ourselves' (Whittle 2007, 377). Critical assessments of radiocarbon dates and chronologies have been made for Southwest Asia (see, e.g., Byrd 2005 and Aurenche *et al.* 2001) but the recognition of the problem has not yet been matched by significant progress to improve the situation. In Britain and Ireland the increasing use of Bayesian modelling of radiocarbon dates, running alongside rigorous assessments of the taphonomy of samples, has changed the nature of our chronologies (Bayliss *et al.* 2007; 2008; Whittle and Bayliss 2007; Whittle *et al.* 2007), with the possibility of generational timescales providing a fundamental challenge to the conceptual models that we operate with to explain change in the past. Bayesian analysis also provides a challenge to the ways in which we interpret all of our dates: for example, the argument that outlying dates are the product of statistical scatter and do not fit within modelled chronologies provides a clear problem for those who seek the earliest date as evidence of origins. The varied extent to which Bayesian methods are applied is demonstrated within this volume: the papers on Britain and Ireland use a variety of approaches to the radiocarbon dates they cite – but rarely formal modelling. Recognition of the potential of Bayesian modelling to the Mesolithic–Neolithic transition is highly regionally variable in Europe (Warren 2009) and is generally rare in Southwest Asia (see Edwards *et al.* 2004 for a recent exception). But to focus on Bayes alone as a solution to our problems with chronologies would miss the more basic point that many sites, in both regions, are very poorly dated by any standards, with too few dates associated with very complex sequences. Key sites such as Jericho were excavated early in the history of radiocarbon dating (e.g. Kenyon 1957, 1960), making their use problematic. A greater emphasis on chronology, with the use of multiple dates from short life samples, will, in conjunction with modelling, profoundly change the nature of our understanding of transitions.

Returning to our volume, Milner, for example, argues that until we can resolve many of our chronological issues it is hard to debate the detail of speed of change or extent of change, as so much depends on chronological scales. At present, archaeologically detectable rapidity is still so poorly defined (several hundred years) that we can barely resolve whether it would have been experienced as rapid by people or if the degree of precision we can obtain actually matches the real rate of change, meaning that it happened at a barely perceptible pace. At the same time, we need to ask which of the climate change effects happened sufficiently rapidly for people to notice what was happening, as this may have a clear impact on behaviour, and hence be archaeologically recognizable. Benz argues that our chronological resolution is simply not good enough to assess stress on hunter-gatherers, which may be cataclysmic over only a few seasons. Although we may not be able to point to a particular year in which change took place, increasing resolution may allow us to associate key changes in different fields of data. Tipping's analysis of climate change attempts to bring various forms of data together, linking climate to archaeology, and to understand how quite specific climate events may have affected people. The close correlation he attempts allows him to present a novel explanation for apparent rhythms of contact and separation and for subsistence change.

What all of these authors do is show how important time remains to us – it is central to how we interpret our data – but that it is not just a matter of increasing precision or the application of methodologies. Instead what is required running alongside this is precision in how we use the chronological information we have, greater and more rigorous correlation of data types, and an understanding that different time scales may be valid in different ways.

Scales of analysis

In addition to different temporal scales there are different geographic scales of analysis. We did not set out to produce any grand narratives at this workshop, rather to consider local evidence and interpretation, and none of our contributors has attempted to create a unified theory of the hunter-gather–farmer transition. Several have, however, taken regional perspectives. Watkins, Asouti and Fairbairn, and Noble in particular, take large, if not continental, canvasses. Robinson *et al.* take a similarly large area for their environmental review. Others take sub-regions, such as the southern Levant of Goring-Morris and Belfer-Cohen or Maher, Bogaard and Isaakidou's series of case studies, or the Northwest Europe of Sturt. Others focus on increasingly small areas, down to the landscape around a single site, taken by Campbell, or the biographies of individual pits considered by Garrow.

These different scales relate in part to the difference between 'bottom-up' or 'top-down' approaches, where we are almost routinely advised that a bottom-up approach should be followed (cf. Sturt, this volume) at a time when the number of synthetic, top-down, books being published (e.g. Barker 2006; Bellwood 2005; Simmons 2007) would seem to suggest that this is the ultimate target for the aspiring academic (reflecting the appeal of the Big Beat referred to by Sturt). Part of the problem would seem to be the difficulty of actually achieving a bottom-up approach. Goring-Morris and Belfer-Cohen illustrate the sheer quantity of data that exists just in the southern Levant, and to attempt to bring this together without becoming swamped in the detail becomes harder and harder. At one level, in Northwest Europe the opposite problem would appear to obtain, with several papers discussing the problems of dealing with a relatively thin database. One result of this is more and more consideration of data brought in from elsewhere, whether from other parts of the world or from the ethnographic present. On the other hand, Sheridan's review demonstrates the depth of reference that is required to make sense of some of the material manifestations of change in the centuries around 4000 BC. That this requires an active engagement with a research tradition conducted in a different language may provide a further challenge for British and Irish archaeologists. In this sense the difference between research in Britain and Ireland, which is primarily conducted by archaeologists based in those countries (although the attraction of Ireland as a location for fieldwork initiated from overseas seems to be greater than it is for Britain), and that in Southwest Asia, which is much more cosmopolitan, is notable.

Sturt discusses the difficulty caused by the tendency of reductionist approaches to break the different strands of evidence up into modern specialist concerns when in fact, in trying to build up models from their component parts, it is their complex interplay that is truly important. A related problem lies in the nature of a fragmentary dataset which often leads to parts being taken as representative of wholes, creating further problems of circularity in the use of familiar analogical reasoning (see below). Our forms of classification constantly establish oppositions and dichotomies, creating boundaries that probably had little relevance in the past; these are not only a side effect of our way of conducting analytical research, but are vital for our more dramatic narratives. In contrast to reductionist approaches, Lefebrve's *rhythmanaytical* approach offers an integrative approach to space as praxis (see also discussion of 'Landscapes of Practice' below). Sturt is surprised that more archaeologists have not taken up this approach as it seems so suited to our endeavours. A number of the papers here are dealing with these same issues – perhaps especially Asouti and Fairbairn, with their focus on linking social behaviour with ecological relationships, but also all those that try to connect patterns of climate change to human activity, or the historical sweep of transformation of hunter-gatherer to farmer to the little keyholes of evidence that we actually have.

Sturt's metaphor of Big Beat and Bebop builds on the rhythmanalytical approach and was taken up enthusiastically during the workshop as a way of discussing different scales of event; it also serves as a metaphor for the grand synthesis (or Big Beat) and the local history (or Bebop). Big Beat narratives are driven by a pleasing, predictable simplicity of beat that appears to incorporate all the significant strands, leaving out the almost irrelevant Bebop fluctuations. The simple rhythm of Big Beat, however, is arguably constructed around reductionism, oppositions and binaries: the appeal of the narrative is founded on the clean rhythms arising from this categorization. In practice, the Big Beat approach can be problematic, ignoring much actual evidence, whether it is fragmentary or contradictory, and omitting much of what may explain transition.

Big Beat is also most easily caught up with cause and effect arguments that, as Robinson *et al.* note, are hard to prove. Equally, as Robinson *et al.* observe, climate phases such as the Younger Dryas are not unique events, but part of a longer history of climate change, and simply noting the longevity of human development does not exclude climate from being an important factor in behavioural choices. Their observation that both archaeological and palaeoclimate datasets are constantly changing and dynamic accords well with Benz's understanding of the relationship as dialectical. To stretch our musical metaphors further, we might suggest that the dynamic character of multiple datasets indicates a radical multitude of rhythms: far from Big Beat, or even Bebop, we may have to consider a Free Jazz approach, characterized by incompatible tempos, discordant melodies and an indeterminate number of players.

Landscapes of practice

Another very strong theme in the workshop was the notion of 'landscapes of practice', most explicitly developed in the archaeobotanical contribution of Asouti and Fairbairn, but present through many other papers. This seems to be a particularly important tool in breaking away from some of our familiar assumptions – whether it be from the stereotypical impression of the transition from wild landscapes to domestic ones (e.g. Foster McCarter 2007), or from the idea of the destruction of the landscape by Southwest Asian mega-site dwellers (Köhler-Rollefson and Rollefson 1990 and references in Campbell, this volume). Considering landscapes as practice also serves as a mechanism for integrating archaeological and palaeoenvironmental data at appropriate scales: rather than black-boxing a 'change' in climate, the emphasis shifts to understanding precisely how this impacted on routine and novel behaviours (cf. Tipping, this volume).

Asouti and Fairbairn provide a robust manifesto for how to understand, in both hunter-gatherer and farming communities, the relationships between people and their resources. Ironically, this may bring Richard Bradley's old joke about hunters having ecological relationships with hazelnuts and farmers social relationships with people full-circle (Bradley 1984, 11), as we turn now to social practice, ecology, and food consumption as a key aspect of our data. Bogaard and Isaakidou, for example, demonstrate that different ways of managing resources are inherently social and ideological and have specific material manifestations. Campbell, in her re-assessment of the exploitation of resources around 'Ain Ghazal, shows how it may have been possible for the community to have sustained their economy on the land immediately available for growing crops and grazing animals. This grounded analysis of praxis has important implications for wider discussions of cause in social change in this region. Garrow's discussion of crop agriculture in the Neolithic of southern England is similar in character, if less extended. Underpinning these narratives is a focus on the materiality and reality of past subsistence practice extended through rigorous archaeobotanical practice and critical use of analogies (see also Bogaard 2004).

Conneller's focus on landscapes of practice comes through the lens of Ingold's 'taskscape' concept (Ingold 1993; 2000), and a database that has often been dominated by lithics. Her assessment suggests that the ways in which Ingold's account has been materialized archaeologically through Northwest European case studies has inadvertently established a structure that reinforces hunter-gatherers as being static and not producing change. Here lies a central difference between Southwest Asia and Northwest Europe. In the European case, whether farming and the Neolithic come in directly with new people or are adopted by indigenous populations, there is no doubt that the nature of the transition is significantly influenced by the presence of communities who are already practising farming: 'the dispersal of farming and the process of Neolithization were embedded in the existing, pre-Neolithic social and historical conditions in each region, in the history of contacts with communities which had already adopted farming' (Zvelebil 2001, 1). Paradoxically, this has facilitated a presentation of European Mesolithic communities as comparatively passive until ousted, influenced or inspired by farmers. In Gamble's terms, '... the Mesolithic, ... has a particular function in mediating the transfer of Neolithic political power to the European continent ... This archaeological space inhabited by the Mesolithic 'folk' was the drum on which the revolutionary roll of agriculture was beaten out' (Gamble 2007, 22).

Ideas of internal and external impetus are highlighted very clearly in discussions of islands and the connections between places, especially in terms of dominant metaphors of isolation. The nature of isolation is considered by McCartney *et al.*, Noble and Kador. All take very different perspectives and, strikingly, those discussing physical islands are more opposed to the idea of isolation than Noble, in his discussion of virtual islands within the transitional European landscape. For all of these papers a key emphasis is on understanding islands as part of landscapes, or perhaps preferably 'seascapes', of practice. Only when this is possible will we be able to adequately assess issues of cause in these geographies. This in turn returns us to Sheridan's discussions of cause: how and what moved, and when, across the sea to Britain and Ireland.

In Southwest Asia the origins of agriculture lie in hunter-gatherers transforming themselves, albeit probably in a very slow process and with our accounts leaving many questions unanswered about what exactly we mean by this self-transformation in terms of degrees of invention, borrowing, contact and transformation within the 'hunter-gatherers' of Southwest Asia. This has huge ramifications for our study, as our generalized ethnographic models for hunter-gatherers tend to draw from a global pool (and are often non-specific in terms of geographies and/or ecologies): thus common themes link discussions of our two research areas: developing complexity, sedentism, specialization and so on. However, in Southwest Asia hunter-gatherers have to be capable of change. Watkins addresses this in his consideration of landscape where, through increasing sedentism and the symbolic construction of a social landscape, people changed themselves.

In understanding landscapes in transition, landscape practice and the interpenetration of subsistence, ideology and routine become of fundamental importance. Asouti and Fairbairn argue that the way to move beyond oppositions between hunter-gatherers and farmers is to abandon such categorization and to consider landscape practices in a continuum that makes the idea of transition redundant. A highly attractive aspect of such an approach is that it allows us to move away from the reinvention of modern anthropological categories in the present to consider these non-analogue societies in their own right. Benz discusses how it is the changing landscape signature created by people that creates the changes that we recognize as the start of the Neolithic. Goring-Morris and Belfer-Cohen argue that for the Early Epipalaeolithic there is no real evidence for anything other than economic criteria being behind site location. By the Natufian this changes, with sites apparently deliberately marking certain points in the landscape while being deliberately secluded within that landscape. At the same time, increasing sedentism will have begun to mark a significant change in landscape practice, with many aspects of the Natufian, such as burial rituals, plaster, flagged floors and massive stone mortars, apparently symbolizing greater permanence of settlement. Goring-Morris and Belfer-Cohen describe the process as 'settling into the local landscape'. They also concur with the idea of a continuum of change.

Parallels in approach, if distinctions in conclusion, are present with Cummings' discussions of connections between people and place over the transition. Cummings takes the settings of Neolithic monuments along the west of the British Isles, and by taking them in their landscape contexts is able to suggest that Early Neolithic people held a similar worldview to that held in the Mesolithic. Monuments here, in contrast to many approaches, cease to reflect a revolutionary change in worldview, but become a new instrument in symbolizing long-held perceptions. Broadly phenomenological in orientation, Cumming's approach is representative of a strong trend in British archaeological practice (exemplified by Tilley 1994), and it will be interesting to see if the specific methodologies and critiques that have arisen for engaging with issues of landscape perception on either side of the transition (among many, see Brück 1998; Cummings and Whittle 2003; 2004; Finlayson and Dennis 2002; Fleming 1999; 2005; 2006) are taken up in Southwest Asia.

Watkins takes on the significance of the built environment, but sees the changes that arose in society as 'revolutionary' rather than as part of the continuum. He understands that the emerging built landscape incorporates symbolic significance, and that this new landscape fundamentally changes us by framing the way we think. As new generations grew up inhabiting this landscape it provided a new and rich social and cultural context allowing for an accelerating pace of change. Watkins sees a new ability to use material culture to create symbolic vocabularies in a novel way, as a revolution as important as language or writing. The key change for us is that, in Watkins' argument, the physical landscape ceases to be as important as the constructed environment. The importance of the built landscape in this argument owes much to Wilson's work (Wilson 1988). What may be seen as contentious here is how radical this change was. It leaves hunter-gatherers as not fully human, and assumes their perception of landscape was dominated by cultural reactions to the existing bio-environment. This is a perspective that most of the other papers that consider hunter-gatherer landscape perceptions would disagree with, as they suggest that the unbuilt landscape inhabited by hunter-gatherers is both created by them and full of symbolic meaning.

Analogies and familiarity

There has been considerable debate regarding the use of analogy in the specific context of the Mesolithic over the years (Zvelebil and Fewster 2001; Jordan 2006). British archaeology has never made as much formal use of ethnographic analogy as American archaeology (at least in its consideration of small-scale societies). However, analogy is probably employed more explicitly in the study of hunter-gatherers than in any other branch of archaeological research. This in part reflects the idea of the hunter-gatherer as 'other', where we require the support of analogy, however actually deployed, in order to understand such different societies. This is in contrast to the relatively rare explicit use of analogy in the study of farming societies, where we tend to oppose the 'other' of hunter-gatherers with an idea of familiarity. Compared to hunter-gatherers it appears that farmers are like us. We move from a terminology of 'sites', 'flint-scatters' and 'optimal foraging' to one of familiarity, of villages, peasants, houses and homes.

The idea of 'farming' has such a great weight of familiarity behind it that it becomes treacherous to use. The concept of the peasant farmer was used by Cauvin as a central plank of his arguments on the transition (Cauvin 2000), and a standardized Eurasian farming package (Barker 2006) becomes the product of the transition. Natufian society (complex hunter-gatherers) sits on the boundary between farming and hunting and gathering, and therefore it is perhaps not surprising that they occupy both 'base camps' and 'villages', together with the bucolic-sounding 'hamlet'. The Southwest Asian Neolithic becomes village society, with frequent references to the 'oriental village', another timeless Orientalist notion. Gamble argues that Childe's Neolithic Revolution was inherently Orientalist as it used the Orient as a theatre, a place of origins which were later transformed in Europe (Gamble 2007, 19–22). Indeed, the European transformation of Neolithic societies is argued to transcend its origins, thus highlighting the superiority of the European to the Orient: '… the West and the East can, for the first time in history, be prised apart. And by prising them apart the Neolithic revolution becomes the origin point of the modern project of European history and culture that gained definition by opposing and containing the Orient' (Gamble 2007, 31).

Our familiarity and comfort in this rural idyll can be seen in the many popular books on the Neolithic that show a well-tilled and ordered landscape opposed to the wild and dangerous hunter-gatherer world. It is also notable that reconstruction drawings often show hunter-gatherers as bearded, but farmers as clean-shaven. We cannot distance ourselves from these 'popular' productions; it is us, the archaeological community, that uses these images to communicate with the wider public and, very often, with our own students, thus perpetuating myths that we may explicitly be trying to counter. The centrality of such a perception is such that it appears to be very close to the debate on the Neolithic of Britain that asks where the villages are, where are these farmers' houses and homes? The importance of this latter debate to notions of the Neolithic in Britain and Ireland suggests a radical epistemological dilemma raised by the absence of familiar images of farmers: if we can't find the villages and houses, where are our peasants? Thomas (1999, 7–11) argues that the importance of agriculture to British identities means that the very appearance of agriculture must be distinguished from the '*un-British*' (1999, 11, original emphasis) hunting and gathering, and that this is set alongside an unwillingness to accept that agricultural practice itself might have changed over time.

This brings us very much to the issue of landscape in both Northwest Europe and Southwest Asia. Campbell, Bogaard and Isaakidou, and Garrow in particular look at the conception of the familiar landscape of the Neolithic. Campbell considers the landscape immediately around 'Ain Ghazal, questioning the still widely accepted model for PPNB mega-site collapse (Köhler-Rollefson and Rollefson 1990 and references in Campbell), a model that was easy to accept given its concordance with contemporary views of unsustainable over-exploitative modern agriculture. Although Campbell argues a strong case, the debate will doubtless continue as evidence for population numbers, climate change and archaeobotany from 'Ain Ghazal and other sites continues to emerge and be correlated.

However, to return to our conventional image of farming, Bogaard has suggested that the decisive moment is represented by the emergence of the full suite of animals and plants that appears at the end of the PPN and comprises the 'full package' that spreads into Greece and across Europe – it really is the intensive mixed farming of the small peasant or yeoman farmer that fuels the spread of the Neolithic. This brings us back to the problem of the mega-site, and Bogaard and Isaakidou express this differently – the problem is not one of explaining their demise, but their longevity. Mega-sites have to be understood in terms of ideologically bound-together communities. Here we are not in new territory: various authors have argued for this, including Watkins and the theatres of memory within revolutionary new Neolithic communities (Watkins 2004), and Kuijt and the forced egalitarianism within what he sees as settlements of hunter-gatherers desperately trying to maintain old social structures within a new context (Kuijt 1994; 1996; 2000). Bogaard and Isaakidou discuss ways in which the mega-site economy could have functioned in ways akin to Campbell's, but for them there still has to be a social mechanism to equalize landscape use.

Conclusion

We began by claiming that the 'useful ambiguity' (Gosden and Head 1994) of landscape may have been problematic in terms of understanding the transition from hunter-gatherer to farmer in our case-study areas. The varied papers included in this volume suggest that this is the case, but that this problem can be productive. The use of the concept of landscape even within this volume is clearly varied, inconsistent and deeply ambiguous. This is, of course, not novel, but as one should expect. The tensions and contrasts evident in our papers are as informative as the synergies. What is perhaps most striking is the degree to which areas of common concern do arise. The various papers presented in this volume all show that we now have a huge wealth of data, but that there is much we still have to learn from our data and from each other. In part learning must include the usual call for more fieldwork but it also includes a far greater precision in chronology, in understanding timescales and in correlating different types of data and measurement. What is achieved here is a demonstration of the beginning of a far more nuanced understanding of the many processes that were taking place during this long sequence of change.

Perhaps partly because most of the authors in this volume do not attempt to span the gap between East and West (only Bogaard and Isaakidou cover the expansion into Europe) the issue of the difference in the nature of the transition in the two areas is not discussed. This is normal. One of the criticisms raised against Hodder's (1990) synthesis of the spread of agriculture into Europe was that it discussed the transition as if there was nothing East of Çatalhöyük. There appeared to be, and still is, something of a gulf between those members of the British archaeological research community interested in the European Mesolithic–Neolithic transition, and those working in a Southwest Asian Epipalaeolithic–Neolithic. To one of us (BF), working in both British and Southwest Asian contexts, this seemed to be partly related to interpretative models that in the West assumed a great divide between hunter-gatherer and farmer, while in Southwest Asia it was essential to identify a developmental path between the two. Developments in recent years have tended to emphasize areas of continuity (see, for example, contributions to Bailey and Spikins 2008), or, especially in the archaeology of the British Mesolithic, have removed some of the disciplinary divisions within the period by increasingly considering Mesolithic society rather than economy (see, for example, contributions to Bevan and Moore 2003; Larsson *et al.* 2003; Milner and Woodman 2005; Conneller and Warren 2006). Understanding hunter-gatherers as capable of innovation and adaptation brings the Mesolithic much closer to our understanding of the Southwest Asian Epipalaeolithic and Neolithic. It is, however, important to recognize that the presence of Neolithic agricultural communities in Europe as neighbours to later Mesolithic communities creates particular kinds of historical possibilities that suggest important differences between the areas. Equally, recognizing the patchwork of communities present in Northwest Europe helps to remind us that a similar situation would have existed in Southwest Asia; hunting and gathering would have persisted in some areas, landscapes were not empty and awaiting colonization by farmers, and the variations between different Early Neolithic communities may have been as significant in the process of transition as those between farmers and foragers. As we have argued above, understanding the wider applicability of models of change requires a more nuanced understanding of scale and temporality than many of our grand narratives have provided.

Unpacking the Neolithic has also taken place within both academic communities, leaving definitions of the Neolithic much less firm and convinced than they used to be. There is an increasing recognition, reflected in the papers in this volume, that our chronological shorthands are problematic: that words such as the 'Neolithic' subsume significant levels of variation and obscure as much as they reveal (see, for example, Sheridan's emphasis on different Neolithic strands). We have argued above that one of the primary problems here in understanding diversity is the insidious familiar analogies that these concepts carry. At the same time, the difficulty of moving beyond this terminology has been repeatedly demonstrated. Attempts to avoid the dichotomies of the Big Beat can fall victim to the difficulties posed by the wealth of disparate material that forms our evidence base: Gabriel Cooney (2007) has highlighted the problems caused by the 'parallel worlds' in which discussions of different sets of evidence can take place. At times a dose of analytical reductionism provides a way of making sense of this material.

While we would not claim that this workshop has developed a new concept of what the Neolithic is or how transitions may have taken place, a number of papers suggest that the dialectic between people and landscape certainly appears to provide common ground for the study of the transformation within both Southwest Asia and Northwest Europe. If we understand landscapes as constituted by varied practices that in turn provide different links between aspects of our data, some of our difficulties with understanding abstracted change are lessened. This is not, of course, to dismiss very similar problems with chronology and the integration of different data sources in the two main case-study areas, but, while these are difficult, recognition of this, and the adoption of a number of relatively straightforward methodological practices, would seem to offer a way forward. The resolution of transition will not come through methodology alone: but changing practices in this sense will facilitate a different scale of analyses; one that in turn will challenge the ways in which we construct wider narratives. It seems to us that such accounts must begin with local histories, based on the relationships between people and particular places and how these change over time. The resulting mosaic will inevitably be a patchwork of differing stories.

Bibliography

Aurenche, O., Galet, P., Régagnon-Caroline E. and Evin, J. (2001) Proto-Neolithic and Neolithic cultures in the Middle East – the birth of agriculture, livestock raising, and ceramics: A calibrated 14C chronology 12,500–5500 cal BC. In H. J. Bruins, I. Carmi and E. Boaretto (eds), Near East chronology: archaeology and environment. *Radiocarbon* 43/3, 1191–202.

Bailey, G. and Spikins, P. (eds) (2008) *Mesolithic Europe*. Cambridge: Cambridge University Press.

Barker, G. (2006) *The Agricultural Revolution in Prehistory: Why did Foragers become Farmers?* Oxford: Oxford University Press.

Bayliss, A., Bronk Ramsey, C., van der Plicht, J. and Whittle, A. (2007) Bradshaw and Bayes: towards a timetable for the Neolithic. *Cambridge Archaeological Journal* 17(S1), 1–28

Bayliss, A., Whittle, A. and Healy, F. (2008) Timing, tempo and temporalities in the early Neolithic of southern Britain. Pp. 25–42 in H. Fokkens, B. J. Coles, A. I. Van Gijn, J. P. Kleijne, H. H. Ponjee and C. G. Slappendel (eds), *Between Foraging and Farming: An Extended Broad Spectrum of Papers Presented to Leendert Louwe Kooijmans*. Leiden: Leiden University (*Analecta Praehistorica Leidensia* 40).

Bellwood, P. (2005) *First Farmers: The Origins of Agricultural Societies*. Oxford: Blackwell.

Bender, B. (1985) Prehistoric developments in the American midcontinent and in Brittany, northwest France. Pp. 21–58 in T. Douglas Price and J. A. Brown (eds), *Prehistoric Hunter-Gatherers: The Emergence of Cultural Complexity*. Studies in Archaeology. Orlando, FL: Academic Press.

Berman, M. (1983) *All That Is Solid Melts into Air: The Experience of Modernity*. London: Verso.

Bevan, L. and Moore, J. (eds) (2003) *Peopling the Mesolithic in a Northern Environment*. British Archaeological Reports, International Series 1157. Oxford: BAR.

Bogaard, A. (2004) *Neolithic Farming in Central Europe*. London: Routledge.

Bradley, R. (1984) *The Social Foundations of Prehistoric Britain*. London: Longman.

Brück, J. (1998) In the footsteps of the ancestors: a critical review of Christopher Tilley's *A Phenomenology of Landscape: Places, Paths, and Monuments*. *Archaeological Review from Cambridge* 15/1, 23–36.

Byrd, B. F. (2005) Reassessing the emergence of village life in the Near East. *Journal of Archaeological Research* 13/3, 231–90

Cauvin, J. (2000) *The Birth of the Gods and the Origins of Agriculture*. Cambridge: Cambridge University Press.

Conneller, C. and Warren, G. M. (eds) (2006) *Mesolithic Britain and Ireland: New Approaches*. Stroud: Tempus.

Cooney, G. (2007) Parallel worlds or multi-stranded identities? Considering the process of 'going over' in Ireland and the Irish Sea zone. Pp. 543–66 in A. Whittle and V. Cummings (eds), *Going Over: The Mesolithic–Neolithic Transition in North-West Europe*. London: British Academy and Oxford University Press.

Cummings, V. and Whittle, A. (2003) Tombs with a view: landscape, monuments and trees. *Antiquity* 77, 255–66.

Cummings, V. and Whittle, A. (2004) *Places of Special Virtue: Megaliths in the Neolithic Landscapes of Wales*. Oxford: Oxbow Books.

Edwards, P. C., Meadows, J., Sayej, G. and Westaway, M. (2004) From the PPNA to the PPNB: new views from the southern levant after excavations at Zahrat adh-Dhra' 2 in Jordan. *Paléorient* 30/2, 21–60.

Finlayson, B. and Dennis, S. (2002) Landscape, archaeology and heritage. *Levant* 34, 219–27.

Fleming, A. (1999) Phenomenology and the megaliths of Wales: a dreaming too far? *Oxford Journal of Archaeology* 18/2, 119–25.

Fleming, A. (2005) Megaliths and post-modernism: the case of Wales. *Antiquity* 79/3, 921–32.

Fleming, A. (2006) Post-processual landscape archaeology: a critique. *Cambridge Archaeological Journal* 16/3, 267–80.

Foster McCarter, S. (2007) *Neolithic*. London: Routledge.

Gamble, C. (2007) *Origins and Revolutions: Human Identity in Earliest Prehistory*. Cambridge: Cambridge University Press.

Gosden, C. and Head, L. (1994) Landscape – a usefully ambiguous concept. *Archaeology in Oceania* 29/3, 113–16.

Hodder, I. (1990) *The Domestication of Europe: Structure and Contingency in Neolithic Societies*. Oxford: Blackwell.

Ingold, T. (1993) The temporality of the landscape. *World Archaeology* 25, 152–74.

Ingold, T. (2000) *The Perception of the Environment*. London: Routledge.

Jordan, P. (2006) Analogy. Pp. 83–100 in C. Conneller and G. M. Warren (eds), *Mesolithic Britain and Ireland: New Approaches*. Stroud: Tempus.

Kenyon, K. (1957) *Digging Up Jericho*. London: Ernst Benn.

Kenyon, K. (1960) *Excavations at Jericho*, vol. I. London: British School of Archaeology.

Köhler-Rollefson, I. and Rollefson, G. O. (1990) The impact of Neolithic subsistence strategies on the environment: the case of 'Ain Ghazal, Jordan. Pp. 3–14 in S. Bottema, G. Entjes-Nieborg and W. van Zeist (eds), *Man's Role in Shaping the Eastern Mediterranean Environment*. Rotterdam: A. A. Balkema/Brookfield.

Kuijt, I. (1994) Pre-Pottery Neolithic A settlement variability: evidence for sociopolitical developments in the Southern Levant. *Journal of Mediterranean Archaeology* 7/2, 165–92.

Kuijt, I. (1996) Negotiating equality through ritual: a consideration of Late Natufian and Pre-Pottery Neolithic A period mortuary practices. *Journal of Anthropological Archaeology* 15/4, 313–36.

Kuijt, I. (2000) When the walls came down: social organization, ideology and the 'collapse' of the Pre-Pottery Neolithic. Pp. 183–99 in H. G. Gebel and H. D. Bienert (eds), *Central Settlements in Neolithic Jordan*. Studies in Early Near Eastern Production, Subsistence, and Environment 5. Berlin: *ex oriente*.

Larsson, L., Kindgen, H., Knutsson, K., Loeffler, D. and Åkerlund, A. (eds) (2003) *Mesolithic on the Move: Papers Presented at the Sixth International Conference on the Mesolithic in Europe, Stockholm 2000*. Oxford: Oxbow Books.

Milner, N. and Woodman, P. C. (eds) (2005) *Mesolithic Studies at the Beginning of the 21st century*. Oxford: Oxbow Books.

Parnell, A. C., Haslett, J., Allen, J. R. M., Buck, C. E. and Huntley, B. (2008) A flexible approach to assessing synchroneity of past events using Bayesian reconstructions of sedimentation history. *Quaternary Science Reviews* 27/19–20, 1872–85.

Price, T. D. and Gebauer, A. B. (1995) New perspectives on the transition to agriculture. Pp. 3–19 in T. D. Price and A. B. Gebauer (eds), *Last Hunters, First Farmers*. Santa Fe, NM: School of American Research Press.

Rowley-Conwy, P. (1983) Sedentary hunters: the Ertebolle example. Pp. 111–26 in G. Bailey (ed.), *Hunter-gatherer Economy*. Cambridge: Cambridge University Press.

Sahlins, M. (1968) Notes on the original affluent society. Pp. 85–9 in R. B. Lee and I. DeVore (eds), *Man the Hunter*. Chicago: Aldine.

Simmons, A. (2007) *The Neolithic Revolution in the Near East*. Tucson, AZ: University of Arizona Press.

Telford, R. J., Heegaard, E. and Birks, H. J. B. (2004) All age-depth models are wrong: but how badly? *Quaternary Science Reviews* 23/1–2, 1–5.

Thomas, J. (1999) *Understanding the Neolithic*. London: Routledge.

Tilley, C. (1994) *A Phenomenology of Landscape*. Oxford: Berg.

Warren, G. M. (2009) How can we understand researchers' perceptions of key research developments? A case study focusing on the adoption of agriculture in Ireland. *World Archaeology* 41/4, 609–25.

Watkins, T. (2004) Architecture and 'theatres of memory' in the Neolithic of Southwest Asia. Pp. 97–106 in E. DeMarrais, C. Gosden and C. Renfrew (eds), *Rethinking Materiality: The Engagement of Mind with the Material World*. Cambridge: McDonald Institute for Archaeological Research.

Whittle, A. (2007) The temporality of transformation: dating the early development of the southern British Neolithic. Pp. 377–98 in A. Whittle and V. Cummings (eds), *Going Over: The Mesolithic–Neolithic Transition in North-West Europe*. London: British Academy and Oxford University Press.

Whittle, A. and Bayliss, A. (2007) The times of their lives: from chronological precision to kinds of history and change. *Cambridge Archaeological Journal* 17/01, 21–8.

Whittle, A., Barclay, A., Bayliss, A., McFadyen, L., Schulting, R. and Wysocki, M. (2007) Building for the dead: events, processes and changing worldviews from the thirty-eighth to the thirty-fourth centuries cal. BC in Southern Britain. *Cambridge Archaeological Journal* 17(S1), 123–47.

Wilson, P. J. (1988) *The Domestication of the Human Species*. New Haven: Yale University Press.

Zvelebil, M. (2001) The agricultural transition and the origins of Neolithic society in Europe. *Documenta Praehistorica* 28, 1–26.

Zvelebil, M. (2002) The invention of hunter-gatherers in seventeenth century Europe? A comment on Plucuennik. *Archaeological Dialogues* 9, 123–9

Zvelebil, M. (2008) Innovating hunter-gatherers: the Mesolithic in the Baltic. Pp. 18–59 in G. Bailey and P. Spikins (eds), *Mesolithic Europe*. Cambridge: Cambridge University Press.

Zvelebil, M. and Fewster, K. J. (2001) Pictures at an exhibition: ethnoarchaeology and hunter-gatherers. Pp. 143–57 in K. J. Fewster and M. Zvelebil (eds), *Ethnoarchaeology and Hunter-Gatherers: Pictures at an Exhibition*. British Archaeological Reports, International Series 955. Oxford: BAR.